Human Immunodeficiency Virus
Reverse Transcriptase

Stuart LeGrice • Matthias Gotte
Editors

# Human Immunodeficiency Virus Reverse Transcriptase

## A Bench-to-Bedside Success

Springer

*Editors*
Stuart LeGrice
HIV Drug Resistance Program
National Cancer Institute
Frederick, MD, USA

Matthias Gotte
McGill University
Montreal, QC, Canada

ISBN 978-1-4899-9953-5 ISBN 978-1-4614-7291-9 (eBook)
DOI 10.1007/978-1-4614-7291-9
Springer New York Heidelberg Dordrecht London



Printed on acid-free paper

Springer is part of Springer Science+Business Media (www.springer.com)

# Preface

The Reverse Transcriptase (RT) of Human Immunodeficiency Virus Type 1 (HIV-1) arguably ranks among one of the most extensively studied retroviral enzymes. Heterologous expression and purification of HIV-1 RT in the early 1980s, approval of the first nucleoside analogue RT inhibitor (NRTI) in 1987, discovery of resistance to RT inhibitors, and approval of the first nonnucleoside analogue RT inhibitor (NNRTI) in 1996 and the various crystal structures of RT with or without bound substrate(s) and/or inhibitors represent only a few of the important milestones that describe a bench-to-bedside success in the continuing effort to combat HIV-1 infection and its consequences. Nucleoside and nonnucleoside RT inhibitors remain important components in frequently used drug regimens to treat the infection. RT inhibitors also play important roles in recently validated strategies to prevent transmission of the virus.

The relevance of HIV-1 RT as a drug target has simultaneously triggered interest in basic research studies aimed at providing a more detailed understanding of interactions between proteins, nucleic acids, and small-molecule ligands in general terms. In light of the ever-growing knowledge on the structure and function of HIV-1 RT, this enzyme serves as a valuable "model system" in efforts to develop novel experimental tools and to explain biochemical processes.

This book is designed to provide an overview of important aspects in past and current HIV-1 RT research, with a focus on mechanistic aspects and translation of knowledge into drug discovery and development. The first section includes chapters with emphasis placed on the coordination of the RT-associated DNA polymerase and ribonuclease H (RNase H) activities. The second covers mechanisms of action and future perspectives associated with NRTIs and NNRTIs, while the third section includes chapters focusing on novel strategies to target the RT enzyme. Chapters of the final part are intended to discuss mechanisms involved in HIV variability and the development of drug resistance. We hope that these contributions will stimulate interest and encourage research aimed at the development of novel RT inhibitors. The lack of bona fide RNase H inhibitors with potent antiviral activity provides an example for challenges and opportunities in the field.

Frederick, MD, USA Stuart LeGrice
Montreal, QC, Canada Matthias Gotte

# In Memoriam: Walter A. Scott

**February 1, 1943–January 28, 2013**

As the publication of this book was nearing completion, the HIV research community was saddened to learn of the sudden passing of one of its contributors, Walter Scott, professor of biochemistry and molecular biology, University of Miami Miller School of Medicine. For those who did not know Walter personally, he will be remembered for his seminal research that unveiled the mechanism through which HIV-1 RT confers resistance to the nucleoside inhibitor AZT. These findings invigorated the field and propelled the careers of many of us. As colleagues who interacted with him personally, we are left with memories of Walter as a quiet and unassuming individual who coupled a passion for his science with a sense of wisdom and humor, and someone who dedicated his life to providing guidance to the many students that passed through his laboratory. The impish smile was rarely absent from Walter's face, regardless of whether he was discussing important research or conjuring up his next joke. Steve Hughes of the National Cancer Institute and a colleague who closely followed Walter's research described him as "in every sense, one of the best people in the field."

Born in Los Angeles and raised in Oregon, Walter received his Ph.D. in physiological chemistry in 1970 from the University of Wisconsin, and following post-doctoral research at the University of California at San Francisco and Johns Hopkins University School of Medicine, he joined the faculty of the University of Miami in 1975. After several years studying gene regulation in SV40, he turned his attention to HIV in 1989, combining his own research program with directing the Pediatric Retrovirology Laboratory of the university's Pediatric AIDS Clinical Trials Unit. During this time, he also coauthored several papers with his wife, Gwen, a faculty member of the Division of Infectious Diseases.

The personal nature of Walter's interaction with colleagues is perhaps best summarized by Peter Meyer, lead author on the 1998 PNAS publication describing the mechanism of AZT resistance, who recalled that he was "like a hand on your shoulder, not someone looking over your shoulder." At times when we feel the pressure of teaching, conducting research and applying for funding, Walter reminded us that there is a lighter, and equally important, side to life, which he most certainly enjoyed to the fullest. His namesake and Scottish novelist once commented that "a sound head, an honest heart, and a humble spirit are the three best guides through time and to eternity." Walter was endowed with each of these attributes, and will be dearly missed.

# Contents

# Chapter 1
# Development of the First AIDS Drugs: AZT and Other Dideoxynueosides

Robert Yarchoan and Hiroaki Mitsuya

## 1.1 Introduction

On March 19, 1987, zidovudine (3′-azido-2′3′dideoxythymidine, azidothymidine, AZT) was approved by the United States Food and Drug Administration (FDA) as the first drug to treat acquired immunodeficiency syndrome (AIDS). The initial development of this drug was the result of a collaboration between scientists in the National Cancer Institute (NCI), Burroughs Wellcome Co., and Duke University. Before this, there was no effective treatment for this devastating disease, and the median survival of AIDS patients was measured in months. Since the development of AZT (Mitsuya et al. 1985; Yarchoan et al. 1986), 25 additional antiretroviral drugs have been approved to treat HIV/AIDS, and combination anti-HIV therapy has converted AIDS from a death sentence to a manageable chronic disease. Several years ago, it was recently estimated that advances in AIDS therapy have already saved over 3 million years of life in the United States alone (Walensky et al. 2006), and we continue to see new benefits from these agents in both the developed world and resource-challenged countries. Moreover, the use of antiviral drugs to prevent HIV transmission was identified as the "2011 Breakthrough of the Year" by *Science* magazine (Cohen 2011). Looking back, the rapidity with which the initial AIDS drugs were introduced into the clinic (Table 1.1) (Mitsuya and Broder 1986, 1987; Mitsuya et al. 1985; Yarchoan and Broder 1987a; Yarchoan et al. 1986, 1988, 1989b) and their impact are even more remarkable in light of the skepticism and pessimism surrounding their initial developmental efforts and the fact that we still do not have an effective AIDS vaccine 28 years after the first isolation of the causative agent.

R. Yarchoan (✉) • H. Mitsuya
HIV and AIDS Malignancy Branch, National Cancer Institute
e-mail: robert.yarchoan@nih.gov

S.F.J. LeGrice and M. Götte (eds.), *Human Immunodeficiency Virus Reverse Transcriptase: A Bench-to-Bedside Success*, DOI 10.1007/978-1-4614-7291-9_1,

**Table 1.1** Timeline of early HIV drug development

| Drug | In vitro activity | Phase I trial started | FDA approval |
|---|---|---|---|
| Zidovudine | February 1985 | July 1985 | March 19, 1987 |
| Zalcitabine | May–June 1985 | October 1986 | June 19, 1992 |
| Didanosine | May–June 1985 | October 1988 | October 9, 1991 |

To get a sense of the backdrop for the development of AZT, it is worthwhile to briefly review the status of the AIDS epidemic and earlier efforts to deal with this new disorder. Although AIDS was identified as a new disease in May 1981 (CDC 1981a, b), it was some time before the scope of this new pandemic was appreciated. The only cases that could be initially tracked were those with full-blown AIDS, and it was not appreciated that many more patients with early prodromes or even asymptomatic HIV infection were part of this same pandemic and, if untreated, would nearly all eventually progress to AIDS and death. Over the next few years, it became apparent that AIDS was caused by an infectious agent, but efforts to identify this causal agent were initially hindered by the multitude of opportunistic infections that AIDS patients harbor as a result of their profound immunodeficiency. In 1983, Barre-Sinoussi, Montagnier, and coworkers first reported isolation of a novel retrovirus from a patient with a condition related to AIDS that was felt to be an AIDS prodrome (Barre-Sinoussi et al. 1983). However, uncertainty lingered as to whether this new agent was indeed the cause of AIDS until a year later, when Gallo and coworkers published a series of papers in *Science* describing the isolation of the a new virus from patients with AIDS, its propagation in a cell line, and development of an antibody test (Gallo et al. 1984; Popovic et al. 1984). After publication of these papers, it became clear to most scientists working in this area that this new virus was indeed the cause of AIDS.

Soon after AIDS was recognized as a new disease, then NCI Director Vincent T. DeVita Jr. urged Dr. Samuel Broder, who had recently been appointed Clinical Director of the National Cancer Institute (NCI) and whose research had been focused at the intersection of cancer and immunology, to spearhead an effort directed at developing a therapy for this new disorder (Harden 2012). As this effort got under way, there were only three research fellows in the Broder laboratory, and the two coauthors of this paper (HM and RY) turned their attention to this project, benefitting from substantial help from others within and outside the NCI. Immediately after publication of these papers by the Gallo laboratory in *Science*, we concentrated on development of antiviral therapy to treat HIV/AIDS. With what we now know, this approach seems straightforward, but at the time, it was rather controversial and, for a number of reasons, was felt to have a very limited chance of success. Prominent among these was the fact that HIV, like other retroviruses, integrates into host cells as a provirus, and because of this, retroviruses were not considered amenable to treatment by antiviral drugs.

A related concern was that it would not be possible to selectively develop effective anti-HIV drugs, since the virus offered few unique targets. All viruses use the host cellular machinery for most of their replicative steps and thus offer very few unique targets for attack. At the time, in contrast to the development of antimicrobial agents, little progress had been made in the antiviral therapy of any disease, and the only effective drug in widespread clinical use was the anti-herpes drug acyclovir, which had been approved in 1982. And HIV, with only nine genes, was considered a much simpler virus than herpesviruses. Also, because HIV was detectable in only a few T cells in AIDS patients, it was felt that most of the T cell depletion and immunodeficiency was through a secondary indirect mechanism that would not be halted by blocking HIV replication, especially if this replication was not completely inhibited. Lastly, there were also concerns that the immunologic damage, once it occurred, would not be reversible.

## 1.2 Initial Developmental Efforts

In spite of the pessimism mentioned above, our group in the NCI felt that it might be feasible to develop effective anti-HIV therapy to treat AIDS. Our thinking was that, unlike other retroviruses that often caused cancer by integrating into the target cell genome and causing expansion of the infected cell population, HIV generally caused AIDS by constantly infecting and destroying new cells. Also, the experience with bone marrow transplantation provided evidence that the immune system (including T cells) could reconstitute itself over a period of months under appropriate conditions. We thus hypothesized that if HIV replication could be halted in patients, we should see evidence of immune stabilization and even reconstitution in a matter of weeks or months.

In considering possible viral targets that could be attacked, we soon focused on reverse transcriptase (RT). This enzyme is essential for replication of retroviruses and hepadnaviruses, but not for eukaryotic cells. In addition, there was earlier literature on agents that blocked the activity of RTs of murine leukemia virus and other retroviruses, and these enzymes had some structural similarity to HIV RT (De Clercq 1979; Furmanski et al. 1980; Ostertag et al. 1974; Ting et al. 1972; Yang et al. 1972). To proceed, we needed an assay to assess drug activity, and one of us (HM), building on his recent studies on human T cell leukemia virus type I (HTLV-1), developed one in which agents could be assessed for their ability to block the cytopathic effect of HIV on an HTLV-1-infected cell line (Mitsuya and Broder 1987; Mitsuya et al. 1984). For a drug to be scored as active in this assay, it had to prevent killing of the target $CD4^+$ T cells by HIV; this assay had the substantial advantage that it could simultaneously assess anti-HIV activity and drug toxicity in the same cell line (Fig. 1.1). Because of the urgency of the epidemic, we focused our initial efforts on drugs currently in clinical use for another indication and thus could be rapidly brought into the clinic. If none of these seemed promising, our next focus

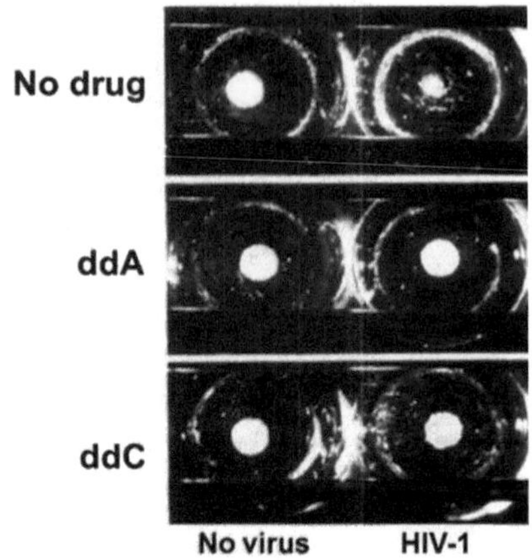


**Fig. 1.1** Cytopathic effect of HIV-1 on clone ATH8 and protection by 2′–3′-deoxyadenosine (ddA) and 2′,3′-dideoxycytidine (ddC). ATH8 cell ($2 \times 10^5$) were exposed to HIV-1 and cultured in test tubes in the presence or absence of dideoxynucleosides. By day 7 of culture, ATH8 cells were almost completely destroyed by the virus. In the absence of drug, the cytopathic effect is seen as a small disrupted pellet, which contains debris of cells (*top right*). In the presence of ddA (40 mM) or ddC (2 mM), ATH8 cells were virtually completely protected and continued to grow, which can be seen as large cell pellets (*middle right* and *bottom right*) comparable to that of HIV-1-unexposed and drug-unexposed ATH8 population (*top left*). ATH8 cells exposed to only drug (*middle left*, *bottom left*) formed large pellets comparable to the virus-unexposed and drug-unexposed cell pellet (*top left*) (Reproduced from Mitsuya et al. 1987b with permission)

was on drugs that we could develop in collaboration with a pharmaceutical drug company with the idea that this would accelerate preclinical development, clinical testing, and manufacture.

One of the first compounds we studied was suramin, an old drug synthesized as an outgrowth of Paul Ehrlich's work on trypanosomiasis that had been developed as a treatment for onchocerciasis. Suramin had been reported by Eric De Clercq to be active against the RT of RNA tumor viruses (De Clercq 1979) and was found to be effective against HIV in the cytopathic screening assay (Mitsuya et al. 1984). Within 3 months of the Gallo *Science* articles on HIV, we had developed an assay to test drugs, identified the in vitro activity of suramin against HIV, obtained approval of an Investigational New Drug (IND) Application from the FDA to test suramin in the clinic, and initiated a clinical trial to test suramin in patients with AIDS or AIDS-related complex (Broder et al. 1985; Mitsuya et al. 1984; Yarchoan et al. 1985). While there was a suggestion that it was more difficult to isolate HIV from some patients on suramin, there was no evidence from this study that it had clinical activity or could reverse HIV-associated immunosuppression (Broder et al. 1985), and we soon turned our attention to other approaches. Nonetheless, the experience in bringing suramin into the clinic and testing it in patients greatly facilitated the subsequent development of AZT.

## 1.3 Preclinical Development of AZT and Other 2′,3′-Dideoxynucleosides

While this work proceeded, we searched for pharmaceutical companies that might have compounds to screen. However, most companies that we approached were unwilling to work on anti-HIV drugs, because there were still relatively few reported cases of AIDS and this was not viewed as a profitable area. One company that did express a willingness to collaborate, however, was Burroughs Welcome Co., whose scientists, although they had considerable expertise in nucleoside analogues and had started working on murine retroviruses as a model for HIV, had no experience or ability to work with HIV. Burroughs Wellcome scientists submitted to us a number of compounds to test, and one, AZT, was identified in February 1985 to be quite potent in the HIV cytopathic assay (Fig. 1.1, Table 1.1) ( Mitsuya et al. 1985). At this stage, Dr. Dani Bolognesi's group in Duke University, which was also working with HIV, joined this collaborative effort.

AZT was not a new drug and in fact was initially synthesized and studied as a possible anticancer drug by Jerome Horwitz on an NCI grant in 1964 (Fig. 1.2) (Horwitz et al. 1964). Although AZT was unsuccessful as a cancer drug, Wolfram Ostertag et al. published a decade later that it interfered with murine retrovirus replication in a complex system consisting of both spleen focus-forming and leukemia viruses (Ostertag et al. 1974). At the same time, however, they found that AZT increased the intracisternal A-type particles in Friend erythroleukemic cells infected with Friend virus, suggesting that it might interfere with viral release (Krieg et al. 1978). While the relevance of this system to HIV was unclear, AZT did in fact prove to have activity against HIV (Mitsuya et al. 1985) and became the first of these drugs that we brought into the clinic (Yarchoan et al. 1986).

Our group continued to test other compounds, based on the postulate that dideoxynucleosides could serve as proviral DNA chain terminators and thereby block HIV infectivity (Fig. 1.2). Such agents had been synthesized some years earlier,

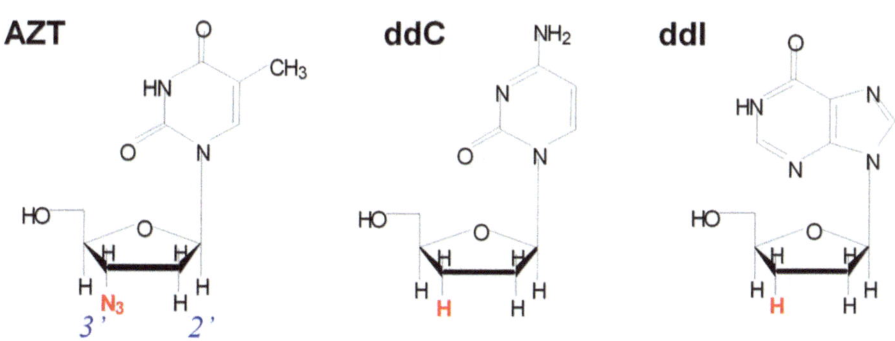


**Fig. 1.2** Structures of AZT (zidovudine), ddC (zalcitabine), and ddI (didanosine)

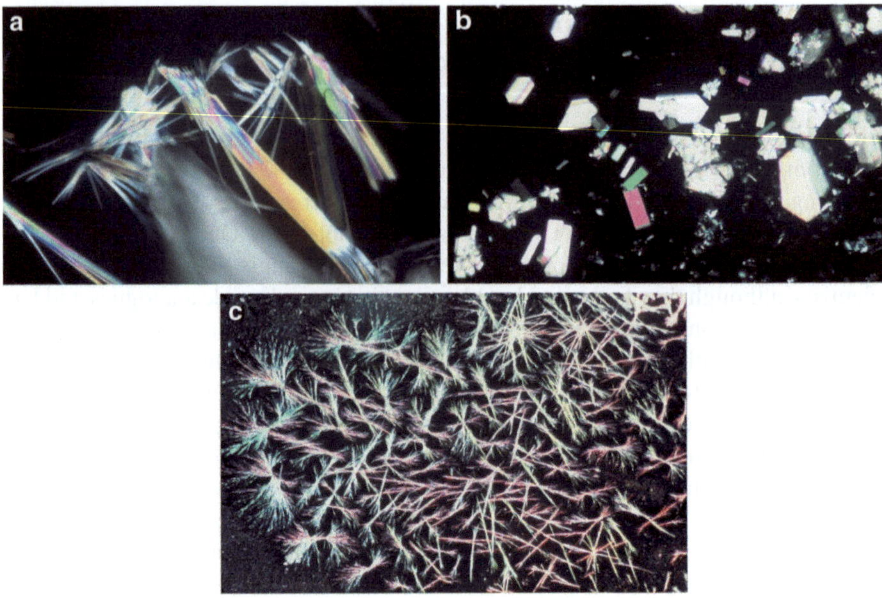


**Fig. 1.3** Crystals of AZT (Panel a), ddC (Panel b), and ddI (Panel c)

e.g., 2′,3′-dideoxycytidine (ddC) had been synthesized by J. Horwitz et al. (Horwitz et al. 1967), and 2′,3′-dideoxyadenosine (ddA) had been synthesized in 1964 (Fig. 1.3) (Robins and Robins 1964). Also, previous reports had suggested that certain of these drugs might have activity against murine retroviruses in cell lines from some, but not other species, and only at high concentrations. In particular, Furmanski et al. had reported that high concentrations (50–100 μM) of dideoxythymidine (ddT) had activity against type C murine leukemia/sarcoma viruses in mouse and human cells, but not in rat cells (Furmanski et al. 1980). Also, Waqar et al. had reported that infection of murine embryonic fibroblasts by murine leukemia virus was inhibited by high concentrations (50 μM or so) of certain dideoxynucleosides (Waqar et al. 1984). In a very short period of time, using the cytopathicity inhibition assay, Mitsuya and Broder found that a number of 2′,3′-dideoxynucleosides, including ddC, ddA, and 2′,3′-dideoxyinosine (ddI), had potent in vitro activity against HIV (Mitsuya and Broder 1986). Subsequent studies revealed a logical pattern of structure-function relationships (Mitsuya and Broder 1987; Mitsuya et al. 1987a, 1990). These agents were not active against HIV by themselves, but had to be phosphorylated to their 5′-triphosphate moiety by cellular enzymes to be effective (Furman et al. 1986; Mitsuya and Broder 1987; Mitsuya et al. 1987a). Phosphorylation of these drugs varied greatly from species to species, which greatly complicated the use of animal models (or even cell lines derived from different species) for activity or toxicity testing. In the triphosphate moiety, they inhibited HIV reverse transcription by acting both as proviral DNA chain terminators and competitive RT inhibitors. In general, active agents had a modification of their 3′-hydroxy group so that once added to a growing chain of proviral DNA by HIV reverse transcriptase, no further 5′->3′ phosphodiester linkages could be formed and the growing proviral DNA chain was terminated (Mitsuya and Broder 1987; Mitsuya et al. 1990; Yarchoan et al. 1989a). They could also act as competitive inhibitors for reverse transcriptase. However, such 2′,3′-dideoxynucleoside-5′-triphosphates were relatively poor substrates for cellular DNA polymerases, especially DNA polymerase α.

ddC, ddA, ddI, and related compounds found to be active in vitro against HIV had been purchased from chemical supply houses for laboratory use, and thus moving them into the clinic would require substantial animal toxicity and other studies. Much of this testing had already been completed for AZT, since Burroughs Wellcome Co. had been considering it as an antibiotic for bacterial infections. It thus made sense to prioritize AZT for clinical testing. Another advantage of testing AZT first was that we already had an established drug company as a partner, which would facilitate the manufacture, large-scale clinical testing, and marketing of the drug if it was found to be promising in initial clinical testing.

## 1.4 Clinical Development of Zidovudine (AZT)

### *1.4.1 Early Clinical Development of AZT*

Following the in vitro observation that AZT had activity against HIV, a concerted effort was made to rapidly move it into clinical trials. There was considerable concern at that time that because AZT, as well as the other dideoxynucleosides, was a DNA chain terminator, thus its use might cause cancer or severe cumulative toxicity. However, because of lethality of AIDS, it was felt that this risk was worth taking. Over the next few months, a phase I clinical protocol, based on the suramin study, was drafted and approved by the relevant Institutional Review Boards, and an IND Application was filed and approved by the FDA. Generation of the protocol was a collaborative effort of the NCI, Burroughs Wellcome Co., and investigators at Duke University. On July 3, 1985, less than 5 months after the initial laboratory observation of AZT activity, the first AIDS patient was administered AZT on the NCI wards of the NIH Clinical Center in Bethesda, MD. As the study was originally designed, patients received a test dose, followed by 2 weeks of intravenous dosing; as experience was gained, the protocol was amended so that patients could receive oral dosing (Yarchoan et al. 1986).

This study showed that AZT was well tolerated at doses that could achieve plasma levels that had been shown to be effective in vitro and that it penetrated into the cerebrospinal fluid (Klecker et al. 1987; Yarchoan et al. 1986). Initially, 19 patients were entered at increasing doses, ranging from 2 mg/kg every 8 h (orally) to 10 mg/kg every 4 h (orally). While it would seem that the best way of assessing a beneficial effect would be to measure its effect on HIV, we had very limited tools at hand, and technologies to measure HIV viral load by RT-PCR or even serum p24 antigen had not been developed. Perhaps the best developed test to assess HIV was the ability to isolate HIV from peripheral blood mononuclear cells; while this was somewhat decreased at higher doses of AZT, results were inconsistent and hard to interpret (Yarchoan et al. 1986). Surprisingly to some, the best evidence that AZT was doing something of benefit came from immunologic and clinical assessments. Fifteen of the 19 patients had increased CD4 counts during the 6 weeks of treatment, and six patients who were anergic at entry developed positive delayed-type hypersensitivity skin test reactions during treatment. Also, two patients had clearing of chronic nail bed infections, one patient reported an increased ability to concentrate, and several patients reported other clinical improvements, and the patients as a group had a mean weight gain of 2.2 kg. Overall, AZT was well tolerated at these doses with the main side effects being headaches and bone marrow suppression (Yarchoan and Broder 1987a; Yarchoan et al. 1986).

### *1.4.2 Randomized Double-Blind Efficacy Trial and Approval of AZT*

There was considerable skepticism at the time whether the increased CD4 count and other findings in this study were indicative of a beneficial effect. In fact, an abstract describing the trial was rejected outright when submitted to the AFCR/ASCI/AAP National Meeting (Yarchoan Personal communication). Even so, many of us were convinced of its activity, and Burroughs Wellcome scientists moved ahead to conduct a multicenter double-blind, placebo-controlled trial in 282 patients to assess the efficacy of oral AZT in patients with AIDS or severe AIDS-related complex (Fischl et al. 1990; Richman et al. 1987). This trial was conducted in 12 centers throughout the United States that had substantial experience with AIDS patients. Patients received 250 mg of AZT every 4 h (145 patients) or placebo (137 patients). The principal endpoints were survival and development of AIDS-related complications. The trial was initiated in June 1986 and terminated by the Data Safety Monitoring Board in September of the same year because of clear evidence that it had clinical activity. At the time of termination, 19 patients on the placebo arm had died, compared to only one patients on the AZT arm, a result that was highly significant ($P<0.001$). In addition, patients receiving AZT had greater CD4 counts, Karnofsky scores, weight, and more frequent reversal of skin test anergy (Fischl et al. 1990; Richman et al. 1987). The principal toxicities seen in the AZT recipients were bone marrow suppression, nausea, myalgia, insomnia, and headaches.

At the time, the design of this randomized trial was highly controversial, with much of this centering on the use of a placebo arm and the deaths that occurred on this arm as one of the principal assessed endpoints (Yarchoan 2012). Nonetheless, the trial greatly accelerated the general acceptance that AZT was in fact active against AIDS, as well as its approval and widespread use. Indeed, from this vantage point, perhaps it would be fair to say that if this trial had not been conducted in this way, many more people would have died because they could not obtain AZT. Even with this trial, many persons held on to the belief for years that AZT toxicity far outweighed the benefits. As an extreme example, a group of so-called AIDS denialists, the most prominent of whom was the scientist Peter Duesberg, claimed that AIDS was not caused by HIV but in fact was caused by a variety of drugs, including AZT (Clayton 1993; Duesberg 1992). This argument has had devastating consequences; it is cited, for example, as being a main contributor to the adoption by the South African government of the viewpoint that HIV was not the cause of AIDS and the opposition of that government to the use of antiretroviral therapy until 2005 (Specter 2007). It has been estimated that as a result of this policy, more than 330,000 lives, or approximately 2.2 million person-years, were lost in South Africa (Chigwedere et al. 2008).

Getting back to AZT's development, on the basis of this phase II study, AZT was almost immediately made widely available to AIDS patients in the United States through a Treatment IND and was approved on March 19, 1987. The time from the initial observation of the in vitro activity of AZT to its approval was 25 months, arguably the fastest time of any drug in the modern era.

While the multicenter trial of AZT was ongoing, our group in the NCI explored higher doses of AZT on a continuation of the phase I trial. We found that the maximum tolerated dose for short-term (6-week) administration was 90 mg/kg/day orally (Yarchoan and Broder 1987b, c). Interestingly, we found that some patients with Kaposi's sarcoma (KS) who received the higher dose of AZT had partial or complete responses of their KS (Yarchoan and Broder 1987b, c). KS has since been shown to be caused by a new herpesvirus, Kaposi's sarcoma-associated herpesvirus (KSHV) (Chang et al. 1994; Moore and Chang 1995), and a lytic gene of this virus has been shown to phosphorylate AZT to a moiety that is toxic to cells (Davis et al. 2007; Gustafson et al. 2000). It is thus possible that the KS responses were in part as a result of AZT's phosphorylation by KSHV.

Also, we had noticed that one patient on the phase I trial had improved cognitive function and we recruited additional patients with AIDS dementia or other neurologic disease to explore this further. We found that each of the three patients with dementia had substantial improvement in their cognitive function on AZT (Yarchoan et al. 1987). The principal cells infected with HIV in the brain are monocytes and related microglial cells, and we found that AZT blocked HIV replication in monocytes (Perno et al. 1988). The claim that AZT could reverse dementia, which was previously felt to be irreversible, was initially met with substantial skepticism by the neurologic community. However, some cognitive testing was done on the multicenter placebo-controlled trial of AZT, and an analysis of those results confirmed that cognitive function did in fact improve in those receiving AZT (Schmitt, et al. 1988). Also, in a subsequent trial of intravenous AZT in HIV-infected children conducted at the NCI, a number had dramatic improvement in HIV-associated neurodevelopmental abnormalities (Pizzo et al. 1988). AZT penetrates into the cerebrospinal fluid better than other AIDS drugs and arguably remains the best drug to treat AIDS-associated neurocognitive disorder.

During early in vitro testing of potential anti-HIV drugs, it was found that the anti-herpes drug acyclovir had little in vitro anti-HIV activity by itself but could potentiate the activity of AZT (Mitsuya and Broder 1987; Mitsuya et al. 1987b). We explored this combination in the clinic and conducted a pilot study of the combination of acyclovir and AZT (Surbone et al. 1988). This small study showed that the regimen was well tolerated and that patients had increased CD4 counts. This study and a second at about the same time of AZT alternating with 2′,3′-dideoxycydine (ddC) (Yarchoan et al. 1988) were the two first combination studies involving effective anti-HIV drugs. Interestingly, there has been renewed interest in the potential of acyclovir as an anti-HIV drug in recent years (McMahon et al. 2011; Vanpouille et al. 2012, 2009).

Finally, during this period, we continued to follow patients on AZT and noticed that their CD4 counts tended to decrease after about 4–6 months on therapy (Yarchoan and Broder 1987c). Several years later, it was found that this was because of the ability of HIV to develop resistance to antiviral drugs (Larder et al. 1989; Larder and Kemp 1989; Mitsuya et al. 1990), which remains one of the principal challenges to prolonged and effective anti-HIV therapy.

Finally, it should be mentioned that when AZT was first approved, the retail price was approximately $8,000 to $12,000 per year, which was at the time one of the highest prices ever for a prescription drug (Zonana 1989). This was called an "inhuman cost" and was highly criticized by AIDS patients, advocates, and government officials. A major factor that helped reduce the price of AZT over the next few years was the anticipation that a competing drug would come onto the market as ddC and ddI were developed.

## 1.5 Development of Zalcitabine (2′,3′-Dideoxycytidine, ddC)

While clinical testing of AZT was ongoing, our group turned to the other dideoxynucleosides that had been shown to have in vitro activity against HIV (Mitsuya and Broder 1986; Mitsuya et al. 1987a, 1990). We initially had no corporate partner for these drugs, and their initial preclinical and early clinical development was completely done within the NCI. The first of these drugs that we chose for clinical development was ddC, which demonstrated potent in vitro activity and a good in vitro activity/toxicity ratio. Animal toxicology was done by Dr. Joseph Tomascewski's group in NCI's Developmental Therapeutics Program, and the IND Application was then filed by Dr. Dale Shoemaker and colleagues in the Regulatory Affairs Branch of the NCI. At the same time, the rights to develop ddC were licensed by the NIH to Hoffmann-La Roche Co., which then became a partner in development of the drug.

The initial ddC phase I trial commenced at the NCI in October 1986 (Yarchoan et al. 1988). By the time this trial was complete, an early version of the HIV p24 antigen assay had been developed, and the evaluable patients receiving ddC at doses of 0.03 mg/kg every 8 h to 0.09 mg/kg every 4 h were found to have a decrease in their serum HIV p24 antigen that was most pronounced at week 2 but was still significantly decreased through week 6 (Yarchoan et al. 1988). The patients also had a slight increase in their CD4 count at week 2, but this did not persist beyond this point. This trial also revealed that ddC was more toxic than AZT at doses that appeared to have anti-HIV activity. Primary toxicities were cutaneous eruptions, fever, mouth sores, arthralgias, thrombocytopenia, and, after several weeks, a painful peripheral neuropathy.

Because the toxicity profile of ddC was different from that of AZT, we reasoned that their combination might yield improved anti-HIV activity with lower toxicity than either drug used alone, and in an extension of the same protocol, we treated an additional six patients with an alternating regimen of AZT and ddC (Yarchoan et al. 1988). An alternating (rather than simultaneous) regimen was utilized because it could be done quickly without the need for combination drug testing in animals. This regimen was well tolerated, and the five patients who completed 9 weeks had a mean increase of 48 CD4 cells/mm$^3$ and tended to have a decrease in p24 antigen. This trial was the first reported study of a combination regimen of effective anti-HIV drugs. Subsequent studies confirmed that the AZT/ddC combination was well tolerated in children and adults (Bozzette and Richman 1990; Pizzo et al. 1990) and provided evidence that the combination of AZT and ddC, administered together, provided better virologic and immunologic activity than AZT alone. ddC (zalcitabine) was ultimately approved by the US FDA as an anti-HIV drug for use in combination with AZT on June 19, 1992. This did not occur until after ddI (didanosine) had been approved, largely because of ddC's relatively weak clinical activity and problematic toxicity profile.

## 1.6 Development of Didanosine (2′,3′-Dideoxyinosine, ddI)

Our group next turned our attention to the purine analogues ddA and ddI. Both of these compounds are phosphorylated in cells to 2′,3′-dideoxyadenosine-5′-triphosphate, which, like AZT-triphosphate and ddC-triphosphate, displayed activity against HIV RT (Ahluwalia et al. 1987, 1997; Cooney et al. 1987; Mitsuya and Broder 1986, 1987; Yarchoan et al. 1989a). Once converted to the 5′-triphosphate moiety, their cellular half-life is more than 12 h, compared to ~3 h for AZT 5′-triphosphate, and could thus potentially be administered only once or twice a day, as a major advantage in treating a chronic infectious disease. Because it seemed more straightforward, clinical development first focused on ddA, and a phase I study of intravenous ddA was initiated in February, 1988 (Yarchoan et al. 1989c).

A barrier to development of these compounds as oral agents was that they rapidly undergo solvolysis under acid conditions to form the free base and dideoxyribose (Cooney et al. 1987); because the stomach is an acid environment, this could reduce their oral absorption. Moreover, it was appreciated that adenine, a product of acid cleavage of ddA that would form in the stomach, could cause renal damage (Linblad et al. 1973); by contrast, the free base of ddI, hypoxanthine, did not have such toxicity. Also, it was recognized that ddA was rapidly converted to ddI in plasma, so that no matter which drug was administered, the form circulating in plasma would be largely ddI. In cells, ddI was then phosphorylated to ddA-5′-triphosphate. For these reasons, the focus of clinical development soon switched to ddI, and a phase I study of this compound was started in October of 1988. To address the problem of acid lability affecting oral absorption, the relatively simple approach was taken of administering ddI with antacids, and when given this way, adequate oral absorption was attained (Hartman et al. 1990; Yarchoan et al. 1989c). As with ddC, the discovery of anti-HIV activity and preclinical development of these drugs was undertaken at the NCI; during its early development, ddI was licensed to Bristol-Myers Squibb Co., which became a partner in its development.

The NCI phase I trial of ddI, which involved 26 patients receiving the drug for up to 42 weeks, showed evidence of anti-HIV activity, especially at the higher doses of 3.2 mg/kg every 12 h orally. In particular, patients had increases in their CD4 cells, improved immunologic function, weight gain, and substantial decreases in HIV p24 antigen (Yarchoan et al. 1990b, 1989b). The principal toxicities were peripheral neuropathy, liver function abnormalities, and rare cases of pancreatitis. Soon after initiation of the NCI study, two others were initiated, one testing it once daily and the other twice daily, and ddI in these trials was found to have similar activity and toxicity profiles to that reported in the initial NCI trial (Cooley et al. 1990; Lambert et al. 1990). In addition, a phase I/II study initiated in children at the NCI showed in addition that it could reverse HIV-associated developmental abnormalities (Butler et al. 1991). Based on these results, several randomized trials were initiated; these trials demonstrated that patients who had been receiving AZT (zidovudine) did better if they were switched to ddI (didanosine) (Delta Coordinating Committee 1996; Kahn et al. 1992; Spruance et al. 1994). Meanwhile, on October 9, 1991, even before these randomized trials were completed, the US FDA approved ddI for both children and adults with advanced AIDS based largely on the phase I data and preliminary CD4 count data from the randomized trials. This approval gave physicians another tool to fight HIV infection, and, as described below, it became possible for physicians to consider combination therapy.

## 1.7 Impact of Anti-HIV Drugs

### *1.7.1 Highly Active Antiretroviral Therapy (HAART)*

Based in part on the experience learned in developing combination chemotherapy for lymphoma and other tumors, our group felt it likely that the treatment of HIV would ultimately require drug combinations and started exploring this in the laboratory and clinic from the outset (Mitsuya et al. 1987b; Nguyen et al. 1993; Surbone et al. 1988; Yarchoan et al. 1988). On an extension of the phase I trial of ddI, we switched patients to a combination of AZT and ddI and found that some with advanced HIV infection or AIDS could survive for 5 years or longer on this combination (Nguyen et al. 1995). In addition, several randomized trials showed that AZT or other dideoxynucleosides used in combination were better than AZT alone and that simultaneous use of these drugs was significantly better than their alternating use (Collier et al. 1993; Hammer et al. 1996; Yarchoan et al. 1994). During this period, work progressed on other HIV targets, and with the development of first protease inhibitors (Kempf et al. 1990; McQuade et al. 1990; Meek et al. 1990;

Miller et al. 1989; Roberts et al. 1990), it was possible to test combinations of three drugs including two dideoxynucleosides and a protease inhibitor (Hammer et al. 1997). The testing of such regimens was greatly aided by development of methodologies to measure the viral load in plasma and to assess HIV resistance (Hirsch et al. 1998; Mellors et al. 1996; Perelson et al. 1996). Certain regimens utilizing three or more drugs, called highly active antiretroviral therapy (HAART), were found to suppress HIV replication to undetectable levels in many patients, so that the development of HIV drug resistance was halted or substantially slowed (Hammer et al. 1997). This approach resulted in substantial immunologic restoration, clinical improvement, and prolonged survival so that AIDS was converted from a rapidly fatal disease to a manageable chronic illness. When HAART regimens were developed, much of the focus was on the contribution of the protease inhibitors. However, it should be remembered that HAART regimens generally included two NRTI such as zidovudine and didanosine and that three-drug regimens utilizing three NRTI were subsequently also found to have excellent activity (Bonjoch et al. 2005), although perhaps not as lasting as regimens utilizing two classes of drugs. It is also noteworthy that despite much attention given to protease and integrase inhibitors, NRTIs have remained the backbone of HIV therapy 25 years after their introduction into the clinic. Contributing to the long-term utility of NRTI are their properties of being relatively inexpensive to synthesize, being co-formulated easily with other drugs, and having relatively minor pharmacokinetic interactions. Also, in many cases, resistance to these drugs developed more slowly than to drugs in other classes.

## 1.8 Prevention of HIV Transmission

Another substantial health benefit of AZT was in preventing mother-to-child and other HIV transmission. A short course of antepartum AZT, intrapartum AZT, and 6-week administration of AZT to the newborn child reduced perinatal transmission by approximately two-thirds in an initial trial (Connor et al. 1994), and adopting this and related anti-HIV therapy around the time of birth has all but eradicated peripartum transmission of HIV. In addition, recent studies have showed that administration of NRTI and other anti-HIV drugs can substantially reduce HIV transmission in discordant couples (Celum and Baeten 2012; Grant et al. 2010), and such preexposure prophylaxis (PrEP) now holds great promise as a strategy to reduce the spread of HIV (Cohen 2011).

## 1.9 Conclusions

When HIV was identified as the causative agent of AIDS, there was hope that this would lead to an effective vaccine in short order, and the development of effective antiviral therapy seemed far less promising. In the April 24, 1984, New York Times report of the press conference given by the Secretary of Health and Human Services to announce the discovery of the cause of AIDS, it was stated that "Finding the cause of AIDS will not necessarily lead to any treatment of the disease soon.... But the finding led the American researchers to express the hope that a vaccine would be developed and ready for testing in 'about 2 years'" (Altman 1984). At that time, patients with HIV infection almost always progressed to AIDS within 10 years or so, and survival of AIDS patients was measured in months. Now, 28 years later, there is no HIV vaccine available, and the most promising lead in this area is a recent clinical trial showing suggestive evidence of a statistically significant but slight protective effect with a prototype vaccine (Rerks-Ngarm et al. 2009). By contrast, effective anti-HIV drugs were rapidly developed and the first approved for use within 3 years; with current combination regimens, the life expectancy of HIV-infected individuals treated with antiretroviral therapy is better than that of individuals with type I diabetes and in some groups is approaching that of the general population (Lewden et al. 2012; Lohse et al. 2007). Moreover, antiretroviral therapy has proven to be effective in preventing mother-to-child transmission and even in preventing transmission of HIV between sexual partners.

During the development of AZT and other dideoxynucleosides, it became apparent early on that these drugs had substantial limitations; at tolerable concentrations, HIV replication was only partially inhibited in vitro, and patients receiving these drugs had only partial and transient clinical improvement, often accompanied by substantial toxicity. We never claimed that these drugs, by themselves, would be the answer to AIDS. Nonetheless, it can be argued that their development led the way to the development of other drugs and that regimens using these drugs have dramatically changed the face of AIDS.

Looking back, the intramural program of the NCI in the 1980s provided an outstanding environment that facilitated the rapid development of early AIDS drugs (Broder 2010a, b). In part because of the "Special Virus-Cancer Program," the NCI had considerable expertise in retroviruses, and soon after the Gallo group developed the ability to isolate and grow HIV, we were able to rapidly transfer this technology to our laboratory. Our group also had considerable immunologic expertise from our previous studies of immunodeficiency disease. The NCI as a whole had substantial expertise in nucleoside chemistry and had an established drug development program that we could tap into for help in animal toxicology, formulation, and regulatory support. Working in an oncologic environment, we were willing to accept and to be attuned for small advances. Also, during this period, we were able to do much of our work as a small, isolated "SWAT team" and were not slowed by a requirement to go through multiple committees for each step of the process. Indeed, because our group encompassed both a laboratory and clinical effort, we could bring drugs to the clinic without a formal hand-off from one group to the other. We should note that the FDA at the time recognized the urgency to develop drugs for AIDS and expeditiously reviewed the applications and protocols for these trials.

During this period of early AIDS drug development, there was considerable controversy that NCI resources were being used in what was viewed as a distraction from the main mission of the NCI – cancer research. However, it is worth noting that these drugs, along with other anti-HIV drugs, dramatically reduced the incidence of AIDS malignancies, especially those associated with low CD4 counts such as Kaposi's sarcoma or primary central nervous system lymphoma (Engels et al. 2006; Herida et al. 2003; Jones et al. 1998). Thus, this work has in fact proven to be extremely important in lowering the morbidity and mortality from AIDS-defining cancers – a goal that is very much within the mission of the NCI.

The NCI continues to contribute to the development of anti-HIV therapy, and Mitsuya and colleagues have more recently developed a fourth approved anti-HIV drug, the protease inhibitor darunavir (Koh et al. 2003). This drug has been licensed royalty free to the Medicines Patent Pool, a UNITAID/WHO initiative that aims to provide new and effective drugs to developing countries. This is a response by the United States Government and the NCI to critical the need for such drugs in resource-limited areas. Along those same lines, in 2003, the United States Government embarked on a program to make antiretroviral therapy available in much of the developing world through the President's Emergency Plan for AIDS Relief (PEPFAR) (Broder 2010a; Walensky and Kuritzkes 2010). Billions of dollars have been committed to PEPFAR, and along with public health measures to reduce transmission, we are finally starting to see a reversal of the AIDS epidemic in many regions of the developing world. It is estimated that more than 1.2 million deaths have been avoided in 5 years by antiretroviral drugs provided by PEPFAR in South Africa alone (Walensky and Kuritzkes 2010). It should also be noted that with advances in the cost-efficient manufacture of antiretroviral drugs and their being made available royalty free, antiretroviral therapy can now be administered in the developing world for less than $350 per person per year (Goosby 2012). This accomplishment is noteworthy in light of the concern raised during the initial development of AZT and related drugs that they were too expensive to be used in the developing world.

*Conflict of Interest* RY, HM, and SB are coinventors on one or more US and/or foreign patents involving dideoxycytidine (zalcitabine), dideoxyadenosine, dideoxyinosine (didanosine), and zidovudine; HM is also coinventor on patents involving darunavir. These inventions were all made as full-time employees of the United States Government under 45 Code of Federal Regulations Part 7. All rights, title, and interest to these patents are assigned to the U.S. Department of Health and Human Services. The government conveys a portion of the royalties it receives to its employee-inventors under the Federal Technology Transfer Act of 1986 (P.L. 99–502).

**Acknowledgments** We thank Dr. Samuel Broder for his thoughtful review of the manuscript.

## References

Ahluwalia G, Cooney DA, Mitsuya H, Fridland A, Flora KP, Hao Z et al (1987) Initial studies on the cellular pharmacology of 2′,3′-dideoxyinosine, an inhibitor of HIV infectivity. Biochem Pharmacol 36(33):3797–3800

Ahluwalia GS, Dedrick RL, Driscoll JS, Morrison PF, Gao W-Y, Johns DG (1997) Decay rates of anti-HIV dideoxynucleotides in tissue culture systems: a simple correction for the effect of cell replication. Drug Metab Dispos 25(7):893–896

Altman LK (1984, April 24, 1984) New U.S. report names virus that may cause AIDS. New York Times. http://partners.nytimes.com/library/national/science/aids/042484sci-aids.html?scp=5&sq=1983%20Gallo%20Heckler&st=cse. Accessed April 24, 2013

Barre-Sinoussi F, Chermann JC, Rey F, Nugeyre T, Charmaret S, Gruest J et al (1983) Isolation of a T-cell lymphotropic virus from a patient at risk for the acquired immunodeficiency syndrome (AIDS). Science 220:868–871

Bonjoch A, Paredes R, Galvez J, Miralles C, Videla S, Martinez E et al (2005) Antiretroviral treatment simplification with 3 NRTIs or 2 NRTIs plus nevirapine in HIV-1-infected patients treated with successful first-line HAART. J Acquir Immune Defic Syndr 39(3):313–316

Bozzette SA, Richman DD (1990) Salvage therapy for zidovudine-intolerant HIV-infected patients with alternating and intermittent regimens of zidovudine and dideoxycytidine. Am J Med 88(5B):24S–26S

Broder S (2010a) The development of antiretroviral therapy and its impact on the HIV-1/AIDS pandemic. Antiviral Res 85(1):1–18

Broder S (2010b) Twenty-five years of translational medicine in antiretroviral therapy: promises to keep. Sci Transl Med 2(39):39ps33

Broder S, Yarchoan R, Collins JM, Lane HC, Markham PD, Klecker RW et al (1985) Effects of suramin on HTLV-III/LAV infection presenting as Kaposi's sarcoma or AIDS-related complex: clinical pharmacology and suppression of viral replication in vitro. Lancet 2:627–630

Butler KM, Husson RN, Balis FM, Brouwers P, Eddy J, El-Amin D et al (1991) Dideoxyinosine (ddI) in symptomatic HIV-infected children: a phase I-II study. N Engl J Med 324:137–144

Celum C, Baeten JM (2012) Tenofovir-based pre-exposure prophylaxis for HIV prevention: evolving evidence. Curr Opin Infect Dis 25(1):51–57

Chang Y, Cesarman E, Pessin M, Lee F, Culpepper J, Knowles DM et al (1994) Identification of herpesvirus-like DNA sequences in AIDS-associated Kaposi's sarcoma. Science 266:1865–1869

Chigwedere P, Seage GR 3rd, Gruskin S, Lee TH, Essex M (2008) Estimating the lost benefits of antiretroviral drug use in South Africa. J Acquir Immune Defic Syndr 49(4):410–415

Clayton J (1993) Duesberg's anti-AZT campaign continues. Nature 363(6431):660

CDC (1981a). MMWR Morb Mortal Wkly Rep 30(25):305–308

CDC (1981b) MMWR Morb Mortal Wkly Rep 30(21):250–252

Cohen J (2011) Breakthrough of the year. HIV treatment as prevention. Science 334(6063):1628

Collier AC, Coombs RW, Fischl MA, Skolnik PR, Northfelt D, Boutin P et al (1993) Combination therapy with zidovudine and didanosine compared with zidovudine alone in HIV-1 infection. Ann Intern Med 119(8):786–793

Connor EM, Sperling RS, Gelber R, Kiselev P, Scott G, O'Sullivan MJ et al (1994) Reduction of maternal-infant transmission of human immunodeficiency virus type 1 with zidovudine treatment. Pediatric AIDS Clinical Trials Group Protocol 076 Study Group. N Engl J Med 331(18):1173–1180

Cooley TP, Kunches LM, Saunders CA, Ritter JK, Perkins CJ, Colin M et al (1990) Once-daily administration of 2′,3′-dideoxyinosine (ddI) in patients with the acquired immunodeficiency syndrome or AIDS-related complex. N Engl J Med 322(19):1430–1345

Cooney DA, Ahluwalia G, Mitsuya H, Fridland A, Johnson M, Hao Z et al (1987) Initial studies on the cellular pharmacology of 2′,3′-dideoxyadenosine, an inhibitor of HTLV-III infectivity. Biochem Pharmacol 36(11):1765–1768

Davis DA, Singer KE, Reynolds IP, Haque M, Yarchoan R (2007) Hypoxia enhances the phosphorylation and cytotoxicity of ganciclovir and zidovudine in Kaposi's sarcoma-associated herpesvirus infected cells. Cancer Res 67(14):7003–7010

De Clercq E (1979) Suramin: a potent inhibitor of the reverse transcriptase of RNA tumor viruses. Cancer Lett 8(1):9–22

Delta Coordinating Committee (1996) DELTA: a randomized double-blind controlled trial comparing combinations of zidovudine plus didanosine or zalcitabine with zidovudine monotherapy in individuals with HIV infection. Lancet 348:293–291

Duesberg PH (1992) AIDS acquired by drug consumption and other noncontagious risk factors. Pharmacol Ther 55(3):201–277

Engels EA, Pfeiffer RM, Goedert JJ, Virgo P, McNeel TS, Scoppa SM et al (2006) Trends in cancer risk among people with AIDS in the United States 1980–2002. AIDS 20(12):1645–1654

Fischl M, Richman DD, Hansen N, Collier AC, Carey JT, Para MF et al (1990) The safety and efficacy of zidovudine (AZT) in the treatment of subjects with mildly symptomatic human immunodeficiency virus type I (HIV) infection. A double-blind, placebo controlled trial. Ann Intern Med 112:727–737

Furman PA, Fyfe JA, St. Clair M, Weinhold K, Rideout JL, Freeman GA et al (1986) Phosphorylation of 3′-azido-3′-deoxythymidine and selective interaction of the 5′-triphosphate with human immunodeficiency virus reverse transcriptase. Proc Natl Acad Sci USA 83:8333–8337

Furmanski P, Bourguignon GJ, Bolles CS, Corombos JD, Das MR (1980) Inhibition by 2′,3′-dideoxythymidine of retroviral infection of mouse and human cells. Cancer Lett 8:307–315

Gallo RC, Salahuddin SZ, Popovic M, Shearer GM, Kaplan M, Haynes BF et al (1984) Frequent detection and isolation of cytopathic retroviruses (HTLV-III) from patients with AIDS and at risk for AIDS. Science 224:500–503

Goosby E (2012) President's budget request reflects strong commitment on global AIDS. DipNote, US Department of State Official Blog, from http://blogs.state.gov/index.php/site/entry/budget_global_aids

Grant RM, Lama JR, Anderson PL, McMahan V, Liu AY, Vargas L et al (2010) Preexposure chemoprophylaxis for HIV prevention in men who have sex with men. N Engl J Med 363(27):2587–2599

Gustafson EA, Schinazi RF, Fingeroth JD (2000) Human herpesvirus 8 open reading frame 21 is a thymidine and thymidylate kinase of narrow substrate specificity that efficiently phosphorylates zidovudine but not ganciclovir. J Virol 74(2):684–692

Hammer S, Katzenstein D, Hughes M, Schooley R, Hirsch M, Merigan T et al (1996) A trial comparing nucleoside monotherapy with combination therapy in HIV-infected adults with CD4 cell counts from 200 to per cubic millimeter. NEJM 335:1081–1090

Hammer SM, Squires KE, Hughes MD, Grimes JM, Demeter LM, Currier JS et al (1997) A controlled trial of two nucleoside analogues plus indinavir in persons with human immunodeficiency virus infection and CD4 cell counts of 200 per cubic millimeter or less. AIDS Clinical Trials Group 320 Study Team. N Engl J Med 337(11):725–733

Harden VA (2012) AIDS at 30: a history, 1st edn. Potomac Books, Dulles

Hartman NR, Yarchoan R, Pluda JM, Thomas RV, Marczyk KS, Broder S et al (1990) Pharmacokinetics of 2′,3′-dideoxyadenosine and 2′,3′-dideoxyinosine in patients with severe HIV infection. Clin Pharmacol Ther 47:647–654

Herida M, Mary-Krause M, Kaphan R, Cadranel J, Poizot-Martin I, Rabaud C et al (2003) Incidence of non-AIDS-defining cancers before and during the highly active antiretroviral therapy era in a cohort of human immunodeficiency virus-infected patients. J Clin Oncol 21(18):3447–3453

Hirsch MS, Conway B, D'Aquila RT, Johnson VA, Brun-Vezinet F, Clotet B et al (1998) Antiretroviral drug resistance testing in adults with HIV infection: implications for clinical management. International AIDS Society–USA Panel [see comments]. JAMA 279(24):1984–1991

Horwitz JP, Chua J, Noel M (1964) Nucleosides. V. The monomesylates of 1-(2′-deoxy-ß-D-lyxofuranosyl)thymidine. J Org Chem 29:2076–2078

Horwitz JP, Chua J, Noel M, Donatti JT (1967) Nucleosides. XI. 2′,3′-dideoxycytidine. J Org Chem 32(3):817–818

Jones JL, Hanson DL, Ward JW (1998) Effect of antiretroviral therapy on recent trends in cancers among HIV-infected. J Acquir Immune Defic Syndr Hum Retrovirol 17(4):A38 (Abstract S33)

Kahn JO, Lagakos SW, Richman DD, Cross A, Pettinetti C, Liou S-H et al (1992) A controlled trial comparing continued zidovudine with didanosine in human immunodeficiency virus infection. N Engl J Med 327(9):581–587

Kempf D, Norbeck D, Codacovi L (1990) Structure-based C2 symmetric inhibitors of HIV protease. J Med Chem 33:2687–2689

Klecker RW Jr, Collins JM, Yarchoan R, Thomas R, Jenkins JF, Broder S et al (1987) Plasma and cerebrospinal fluid pharmacokinetics of 3′-azido-3′-deoxythymidine: a novel pyrimidine analog with potential application for the treatment of patients with AIDS and related diseases. Clin Pharmacol Ther 41:407–412

Koh Y, Nakata H, Maeda K, Ogata H, Bilcer G, Devasamudram T et al (2003) Novel bis-tetrahydrofuranylurethane-containing nonpeptidic protease inhibitor (PI) UIC-94017 (TMC114) with potent activity against multi-PI-resistant human immunodeficiency virus in vitro. Antimicrob Agents Chemother 47(10):3123–3129

Krieg CJ, Ostertag W, Clauss U, Pragnell IB, Swetly P, Roesler G et al (1978) Increase in intracisternal A-type particles in Friend cells during inhibition of Friend virus (SFFV) release by interferon or azidothymidine. Exp Cell Res 116:21–29

Lambert JS, Seidlin M, Reichman RC, Plank CS, Laverty M, Morse GD et al (1990) 2′,3′-Dideoxyinosine (ddI) in patients with the acquired immunodeficiency syndrome or the AIDS-related complex. A phase I trial. N Engl J Med 322:1333–1340

Larder BA, Kemp SD (1989) Multiple mutations in HIV-1 reverse transcriptase confer high-level resistance to zidovudine (AZT). Science 246:1155–1158

Larder BA, Darby G, Richman DD (1989) HIV with reduced sensitivity to zidovudine (AZT) isolated during prolonged therapy. Science 243:1731–1734

Lewden C, Bouteloup V, De Wit S, Sabin C, Mocroft A, Wasmuth JC et al (2012) All-cause mortality in treated HIV-infected adults with CD4>=500/$mm^3$ compared with the general population: evidence from a large European observational cohort collaboration{dagger}. Int J Epidemiol 41(2):433–445

Linblad G, Johnsson G, Falk J (1973) Adenine toxicity; a three week intravenous study in dogs. Acta Pharmacol Toxicol 32:246–256

Lohse N, Hansen AB, Pedersen G, Kronborg G, Gerstoft J, Sorensen HT et al (2007) Survival of persons with and without HIV infection in Denmark, 1995–2005. Ann Intern Med 146(2):87–95

McMahon MA, Parsons TL, Shen L, Siliciano JD, Siliciano RF (2011) Consistent inhibition of HIV-1 replication in CD4+ T cells by acyclovir without detection of human herpesviruses. J Virol 85(9):4618–4622

McQuade TJ, Tomasselli AG, Liu L, Karacostas V, Moss B, Sawyer TK et al (1990) A synthetic HIV-1 protease inhibitor with antiviral activity arrests HIV-like particle maturation. Science 247:454–456

Meek TD, Lambert DM, Dreyer GB, Carr TJ, Tomaszek TA Jr, Moore ML et al (1990) Inhibition of HIV-1 protease in infected T-lymphocytes by synthetic peptide analogues. Nature 343:90–92

Mellors JW, Rinaldo CR Jr, Gupta P, White RM, Todd JA, Kingsley LA (1996) Prognosis in HIV-1 infection predicted by the quantity of virus in plasma. Science 272(5265):1167–1170

Miller M, Schneider J, Sathyanarayana BK, Toth MV, Marshall GR, Clawson L et al (1989) Structure of a complex of synthetic HIV-1 protease with a substrate-based inhibitor at 2.3 A resolution. Science 246:1149–1152

Mitsuya H, Broder S (1986) Inhibition of the in vitro infectivity and cytopathic effect of human T-lymphotropic virus type III/lymphadenopathy virus-associated virus (HTLV-III/LAV) by 2′,3′-dideoxynucleosides. Proc Natl Acad Sci USA 83:1911–1915

Mitsuya H, Broder S (1987) Strategies for antiviral therapy in AIDS. Nature 325:773–778

Mitsuya H, Popovic M, Yarchoan R, Matsushita S, Gallo RC, Broder S (1984) Suramin protection of T cells in vitro against infectivity and cytopathic effect of HTLV-III. Science 226:172–174

Mitsuya H, Weinhold KJ, Furman PA, St. Clair MH, Nusinoff Lehrman S, Gallo RC et al (1985) 3′-Azido-3′-deoxythymidine (BW A509U): an antiviral agent that inhibits the infectivity and cytopathic effect of human T-lymphotropic virus type III/lymphadenopathy-associated virus in vitro. Proc Natl Acad Sci USA 82:7096–7100

Mitsuya H, Jarrett RF, Matsukura M, di Marzo Veronese F, deVico AL, Sarngadharan MG et al (1987a) Long-term inhibition of human T-lymphotropic virus type III/lymphadenopathy-associated virus (human immunodeficiency virus) DNA synthesis and RNA expression in T cells protected by 2′,3′-dideoxynucleosides in vitro. Proc Natl Acad Sci USA 84:2033–2037

Mitsuya H, Matsukura M, Broder S (1987b) Rapid in vitro systems for assessing activity of agents against HTLV-III/LAV. In: Broder S (ed) AIDS: modern concepts and therapeutic challenges. Marcel Dekker, New York, pp. 303–333

Mitsuya H, Yarchoan R, Broder S (1990) Molecular targets for AIDS therapy. Science 249: 1533–1544

Moore PS, Chang Y (1995) Detection of herpesvirus-like DNA sequences in Kaposi's sarcoma in patients with and without HIV infection. N Engl J Med 332(18):1181–1185

Nguyen B-Y, Shay LE, Wyvill KM, Pluda JM, Brawley O, Cohen RB et al (1993) A pilot study of sequential therapy with zidovudine plus acyclovir, dideoxyinosine, and dideoxycytidine in patients with severe human immunodeficiency virus infection. J Infect Dis 168:810–817

Nguyen B-Y, Yarchoan R, Wyvill KM, Venzon DJ, Pluda JM, Mitsuya H et al (1995) Five-year follow-up of a phase I study of didanosine in patients with advanced human immunodeficiency virus infection. J Infect Dis 171:1180–1189

Ostertag W, Roesler G, Krieg CJ, Kind J, Cole T, Crozier T et al (1974) Induction of endogenous virus and of thymidine kinase by bromodeoxyuridine in cell cultures transformed by friend virus. Proc Natl Acad Sci USA 71(12):4980–4985

Perelson AS, Neumann AU, Markowitz M, Leonard JM, Ho DD (1996) HIV-1 dynamics in vivo: virion clearance rate, infected cell life-span, and viral generation time. Science 271(5255): 1582–1586

Perno CF, Yarchoan R, Cooney DA, Hartman NR, Gartner S, Popovic M et al (1988) Inhibition of human immunodeficiency virus (HIV-1/HTLV-$III_{Ba\text{-}L}$) replication in fresh and cultured human peripheral blood monocytes/macrophages by azidothymidine and related 2′,3′-dideoxynucleosides. J Exp Med 168:1111–1125

Pizzo PA, Eddy J, Falloon J, Balis FM, Murphy RF, Moss H et al (1988) Effect of continuous intravenous infusion zidovudine (AZT) in children with symptomatic HIV infection. N Engl J Med 319:889–896

Pizzo PA, Butler K, Balis F, Brouwers P, Hawkins M, Eddy J et al (1990) Dideoxycytidine alone and in an alternating schedule with zidovudine (AZT) in children with symptomatic human immunodeficiency virus infection. J Pediatr 117:799–808

Popovic M, Sarngadharan MG, Read E, Gallo RC (1984) Detection, isolation, and continuous production of cytopathic retroviruses (HTLV-III) from patients with AIDS and pre-AIDS. Science 224:497–500

Rerks-Ngarm S, Pitisuttithum P, Nitayaphan S, Kaewkungwal J, Chiu J, Paris R et al (2009) Vaccination with ALVAC and AIDSVAX to prevent HIV-1 infection in Thailand. N Engl J Med 361(23):2209–2220

Richman DD, Fischl MA, Grieco MH, Gottlieb MS, Volberding PA, Laskin OL et al (1987) The toxicity of azidothymidine (AZT) in the treatment of patients with AIDS and AIDS-related complex: a double-blind, placebo-controlled trial. N Engl J Med 317:192–197

Roberts NA, Martin JA, Kinchington D, Broadhurst AV, Craig JC, Duncan IB et al (1990) Rational design of peptide-based HIV proteinase inhibitors. Science 248:358–361

Robins MJ, Robins RK (1964) The synthesis of 2′,3′-dideoxyadenosine from 2′-deoxyadenosine. J Am Chem Soc 86(17):3585–3586

Schmitt FA, Bigley JW, McKinnis R, Logue PE, Evans RW, Drucker JL et al (1988) Neuropsychological outcome of zidovudine (AZT) treatment of patients with AIDS and AIDS-related complex. N Engl J Med 319(24):1573–1578

Specter M (2007) The denialists: the dangerous attacks on the consensus about H.I.V. and AIDS. New Yorker 32–38

Spruance SL, Pavia AT, Peterson D, Berry A, Pollard R, Patterson TF et al (1994) Didanosine compared with continuation of zidovudine in HIV-infected patients with signs of clinical deterioration while receiving zidovudine. A randomized, double-blind clinical trial. Ann Intern Med 120:360–368

Surbone A, Yarchoan R, McAtee N, Blum R, Allain J-P, Thomas RV et al (1988) Treatment of acquired immunodeficiency syndrome (AIDS) and AIDS-related complex with a regimen of 3′-azido-2′,3′-dideoxythymidine (azidothymidine or zidovudine) and acyclovir. Ann Intern Med 108(4):534–540

Ting RC, Yang SS, Gallo RC (1972) Reverse transcriptase, RNA tumour virus transformation and derivatives of rifamycin SV. Nat New Biol 236(67):163–166

Vanpouille C, Lisco A, Margolis L (2009) Acyclovir: a new use for an old drug. Curr Opin Infect Dis 22(6):583–587

Vanpouille C, Lisco A, Introini A, Grivel JC, Munawwar A, Merbah M et al (2012) Exploiting the Anti-HIV-1 activity of acyclovir: the suppression of primary and drug-resistant HIV isolates and its potentiation by Ribavirin. Antimicrob Agents Chemother 56:2604–2611

Walensky RP, Kuritzkes DR (2010) The impact of the president's emergency plan for AIDS relief (PEPfAR) beyond HIV and why it remains essential. Clin Infect Dis 50(2):272–275

Walensky RP, Paltiel AD, Losina E, Mercincavage LM, Schackman BR, Sax PE et al (2006) The survival benefits of AIDS treatment in the United States. J Infect Dis 194(1):11–19

Waqar MA, Evans MJ, Manly KF, Hughes RG, Huberman JA (1984) Effects of 2′,3′-dideoxynucleosides on mammalian cells and viruses. J Cell Physiol 121:402–408

Yang SS, Herrera FM, Smith RG, Reitz MS, Lancini G, Ting RC et al (1972) Rifamycin antibiotics: inhibitors of Rauscher murine leukemia virus reverse transcriptase and of purified DNA polymerases from human normal and leukemic lymphoblasts. J Natl Cancer Inst 49(1):7–25

Yarchoan M (2012) The story of AZT: partnership and conflict. Downloaded from Scribd.com on 12 May 2012: Scribd

Yarchoan R, Broder S (1987a) Development of antiretroviral therapy for the acquired immunodeficiency syndrome and related disorders. A progress report. New Engl J Med 316:557–564

Yarchoan R, Broder S (1987b) Preliminary results on the use of dideoxynucleosides in the therapy of AIDS. In: Chanock RM, Lerner RA, Brown F, Ginsberg H (eds) Vaccines 87: modern approaches to new vaccines: prevention of AIDS and other viral bacterial, and parasitic diseases. Cold Spring Harbor Laboratory Press, Cold Spring Harbor, pp. 214–224

Yarchoan R, Broder S (1987c) Progress in the development of antiviral therapy for HTLV-III-associated diseases. In: DeVita VT, Hellman S, Rosenberg SA (eds) Important advances in oncology 1987. J.B. Lippincott Co, Philadelphia, pp. 293–311

Yarchoan R, Mitsuya H, Matsushita S, Broder S (1985) Implications of the discovery of HTLV-III for the treatment of AIDS. Cancer Res 45:4685s–4688s

Yarchoan R, Klecker RW, Weinhold KJ, Markham PD, Lyerly HK, Durack DT et al (1986) Administration of 3′-azido-3′-deoxythymidine, an inhibitor of HTLV-III/LAV replication, to patients with AIDS or AIDS-related complex. Lancet 1:575–580

Yarchoan R, Berg G, Brouwers P, Fischl MA, Spitzer AR, Wichman A et al (1987) Response of human-immunodeficiency-virus-associated neurological disease to 3′-azido-3′-deoxythymidine. Lancet 1:132–135

Yarchoan R, Perno CF, Thomas RV, Klecker RW, Allain J-P, Wills RJ et al (1988) Phase I studies of 2′,3′-dideoxycytidine in severe human immunodeficiency virus infection as a single agent and alternating with zidovudine (AZT). Lancet 1:76–81

Yarchoan R, Mitsuya H, Myers CE, Broder S (1989a) Clinical pharmacology of 3′-azido-2′,3′-dideoxythymidine (zidovudine) and related dideoxynucleosides. N Engl J Med 321:726–738

Yarchoan R, Mitsuya H, Thomas RV, Pluda JM, Hartman NR, Perno C-F et al (1989b) In vivo activity against HIV and favorable toxicity profile of 2′,3′-dideoxyinosine. Science 245:412–415

Yarchoan R, Thomas RV, Mitsuya H, Perno C-F, Pluda JM, Hartman NR et al (1989) Initial clinical studies of 2′,3′-dideoxyadenosine (ddA) and 2′,3′-dideoxyinosine (ddI) in patients with AIDS or AIDS-related complex (ARC). J Cell Biochem (41:13B):313

Yarchoan R, Mitsuya H, Pluda J, Marczyk KS, Thomas RV, Hartman NR et al (1990a) The National Cancer Institute phase I study of ddI administration in adults with AIDS or AIDS-related complex: analysis of activity and toxicity profiles. Rev Infect Dis 12(5):S522–S533

Yarchoan R, Pluda JM, Thomas RV, Mitsuya H, Prouwers P, Wyvill KM et al (1990b) Long-term toxicity/activity profile of 2′,3′-dideoxyinosine in AIDS or AIDS-related complex. Lancet 2:526–529

Yarchoan R, Lietzau JA, Nguyen B-Y, Brawley OW, Pluda JM, Saville MW et al (1994) A randomized pilot study of alternating or simultaneous zidovudine and didanosine therapy in patients with symptomatic immunodeficiency virus infection. J Infect Dis 169:9–17

Zonana VF (1989, August 31, 1989) AIDS Groups Urge Firm to Lower AZT Price. Los Angeles Times

# Part I
# Structure and Function of HIV RT

# Chapter 2
# Proviral DNA Synthesis in HIV: Background

**Dorota Piekna-Przybylska and Robert A. Bambara**

## 2.1 Introduction

Human immunodeficiency virus (HIV) stores its genetic information in the form of RNA. This RNA genome is introduced into the target cell during infection. The virus belongs to the family of retroviruses, as it is able to reverse the normal flow of genetic information from DNA to RNA by copying its RNA genome into DNA using the viral enzyme reverse transcriptase (RT). Each viral particle contains two copies of positive-strand RNA genome enclosed by a core composed of 2,000 copies of the capsid (CA) protein. The RNA genome is tightly bound to nucleocapsid proteins (NC) and other enzymes needed for the early steps of viral infection and is protected by a capsid surrounded by a shell composed of matrix (MA) proteins. The shell is located underneath the virion envelope, a plasma membrane of host-cell origin. During infection, the viral envelope fuses with the cell membrane, releasing the capsid into the cytoplasm. Thereafter, the RNA genome undergoes a multistep process of conversion into DNA within the reverse transcription complex (RTC) (Fig. 2.1). After reverse transcription, the newly synthesized DNA is integrated into the host nuclear genome and permanently linked with their target cells.

## 2.2 The Viral RNA Genome and DNA Provirus

In all retroviruses, reverse transcription of the RNA genome is a discontinuous process, in which short single-stranded DNA products initiated from RNA primers can be fully elongated only after they are moved from the original site of synthesis to

D. Piekna-Przybylska, Ph.D. • R.A. Bambara, Ph.D. (✉)
Department of Microbiology and Immunology, School of Medicine and Dentistry, University of Rochester, 601 Elmwood Ave, Box 672, Rochester, NY, USA
e-mail: Dorota_Piekna@urmc.rochester.edu; Robert_Bambara@urmc.rochester.edu

S.F.J. LeGrice and M. Götte (eds.), *Human Immunodeficiency Virus Reverse Transcriptase: A Bench-to-Bedside Success*, DOI 10.1007/978-1-4614-7291-9_2,

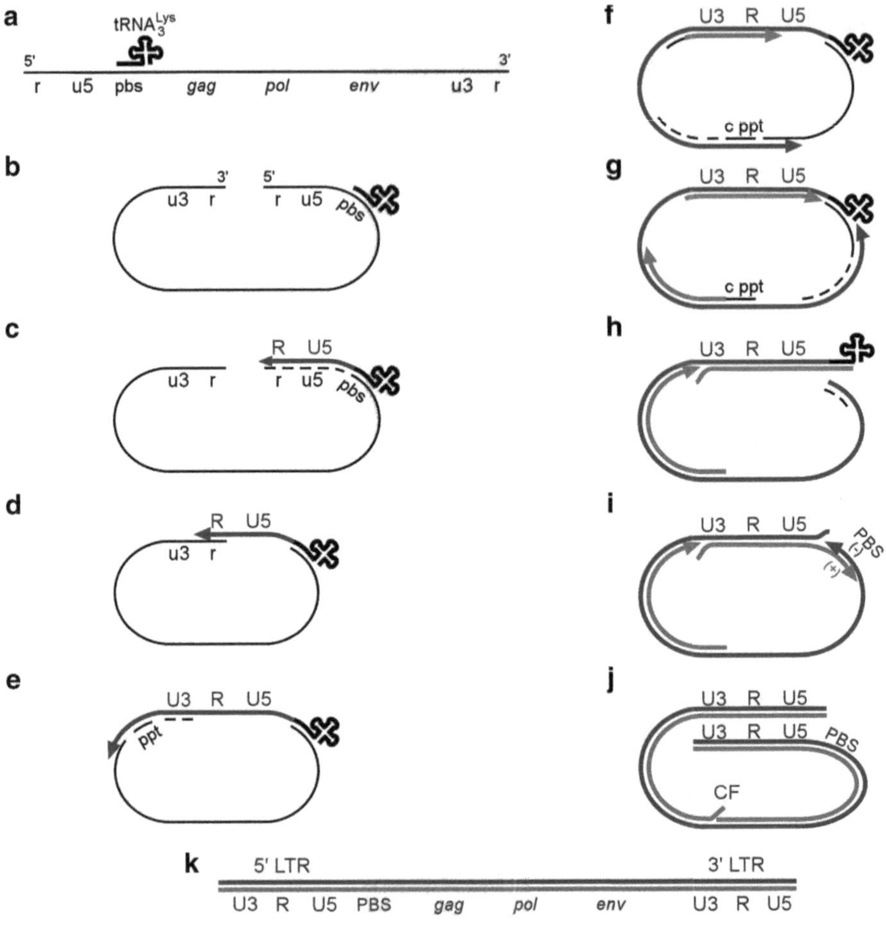


**Fig. 2.1** HIV-1 proviral DNA synthesis. (**a**) After virus entry into the target cell, the RNA genome (*black thin line*) undergoes a multistep process of conversion into double-stranded DNA. tRNA3Lys bound to the PBS near the 5′ end of RNA genome serves as a primer to initiate reverse transcription. (**b**) Genome circularization brings the 5′ and 3′ ends of the RNA genome into proximity. (**c**) While minus-strand strong-stop DNA is synthesized (*dark gray thick line*), the RNA template is degraded (*dashed black line*) by the RNase H activity of RT. (**d**) In order to continue synthesis of minus-strand DNA, complementary sequences of R elements in the cDNA and at the 3′ end of the RNA genome interact, allowing the first (minus)-strand transfer. (**e**) When RT passes the U3 region, the first U3-R-U5 sequence of the LTR is created. During RNA genome degradation, a short fragment (3′PPT) is not removed, being resistant to RNase H cleavage. (**f**) The 3′PPT serves as a primer to start synthesis of plus-strand DNA (*light gray thick line*). The second RNA primer, called the central PPT (cPPT), is created in the middle of the viral genome. (**g**) cPPT extension allows synthesis of plus-strand DNA in the 3′ half of the viral genome. (**h**) Extension of plus-strand DNA (from the 3′PPT) and minus-strand DNA leads to displacement of the tRNA primer from the RNA genome and copying of sequences of the 3′ end in the tRNA primer and PBS region, respectively. (**i**) Complementary sequences of the (–) and (+) PBS interact to promote the second (plus)-strand DNA transfer reaction. (**j**) Further extension of minus-strand DNA and plus-strand DNA initiated at the cPPT leads to strand displacement within the U3-R-U5 DNA duplex, creating two copies of this region at each genomic end. Plus-strand synthesis initiated at the 3′PPT is terminated at the CTS, where the central DNA flap (CF) is created. (**k**) The final product is a linear double-stranded DNA, designated proviral DNA, flanked by LTR sequences

another location on the viral genome. The process is called strand transfer and allows synthesis of full-length proviral DNA. The main reason why retroviruses exploit discontinuous DNA synthesis and strand transfer is to produce two identical long terminal repeats (LTRs) at each end of the viral DNA genome, which are needed for successful integration. Most importantly, the term retrovirus refers to the conversion of a messenger RNA-like genome back to the gene that created it. The gene has regulatory regions for transcription that flank the mRNA product coding sequence. The only way to produce these flanking sequences is discontinuous sequence duplication mechanism of reverse transcription. The LTR is about 650-nt long and consists of two unique sequences designated U3 and U5, separated by the repeat (R) element. The LTR does not encode proteins but contains regulatory elements important for replication and integration into the host chromosome. The LTR flanking the 5′ end of the proviral genome acts as a promoter and enhancer of viral RNA transcription, whereas the 3′ LTR acts as a transcription termination and polyadenylation site (Guntaka 1993; Pereira et al. 2000; Ramirez de Arellano et al. 2006).

The coding region of the viral RNA is flanked by sequences of R-U5 at the 5′ end and U3-R at the 3′ end. The presence of identical sequences of the R region at both ends is crucial for continuous synthesis of DNA and reconstruction/duplication of the U3-R-U5 LTR sequence during reverse transcription. In HIV-1, the 97-nt R element folds into two conserved stem-loop structures, the trans-activation response (TAR) element, and the PolyA hairpin. TAR provides a binding site for the viral Tat protein, which stimulates transcription from the 5′ LTR promoter (Hauber and Cullen 1988; Muesing et al. 1987; Selby et al. 1989). The PolyA hairpin encompasses the AAUAAA polyadenylation signal recognized by the cleavage polyadenylation specificity factor (CPSF), resulting in addition of a poly(A) tail to viral transcripts (Gilmartin et al. 1995). Although both elements are present at the 5′ and 3′ ends of the viral genome, TAR exerts its function only within the 5′ LTR, whereas the polyadenylation signal is recognized exclusively in the 3′ LTR.

## 2.3 Location of the Reverse Transcription Process

After infection, the viral core containing the RTC is released into the cytoplasm. During the process of reverse transcription, the core-RTC traverses the cytoplasm toward the nucleus utilizing the microtubule network (McDonald et al. 2002). The HIV-1 core has a cone-shaped morphology (Ganser et al. 1999), and in addition to two copies of genomic RNA, it contains the tRNA3Lys primer and structural (NC, MA), enzymatic (PR, RT and IN), and accessory proteins (Nef and Vpr) (Bukrinsky et al. 1993; Heinzinger et al. 1994; Vogt 1997). Loss of the viral capsid occurs in the cytoplasm and before nuclear import. It was previously believed that capsid disassembly, also called uncoating, occurs shortly after entry into infected cells following fusion with the plasma membrane (Bukrinsky 2004; Dvorin and Malim 2003; Suzuki and Craigie 2007). However, many studies indicate that the RTC remains associated with the capsid after entry (Forshey and Aiken 2003; Forshey et al. 2002; McDonald et al. 2002). Moreover, scanning electron microscopy imaging revealed that intact HIV-1 capsids containing viral DNA appear in close proximity to the nuclear membrane and the nuclear pore (Arhel et al. 2007). This implies that HIV-1 reverse transcription occurs within an intact capsid core, which remains associated with the viral genome from entry of the cell to the time when the complex moves to the nuclear pore (Arhel et al. 2007). It is believed that association of the capsid with the RTC is weak, but necessary to support its integrity and protect the RNA genome and reverse transcription intermediates from degradation. After completion of DNA synthesis, the RTC converts into a pre-integration complex (PIC), which is capable of translocation through the nuclear pore and then integrating into the host chromosome (Arhel 2010).

## 2.4 HIV-1 Reverse Transcriptase

The key player in the HIV-1 replication cycle is the viral reverse transcriptase (RT), which is an RNA-dependent and DNA-dependent DNA polymerase with RNase H activity. It is responsible for copying the single-stranded RNA into double-stranded DNA. The enzyme is encoded downstream from the *gag* gene within the coding region of the *pol* gene, which also contains sequences of protease (PR) and integrase (IN). However, the enzymes of the *pol* gene are translated into proteins only as a result of ribosomal frameshifting near the 3′ end of *gag*. Most ribosomes synthesizing the gag polyprotein terminate elongation before the *pol* coding region. A minority of ribosomes (about 5 %) undergo a frameshift that allows bypass of the *gag* termination codon and synthesis of the large gal-pol fusion protein (Wills et al. 1994). The gag-pol precursor is subsequently processed into separate enzymes and proteins by the viral PR.

Mature HIV-1 RT is a heterodimer composed of two subunits, p66 and p51. The p66 subunit contains four subdomains named fingers, palm, thumb, and connection, based on the structural similarity to a hand (Arnold et al. 1992; Jacobo-Molina et al. 1991; Kohlstaedt et al. 1992) and a C-terminal RNase H domain. The p51 subunit is a product of the same gene as p66, but is devoid of 120 amino acids of the RNase H domain as a result of proteolytic cleavage (Le Grice and Gruninger-Leitch 1990). The polymerization and RNase H activities of RT are present within the p66 subunit, whereas p51 serves as a structural support (Hostomsky et al. 1992; Le Grice et al. 1991). Polymerase domain and RNase H domain are important for RT to bind to RNA-DNA, RNA-RNA, or DNA-DNA hybrids.

The DNA polymerase active site at the primer 3′ terminus and the RNase H active site accommodate 18 nucleotides of RNA-DNA hybrid and 17 nucleotides for duplex DNA (Gopalakrishnan et al. 1992; Jacobo-Molina et al. 1993; Sarafianos et al. 2001). This configuration allows RT to cleave the plus-strand genomic RNA template during synthesis of minus-strand DNA. The process has been designated polymerization-dependent RNase H activity and initiates the degradation of the RNA genome. Since the rate of DNA synthesis is greater than that of hydrolysis, the excess RT molecules in the virion (~50) participate in polymerization-independent cleavages for complete removal of genomic RNA (DeStefano et al. 1991; Julias et al. 2001; Schultz and Champoux 2008). The 5′ and 3′ ends of the RNA molecule in RNA-DNA duplexes can direct binding of one or more RTs not involved in synthesis. RNA 5′ end-directed RT binding occurs when the enzyme aligns the polymerase active site near the 5′ end of the RNA. The primary cut appears approximately 18 nt away from this end (DeStefano et al. 1993; Palaniappan et al. 1996). The enzyme subsequently slides toward the RNA 5′ end or rebinds to make a secondary cut at a distance of 8–9 nt from the 5′ end of the RNA (Wisniewski et al. 2000a, b). For RNA 3′ end-directed binding, the RT positions its RNase H active site approximately 5 nt from the RNA 3′ end for a secondary-type cleavage (Wisniewski et al. 2000a, b). In addition, the enzyme can bind RNA-DNA duplexes away from both RNA termini and perform internal cuts (Schultz et al. 2004).

## 2.5 Selection of the tRNA Primer for HIV-1

The tRNA primer used by HIV-1 RT to initiate reverse transcription is the host tRNA3Lys. Conversion of the RNA genome into double-stranded DNA begins with the extension of tRNA3Lys annealed to an 18-nucleotide primer binding site (PBS) sequence located near the 5′ end of the viral RNA. RT has strong affinity for this primer, but these interactions are not specific (Thrall et al. 1996). tRNA3Lys is selectively incorporated into virions along with two other tRNALys isoacceptors and human lysyl-tRNA synthetase (LysRS) (Jiang et al. 1993). The mechanism of this process is not yet well understood, but it occurs independent of genome encapsidation, indicating that RNA-RNA contacts are not driving the incorporation (Mak et al. 1994). During formation of the viral particle, tRNA3Lys and two major isoacceptors are selectively concentrated at the site of assembly. In the virion these three tRNAs represent 60 % of the incorporated low-molecular-weight RNAs, while in the cytoplasm their proportion is only 6 % (Mak et al. 1994). A single viral particle contains eight molecules of tRNA3Lys and twelve molecules of tRNA1,2Lys for each genomic RNA dimer (Huang et al. 1994). While tRNA3Lys is used to prime reverse transcription, the role of two other isoacceptors is not known. The tRNA1,2Lys molecules differ in sequence from tRNA3Lys and cannot serve as initiation primers. However, they may be related to an import of the PIC into the nucleus, based on studies showing that tRNA1,2Lys lacking a 3′ CCA tail efficiently induced nuclear import (Zaitseva et al. 2006).

In the cytoplasm, LysRS is a major tRNA-binding protein for tRNALys isoacceptors and normally is responsible for tRNA aminoacylation. About 25 molecules of LysRS are present in each virion, similar to the number of packaged tRNALys molecules (Cen et al. 2002). The interaction between tRNALys and LysRS (Javanbakht et al. 2002), but not tRNA aminoacylation, is required for tRNA incorporation (Cen et al. 2004). Gag and gag-pol can interact with LysRS (Halwani et al. 2004) and tRNALys (Khorchid et al. 2000), respectively, but only gag is believed to guide their packaging into the virions (Cen et al. 2001).

Recently, it was reported that the mitochondrial LysRS, and not the cytoplasmic LysRS, is a target for incorporation into viral particles (Kaminska et al. 2007). However, several reports demonstrated that exogenous cytoplasmic LysRS expressed in infected cells results in its incorporation, with increased tRNALys packaging and viral infectivity (Cen et al. 2004; Gabor et al. 2002; Guo et al. 2005). It was suggested that cytoplasmic LysRS could not be detected in virions when using antibodies against the labile N-terminus of this protein (Cen et al. 2001; Kleiman et al. 2010). Antibodies recognizing a full-length LysRS can also detect truncated species (Kaminska et al. 2007). Thus, it is likely that both forms of LysRS are packed into the virus (Kleiman et al. 2010).

## 2.6 Placement of tRNA3Lys onto the Viral Genome

It is not known whether tRNA3Lys is annealed to the PBS before or after formation of the viral particle. Eighteen nucleotides at the 3′ end of tRNA3Lys, which are used by RT to prime reverse transcription, are part of the amino-acid accepting stem (tRNA nucleosides 66–76) and the 3′ side of the TΨC stem-loop arm domain (nucleosides 59–65) (Fig. 2.2a). The TΨC and D loops are involved in tertiary interactions causing the tRNA to assume a stable L-shaped configuration (Benas et al. 2000). Thus, annealing to the PBS would require destabilizing of the structure and melting of two helices. Hybrid formation between the tRNA and PBS may initiate via the unpaired 3′ CCA in the tRNA and/or from the bases at the bottom of the acceptor-TΨC stems (Barraud et al. 2007; Hargittai et al. 2004; Tisne et al. 2004). In the presence of the PBS complementary sequence, both NC and NC of the precursor gag protein facilitate tRNA annealing by destabilizing the structure and unwinding the acceptor-TΨC stem (Chan et al. 1999; Guo et al. 2009; Hargittai et al. 2001; Huang et al. 1997). However, tRNA3Lys placement by gag is less effective and RT initiates reverse transcription with less efficiency, when compared to the tRNA3Lys annealing in the presence of NC (Cen et al. 2000; Guo et al. 2009;

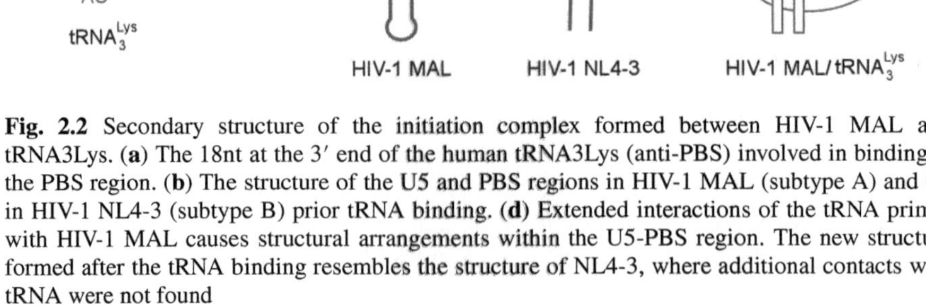


**Fig. 2.2** Secondary structure of the initiation complex formed between HIV-1 MAL and tRNA3Lys. (**a**) The 18nt at the 3′ end of the human tRNA3Lys (anti-PBS) involved in binding at the PBS region. (**b**) The structure of the U5 and PBS regions in HIV-1 MAL (subtype A) and (**c**) in HIV-1 NL4-3 (subtype B) prior tRNA binding. (**d**) Extended interactions of the tRNA primer with HIV-1 MAL causes structural arrangements within the U5-PBS region. The new structure formed after the tRNA binding resembles the structure of NL4-3, where additional contacts with tRNA were not found

Saadatmand et al. 2009). Interestingly, HIV-1 Vif was also found to promote tRNA3Lys annealing to the PBS sequence. However, Vif inhibited formation of tRNALys3/PBS hybrids induced by NC (Henriet et al. 2007), suggesting that tRNA3Lys placement is initiated by Vif and/or gag at the site of viral assembly and completed by NC after gag processing (Henriet et al. 2007; Kleiman et al. 2010).

Recent studies conducted in vitro and in cells revealed that LysRS could target the tRNA primer to the PBS region (Levin et al. 2010), exploiting the tRNA3Lys-like element (TLE) present upstream of the PBS, i.e., closer to the 5′ end of the genome. The TLE resembles the anticodon loop of tRNA3Lys and has affinity for LysRS, suggesting the TLE/LysRS interaction facilitates primer placement by directing the tRNA3Lys/LysRS complex to the PBS and subsequent tRNA release and annealing by binding LysRS to the TLE (Levin et al. 2010).

The specificity of tRNA3Lys usage as a primer is not determined solely by complementarity of 18 nt at its 3′ end with the PBS. Mutant PBS sequences allowing only partial annealing of tRNA3Lys can revert to the wild-type sequence in a single round of virus replication cycle (Das et al. 1994; Rhim et al. 1991). When the wild-type PBS was replaced with sequences complementary to other tRNAs (tRNA1,2Lys, tRNAPro, tRNATrp, tRNAIle, tRNAPhe), the new primers were utilized but resulted in a lower rate of viral replication. Moreover, after prolonged culturing, the new PBS sequences reverted to the wild-type PBS specific for tRNA3Lys (Das et al. 1995; Li et al. 1994; Wakefield et al. 1995). Therefore, factors other than the PBS sequence are also important for preferential use of tRNA3Lys as a primer.

Additional contacts between the tRNA primer and viral genome formed during initiation of reverse transcription were indicated as possible determinants of primer specificity. The sequences involved in additional interaction are located within the 5′ UTR proximal to the PBS (Fig. 2.2). However, despite the fact that the 5′ UTR of HIV-1 is the most conserved part of the viral genome, regions identified to form extended interactions with tRNA3Lys in subtype A isolates were not observed for subtype B (Goldschmidt et al. 2004; Isel et al. 1995). In HIV-1 MAL (subtype A), sequences involved in binary complex formation with primer tRNA are located in the U5 region, upstream of PBS (Fig. 2.2). The stability of this complex depends on nucleotide modifications in tRNA3Lys (Isel et al. 1993; Skripkin et al. 1996) and complementarity between the adenosine-rich (A-rich) loop located in the viral RNA and anticodon loop of the tRNA primer (Isel et al. 1995, 1998).

Structural probing of the initiation complex formed in NL4-3 and HxB2 (subtype B) did not reveal additional interactions outside of the PBS region (Goldschmidt et al. 2004). A comparison of structural arrangements of the 5′ UTR in NL4-3/HxB2 and for binary complex formed between tRNA3Lys and MAL might explain why extended interactions are absent in subtype B isolates (Fig. 2.2c, d). It was proposed that structural elements found in the MAL/tRNA3Lys complex already exist in NL4-3/HxB2 RNAs prior to tRNA3Lys annealing (Goldschmidt et al. 2002, 2004). In MAL, structural rearrangement of sequences proximal to the PBS region might be necessary for correct placement of the helix formed between the PBS and the 3′ end of the tRNA primer into the RT substrate binding cleft (Goldschmidt et al. 2002; Isel et al. 2010).

The A-rich loop of HIV-1 MAL that is complementary to the tRNA3Lys U-rich anticodon was proposed as a possible structural element required in selection of that specific tRNA primer. Simultaneous conversion of nucleotides in the A-rich loop and PBS region to sequences complementary to the anticodon and 3′-terminal region of tRNAHis or tRNAMet resulted in stable maintenance of the PBS complementary to the new tRNA primer during prolonged culturing (Kang et al. 1997; Li et al. 1997; Zhang et al. 1996). This suggests key roles for both viral RNA sequences in priming specificity. However, further studies showed that prolonged maintenance of nonself primer binding sites correlated with reduced replication capacity of these new mutant viruses (Moore-Rigdon et al. 2005; Wei et al. 2005). Replication efficiency improved for mutant virus with PBS and A-rich sequences adapted to use tRNAHis, but only after generation of additional mutations (Wakefield et al. 1996; Zhang et al. 1996). Moreover, the A-rich loop is not present in many viruses that utilize tRNA3Lys as a primer; thus, the putative A-rich/anticodon interactions are not conserved.

## 2.7 Initiation of Reverse Transcription

In HIV-1 and HIV-2, it was suggested that initiation of reverse transcription is regulated by a specific motif called the primer activation sequence (PAS) (Fig. 2.3) (Beerens et al. 2001; Berkhout and Schoneveld 1993; Freund et al. 2001). The sequence, located upstream of the PBS region, is complementary to eight nucleotides of the TΨC stem of the tRNA primer. A similar motif was identified in the Rous sarcoma virus (RSV) and is complementary to the TΨC stem of the corresponding tRNAPro primer (Aiyar et al. 1992; Cobrinik et al. 1991, 1988). Phylogenetic analysis of different retrovirus genera demonstrated that despite diversity in tRNA primer usage, the tRNA/PAS interactions appear to be conserved in evolution (Beerens and Berkhout 2002).

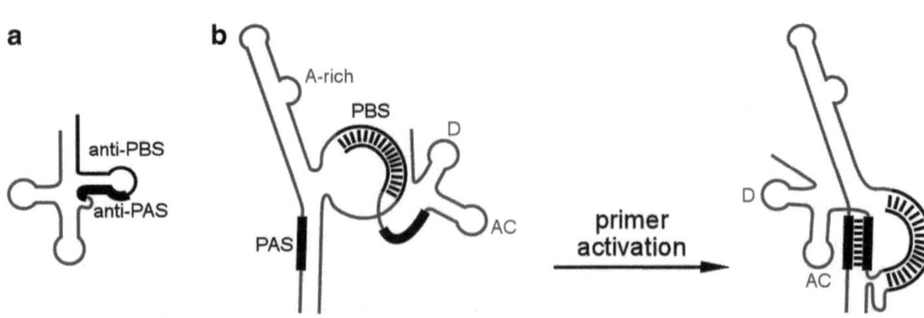


**Fig. 2.3** Primer activation signal (PAS) for initiation of reverse transcription. (**a**) Scheme of the tRNA3Lys primer. (**b**) The anti-PBS of the tRNA3Lys primer anneals to the PBS (*black thin lines*). Before initiation of reverse transcription, the anti-PAS sequence in tRNA3Lys must anneal to the PAS in U5 (*black thick lines*)

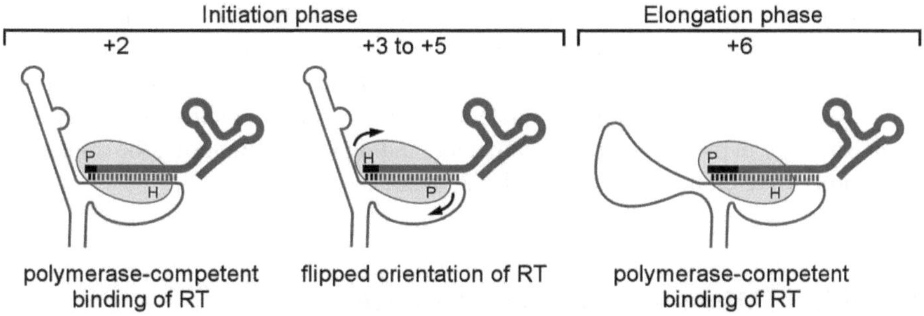


**Fig. 2.4** Dynamics of initiation complex in early phases of reverse transcription. During the initiation phase of reverse transcription, the RT (*oval*) is mostly bound in an orientation that allows addition of the first couple of deoxyribonucleotides (*black*) to the tRNA primer (*thick gray line*). The stem-loop structure formed upstream of the PBS (*thin gray line*) forces RT to bind predominantly in the flipped orientation causing arrest of DNA synthesis. Addition of the sixth deoxyribonucleotide causes unfolding of the stem-loop structure. The enzyme can eventually reorient into the polymerase-competent binding mode and enter the elongation phase of DNA synthesis. *P* polymerization active site, *H* RNase H active site

How would PAS exert its control? The mechanism of regulation requires the tRNA primer to be annealed to the PBS. Hybrid formation with the PBS disrupts annealing of nucleotides in the 5′ stem of the TΨC arm of the primer tRNA, termed the anti-PAS sequence (Fig. 2.3). Since nucleotides of the PAS motif form a stable helix with complementary sequence downstream of the PBS, structural rearrangements within the 5′ UTR are necessary to allow interactions between PAS and the tRNA primer. Initially, the tRNA primer annealed to the PBS is inactive, but once the PAS/anti-PAS interactions are established, reverse transcription is activated. Conformational changes of the 5′ UTR could be mediated by NC protein, released from the gag precursor during maturation of viral particles, and shown to be important for efficient tRNA extension. Therefore, PAS interactions may serve to prevent premature reverse transcription, so that maturation of NC can then trigger initiation of minus-strand DNA synthesis.

In vitro experiments showed that synthesis of minus-strand DNA from the tRNA primer consists of two steps (Arts et al. 1996; Isel et al. 1996; Lanchy et al. 1996, 1998). The first, or initiation phase, is characterized by slow distributive addition of the first six nucleotides, during which RT frequently dissociates. Interestingly, this step was not observed when AMV and MLV RTs were used. However, nonhomologous RTs were not able to extend tRNA3Lys primer efficiently, indicating that HIV-1 RT specifically recognizes the cognate tRNA3Lys/viral RNA complex. After addition of the sixth nucleotide, reverse transcription transits to the elongation phase, during which minus-strand DNA synthesis proceeds rapidly with rare enzyme dissociation. Monitoring of the dynamics of several initiation complexes by using single-molecule spectroscopy revealed that RT binds the initiation complex in two orientations, a polymerase-competent orientation and a flipped orientation, which precludes tRNA extension (Fig. 2.4) (Liu et al. 2010). A polymerase-competent orientation was mostly observed for the addition of the first two nucleotides. However, at position +3, RT encounters a stem-loop structure in U5, which forces the enzyme to bind predominantly in the flipped polymerase-inactive orientation. At this stage, addition of subsequent nucleotides is slow until position +6, when RT again is mostly in the polymerase-competent binding mode, triggering the transition to the processive elongation phase of DNA synthesis.

## 2.8 First (Minus)-Strand DNA Transfer

Extension of the 3′ end of the PBS-bound tRNA primer allows copying the U5 and R elements into a 181-nt long cDNA, designated minus strong-stop DNA ((−) ssDNA). Synthesis of this DNA fragment is continued after its transfer to the 3′ end of the viral genome. The transfer reaction is facilitated by annealing of complementary sequences of R, one present at the 3′ end of the RNA genome and the other in the (−) ssDNA synthesized at the 5′ end (Fig. 2.1c). Studies showed that the length of complementarity and secondary structure of the R sequence are important determinants for efficient strand transfer (Berkhout et al. 1995, 2001; Negroni and Buc 2000; Song et al. 2008).

During synthesis of (−)ssDNA, RT pauses at the TAR hairpin, promoting RNase H cleavage approximately 14–20 nucleotides downstream from the pause site within the polyA hairpin in the copied template (Fig. 2.5) (Chen et al. 2003a, b; Kim et al. 1997). The structure is destabilized with the help of NC (Klasens et al. 1999). A short gap created by RNase H activity exposes the (−)ssDNA sequence that serves as an initiation site for an invasion-driven transfer. Here, the complementary sequence in R at the genomic 3′ end invades and anneals to the cDNA through an interaction stabilized by NC. The RNA-DNA hybrid starts to propagate displacing adjacent RNA template fragments of the genomic 5′ end. Successive base pairing of (−)ssDNA with invading 3′ R proceeds until it reaches the 3′ terminus of synthesized (−)ssDNA, completing minus-strand DNA transfer (Chen et al. 2003a, b).

Since the 3′ end of (−)ssDNA is exactly complementary to TAR, the sequence can fold back to form a stem-loop primer-template, which can be elongated and prevent minus-strand transfer. However, in the presence of NC, the 5′-most fragment of the RNA template remaining after polymerase-dependent RNase H cleavage is not removed until it can be strand-exchanged by the equivalent sequence of the genomic 3′ end (Guo et al. 1997; Lapadat-Tapolsky et al. 1997). First-strand transfer allows minus-strand DNA synthesis to be continued into the U3 region and beyond. After RT completes copying the U3 region, the first U3-R-U5 sequence of the 3′ LTR has been created.

The presence of two RNA genome molecules in a single virion particle allows minus-strand transfer to occur intra- or intermolecularly. It is believed that the process is facilitated by genome circularization that brings in proximity the R elements from the 5′ and 3′ ends, otherwise separated by over 9,000 nucleotides in the linear form of the HIV-1 RNA genome. The interactions that mediate HIV-1 genome

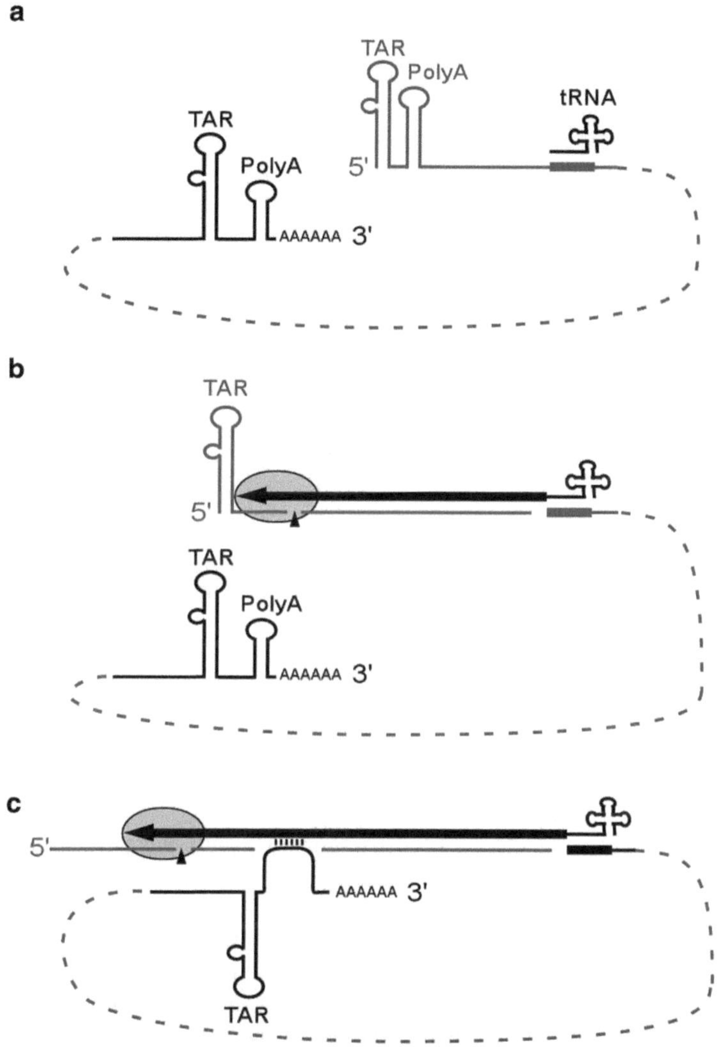


**Fig. 2.5** Minus-strand strong-stop DNA transfer. (**a**) Sequences of R elements at the 5′ end (*thin line* in *gray*) and the 3′ end (*thin line* in *black*) are involved in the first-strand transfer. (**b**) The stable stem-loop structure of the TAR hairpin can pause RT (*oval*) during synthesis of minus-strand strong-stop DNA (*black thick line*). RNase H activity of RT makes cuts (*black triangle*) within the RNA template and helps to create a gap exposing the cDNA sequence. (**c**) The complementary sequence of the R at the 3′ end can interact with cDNA within the gap region. Displacement of the RNA segments and strand exchange follows extension of the synthesized (−)ssDNA until its 3′ terminus is transferred (not shown). A *dashed line* designates the bulk of the genome

circularization and facilitate minus-strand transfer involve a 5′ end sequence of *gag*, located over 600 nucleotides from the 5′ end of the genomic RNA, and a region of U3 and R at the 3′ end (Fig. 2.6a) (Beerens and Kjems 2010; Ooms et al. 2007).

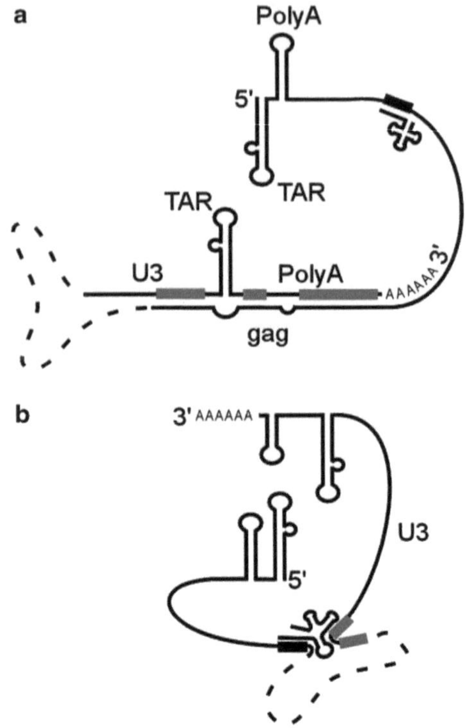


**Fig. 2.6** RNA genome circularization in HIV-1. (**a**) RNA base-pairing model of interactions between sequences of *gag* and U3-R at the 3′ end of the RNA genome. (**b**) Proposed tRNA3Lys-U3 interactions known to facilitate minus-strand transfer in HIV-1. Sequences of the genomic 3′ end involved in the interactions are shown as the *thick gray lines*

Similar RNA-RNA contacts between the genomic ends were described in plant viruses (Miller and White 2006), flaviviruses (Corver et al. 2003; Hahn et al. 1987; Shurtleff et al. 2001), hepatitis C virus (Romero-Lopez and Berzal-Herranz 2009), FMDV (foot-and-mouth disease virus) (Serrano et al. 2006), and the LTR Ty1 retrotransposon (Cristofari et al. 2002). In addition to this gag/U3-R interaction, juxtaposing of R regions at both ends for minus-strand transfer could be mediated by TAR/TAR interactions. Here, interaction could involve palindromic sequences of ten nucleotides in a stem-loop, similar to the contact described for DIS hairpins (Beerens and Kjems 2010; Piekna-Przybylska and Bambara 2011). Interestingly, for many plant viruses stem-loops and loop-loop kissing structures formed by sequences at both ends of the RNA genome were demonstrated to participate in genome circularization (Miller and White 2006).

Other nucleic acid molecules, including the tRNA primer and the (−)ssDNA, were also proposed to support a circular genome conformation during minus-strand transfer. Primer tRNA used in reverse transcription was described as a bridging factor in genome circularization in LTR retrotransposons (Gabus et al. 1998), endogenous LTR retroviruses (Gabus et al. 2006), and also HIV-1 (Brule et al. 2000; Piekna-Przybylska et al. 2010). In HIV-1, the tRNA3Lys placed on the PBS may interact with complementary nucleotides of the tRNA gene-like sequence embedded in the U3-R region at the 3′ end of the genome (Fig. 2.6b) (Brule et al. 2000; Piekna-Przybylska et al. 2010). Mutation analyses in vitro and in vivo have demonstrated that complementary sequences in U3 stimulate minus-strand transfer (Brule et al. 2000; Piekna-Przybylska et al. 2010, 2011). In the case of interactions involving (−)ssDNA, proximity of sequences participating in strand transfer could be achieved by contacts formed during invasion-driven transfer between the polyA hairpin at the 3′ end and the invasion site in (−)ssDNA, described above (Piekna-Przybylska et al. 2011). Moreover, Berkhout and coworkers proposed that the TAR sequence in (−)ssDNA could interact with TAR at the 3′ end of the RNA genome via kissing loop contact (Berkhout et al. 2001).

## 2.9 Recombination Between the Two Copies of the RNA Genome

Packaging of two copies of single-stranded RNA viral genome is important for viral fitness and infectivity. It has been proposed that the availability of two copies of the viral genome ensures that the process of DNA synthesis is not interrupted in case of RNA damage (Coffin 1979). It is known that the extending DNA terminus and RT can relocate from one RNA strand to the other within internal genomic regions during reverse transcription. The mechanism of template switching is similar to minus-strand transfer, although stable hairpins are not the only possible places where strand transfer occurs. Recent studies have shown that frequent recombination also takes place when RT encounters sequences with several nearby G-runs that can form G-quartets (Shen et al. 2009) and are also capable of pausing RT such that RNase H cuts are concentrated. Moreover, pausing is not the only factor that can trigger strand transfer, since template switching was also observed in regions that do not correlate with pauses in reverse transcription (Negroni and Buc 2000). A slower rate of polymerization caused by unfavorable template sequence, low dNTP concentration, bound protein, or other factors may also promote strand transfer. Here, template switching is driven by increased frequency of polymerase-dependent RNase H cleavage, facilitating formation of gaps that invite invasion by the second template (Svarovskaia et al. 2000). Most recombination events occur during minus-strand synthesis (Zhang et al. 2000).

Transfers between two copies of HIV RNA genomes occur with a frequency of about 3.3 crossovers for every 1,000 base pairs transcribed, predicting 30 recombination events during each cycle of reverse transcription. Interestingly, the crossover frequency depends on the cell type (Jetzt et al. 2000; Levy et al. 2004; Zhuang et al. 2002). Since native HIV made in cells simultaneously infected with two genetically distinct viral strains can be heterozygotic, the consequence of recombination is that progeny provirus may have combined mutations initially present only in one of the two copies of the RNA genome. In general, when RNA genomes are not identical, intermolecular strand transfer causes diversity in genome sequence, promoting evolution, which is the key factor in developing resistance against antiviral host responses.

## 2.10 Plus-Strand DNA Synthesis

Lentiviruses, some spumaviruses, and the retrotransposon Ty1 initiate plus-strand synthesis at two sites, one in proximity to the 3′ end and the second in the middle of the genome (Charneau et al. 1992; Harris et al. 1981; Heyman et al. 1995; Stetor et al. 1999; Tobaly-Tapiero et al. 1991). As with initiation of minus-strand DNA synthesis, HIV RT requires a primer to initiate synthesis of plus-strand DNA. During the course of RNA genome degradation associated with synthesis of minus-strand DNA, two short RNA fragments are resistant to RNase H cleavage. These are 15-nt sequences, called polypurine tracts (PPTs), containing only purine bases (5′-AAA AGA AAA GGG GGG-3′) (Fig. 2.1e, f) (Rausch and Le Grice 2004) and are used as primers for plus-strand DNA synthesis. The 3′PPT is located adjacent to the 3′ LTR of the RNA genome and delineates the 5′ boundary of the U3 region. The central PPT (cPPT) is located nearly in the center of the genome, in the IN open reading frame (ORF). Precise RNase H-mediated cuts in DNA-RNA duplexes within PPT regions are required to create each PPT sequence (Huber and Richardson 1990; Pullen et al. 1993; Wohrl and Moelling 1990). Mutational analysis determined that the stretch of G residues within the PPT and its helical structure are important for specific cleavage and extension by RT (Jones and Hughes 2007; Miles et al. 2005; Powell and Levin 1996). In most retroviruses, a U-rich region upstream of the PPT sequence is also required for efficient formation of the RNA primer (Rausch and Le Grice 2004).

The 3′PPT serves as a main primer for plus-strand DNA synthesis. Shortly after minus-strand strong-stop DNA transfer, synthesis of plus-strand DNA begins. This takes place after RT passes the U3 region of the RNA genome while extending minus-strand DNA. Template RNA in the RNA-DNA duplex formed by synthesis of the minus-strand DNA is efficiently degraded by RNase H. The synthesis of plus-strand DNA can now begin from the 3′PPT primer. DNA synthesis copies the U3, R, and U5 sequences from the minus-strand DNA, now used as a template. Human tRNA3Lys, which is still linked to minus-strand DNA, becomes the template for reconstitution of the PBS by copying of the first 18 nt at its 3′ end. The resulting DNA product is termed plus-strand strong-stop DNA ((+) ssDNA) and contains at its 3′ end the regenerated PBS sequence, designated (+) PBS. Synthesis of the (+) PBS would require the displacement of the RNA sequence of the original genomic PBS region still annealed to the tRNA primer (Fig. 2.1g). While is not known when PBS-tRNA displacement occurs, studies in vitro showed that both an extension of the (+) ssDNA and an extension of the long minus-strand DNA, coming from the other direction, can partially displace the tRNA3Lys primer (Ben-Artzi et al. 1996).

The modified base of adenosine at position 58 (methyladenosine 58, m1A58) in tRNA3Lys serves as a pause or termination signal for plus-strand strong-stop DNA synthesis (Gilboa et al. 1979b; Renda et al. 2001). Lack of modification at this position in some tRNA3Lys molecules allows the RT to copy the tRNA sequence beyond residue 58. Such extension would interfere with subsequent plus-strand transfer (see "Plus (second)-strand DNA Transfer" below). Mutation of the methylated tRNA3Lys residue A58 disrupts reverse transcription and inhibits replication of HIV-1 (Renda et al. 2001, 2004). However, in reconstituted systems it was observed that termination does not occur exclusively at methyl A58 but also at pseudouridine 55 (Ψ55) or even hyper-modified adenosine 37 ($ms^2t^6A$-37) in the anticodon loop (Auxilien et al. 1999; Wu et al. 1999). In some HIV-1 strains, the first three bases downstream of the PBS are complementary to nucleotides 56–58 in tRNA3Lys. Thus, bypass of the methyl A at position 58, but termination at position 55 in the tRNA does not interfere with the base pairing needed for successful plus-strand transfer in these viruses. However, plus-strand transfer will not take place if tRNA3Lys is completely unmodified and the entire molecule is reverse transcribed (Ben-Artzi et al. 1996; Burnett and McHenry 1997; Yusupova et al. 1996).

## 2.11 tRNA Primer Removal from the (+)PBS

tRNA removal from the RNA-DNA hybrid is necessary to expose the (+) PBS sequence that can be annealed with the (−) PBS for the second-strand transfer reaction. The (−) PBS is created by extension of the minus-strand DNA through the PBS region originally occupied by the tRNA3Lys primer. RNase H activity of RT participates in removing the tRNA primer from the RNA-DNA hybrid. Initial cleavage is not at the RNA-DNA junction but instead between the last two ribonucleotides, the 3′-terminal rA and penultimate rC in tRNA3Lys. The product then has an rA residue at the 5′ end of the minus-strand DNA, following 17 nt of RNA-DNA hybrid made from the tRNA extended from the (+) PBS in plus-strand DNA. This rA residue will appear at the end of the 3′ LTR and be removed during integration into host DNA. Reconstituted systems suggest that after primary cleavage, the tRNA is unable to completely dissociate from the (+) PBS. Secondary cuts and the nucleic acid chaperone activity of NC are both required for efficient primer removal (Smith et al. 1999; Wu et al. 1999). Once annealing between the PBS regions in the plus and minus strand occurs, the further elongation of (+) ssDNA continues until RT reaches the cPPT (see "Priming and termination of DNA synthesis within the cPPT region" below).

## 2.12 Plus (Second)-Strand DNA Transfer

As with minus-strand DNA synthesis, plus-strand DNA elongation is not continuous and requires interactions between complementary sequences. In this case, however, the genomic 5′ and 3′ ends are already in proximity after the first-strand transfer. Contact is mediated by base pairing of complementary (+) PBS and (−) PBS sequences present at the 3′ ends of (+) ssDNA and minus-strand DNA, respectively (Fig. 2.1i), and is enhanced in the presence of NC (Muthuswami et al. 2002; Wu et al. 1999).

The mechanism of annealing requires destabilization of short stem-loop hairpins formed by 18-nt long sequences of the (–) and (+) PBS. NMR studies have demonstrated that the (–) PBS sequence at the 3′ end of the minus-strand DNA can form a stable hairpin structure with a pentanucleotide loop and terminal single-stranded 4-nt overhangs (Johnson et al. 2000). A similar structure could also be formed by the (+) PBS sequence (Egele et al. 2004). Studies by Ramalanjaona and coworkers showed that in the absence of NC protein, annealing was mediated by nucleation through base pairing of the PBS single-stranded overhangs (Ramalanjaona et al. 2007). However, both the (–) PBS and (+) PBS sequences were found to bind with three NC molecules, which induced a limited secondary structure destabilization in both PBS hairpins (Egele et al. 2004). Moreover, Egele et al. suggest NC can activate formation of kissing complexes between the (+) PBS and (–) PBS within the loops, promoting chaperoning of plus-strand transfer (Egele et al. 2004).

As a consequence of plus-strand transfer, the U3-R-U5 sequences in plus-strand strong-stop DNA can be duplicated to form the double-stranded 5′ LTR by extension of the (–) PBS terminus of minus-strand DNA. Extension of the (+) PBS allows the synthesis of the second DNA strand for completion of the rest of the genome.

## 2.13 Priming and Termination of DNA Synthesis Within the cPPT Region

The presence of an additional PPT primer in the middle of the viral genome allows synthesis of plus-strand DNA as two distinct segments. The 3′PPT primer is used for synthesis of plus-strand DNA within the 5′ half of the viral genome. The copied sequences of U3, R, and U5 become the 5′ LTR. Elongation from the cPPT primer results in the synthesis of plus-strand DNA within the 3′ half of the genome, including the 3′ LTR sequence made after the displacement of the plus-strand DNA of the 5′ LTR (initiated from the 3′PPT). Synthesis of the plus-strand DNA in the 5′ half of the genome terminates 90–100 nucleotides beyond the cPPT, within a region known as the central termination sequence (CTS) (Fig. 2.1j) (Charneau et al. 1994). The CTS blocks RT, which engages in a termination process involving elongation of plus-strand DNA initiated at the 3′PPT while displacing a 5′flap of plus strand initiated from the cPPT. The mechanism of CTS-mediated termination involves two sites, Ter1 and Ter2, located at the 3′ ends of $A_nT_n$ motifs (Lavigne et al. 1997). These motifs generate a global curvature of the DNA helix compressing the DNA minor groove, a structure apparently responsible for terminating DNA synthesis (Lavigne and Buc 1999). The consequence of cPPT priming is that the unintegrated linear HIV-1 DNA has two overlapping short plus-strand segments. The displaced fragment of the plus-strand DNA initiated from cPPT is termed the central DNA flap. Studies in vitro demonstrated that flap removal and gap repair can be catalyzed by the human enzymes, FEN1 and DNA ligase, respectively (Rumbaugh et al. 1998).

The cPPT is an important cis-active sequence for the HIV-1 life cycle, since its mutagenesis affects growth of mutant viruses and their integration into the host genome, with loss of viral infectivity (Charneau et al. 1992; Hungnes et al. 1992). It was found that mutations in the cPPT and CTS sequences could alter the structure of the central DNA flap (Charneau et al. 1992, 1994). In the absence of a central DNA flap, a linear DNA remains uncoated and trapped within the intact capsid (Arhel et al. 2007). Consequently, the RTC does not mature into a PIC, and viral DNA cannot cross the nuclear membrane for integration into the host nuclear DNA (Arhel et al. 2006).

## 2.14 Completion of the Viral DNA Genome

Completion of both the minus and plus DNA strands requires strand displacement synthesis through the DNA hybrid regions in order to generate linear integration-competent proviral DNA with an LTR sequence at each end (Fig. 2.1i). During plus-strand transfer, annealing of the (−) PBS and (+) PBS sequences allows maintenance of a circular intermediate. Synthesis of the plus strand in the 5′ half of the genome is then continued, and the extension of the minus strand from the (−) PBS sequence leads to displacement of minus-strand DNA originally extended from tRNA3Lys primer. Plus-strand DNA initiated from 3′PPT becomes the template for regeneration of the 5′ LTR sequence. In the meantime, extension of plus-strand DNA from the cPPT primer displaces plus-strand DNA initiated from the 3′PPT. Here, minus-strand DNA extended from tRNA3Lys is the template for restoration of the 3′ LTR. The final product of reverse transcription is a blunt-ended linear double-stranded proviral DNA.

## 2.15 Integration of the HIV-1 Genome into the Host Genome

The final step in the HIV-1 replication cycle is integration of a DNA copy of its genome into host nuclear DNA. Integrated proviruses persist in the host cell and serve as templates for viral replication and gene expression needed to generate new viral particles. Integration involves processing of the LTR termini, their insertion into the chromosome, and finally DNA repair within the junction strands between the viral DNA genome and host DNA. All steps require the enzymatic activity of retroviral IN.

After conversion of the RNA genome into the linear provirus, IN binds to the terminal sequences over a distance of about 200 nt within the U3 and U5 of the LTR and forms the PIC, which also includes RT, MA, Vpr, NC, and cellular proteins (Bukrinsky et al. 1993; Farnet and Bushman 1997; Farnet and Haseltine 1991b; Gallay et al. 1995; Miller et al. 1997; Nermut and Fassati 2003). The necessary regions of the LTR for specific and efficient IN binding are attachment (att) sites located 7 to 13 bp from the highly conserved terminal 5′-CA dinucleotide (Leavitt et al. 1992; Sherman and Fyfe 1990; Vink et al. 1991). Linear viral DNA is cleaved

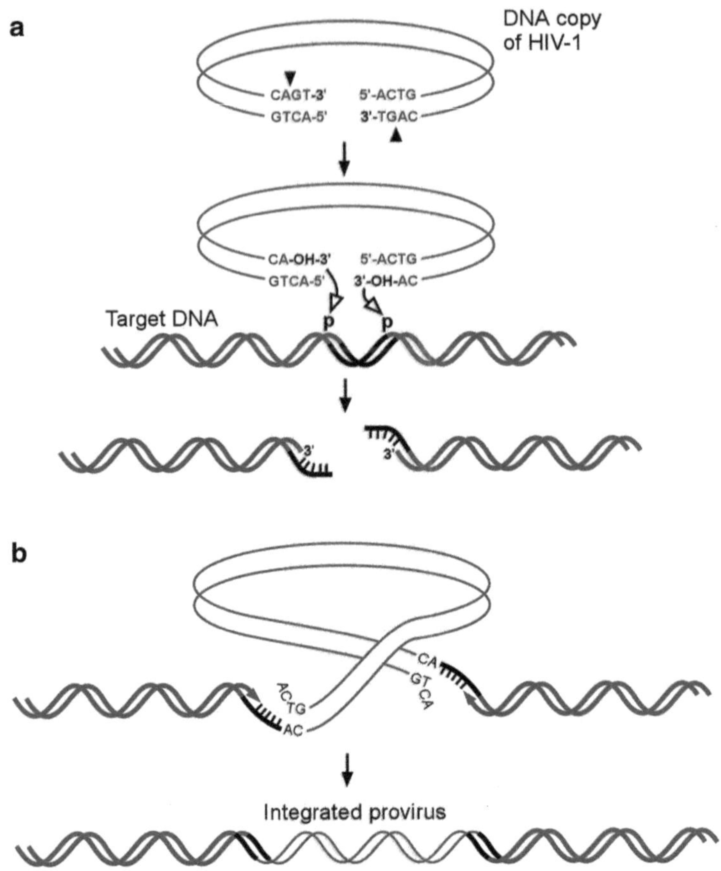


**Fig. 2.7** Steps in the integration of viral DNA. (**a**) During 3′-processing of viral DNA (*black triangle*), IN removes a pGT dinucleotide at the 3′ end of each LTR sequence. After nuclear import, IN mediates nucleophilic attack on the phosphodiester bridges (p) in the target DNA by the 3′-hydroxyl residues of the viral DNA. (**b**) Only 3′-OH viral ends are ligated to 5′-*O*-phosphate ends separated by five base pairs (in *black*) within the integration region. The 5′-CA dinucleotide in the viral DNA is removed and the 5-nucleotide single-stranded gap is repaired. The integrated provirus (*thin gray lines*) is flanked by the 5-nucleotide sequence of integration site

by IN at both 3′ ends by removal of the terminal GT, which exposes the adjacent conserved CA dinucleotides (Fig. 2.7) (Chen and Engelman 2001; Engelman et al. 1991; Masuda et al. 1998). This reaction takes place in the cytoplasm.

Following nuclear import, cleaved DNA with two recessed 3′ ends is used as a substrate for strand transfer (unrelated mechanistically to strand transfer during reverse transcription) at the target site in the chromosome. During this process, IN probably links the two viral DNA ends to keep them in proximity (Miller et al. 1997; Wei et al. 1998). In the strand transfer reaction, the 3′ hydroxyl groups of the processed viral DNA attack a pair of phosphodiester bonds in the target DNA.

The 5′ overhangs in the viral DNA are removed, defining the boundaries of integrated DNA at the conserved CA dinucleotide. Integration is completed by repair of the single-stranded gaps created between the target and viral DNA, resulting in duplication of a 5-nt sequence that was initially present in the chromosome as a single copy, but now flanks the integrated provirus (Ellison et al. 1990; Muesing et al. 1985).

## 2.16 Unintegrated Viral DNA

Generation of proviral DNA and its integration into the host chromosome are critical steps for long-term HIV survival. However, despite the importance of integration, the majority of viral genomes in the nucleus exist as circular DNA with one or two LTRs or as a linear DNA. In infected cells, the proportion of non-integrated HIV can reach the surprising level of 99 % of total viral DNA (Chun et al. 1997). The linear form of proviral DNA is a precursor for the population of 1- and 2-LTR circles, and among them there are forms with non-defective and defective genomes.

1-LTR DNA circles are the most abundant and could be generated by homologous recombination between the two LTRs of a linear DNA (Farnet and Haseltine 1991a; Shank et al. 1978) or from a circular intermediate formed in reverse transcription (Dina and Benz 1980; Gilboa et al. 1979a; Junghans et al. 1982). 2-LTR circles with non-defective genomes are presumably generated by ligation of the linear DNA ends, in reactions catalyzed by host-cell enzymes (Li et al. 2001). Analysis of 2-LTR-circle junctions showed that about a half of the population are products of blunt-end ligation of full-length linear DNA genomes that retained the GT dinucleotide (Svarovskaia et al. 2004), while the remainder contained deletions or insertions. The insertions were derived from viral sequences around the PPT and PBS regions, as a result of aberrant RNase H cleavage (Julias et al. 2002; Svarovskaia et al. 2004). In addition, 2-LTR circles can arise from auto-integration, in which the integration took place in another site in the viral DNA (Farnet and Haseltine 1991a). The linear DNA forms are presumably molecules not recognized by IN because of defective ends resulting from host nuclease cleavages or errors in reverse transcription.

## 2.17 Conclusion

Intensive research on the HIV-1 replication cycle has added much to our knowledge of both the reverse transcription process and the structure and functions of the viral polymerase. Detailed studies of different steps in reverse transcription reveal how the low fidelity of RT during polymerization and frequent recombination events contribute to viral diversity (see Chap. 6). Evolved viruses have new mutations, which help to overcome host immuno-defense and drug treatment therapies. Thus, at present, anti-HIV therapies require a strict regimen of medication to minimize the potential for viral mutations that allow it to escape inhibition. The treatment usually includes a combination of three or more anti-HIV drugs, such as inhibitors of RT, PR, and IN, plus fusion or entry inhibitors.

Reverse transcription presents additional targets for future therapy. RT inhibitors are nucleoside and non-nucleoside agents that work by interfering with polymerase function (see Chaps. 3 and 4). Although inhibiting polymerization activity of RT is effective, the emergence of viruses resistant to current drugs constantly forces development of new antiretroviral agents. The replication of HIV also requires the RT to have efficient RNase H activity (see Chap. 8). Detailed analyses of reverse transcription showed that RNase H activity is absolutely necessary for generation of PPTs, removal of the tRNA primer, recombination, and clearing the cDNA of RNA genome template fragments to allow plus-strand synthesis. Altering of RNase H activity in RT can affect any of these steps resulting in inhibition of reverse transcription. Thus, the RNase H active site in RT is currently considered as a target in the identification and development of novel antiretroviral agents (see Chap. 9). Other steps in proviral DNA synthesis, discussed above, are also potential therapeutically susceptible points in the replication mechanism. Very likely, combinations of inhibitors targeting different steps of reverse transcription have the best potential for disrupting the HIV life cycle before viral integration into the host chromosome (see Chaps. 9, 11, 12, 13, and 16).

## References


Aiyar A, Cobrinik D, Ge Z, Kung HJ, Leis J (1992) Interaction between retroviral U5 RNA and the T psi C loop of the tRNA(Trp) primer is required for efficient initiation of reverse transcription. J Virol 66(4):2464–2472

Arhel N (2010) Revisiting HIV-1 uncoating. Retrovirology 7:96

Arhel N, Munier S, Souque P, Mollier K, Charneau P (2006) Nuclear import defect of human immunodeficiency virus type 1 DNA flap mutants is not dependent on the viral strain or target cell type. J Virol 80(20):10262–10269

Arhel NJ, Souquere-Besse S, Munier S, Souque P, Guadagnini S, Rutherford S et al (2007) HIV-1 DNA Flap formation promotes uncoating of the pre-integration complex at the nuclear pore. EMBO J 26(12):3025–3037

Arnold E, Jacobo-Molina A, Nanni RG, Williams RL, Lu X, Ding J et al (1992) Structure of HIV-1 reverse transcriptase/DNA complex at 7 A resolution showing active site locations. Nature 357(6373):85–89

Arts EJ, Stetor SR, Li X, Rausch JW, Howard KJ, Ehresmann B et al (1996) Initiation of (−) strand DNA synthesis from tRNA(3Lys) on lentiviral RNAs: implications of specific HIV-1 RNA-tRNA(3Lys) interactions inhibiting primer utilization by retroviral reverse transcriptases. Proc Natl Acad Sci USA 93(19):10063–10068

Auxilien S, Keith G, Le Grice SF, Darlix JL (1999) Role of post-transcriptional modifications of primer tRNALys,3 in the fidelity and efficacy of plus strand DNA transfer during HIV-1 reverse transcription. J Biol Chem 274(7):4412–4420

Barraud P, Gaudin C, Dardel F, Tisne C (2007) New insights into the formation of HIV-1 reverse transcription initiation complex. Biochimie 89(10):1204–1210

Beerens N, Berkhout B (2002) Switching the in vitro tRNA usage of HIV-1 by simultaneous adaptation of the PBS and PAS. RNA 8(3):357–369

Beerens N, Kjems J (2010) Circularization of the HIV-1 genome facilitates strand transfer during reverse transcription. RNA 16(6):1226–1235

Beerens N, Groot F, Berkhout B (2001) Initiation of HIV-1 reverse transcription is regulated by a primer activation signal. J Biol Chem 276(33):31247–31256

Ben-Artzi H, Shemesh J, Zeelon E, Amit B, Kleiman L, Gorecki M et al (1996) Molecular analysis of the second template switch during reverse transcription of the HIV RNA template. Biochemistry 35(32):10549–10557

Benas P, Bec G, Keith G, Marquet R, Ehresmann C, Ehresmann B et al (2000) The crystal structure of HIV reverse-transcription primer tRNA(Lys,3) shows a canonical anticodon loop. RNA 6(10):1347–1355

Berkhout B, Schoneveld I (1993) Secondary structure of the HIV-2 leader RNA comprising the tRNA-primer binding site. Nucleic Acids Res 21(5):1171–1178

Berkhout B, van Wamel J, Klaver B (1995) Requirements for DNA strand transfer during reverse transcription in mutant HIV-1 virions. J Mol Biol 252(1):59–69

Berkhout B, Vastenhouw NL, Klasens BI, Huthoff H (2001) Structural features in the HIV-1 repeat region facilitate strand transfer during reverse transcription. RNA 7(8):1097–1114

Brule F, Bec G, Keith G, Le Grice SF, Roques BP, Ehresmann B et al (2000) In vitro evidence for the interaction of tRNA(3)(Lys) with U3 during the first strand transfer of HIV-1 reverse transcription. Nucleic Acids Res 28(2):634–640

Bukrinsky M (2004) A hard way to the nucleus. Mol Med 10(1–6):1–5

Bukrinsky MI, Sharova N, McDonald TL, Pushkarskaya T, Tarpley WG, Stevenson M (1993) Association of integrase, matrix, and reverse transcriptase antigens of human immunodeficiency virus type 1 with viral nucleic acids following acute infection. Proc Natl Acad Sci USA 90(13):6125–6129

Burnett BP, McHenry CS (1997) Posttranscriptional modification of retroviral primers is required for late stages of DNA replication. Proc Natl Acad Sci USA 94(14):7210–7215

Cen S, Khorchid A, Gabor J, Rong L, Wainberg MA, Kleiman L (2000) Roles of Pr55(gag) and NCp7 in tRNA(3)(Lys) genomic placement and the initiation step of reverse transcription in human immunodeficiency virus type 1. J Virol 74(22):10796–10800

Cen S, Khorchid A, Javanbakht H, Gabor J, Stello T, Shiba K et al (2001) Incorporation of lysyl-tRNA synthetase into human immunodeficiency virus type 1. J Virol 75(11):5043–5048

Cen S, Javanbakht H, Kim S, Shiba K, Craven R, Rein A et al (2002) Retrovirus-specific packaging of aminoacyl-tRNA synthetases with cognate primer tRNAs. J Virol 76(24):13111–13115

Cen S, Javanbakht H, Niu M, Kleiman L (2004) Ability of wild-type and mutant lysyl-tRNA synthetase to facilitate tRNA(Lys) incorporation into human immunodeficiency virus type 1. J Virol 78(3):1595–1601

Chan B, Weidemaier K, Yip WT, Barbara PF, Musier-Forsyth K (1999) Intra-tRNA distance measurements for nucleocapsid proteindependent tRNA unwinding during priming of HIV reverse transcription. Proc Natl Acad Sci USA 96(2):459–464

Charneau P, Alizon M, Clavel F (1992) A second origin of DNA plus-strand synthesis is required for optimal human immunodeficiency virus replication. J Virol 66(5):2814–2820

Charneau P, Mirambeau G, Roux P, Paulous S, Buc H, Clavel F (1994) HIV-1 reverse transcription. A termination step at the center of the genome. J Mol Biol 241(5):651–662

Chen H, Engelman A (2001) Asymmetric processing of human immunodeficiency virus type 1 cDNA in vivo: implications for functional end coupling during the chemical steps of DNA transposition. Mol Cell Biol 21(20):6758–6767

Chen Y, Balakrishnan M, Roques BP, Bambara RA (2003a) Steps of the acceptor invasion mechanism for HIV-1 minus strand strong stop transfer. J Biol Chem 278(40):38368–38375

Chen Y, Balakrishnan M, Roques BP, Fay PJ, Bambara RA (2003b) Mechanism of minus strand strong stop transfer in HIV-1 reverse transcription. J Biol Chem 278(10):8006–8017

Chun TW, Carruth L, Finzi D, Shen X, DiGiuseppe JA, Taylor H et al (1997) Quantification of latent tissue reservoirs and total body viral load in HIV-1 infection. Nature 387(6629): 183–188

Cobrinik D, Soskey L, Leis J (1988) A retroviral RNA secondary structure required for efficient initiation of reverse transcription. J Virol 62(10):3622–3630

Cobrinik D, Aiyar A, Ge Z, Katzman M, Huang H, Leis J (1991) Overlapping retrovirus U5 sequence elements are required for efficient integration and initiation of reverse transcription. J Virol 65(7):3864–3872

Coffin JM (1979) Structure, replication, and recombination of retrovirus genomes: some unifying hypotheses. J Gen Virol 42(1):1–26

Corver J, Lenches E, Smith K, Robison RA, Sando T, Strauss EG et al (2003) Fine mapping of a cis-acting sequence element in yellow fever virus RNA that is required for RNA replication and cyclization. J Virol 77(3):2265–2270

Cristofari G, Bampi C, Wilhelm M, Wilhelm FX, Darlix JL (2002) A 5′-3′ long-range interaction in Ty1 RNA controls its reverse transcription and retrotransposition. EMBO J 21(16):4368–4379

Das AT, Koken SE, Essink BB, van Wamel JL, Berkhout B (1994) Human immunodeficiency virus uses tRNA(Lys,3) as primer for reverse transcription in HeLa-CD4+ cells. FEBS Lett 341(1):49–53

Das AT, Klaver B, Berkhout B (1995) Reduced replication of human immunodeficiency virus type 1 mutants that use reverse transcription primers other than the natural tRNA(3Lys). J Virol 69(5):3090–3097

DeStefano JJ, Buiser RG, Mallaber LM, Myers TW, Bambara RA, Fay PJ (1991) Polymerization and RNase H activities of the reverse transcriptases from avian myeloblastosis, human immunodeficiency, and Moloney murine leukemia viruses are functionally uncoupled. J Biol Chem 266(12):7423–7431

DeStefano JJ, Bambara RA, Fay PJ (1993) Parameters that influence the binding of human immunodeficiency virus reverse transcriptase to nucleic acid structures. Biochemistry 32(27):6908–6915

Dina D, Benz EW Jr (1980) Structure of murine sarcoma virus DNA replicative intermediates synthesized in vitro. J Virol 33(1):377–389

Dvorin JD, Malim MH (2003) Intracellular trafficking of HIV-1 cores: journey to the center of the cell. Curr Top Microbiol Immunol 281:179–208

Egele C, Schaub E, Ramalanjaona N, Piemont E, Ficheux D, Roques B et al (2004) HIV-1 nucleocapsid protein binds to the viral DNA initiation sequences and chaperones their kissing interactions. J Mol Biol 342(2):453–466

Ellison V, Abrams H, Roe T, Lifson J, Brown P (1990) Human immunodeficiency virus integration in a cell-free system. J Virol 64(6):2711–2715

Engelman A, Mizuuchi K, Craigie R (1991) HIV-1 DNA integration: mechanism of viral DNA cleavage and DNA strand transfer. Cell 67(6):1211–1221

Farnet CM, Bushman FD (1997) HIV-1 cDNA integration: requirement of HMG I(Y) protein for function of preintegration complexes in vitro. Cell 88(4):483–492

Farnet CM, Haseltine WA (1991a) Circularization of human immunodeficiency virus type 1 DNA in vitro. J Virol 65(12):6942–6952

Farnet CM, Haseltine WA (1991b) Determination of viral proteins present in the human immunodeficiency virus type 1 preintegration complex. J Virol 65(4):1910–1915

Forshey BM, Aiken C (2003) Disassembly of human immunodeficiency virus type 1 cores in vitro reveals association of Nef with the subviral ribonucleoprotein complex. J Virol 77(7): 4409–4414

Forshey BM, von Schwedler U, Sundquist WI, Aiken C (2002) Formation of a human immunodeficiency virus type 1 core of optimal stability is crucial for viral replication. J Virol 76(11): 5667–5677

Freund F, Boulme F, Litvak S, Tarrago-Litvak L (2001) Initiation of HIV-2 reverse transcription: a secondary structure model of the RNA-tRNA(Lys3) duplex. Nucleic Acids Res 29(13): 2757–2765

Gabor J, Cen S, Javanbakht H, Niu M, Kleiman L (2002) Effect of altering the tRNA(Lys)(3) concentration in human immunodeficiency virus type 1 upon its annealing to viral RNA, GagPol incorporation, and viral infectivity. J Virol 76(18):9096–9102

Gabus C, Ficheux D, Rau M, Keith G, Sandmeyer S, Darlix JL (1998) The yeast Ty3 retrotransposon contains a 5′-3′ bipartite primer-binding site and encodes nucleocapsid protein NCp9 functionally homologous to HIV-1 NCp7. EMBO J 17(16):4873–4880
Gabus C, Ivanyi-Nagy R, Depollier J, Bucheton A, Pelisson A, Darlix JL (2006) Characterization of a nucleocapsid-like region and of two distinct primer tRNALys,2 binding sites in the endogenous retrovirus Gypsy. Nucleic Acids Res 34(20):5764–5777
Gallay P, Swingler S, Song J, Bushman F, Trono D (1995) HIV nuclear import is governed by the phosphotyrosine-mediated binding of matrix to the core domain of integrase. Cell 83(4):569–576
Ganser BK, Li S, Klishko VY, Finch JT, Sundquist WI (1999) Assembly and analysis of conical models for the HIV-1 core. Science 283(5398):80–83
Gilboa E, Goff S, Shields A, Yoshimura F, Mitra S, Baltimore D (1979a) In vitro synthesis of a 9 kbp terminally redundant DNA carrying the infectivity of Moloney murine leukemia virus. Cell 16(4):863–874
Gilboa E, Mitra SW, Goff S, Baltimore D (1979b) A detailed model of reverse transcription and tests of crucial aspects. Cell 18(1):93–100
Gilmartin GM, Fleming ES, Oetjen J, Graveley BR (1995) CPSF recognition of an HIV-1 mRNA 3′-processing enhancer: multiple sequence contacts involved in poly(A) site definition. Genes Dev 9(1):72–83
Goldschmidt V, Rigourd M, Ehresmann C, Le Grice SF, Ehresmann B, Marquet R (2002) Direct and indirect contributions of RNA secondary structure elements to the initiation of HIV-1 reverse transcription. J Biol Chem 277(45):43233–43242
Goldschmidt V, Paillart JC, Rigourd M, Ehresmann B, Aubertin AM, Ehresmann C et al (2004) Structural variability of the initiation complex of HIV-1 reverse transcription. J Biol Chem 279(34):35923–35931
Gopalakrishnan V, Peliska JA, Benkovic SJ (1992) Human immunodeficiency virus type 1 reverse transcriptase: spatial and temporal relationship between the polymerase and RNase H activities. Proc Natl Acad Sci USA 89(22):10763–10767
Guntaka RV (1993) Transcription termination and polyadenylation in retroviruses. Microbiol Rev 57(3):511–521
Guo J, Henderson LE, Bess J, Kane B, Levin JG (1997) Human immunodeficiency virus type 1 nucleocapsid protein promotes efficient strand transfer and specific viral DNA synthesis by inhibiting TAR-dependent self-priming from minus-strand strong-stop DNA. J Virol 71(7):5178–5188
Guo F, Gabor J, Cen S, Hu K, Mouland AJ, Kleiman L (2005) Inhibition of cellular HIV-1 protease activity by lysyl-tRNA synthetase. J Biol Chem 280(28):26018–26023
Guo F, Saadatmand J, Niu M, Kleiman L (2009) Roles of Gag and NCp7 in facilitating tRNA(Lys) (3) Annealing to viral RNA in human immunodeficiency virus type 1. J Virol 83(16):8099–8107
Hahn CS, Hahn YS, Rice CM, Lee E, Dalgarno L, Strauss EG et al (1987) Conserved elements in the 3′ untranslated region of flavivirus RNAs and potential cyclization sequences. J Mol Biol 198(1):33–41
Halwani R, Cen S, Javanbakht H, Saadatmand J, Kim S, Shiba K et al (2004) Cellular distribution of Lysyl-tRNA synthetase and its interaction with Gag during human immunodeficiency virus type 1 assembly. J Virol 78(14):7553–7564
Hargittai MR, Mangla AT, Gorelick RJ, Musier-Forsyth K (2001) HIV-1 nucleocapsid protein zinc finger structures induce tRNA(Lys,3) structural changes but are not critical for primer/template annealing. J Mol Biol 312(5):985–997
Hargittai MR, Gorelick RJ, Rouzina I, Musier-Forsyth K (2004) Mechanistic insights into the kinetics of HIV-1 nucleocapsid protein-facilitated tRNA annealing to the primer binding site. J Mol Biol 337(4):951–968
Harris JD, Scott JV, Traynor B, Brahic M, Stowring L, Ventura P et al (1981) Visna virus DNA: discovery of a novel gapped structure. Virology 113(2):573–583
Hauber J, Cullen BR (1988) Mutational analysis of the trans-activation-responsive region of the human immunodeficiency virus type I long terminal repeat. J Virol 62(3):673–679

Heinzinger NK, Bukinsky MI, Haggerty SA, Ragland AM, Kewalramani V, Lee MA et al (1994) The Vpr protein of human immunodeficiency virus type 1 influences nuclear localization of viral nucleic acids in nondividing host cells. Proc Natl Acad Sci USA 91(15):7311–7315

Henriet S, Sinck L, Bec G, Gorelick RJ, Marquet R, Paillart JC (2007) Vif is a RNA chaperone that could temporally regulate RNA dimerization and the early steps of HIV-1 reverse transcription. Nucleic Acids Res 35(15):5141–5153

Heyman T, Agoutin B, Friant S, Wilhelm FX, Wilhelm ML (1995) Plus-strand DNA synthesis of the yeast retrotransposon Ty1 is initiated at two sites, PPT1 next to the 3′LTR and PPT2 within the pol gene. PPT1 is sufficient for Ty1 transposition. J Mol Biol 253(2):291–303

Hostomsky Z, Hostomska Z, Fu TB, Taylor J (1992) Reverse transcriptase of human immunodeficiency virus type 1: functionality of subunits of the heterodimer in DNA synthesis. J Virol 66(5):3179–3182

Huang Y, Mak J, Cao Q, Li Z, Wainberg MA, Kleiman L (1994) Incorporation of excess wild-type and mutant tRNA(3Lys) into human immunodeficiency virus type 1. J Virol 68(12): 7676–7683

Huang Y, Wang J, Shalom A, Li Z, Khorchid A, Wainberg MA et al (1997) Primer tRNA3Lys on the viral genome exists in unextended and two-base extended forms within mature human immunodeficiency virus type 1. J Virol 71(1):726–728

Huber HE, Richardson CC (1990) Processing of the primer for plus strand DNA synthesis by human immunodeficiency virus 1 reverse transcriptase. J Biol Chem 265(18):10565–10573

Hungnes O, Tjotta E, Grinde B (1992) Mutations in the central polypurine tract of HIV-1 result in delayed replication. Virology 190(1):440–442

Isel C, Marquet R, Keith G, Ehresmann C, Ehresmann B (1993) Modified nucleotides of tRNA(3Lys) modulate primer/template loop-loop interaction in the initiation complex of HIV-1 reverse transcription. J Biol Chem 268(34):25269–25272

Isel C, Ehresmann C, Keith G, Ehresmann B, Marquet R (1995) Initiation of reverse transcription of HIV-1: secondary structure of the HIV-1 RNA/tRNA(3Lys) (template/primer). J Mol Biol 247(2):236–250

Isel C, Lanchy JM, Le Grice SF, Ehresmann C, Ehresmann B, Marquet R (1996) Specific initiation and switch to elongation of human immunodeficiency virus type 1 reverse transcription require the post-transcriptional modifications of primer tRNA3Lys. EMBO J 15(4):917–924

Isel C, Keith G, Ehresmann B, Ehresmann C, Marquet R (1998) Mutational analysis of the tRNA3Lys/HIV-1 RNA (primer/template) complex. Nucleic Acids Res 26(5):1198–1204

Isel C, Ehresmann C, Marquet R (2010) Initiation of HIV reverse transcription. Viruses 2:213–243

Jacobo-Molina A, Clark AD Jr, Williams RL, Nanni RG, Clark P, Ferris AL et al (1991) Crystals of a ternary complex of human immunodeficiency virus type 1 reverse transcriptase with a monoclonal antibody Fab fragment and double-stranded DNA diffract x-rays to 3.5-A resolution. Proc Natl Acad Sci USA 88(23):10895–10899

Jacobo-Molina A, Ding J, Nanni RG, Clark AD Jr, Lu X, Tantillo C et al (1993) Crystal structure of human immunodeficiency virus type 1 reverse transcriptase complexed with double-stranded DNA at 3.0 A resolution shows bent DNA. Proc Natl Acad Sci USA 90(13):6320–6324

Javanbakht H, Cen S, Musier-Forsyth K, Kleiman L (2002) Correlation between tRNALys3 aminoacylation and its incorporation into HIV-1. J Biol Chem 277(20):17389–17396

Jetzt AE, Yu H, Klarmann GJ, Ron Y, Preston BD, Dougherty JP (2000) High rate of recombination throughout the human immunodeficiency virus type 1 genome. J Virol 74(3):1234–1240

Jiang M, Mak J, Ladha A, Cohen E, Klein M, Rovinski B et al (1993) Identification of tRNAs incorporated into wild-type and mutant human immunodeficiency virus type 1. J Virol 67(6):3246–3253

Johnson PE, Turner RB, Wu ZR, Hairston L, Guo J, Levin JG et al (2000) A mechanism for plus-strand transfer enhancement by the HIV-1 nucleocapsid protein during reverse transcription. Biochemistry 39(31):9084–9091

Jones FD, Hughes SH (2007) In vitro analysis of the effects of mutations in the G-tract of the human immunodeficiency virus type 1 polypurine tract on RNase H cleavage specificity. Virology 360(2):341–349

Julias JG, Ferris AL, Boyer PL, Hughes SH (2001) Replication of phenotypically mixed human immunodeficiency virus type 1 virions containing catalytically active and catalytically inactive reverse transcriptase. J Virol 75(14):6537–6546

Julias JG, McWilliams MJ, Sarafianos SG, Arnold E, Hughes SH (2002) Mutations in the RNase H domain of HIV-1 reverse transcriptase affect the initiation of DNA synthesis and the specificity of RNase H cleavage in vivo. Proc Natl Acad Sci USA 99(14):9515–9520

Junghans RP, Boone LR, Skalka AM (1982) Products of reverse transcription in avian retrovirus analyzed by electron microscopy. J Virol 43(2):544–554

Kaminska M, Shalak V, Francin M, Mirande M (2007) Viral hijacking of mitochondrial lysyl-tRNA synthetase. J Virol 81(1):68–73

Kang SM, Zhang Z, Morrow CD (1997) Identification of a sequence within U5 required for human immunodeficiency virus type 1 to stably maintain a primer binding site complementary to tRNA(Met). J Virol 71(1):207–217

Khorchid A, Javanbakht H, Wise S, Halwani R, Parniak MA, Wainberg MA et al (2000) Sequences within Pr160gag-pol affecting the selective packaging of primer tRNA(Lys3) into HIV-1. J Mol Biol 299(1):17–26

Kim JK, Palaniappan C, Wu W, Fay PJ, Bambara RA (1997) Evidence for a unique mechanism of strand transfer from the transactivation response region of HIV-1. J Biol Chem 272(27): 16769–16777

Klasens BI, Huthoff HT, Das AT, Jeeninga RE, Berkhout B (1999) The effect of template RNA structure on elongation by HIV-1 reverse transcriptase. Biochim Biophys Acta 1444(3):355–370

Kleiman L, Jones CP, Musier-Forsyth K (2010) Formation of the tRNALys packaging complex in HIV-1. FEBS Lett 584(2):359–365

Kohlstaedt LA, Wang J, Friedman JM, Rice PA, Steitz TA (1992) Crystal structure at 3.5 A resolution of HIV-1 reverse transcriptase complexed with an inhibitor. Science 256(5065): 1783–1790

Lanchy JM, Ehresmann C, Le Grice SF, Ehresmann B, Marquet R (1996) Binding and kinetic properties of HIV-1 reverse transcriptase markedly differ during initiation and elongation of reverse transcription. EMBO J 15(24):7178–7187

Lanchy JM, Keith G, Le Grice SF, Ehresmann B, Ehresmann C, Marquet R (1998) Contacts between reverse transcriptase and the primer strand govern the transition from initiation to elongation of HIV-1 reverse transcription. J Biol Chem 273(38):24425–24432

Lapadat-Tapolsky M, Gabus C, Rau M, Darlix JL (1997) Possible roles of HIV-1 nucleocapsid protein in the specificity of proviral DNA synthesis and in its variability. J Mol Biol 268(2):250–260

Lavigne M, Buc H (1999) Compression of the DNA minor groove is responsible for termination of DNA synthesis by HIV-1 reverse transcriptase. J Mol Biol 285(3):977–995

Lavigne M, Roux P, Buc H, Schaeffer F (1997) DNA curvature controls termination of plus strand DNA synthesis at the centre of HIV-1 genome. J Mol Biol 266(3):507–524

Le Grice SF, Gruninger-Leitch F (1990) Rapid purification of homodimer and heterodimer HIV-1 reverse transcriptase by metal chelate affinity chromatography. Eur J Biochem 187(2):307–314

Le Grice SF, Naas T, Wohlgensinger B, Schatz O (1991) Subunit-selective mutagenesis indicates minimal polymerase activity in heterodimer-associated p51 HIV-1 reverse transcriptase. EMBO J 10(12):3905–3911

Leavitt AD, Rose RB, Varmus HE (1992) Both substrate and target oligonucleotide sequences affect in vitro integration mediated by human immunodeficiency virus type 1 integrase protein produced in Saccharomyces cerevisiae. J Virol 66(4):2359–2368

Levin JG, Mitra M, Mascarenhas A, Musier-Forsyth K (2010) Role of HIV-1 nucleocapsid protein in HIV-1 reverse transcription. RNA Biol 7(6):754–774

Levy DN, Aldrovandi GM, Kutsch O, Shaw GM (2004) Dynamics of HIV-1 recombination in its natural target cells. Proc Natl Acad Sci USA 101(12):4204–4209

Li X, Mak J, Arts EJ, Gu Z, Kleiman L, Wainberg MA et al (1994) Effects of alterations of primer-binding site sequences on human immunodeficiency virus type 1 replication. J Virol 68(10): 6198–6206

Li Y, Zhang Z, Wakefield JK, Kang SM, Morrow CD (1997) Nucleotide substitutions within U5 are critical for efficient reverse transcription of human immunodeficiency virus type 1 with a primer binding site complementary to tRNA(His). J Virol 71(9):6315–6322

Li L, Olvera JM, Yoder KE, Mitchell RS, Butler SL, Lieber M et al (2001) Role of the non-homologous DNA end joining pathway in the early steps of retroviral infection. EMBO J 20(12):3272–3281

Liu S, Harada BT, Miller JT, Le Grice SF, Zhuang X (2010) Initiation complex dynamics direct the transitions between distinct phases of early HIV reverse transcription. Nat Struct Mol Biol 17(12):1453–1460

Mak J, Jiang M, Wainberg MA, Hammarskjold ML, Rekosh D, Kleiman L (1994) Role of Pr160gag-pol in mediating the selective incorporation of tRNA(Lys) into human immunodeficiency virus type 1 particles. J Virol 68(4):2065–2072

Masuda T, Kuroda MJ, Harada S (1998) Specific and independent recognition of U3 and U5 att sites by human immunodeficiency virus type 1 integrase in vivo. J Virol 72(10):8396–8402

McDonald D, Vodicka MA, Lucero G, Svitkina TM, Borisy GG, Emerman M et al (2002) Visualization of the intracellular behavior of HIV in living cells. J Cell Biol 159(3):441–452

Miles LR, Agresta BE, Khan MB, Tang S, Levin JG, Powell MD (2005) Effect of polypurine tract (PPT) mutations on human immunodeficiency virus type 1 replication: a virus with a completely randomized PPT retains low infectivity. J Virol 79(11):6859–6867

Miller WA, White KA (2006) Long-distance RNA-RNA interactions in plant virus gene expression and replication. Annu Rev Phytopathol 44:447–467

Miller MD, Farnet CM, Bushman FD (1997) Human immunodeficiency virus type 1 preintegration complexes: studies of organization and composition. J Virol 71(7):5382–5390

Moore-Rigdon KL, Kosloff BR, Kirkman RL, Morrow CD (2005) Preferences for the selection of unique tRNA primers revealed from analysis of HIV-1 replication in peripheral blood mononuclear cells. Retrovirology 2:21

Muesing MA, Smith DH, Cabradilla CD, Benton CV, Lasky LA, Capon DJ (1985) Nucleic acid structure and expression of the human AIDS/lymphadenopathy retrovirus. Nature 313(6002):450–458

Muesing MA, Smith DH, Capon DJ (1987) Regulation of mRNA accumulation by a human immunodeficiency virus trans-activator protein. Cell 48(4):691–701

Muthuswami R, Chen J, Burnett BP, Thimmig RL, Janjic N, McHenry CS (2002) The HIV plus-strand transfer reaction: determination of replication-competent intermediates and identification of a novel lentiviral element, the primer over-extension sequence. J Mol Biol 315(3):311–323

Negroni M, Buc H (2000) Copy-choice recombination by reverse transcriptases: reshuffling of genetic markers mediated by RNA chaperones. Proc Natl Acad Sci USA 97(12):6385–6390

Nermut MV, Fassati A (2003) Structural analyses of purified human immunodeficiency virus type 1 intracellular reverse transcription complexes. J Virol 77(15):8196–8206

Ooms M, Abbink TE, Pham C, Berkhout B (2007) Circularization of the HIV-1 RNA genome. Nucleic Acids Res 35(15):5253–5261

Palaniappan C, Fuentes GM, Rodriguez-Rodriguez L, Fay PJ, Bambara RA (1996) Helix structure and ends of RNA/DNA hybrids direct the cleavage specificity of HIV-1 reverse transcriptase RNase H. J Biol Chem 271(4):2063–2070

Pereira LA, Bentley K, Peeters A, Churchill MJ, Deacon NJ (2000) A compilation of cellular transcription factor interactions with the HIV-1 LTR promoter. Nucleic Acids Res 28(3): 663–668

Piekna-Przybylska D, Bambara RA (2011) Requirements for efficient minus strand strong-stop DNA transfer in human immunodeficiency virus 1. RNA Biol 8(2):230–236

Piekna-Przybylska D, DiChiacchio L, Mathews DH, Bambara RA (2010) A sequence similar to tRNA 3 Lys gene is embedded in HIV-1 U3-R and promotes minus-strand transfer. Nat Struct Mol Biol 17(1):83–89

Piekna-Przybylska D, Dykes C, Demeter LM, Bambara RA (2011) Sequences in the U3 region of human immunodeficiency virus 1 improve efficiency of minus strand transfer in infected cells. Virology 410(2):368–374

Powell MD, Levin JG (1996) Sequence and structural determinants required for priming of plus-strand DNA synthesis by the human immunodeficiency virus type 1 polypurine tract. J Virol 70(8):5288–5296
Pullen KA, Rattray AJ, Champoux JJ (1993) The sequence features important for plus strand priming by human immunodeficiency virus type 1 reverse transcriptase. J Biol Chem 268(9):6221–6227
Ramalanjaona N, de Rocquigny H, Millet A, Ficheux D, Darlix JL, Mely Y (2007) Investigating the mechanism of the nucleocapsid protein chaperoning of the second strand transfer during HIV-1 DNA synthesis. J Mol Biol 374(4):1041–1053
Ramirez de Arellano E, Soriano V, Alcamil J, Holguin A (2006) New findings on transcription regulation across different HIV-1 subtypes. AIDS Rev 8(1):9–16
Rausch JW, Le Grice SF (2004) 'Binding, bending and bonding': polypurine tract-primed initiation of plus-strand DNA synthesis in human immunodeficiency virus. Int J Biochem Cell Biol 36(9):1752–1766
Renda MJ, Rosenblatt JD, Klimatcheva E, Demeter LM, Bambara RA, Planelles V (2001) Mutation of the methylated tRNA(Lys)(3) residue A58 disrupts reverse transcription and inhibits replication of human immunodeficiency virus type 1. J Virol 75(20):9671–9678
Renda MJ, Bradel-Tretheway B, Planelles V, Bambara RA, Dewhurst S (2004) Inhibition of HIV type 1 replication using lentiviral-mediated delivery of mutant tRNA(Lys3)A58U. AIDS Res Hum Retroviruses 20(12):1324–1334
Rhim H, Park J, Morrow CD (1991) Deletions in the tRNA(Lys) primer-binding site of human immunodeficiency virus type 1 identify essential regions for reverse transcription. J Virol 65(9):4555–4564
Romero-Lopez C, Berzal-Herranz A (2009) A long-range RNA-RNA interaction between the 5′ and 3′ ends of the HCV genome. RNA 15(9):1740–1752
Rumbaugh JA, Fuentes GM, Bambara RA (1998) Processing of an HIV replication intermediate by the human DNA replication enzyme FEN1. J Biol Chem 273(44):28740–28745
Saadatmand J, Niu M, Kleiman L, Guo F (2009) The contribution of the primer activation signal to differences between Gag- and NCp7-facilitated tRNA(Lys3) annealing in HIV-1. Virology 391(2):334–341
Sarafianos SG, Das K, Tantillo C, Clark AD Jr, Ding J, Whitcomb JM et al (2001) Crystal structure of HIV-1 reverse transcriptase in complex with a polypurine tract RNA:DNA. EMBO J 20(6):1449–1461
Schultz SJ, Champoux JJ (2008) RNase H activity: structure, specificity, and function in reverse transcription. Virus Res 134(1–2):86–103
Schultz SJ, Zhang M, Champoux JJ (2004) Recognition of internal cleavage sites by retroviral RNases H. J Mol Biol 344(3):635–652
Selby MJ, Bain ES, Luciw PA, Peterlin BM (1989) Structure, sequence, and position of the stem-loop in tar determine transcriptional elongation by tat through the HIV-1 long terminal repeat. Genes Dev 3(4):547–558
Serrano P, Pulido MR, Saiz M, Martinez-Salas E (2006) The 3′ end of the foot-and-mouth disease virus genome establishes two distinct long-range RNA-RNA interactions with the 5′ end region. J Gen Virol 87(Pt 10):3013–3022
Shank PR, Hughes SH, Kung HJ, Majors JE, Quintrell N, Guntaka RV et al (1978) Mapping unintegrated avian sarcoma virus DNA: termini of linear DNA bear 300 nucleotides present once or twice in two species of circular DNA. Cell 15(4):1383–1395
Shen W, Gao L, Balakrishnan M, Bambara RA (2009) A recombination hot spot in HIV-1 contains guanosine runs that can form a G-quartet structure and promote strand transfer in vitro. J Biol Chem 284(49):33883–33893
Sherman PA, Fyfe JA (1990) Human immunodeficiency virus integration protein expressed in *Escherichia coli* possesses selective DNA cleaving activity. Proc Natl Acad Sci USA 87(13):5119–5123
Shurtleff AC, Beasley DW, Chen JJ, Ni H, Suderman MT, Wang H et al (2001) Genetic variation in the 3′ non-coding region of dengue viruses. Virology 281(1):75–87

Skripkin E, Isel C, Marquet R, Ehresmann B, Ehresmann C (1996) Psoralen crosslinking between human immunodeficiency virus type 1 RNA and primer tRNA3(Lys). Nucleic Acids Res 24(3): 509–514

Smith CM, Smith JS, Roth MJ (1999) RNase H requirements for the second strand transfer reaction of human immunodeficiency virus type 1 reverse transcription. J Virol 73(8):6573–6581

Song M, Basu VP, Hanson MN, Roques BP, Bambara RA (2008) Proximity and branch migration mechanisms in HIV-1 minus strand strong stop DNA transfer. J Biol Chem 283(6): 3141–3150

Stetor SR, Rausch JW, Guo MJ, Burnham JP, Boone LR, Waring MJ et al (1999) Characterization of (+) strand initiation and termination sequences located at the center of the equine infectious anemia virus genome. Biochemistry 38(12):3656–3667

Suzuki Y, Craigie R (2007) The road to chromatin – nuclear entry of retroviruses. Nat Rev Microbiol 5(3):187–196

Svarovskaia ES, Delviks KA, Hwang CK, Pathak VK (2000) Structural determinants of murine leukemia virus reverse transcriptase that affect the frequency of template switching. J Virol 74(15):7171–7178

Svarovskaia ES, Barr R, Zhang X, Pais GC, Marchand C, Pommier Y et al (2004) Azido-containing diketo acid derivatives inhibit human immunodeficiency virus type 1 integrase in vivo and influence the frequency of deletions at two-long-terminal-repeat-circle junctions. J Virol 78(7): 3210–3222

Thrall SH, Reinstein J, Wohrl BM, Goody RS (1996) Evaluation of human immunodeficiency virus type 1 reverse transcriptase primer tRNA binding by fluorescence spectroscopy: specificity and comparison to primer/template binding. Biochemistry 35(14):4609–4618

Tisne C, Roques BP, Dardel F (2004) The annealing mechanism of HIV-1 reverse transcription primer onto the viral genome. J Biol Chem 279(5):3588–3595

Tobaly-Tapiero J, Kupiec JJ, Santillana-Hayat M, Canivet M, Peries J, Emanoil-Ravier R (1991) Further characterization of the gapped DNA intermediates of human spumavirus: evidence for a dual initiation of plus-strand DNA synthesis. J Gen Virol 72(Pt 3):605–608

Vink C, van Gent DC, Elgersma Y, Plasterk RH (1991) Human immunodeficiency virus integrase protein requires a subterminal position of its viral DNA recognition sequence for efficient cleavage. J Virol 65(9):4636–4644

Vogt VM (1997) Retroviral virions and genomes. In: Coffin JM, Hughes SM, Varmus HE (eds) Retroviruses (25 March 2011 ed.). Cold Spring Harbor Laboratory Press, Cold Spring Harbor, NY, pp. 27–69

Wakefield JK, Wolf AG, Morrow CD (1995) Human immunodeficiency virus type 1 can use different tRNAs as primers for reverse transcription but selectively maintains a primer binding site complementary to tRNA(3Lys). J Virol 69(10):6021–6029

Wakefield JK, Kang SM, Morrow CD (1996) Construction of a type 1 human immunodeficiency virus that maintains a primer binding site complementary to tRNA(His). J Virol 70(2): 966–975

Wei SQ, Mizuuchi K, Craigie R (1998) Footprints on the viral DNA ends in Moloney murine leukemia virus preintegration complexes reflect a specific association with integrase. Proc Natl Acad Sci USA 95(18):10535–10540

Wei M, Cen S, Niu M, Guo F, Kleiman L (2005) Defective replication in human immunodeficiency virus type 1 when non-primers are used for reverse transcription. J Virol 79(14):9081–9087

Wills NM, Gesteland RF, Atkins JF (1994) Pseudoknot-dependent read-through of retroviral gag termination codons: importance of sequences in the spacer and loop 2. EMBO J 13(17): 4137–4144

Wisniewski M, Balakrishnan M, Palaniappan C, Fay PJ, Bambara RA (2000a) The sequential mechanism of HIV reverse transcriptase RNase H. J Biol Chem 275(48):37664–37671

Wisniewski M, Balakrishnan M, Palaniappan C, Fay PJ, Bambara RA (2000b) Unique progressive cleavage mechanism of HIV reverse transcriptase RNase H. Proc Natl Acad Sci USA 97(22):11978–11983

Wohrl BM, Moelling K (1990) Interaction of HIV-1 ribonuclease H with polypurine tract containing RNA-DNA hybrids. Biochemistry 29(44):10141–10147

Wu T, Guo J, Bess J, Henderson LE, Levin JG (1999) Molecular requirements for human immunodeficiency virus type 1 plus-strand transfer: analysis in reconstituted and endogenous reverse transcription systems. J Virol 73(6):4794–4805

Yusupova G, Lanchy JM, Yusupov M, Keith G, Le Grice SF, Ehresmann C et al (1996) Primer selection by HIV-1 reverse transcriptase on RNA-tRNA(3Lys) and DNA-tRNA(3Lys) hybrids. J Mol Biol 261(3):315–321

Zaitseva L, Myers R, Fassati A (2006) tRNAs promote nuclear import of HIV-1 intracellular reverse transcription complexes. PLoS Biol 4(10):e332

Zhang Z, Kang SM, LeBlanc A, Hajduk SL, Morrow CD (1996) Nucleotide sequences within the U5 region of the viral RNA genome are the major determinants for an human immunodeficiency virus type 1 to maintain a primer binding site complementary to tRNA(His). Virology 226(2):306–317

Zhang J, Tang LY, Li T, Ma Y, Sapp CM (2000) Most retroviral recombinations occur during minus-strand DNA synthesis. J Virol 74(5):2313–2322

Zhuang J, Jetzt AE, Sun G, Yu H, Klarmann G, Ron Y et al (2002) Human immunodeficiency virus type 1 recombination: rate, fidelity, and putative hot spots. J Virol 76(22):11273–11282

# Chapter 3
# The RNase H Domain: Structure, Function and Mechanism

**Marcin Nowotny and Małgorzata Figiel**

## 3.1 Overview

An essential step of proliferation of retroviruses and transposition of long terminal repeat-containing retrotransposons is conversion of their single-stranded RNA genome into integration-competent, double-stranded proviral DNA by the multifunctional reverse transcriptase (RT) (Gilboa et al. 1979). RT is an enzyme with two activities. RNA-dependent DNA polymerase activity is first used to synthesize minus (−) strand DNA from the positive-stranded RNA genome, resulting in an RNA/DNA replication intermediate. The RNA strand of these hybrids is degraded by the RNase H activity to allow DNA-dependent synthesis of (+) strand DNA. RNase H activity is used not only to nonspecifically remove the RNA but also to specifically generate and remove RNA primers required to initiate synthesis of both DNA strands. In this chapter, we describe the current understanding of the HIV-1 RNase H domain and its cellular counterparts – RNases H1 – with particular focus on structural data which, together with biochemical and computational studies, have revealed a detailed picture of the mechanism of action of this important and clinically significant enzyme. We will also discuss how the RNase H domain functions in the context of the dimeric HIV-1 RT.

Ribonucleases H (RNases H) are defined as nucleases that hydrolyze the RNA strand in RNA/DNA hybrids. They catalyze this hydrolytic reaction in a sequence-nonspecific manner. Two types of cellular RNases H have been described – RNases H1 and RNases H2 (or type 1 and type 2). The RNase H domain from HIV-1 RT resembles very closely the catalytic domains of cellular RNases H1, and we therefore focus here on type 1 enzymes.

M. Nowotny (✉) • M. Figiel
International Institute of Molecular and Cell Biology, Warsaw, Poland
e-mail: mnowotny@iimcb.gov.pl

S.F.J. LeGrice and M. Götte (eds.), *Human Immunodeficiency Virus Reverse Transcriptase: A Bench-to-Bedside Success*, DOI 10.1007/978-1-4614-7291-9_3,

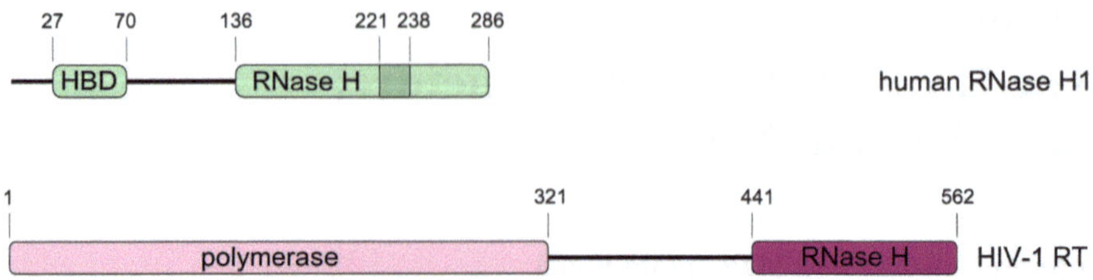


**Fig. 3.1** *Domain representation of human RNase H1 (green) and HIV-1 RT (pink)*. Basic protrusion of human RNase H1 is colored in *dark green*. Residue numbers at the domain boundaries are given. HBD denotes the hybrid-binding domain

RNases H1 are present in bacteria, eukaryotes, and some archaea (Ohtani et al. 1999a). Their catalytic domain comprises approximately 150 residues. All eukaryotic and some bacterial RNases H1 contain an additional N-terminal hybrid-binding domain of ~50 residues (Cerritelli et al. 1998) (Fig. 3.1). The two domains are connected by a flexible linker of variable length and non-conserved sequence. The proposed role of the N-terminal domain is to increase processivity of the enzyme (Nowotny et al. 2008; Gaidamakov et al. 2005). It is thought to tether the protein to the RNA/DNA substrate which allows the catalytic domain to execute numerous cuts in a defined region of the hybrid, resulting in formation of small RNA fragments that can dissociate from the DNA more easily.

Several functions of cellular RNases H1 have been proposed. They are involved in the processing of R-loops which form during translation when nascent mRNA hybridizes with the complementary strand of the DNA (Broccoli et al. 2004; Drolet et al. 1995). RNases H are also implicated in DNA repair (Arudchandran et al. 2000). Recent reports have demonstrated that RNA/DNA hybrids can lead to genomic instability, and by removal of such hybrids, RNases H help protect genetic information (Wahba et al. 2011). Both type 1 and type 2 RNases H also repair DNA via excision of single ribonucleotides misincorporated into the DNA by replicative DNA polymerases (Lazzaro et al. 2012). In higher organisms, RNases H1 are localized predominantly in nucleus and mitochondria (Cerritelli et al. 2003). RNase H1 knockout mice die at embryonic stage due to defects in replication of mitochondrial DNA which points to another function of this enzyme and demonstrates that it is essential in animals (Cerritelli et al. 2003).

## 3.2 Structures of RNases H1

Crystal structures are available for RNases H1 from eight organisms (reviewed in (Ohtani et al. 1999a)). The first structures of RNase H1 were solved for the *E. coli* enzyme which only contains the catalytic domain. These structures, published in 1990 (Yang et al. 1990; Katayanagi et al. 1990), revealed a fold that was novel at the time and was named the "RNase H fold" (Fig. 3.2). This topology has since been

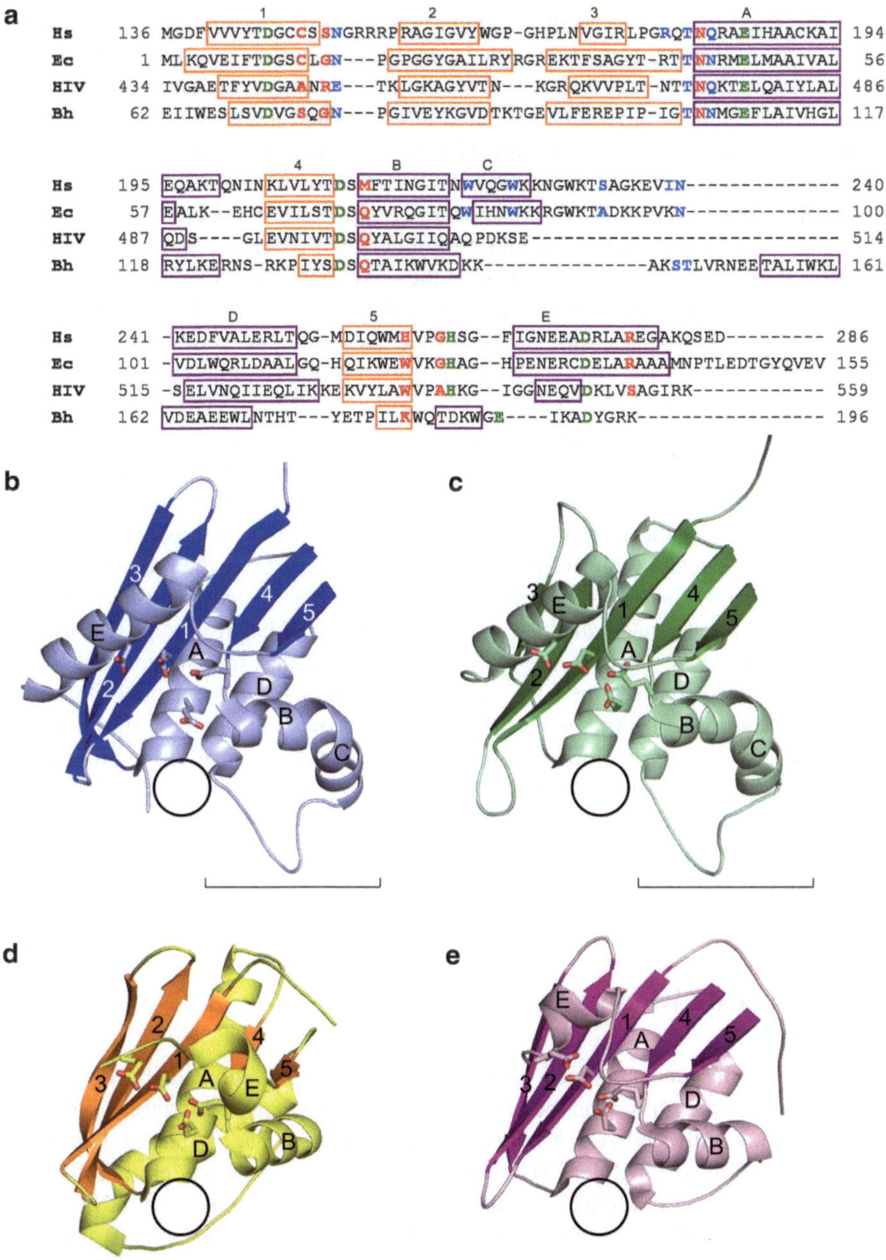


**Fig. 3.2** *Structures of RNases H1*. (**a**) Sequence alignment of RNase H domains. Secondary structure elements are marked by *boxes* (α-helices, *purple*; β-strands, *orange*) and labeled. Important residues are shown in colors: active site residues, *green*; DNA-interacting residues, *blue*; RNA-interacting residues, *red*. Hs, human; Ec, *E. coli*; HIV, human immunodeficiency virus type 1; Bh, *B. halodurans*. (**b–e**) Ribbon representations of RNase H1 structures: *E. coli* RNase H1 (**b**), catalytic domain of human RNase H1 (**c**), catalytic domain of *B. halodurans* RNase H1 (**d**), and RNase H domain from HIV-1 RT (**e**). Secondary structure elements are labeled as in 2A, and active site residues are shown as *sticks*. *Brackets* indicate the position of the basic protrusion. Circles indicate the position of the phosphate-binding pocket

found in catalytic cores of many enzymes involved in the processing of nucleic acids – retroviral integrases, transposases, Holliday junction resolvases, and RNAi enzyme Argonaute (Nowotny 2009). Two out of four enzymatic activities encoded by HIV-1 rely on this fold – RT-associated RNase H and integrase.

The central element of the RNase H fold comprises a β-sheet of five strands numbered 1–5 and arranged in the order 5–4–1–2–3, with strand 2 running antiparallel to the others (Fig. 3.2b–e). Strands 1, 2, and 3 are connected by loops, run without any additional insertions, and are longer than strands 4 and 5. The central β-sheet is flanked by α-helices. The first of these – α-helix A – is located after strand 3 and reinforces the structure of the β-sheet (Fig. 3.2a). Strand 4 is followed by helices located on the same side of the β-sheet as α-helix A. Based on this part of the structure, RNases H1 can be divided into two groups. Some enzymes, such as the RNase H domain of HIV-1 RT or *B. halodurans* RNase H1 (Bh-RNase H1), only have two helices after strand 4. Others, i.e., eukaryotic enzymes, *E. coli* RNase H1 (Ec-RNase H1), and the RNase H domain of Moloney murine leukemia virus (MMLV) RT, have an additional α-helix C which, together with the following loop, forms an element of the structure often referred to as "basic protrusion" (Fig. 3.2). Scanning mutagenesis studies of *E. coli* RNase H1 showed that the protrusion plays an important role in substrate binding (Kanaya et al. 1991). A deletion of 18 residues from this region inactivated the enzyme and reduced its affinity for RNA/DNA hybrids (Haruki et al. 1997). Upon deletion of 13 residues, the enzyme lost $Mg^{2+}$-dependent activity but retained the ability to perform $Mn^{2+}$-dependent cleavages (Keck and Marqusee 1996). In MMLV RT, the removal of α-helix C dramatically reduced RNase H activity, and substitutions of several of its residues led to defects in RT functions and virus replication (Lim et al. 2002). In most RNases H1 (Bh-RNase H1 is an exception), the C-terminal part of the structure, after strand 5, contains another long α-helix E located on the opposite side of the central β-sheet relative to the other helices (Fig. 3.2).

## 3.3 Substrate Binding

The details of the mechanism of substrate recognition and catalysis by RNasesH1 have been elucidated based on the crystal structures of catalytic domains of RNases H1 in complex with RNA/DNA hybrid substrates (Fig. 3.3a, b). The first complex structure was solved for RNase H1 from *B. halodurans* (Nowotny et al. 2005) and was followed by a complex structure of the human enzyme (Hs-RNase H1) (Nowotny et al. 2007). The active site of RNase H1 approaches the substrate from the minor groove side. Base edges in this groove form contacts with several conserved residues – Asn151, Asn182, Gln183 in Hs-RNase H1 and Asn77, Asn105, and Asn106 in Bh-RNase H1. Their equivalents in HIV-1 RT are Glu449, Asn474, and Gln475 (Fig. 3.2a). Two of these are critical elements of a motif of HIV-1 RT designated the "RNase H primer grip," which is discussed in greater detail below.

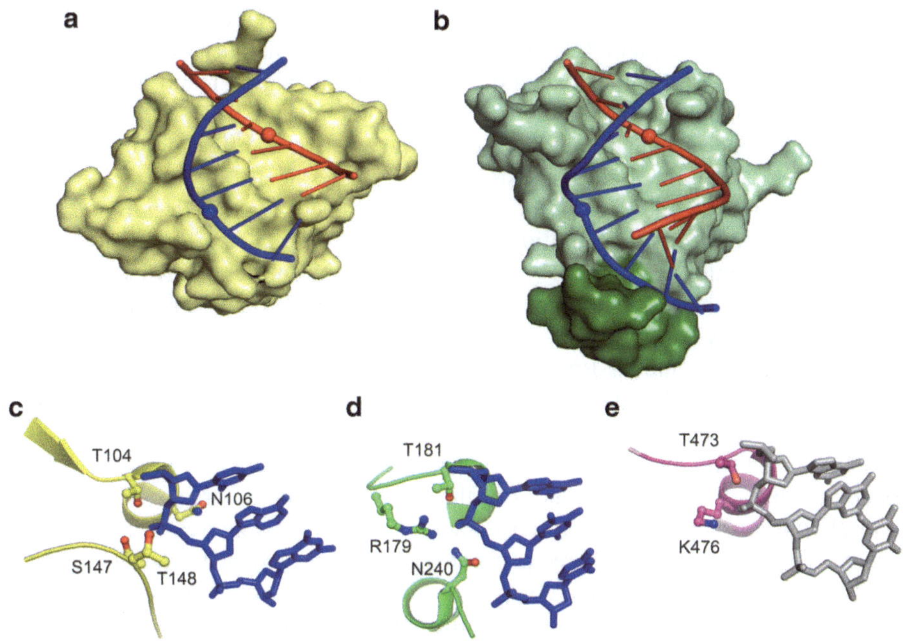


**Fig. 3.3** *Substrate binding by RNase H1.* (**a–b**) Substrate complexes of Bh-RNase H1 (**a**) and Hs-RNase H1 (**b**). RNA is shown in *red*, and DNA in *blue*; scissile phosphate and the phosphate bound in the phosphate-binding pocket are shown as *spheres*. The basic protrusion of Hs-RNase H1 is colored in *dark green*. (**c–e**) Close-up of the phosphate-binding pockets of Bh-RNase H1 (**c**), Hs-RNase H1 (**d**), and HIV-1 RT (E). Residues forming the phosphate-binding pocket are shown in *ball* and *stick* and *labeled*. The DNA strand interacting with the protein is shown in *blue* for crystal structures and in *gray* for the model of HIV-1 RNase H-substrate complex

The binding of the sugar-phosphate backbone of the RNA strand is conserved in bacterial and human RNases H1 – 2′-OH groups of the RNA form extensive interactions with the protein, which serves for the specific recognition of the ribonucleotides. These interactions are mainly mediated by the backbone of the protein which means that they are determined by the overall shape of the fold rather than the positions and the identity of side chains. On the RNA side, the interactions are formed by four consecutive ribonucleotides – two on each side of the scissile phosphate, which explains why RNases H1 require a stretch of at least four ribonucleotides in a chimeric DNA–RNA–DNA/DNA substrate and cleave such a minimal substrate preferentially in the middle of RNA sequence (Ohtani et al. 1999b). As determined by sugar puckers, the RNA strand of the hybrid bound by RNase H1 assumes an A-form conformation, the only one allowed for RNA.

For binding of the DNA strand by RNase H1, the most prominent feature is the "phosphate-binding pocket" (Fig. 3.3c–e), located close to the N-terminus of α-helix A. Its positioning relative to the active site is very well conserved among RNases H1 (Fig. 3.2b–e). The pocket comprises one absolutely conserved threonine residue

(Thr104, Thr181, and Thr473 in Bh-RNase H1, Hs-RNase H1, and HIV-1 RNase H, respectively, Figs. 3.2a and 3.3c–e), which is critical for activity. Substituting this residue in Ec-RNase H1 inactivated the enzyme (Kanaya 1998), and in HIV-1 RT, the equivalent mutation disrupted viral replication (Julias et al. 2002). Other residues forming the phosphate-binding pocket vary between different enzymes. For example, three additional amino acids form the pocket in Bh-RNase H1 – Asn106 located in α-helix A and Ser147 and Thr148 located between α-helices B and D. In HIV-1 RT, there is only one additional residue Lys476 (Fig. 3.3e). The fact that the phosphate-binding pocket of HIV-1 RNase H domain only contains two residues may account for the observation that the isolated RNase H domain is inactive (vide infra) (Hostomsky et al. 1991).

In Bh-RNase H1 and Hs-RNase H1-substrate complexes, the phosphate-binding pocket forms very tight interactions with the phosphate group of DNA located two base pairs from the scissile phosphate of the RNA strand (Nowotny et al. 2005, 2007). In both structures, this interaction results in deformation of the DNA backbone, manifested by unusual values of the alpha and gamma torsion angles of the phosphodiester backbone. Such deformation of the DNA requires that the minor groove of the duplex, as measured between the active site and the phosphate-binding pocket, is narrow – approximately 7–8 Å – which corresponds to B-like form of nucleic acid. This is possible only when the non-cleaved strand adopts B-form conformation. Since such a conformation is only permissible for DNA, this provides a means for the enzyme to recognize that the non-cleaved strand of the substrate is the DNA. In other words, RNase H exploits the ability of DNA to adopt B-form conformation and its flexibility to differentiate it from RNA. This is also demonstrated by biochemical studies – when modified nucleotides that preferentially adopt A-form conformation were introduced into the DNA strand of the hybrid, such duplexes were poor substrates for RNase H1 (Lima et al. 2007a, b).

The importance of basic protrusion in substrate binding first implied in biochemical experiments is further confirmed by analyses of the crystal structure of Hs-RNase H1 (Haruki et al. 1997; Keck and Marqusee 1996). In the human enzyme, a tight channel is formed between the basic protrusion and the remainder of the fold (Fig. 3.3b) which accommodates the DNA strand of the substrate (Nowotny et al. 2005). A tryptophan residue (Trp221) located within the channel forms a stacking interaction with a deoxyribose ring of the DNA. Since the presence of 2′-OH group in the ribose would disrupt this interaction, resulting in weaker binding, this is an additional mechanism that leads to the preference for DNA.

## 3.4 Catalysis

The active site of RNases H1 is composed of four carboxylates forming a conserved –D–E–D–D– motif coordinating the divalent metal ions (Fig. 3.2a). The first residue of the –D–E–D–D– motif is located in the middle of β-strand 1, the second in α-helix A, and the third in the C-terminus of β-strand 4. These three residues are absolutely conserved among RNases H1, and their substitutions lead to a complete

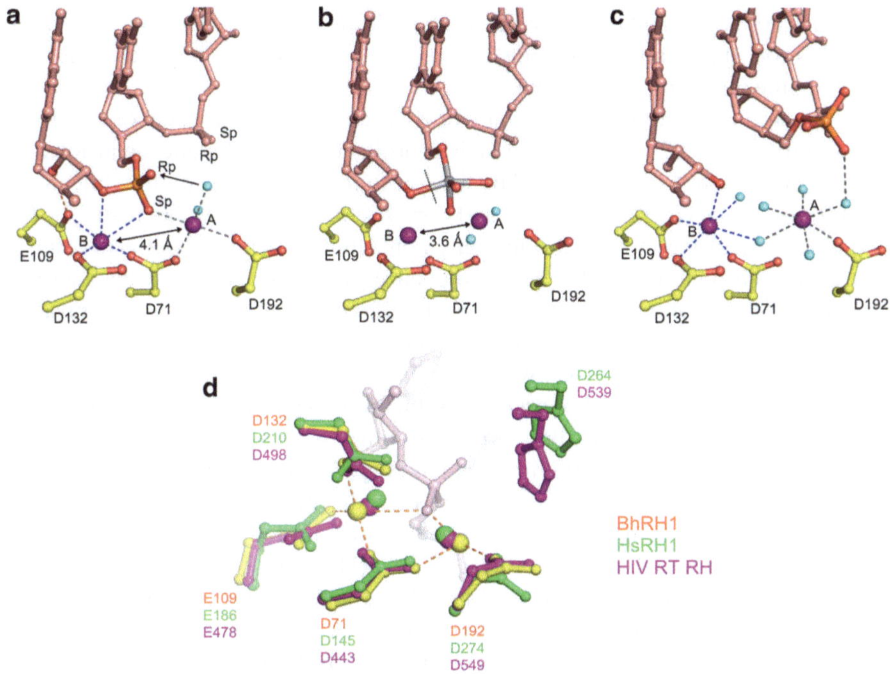


**Fig. 3.4** *Catalysis by RNase H1*. (**a–c**). Active site of Bh-RNase H1 at different steps of the hydrolysis reaction: substrate complex (**a**), reaction intermediate (**b**), reaction product (**c**). Active site residues are shown in *ball* and *stick* and *labeled*. $Mg^{2+}$ ions are shown as *purple spheres*, and water as *cyan spheres*. RNA is depicted as *pink sticks*, with the scissile phosphate in *red* and *orange* (when determined from structure) or *red* and *gray* (when modeled). (**d**) Superposition of the active site residues from the substrate complexes of Bh-RNase H1 (*yellow*) and Hs-RNase H1 (*green*) and the inhibitor complex of HIV-1 RNase H (*magenta*). Metal ions ($Mg^{2+}$ for Bh-RNase H1, $Ca^{2+}$ for Hs-RNase H1, and $Mn^{2+}$ for HIV-1 RNase H) are shown as *spheres*. RNA strand from the Bh-RNase H1-substrate complex is shown in *pink*. The inhibitor from the HIV-1 RNase H structure is omitted for clarity

loss of activity (Kanaya et al. 1990; Haruki et al. 1994). The fourth residue is located in the C-terminal α-helix E. In some enzymes, substitutions of this last residue of the active site leave the enzyme partially active (Nowotny et al. 2005; Rausch and Le Grice 1997).

RNase H-catalyzed hydrolysis generates a product with 5′-phosphate and 3′-OH groups (Krug and Berger 1989; Miller et al. 1973) and the reaction occurs by $S_N2$-like mechanism with inversion of stereoconfiguration of the phosphorus atom (Krakowiak et al. 2002). Catalysis requires divalent metal ions coordinated by the carboxylate-rich active site; $Mg^{2+}$ is the preferred divalent cation but others, for example, $Mn^{2+}$, can also support hydrolysis. RNase H1 activity is inhibited in presence of calcium ions (Nowotny and Yang 2006) or high concentrations of $Mg^{2+}$ and $Mn^{2+}$ (Goedken and Marqusee 2001; Keck et al. 1998). In the crystal structures of substrate complexes of RNases H1, two metal ions, designated A and B, are observed at the active site (Fig. 3.4), supporting the two-metal-ion mechanism of catalysis first proposed by Steitz and Steitz (1993). In this mechanism, the two metal ions are positioned on either side of the scissile phosphate – one is stabilizing the transition state and the leaving group, while the second is positioning and activating the attacking nucleophile.

The central element of the active site of RNases H1 is the first Asp of the -D-E-D-D- motif (Figs. 3.2a and 3.4a), which coordinates both divalent metal ions. In addition, metal ion A is coordinated by the fourth residue of this motif, while metal ion B forms contacts with the second and third residue (Fig. 3.4a). Interestingly, the second residue of the motif (Glu) not only contacts the metal ion but also forms a hydrogen bond with a 2′-OH group of the substrate. This residue is characteristic for RNases H1 and absent from the active sites of the related enzymes. Conceivably, it may have evolved to serve as an additional RNA specificity check by coupling an interaction with the 2′-OH group with formation of the active site.

Metal ions A and B form important coordination contacts with the scissile phosphate – both interact with the pro-$S_p$ non-bridging oxygen, and in addition, metal ion B interacts with the 3′ bridging oxygen (Fig. 3.4a). The importance of these coordination contacts is vividly demonstrated by the structure of one of the human RNase H1-substrate complexes, which contains two protein molecules interacting with a 14-mer RNA/DNA hybrid (PDB: 2QKK) (Nowotny et al. 2005). The active site of one protein subunit is properly formed, but within the active site of the other subunit, the substrate is deformed due to crystal packing, and the scissile phosphate is displaced from the active site by 3 Å. While the properly formed active site contains two metal ions ($Ca^{2+}$) in the predicted positions, the deformed one contains only one, and its position does not correspond to either A or B binding sites. The crucial role of substrate in metal-ion coordination was also demonstrated by the crystal structure of Bh-RNase H1 in complex with dsDNA (Pallan and Egli 2008). In this structure, the non-cleaved DNA strand interacts with the phosphate-binding pocket in a manner similar to the DNA of the correct RNA/DNA substrate. In contrast, the DNA corresponding to the cleaved strand of the duplex is shifted at the active site by only 1.6 Å. In addition, Glu109, which normally interacts with 2′-OH of the substrate, swings away. The displacement of the substrate precludes the coordination of the metal ions and they are missing from the active site. These results demonstrate that substrate binding and metal ion coordination at the active site are tightly coupled, ensuring that catalysis occurs only when the proper substrate is bound (Yang et al. 2006). Further support for this notion comes from the fact that human RNase H1 can bind dsRNA (albeit more weakly than RNA/DNA) but is not able to cleave it (Nowotny et al. 2005).

The two-metal-ion mechanism of catalysis involves an attack of a nucleophile on the scissile phosphate in $S_N2$-like reaction. In the case of hydrolysis, such as catalyzed by RNase H1, the nucleophile is a water molecule or hydroxide ion. In the Bh-RNase H1 complex structures, metal ion A coordinates two water molecules (Fig. 3.4a), one of which is appropriately positioned to function as the attacking nucleophile. It is located 3.5 Å from the phosphorus, roughly in line with the 3′O–P bond to be hydrolyzed. Upon nucleophilic attack, a phosphorane transition state with pentavalent phosphorus is formed in which the attacking and leaving group (3′ oxygen) are equidistant (Fig. 3.4b). The bond between 3′ oxygen and phosphorus is next broken, leading to formation of the product with 3′-OH and 5′-phosphate groups with concomitant inversion of chirality of the phosphorus atom (Fig. 3.4c) (Krug and Berger 1989; Miller et al. 1973).

Further insights into the catalytic mechanism of RNase H1 have been gleaned from crystal structures of *B. halodurans* RNase H1 in complex with RNA/DNA hybrids specially designed to place nicks in the RNA strand at the active site upon crystal formation (Nowotny and Yang 2006). One of these contained non-phosphorylated nick for modeling of the transition state, while the second contained a phosphorylated 5′ terminus corresponding to the product of the reaction. These structures revealed that the metal ions move closer together when the transition state is formed – the distance in the substrate complex was 4.1 Å and in the transition-state mimic 3.6 Å. Larger ions such as $Ca^{2+}$ cannot move so close to each other, which would explain why they fail to support the RNase H1 activity. Following hydrolysis, the resulting 5′ phosphate is removed from the active site, but the metal ions remain bound. Metal ion B changes its conformation from an irregular configuration with five ligands observed in the substrate complex to a regular octahedral one with six (two of which are new water molecules). This octahedral configuration is preferred by $Mg^{2+}$ ions, and the transition between irregular (high energy) and regular (low energy) coordination likely provides an additional driving force for the reaction and assures its irreversibility. It also explains why $Mg^{2+}$ is the preferred metal ion for RNases H1.

One important aspect of RNase H-mediated hydrolysis that cannot be addressed based on crystal structures is proton transfer events. It is assumed that deprotonated water (hydroxide ion) is a more efficient nucleophile, and one of the early steps in the reaction is the deprotonation of the nucleophilic water. Following strand scission, another proton transfer must occur in order to protonate the leaving 3′ oxygen. The issue of proton transfers has been addressed by several computational studies. Quantum mechanical/molecular mechanical (QM/MM) simulations showed several possible pathways of proton transfer. One outcome of such simulations was transfer of the proton from the attacking nucleophile to the pro-Rp oxygen of the scissile phosphate and then to the leaving group (De Vivo et al. 2008; Elsasser and Fels 2010). Both transfers were mediated by water molecules. Another possible proton transfer pathway is from the attacking water to the phosphate group downstream of the scissile phosphate (Rosta et al. 2011). In substrate complex crystal structures of Bh-RNase H1, the pro-Rp non-bridging phosphorus atom of this downstream phosphate is located ~3 Å from the attacking water – a distance suitable for proton transfers. Involvement of this phosphorus in the reaction is supported by the fact that its stereoselective substitution with sulfur inhibited RNase H1 cleavage, while a similar pro-Rp substitution at the scissile phosphate had no effect (Haruki et al. 2000). Simulations by Rosta et al. (2011) also showed that protonation of the leaving group could be achieved by transfer from the third Asp of the -D-E-D-D- active site motif (Asp132 in Bh-RNase H1).

RNases H1 contain an additional conserved His in the vicinity of the active site located in a flexible loop between β-strand 5 and the C-terminal α-helix E. In Hs-RNase H1, this is His264 and the counterpart of HIV-1 RNase H is His539

(Fig. 3.4d), substitution of which leads to a significant decrease in enzymatic activity (Oda et al. 1993; Tisdale et al. 1991). In Bh-RNase H1, the region after β-strand 5 has a different structure, but a functional equivalent of the His residue has been identified – Glu188 – which in some structures participates in coordinating metal ion A (Nowotny et al. 2005). In contrast, the His residue in human RNase H1 does not directly contact the substrate or the metal ions (Nowotny et al. 2007). It does, however, clash with the 5′ phosphate removed from the active site in the modeled post-cleavage state. It has been thus proposed that this residue facilitates the release of the product from the enzyme following strand scission (Nowotny et al. 2007).

## 3.5 Catalytic Mechanism of HIV-1 RNase H

As already mentioned, HIV-1 RT possesses two enzymatic activities – DNA polymerase and RNase H (Hansen et al. 1987). The DNA polymerase domain comprises the N-terminal domain and is separated from the C-terminal RNase H domain by a connection subdomain (Hansen et al. 1988) (Fig. 3.1). RTs can be monomeric, an example of which is the MMLV enzyme (Cote and Roth 2008). Lentiviral RTs, including that of HIV-1, are heterodimers of a full-length active subunit (p66 in HIV-1 RT) and a smaller one from which the RNase H domain has been proteolytically processed (p51 in HIV-1 RT) (Fig. 3.5). The conformation of p51 is significantly different (Wang et al. 1994) and it has been proposed to provide an architectural platform that stabilizes the catalytic subunit. One of the key

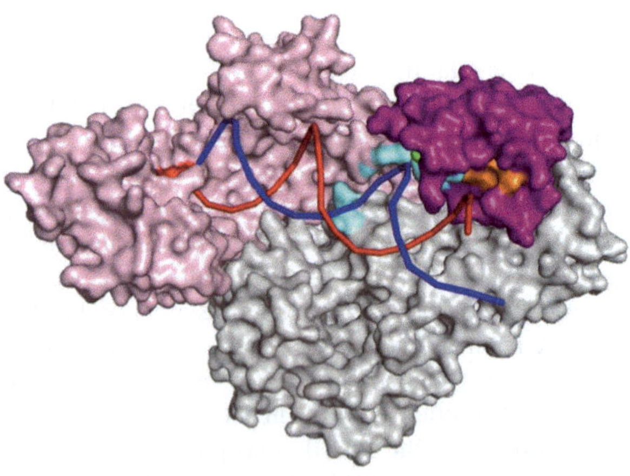

**Fig. 3.5** *Structure of HIV-1 RT in complex with RNA/DNA hybrid* (PDB ID: 1HYS). The p66 subunit is shown in *pink* (DNA polymerase and connection subdomains) and *purple* (RNase H domain) and the p51 subunit in *gray*. The RNA/DNA hybrid is shown in ribbon representation (*red*, RNA; *blue*, DNA). DNA polymerase and RNase H active sites are indicated with *red* and *orange*, respectively. The RNase H primer grip is colored in *cyan*, the phosphate-binding pocket is colored in *green*

differences between dimeric and monomeric RTs is the fact that the RNase H domain in the latter is linked with the connection domain via a flexible linker and is very mobile in the absence of the substrate (Nowak et al. 2013). In contrast, in the dimeric RTs such as HIV-1 enzyme the RNase H domain is relatively immobile and rigidly placed within the dimer.

The interaction of HIV-1 RT heterodimer with nucleic acid has been elucidated thanks to numerous crystal structures. Most of these structures contain dsDNA as the nucleic acid and until recently only one structure of HIV-1 RT bound to an RNA/DNA hybrid was available (Sarafianos et al. 2001). In this structure, the substrate forms interactions with both the DNA polymerase and RNase H domains but does not make contact with the RNase H active site, with the nearest phosphate group located 3.7 Å form the expected position of the scissile phosphate. Given the very high structural similarity of HIV-1 RNase H domain and its cellular counterparts, implying a similar mechanism of substrate binding and catalysis, a theoretical model of a productive HIV-1 RNase H-substrate complex was proposed (Nowotny et al. 2007). It suggested that in order to retain required interactions with both the polymerase and RNase H domains, the substrate needs to bend by at least 35° in the region located 4-6 bp from the RNase H active site.

Recently, crystal structures of a ternary complex of HIV-1 RT with RNA/DNA and non-nucleoside reverse transcriptase-inhibitors (NNRTIs) have been reported (Lapkouski et al. 2013). Binding of NNRTIs results in a displacement of the substrate from the polymerase active site which in turn increases the likelihood of binding of the substrate at the active site of the RNase H. In these structures a phosphate group of the RNA strand of the substrate is located ~2 Å away from the expected position of the scissile phosphate at the RNase H active site and therefore these complexes can be assumed to represent the catalysis-competent state. The substrate contains a kink right before the entrance to the RNase H domain, accompanied by a widening of the major groove due to unwinding of the duplex. In addition to the changes in the nucleic acid conformation, also the relative positions of HIV-1 RT domains change. The distance between the two subunits increases and the fingers and palm domains of p66 move with respect to the rest of the subunit leading to a more open conformation of the heterodimer.

The catalytic mechanism is very likely shared between cellular and HIV-1 RNases H-1. All require divalent metal ions for the activity (Cirino et al. 1995; Starnes and Cheng 1989), and the superposition of the structures of *B. halodurans*, human, and HIV-1 RNases H1 reveals excellent conservation of the active site (Fig. 3.4d). Moreover, similar to other RNases H1, HIV-1 RT residues forming the -D-E-D-D- active site motif (Asp443, Glu478, Asp498, Asp549) and the conserved His located nearby (His539) are necessary for the activity (Davies et al. 1991; Schatz et al. 1989; Mizrahi et al. 1990, 1994). Recently, crystal structures with several inhibitors interacting with the active site through metal ion coordination have been reported (Himmel et al. 2009). Two of these have been solved for the natural product hydroxytropolones β-thujaplicinol and manicol, both of which contain a seven-membered ring with three oxygens which mimics the scissile phosphate. The structure of these complexes with HIV-1 RNase H contains two $Mn^{2+}$ ions at the active site that occupy nearly identical positions to metal ions in the substrate complex structures of Bh-RNase H1 and Hs-RNase H1 (Himmel et al. 2009; Chung et al. 2011) (Fig. 3.4d). This further supports the idea that the catalytic mechanism established for cellular RNases H1 is conserved in their HIV-1 counterpart.

One significant difference between cellular and HIV-1 RNases H is that the isolated HIV-1 domain has minimal $Mg^{2+}$-dependent activity (Hostomsky et al. 1991; Smith and Roth 1993), which can, however, be restored by various modifications – for example, inclusion of an N- or C-terminal poly-His extension (Hansen et al. 1988; Smith and Roth 1993; Evans et al. 1991). Similar restoration of $Mn^{2+}$-dependent activity, but to a greater extent, has been observed upon insertion of the basic protrusion from *E. coli* RNase H into the structure of HIV-1 RNase H (Stahl et al. 1994; Keck and Marqusee 1995). It has also been reported that $Mg^{2+}$-dependent activity can be reconstituted by the addition of p51 subunit of HIV-1 RT (Hostomsky et al. 1991; Smith and Roth 1993; Evans et al. 1991). Progressive addition of HIV-1 RT sequence to the N-terminus of the RNase H domain also increased $Mn^{2+}$-dependent activity (Smith et al. 1994), while an engineered thumb-connection-RNase H fragment had low levels of $Mg^{2+}$-dependent RNase H activity that could be enhanced by reconstitution with the p51 subunit (Smith et al. 1994). Moreover, the HIV-1 connection subdomain could stimulate activity *in trans* (Smith et al. 1994). $Mn^{2+}$ ions are known to relax specificity of catalysis for many enzymes that utilize two-metal-ion mechanism and allow cleavage of nonideal or slightly misplaced substrates (Yang et al. 2006). HIV-1 RT can, for example, cleave dsRNA substrates in the presence of $Mn^{2+}$ (Ben-Artzi et al. 1992; Hostomsky et al. 1994), although this so-called RNase H* activity is not likely to be of physiological significance. The fact that many of the HIV-1 RNase H constructs described above exhibit only $Mn^{2+}$-dependent activity likely means that the substrate positioning at the active site of these fusions and fragments is suboptimal and that proper positioning requires a contribution from other portions of the RT heterodimer. Inactivity of the isolated HIV-1 RNase H domain likely reflects a combination of two factors, namely, (1) lack of the basic protrusion and (2) the observation that the phosphate-binding pocket only contains two residues, which reduces affinity for the substrate. This notion is further supported by the fact that the isolated RNase H domain from the gammaretroviral MMLV RT, which contains both the basic protrusion and a phosphate-binding pocket with four residues, retains some activity (Schultz and Champoux 1996; Zhan and Crouch 1997; Tanese and Goff 1988; Telesnitsky et al. 1992). For HIV-1 RT, both the protrusion and the pocket can be substituted by the residues from the RNase H primer grip described below.

## 3.6 The Role of RNase H Activity in Reverse Transcription

RNase H activity is essential for reverse transcription and consequently viral replication, i.e., mutations that inhibit RNase H block viral infectivity (Tisdale et al. 1991). The primary role of RNase H is to remove the original genomic viral (+) RNA from the RNA/DNA replication intermediate. This is achieved by multiple nonspecific cuts throughout the RNA strand which are similar to the mode of action of cellular RNases H1 that cleave the substrate in a nondiscriminate fashion with only limited sequence preference. HIV-1 RNase H also has the ability to perform specific cleavages of the tRNA primer and the polypurine tract (PPT) that are required for proper execution of reverse transcription (reviewed in (Schultz and Champoux 2008; Rausch and Le Grice 2004)). This ability distinguishes the retroviral enzyme from its cellular counterparts. While the basic features and mechanism of HIV-1 RNase H are most likely conserved compared to the cellular enzymes, the overall mode of action of HIV-1 RNase H domain is heavily influenced by the environment of the RT heterodimer. Although a significant amount of data is available for several retroviral RNases H, in particular from MMLV, here we mainly focus on the domain from HIV-1 RT and briefly summarize its most important biochemical properties.

Structural studies showed that the substrate bound by HIV-1 RT spans the entire length of the enzyme and interacts with both the DNA polymerase and RNaseH domains (Fig. 3.5). The distance between the active sites of polymerase and RNaseH is approximately 60 Å which corresponds to 17 base pairs (bp) on dsDNA or 18 bp on RNA/DNA (Sarafianos et al. 2001; Jacobo-Molina et al. 1993; Huang et al. 1998; Lapkouski et al. 2013). This mode of substrate binding predicts that extensive interactions between nucleic acid and the DNA polymerase domain heavily influence RNase H activity. This results in three modes of RNA/DNA cleavages observed for HIV-1 RT: DNA 3′-end-directed, RNA 5′-end-directed, and internal cleavages (Fig. 3.6a–c). For 3′-end-directed cleavage, RNA/DNA is bound by the DNA polymerase domain on the recessed 3′-end of the DNA (Wohrl and Moelling 1990; Gopalakrishnan et al. 1992; Gotte et al. 1998) (Fig. 3.6a). This configuration allows the enzyme to extend the DNA primer using RNA as the template and promotes RNase H cleavages in the RNA strand 15–20 nucleotides from the recessed DNA end (Gopalakrishnan et al. 1992; Furfine and Reardon 1991a; Kati et al. 1992). In fact, when the primer is extended, resulting in translocation, the sites of RNase H cuts are also relocated accordingly (Furfine and Reardon 1991a). The window in which cleavages are observed is wider by 2–3 bp in each direction than the actual distance between the DNA polymerase and RNase H active sites. Since the overall arrangement of these RT domains is relatively rigid and the distance between the active sites is unlikely to change significantly, the shift of the position of the cut probably reflects translocation of the substrate within the binding cleft. 3′-end-directed cleavages can be either static when the 3′ end of the DNA is stably bound by the DNA polymerase domain (polymerization independent) or concurrent with DNA polymerization (polymerization dependent). Polymerization-dependent cleavages are not strictly coupled with DNA synthesis, i.e., they occur less frequently than nucleotide addition to nascent DNA (DeStefano et al. 1991). When RNase H and DNA polymerase activities were measured simultaneously, the former was approximately sevenfold slower (Kati et al. 1992). In agreement with this notion, experiments in which single binding events were monitored for HIV-1 RT showed that during processive DNA synthesis, RNase H activity degraded around 80 % of the RNA in nascent hybrid to fragments of an average length of 7 nt, while 20 % of

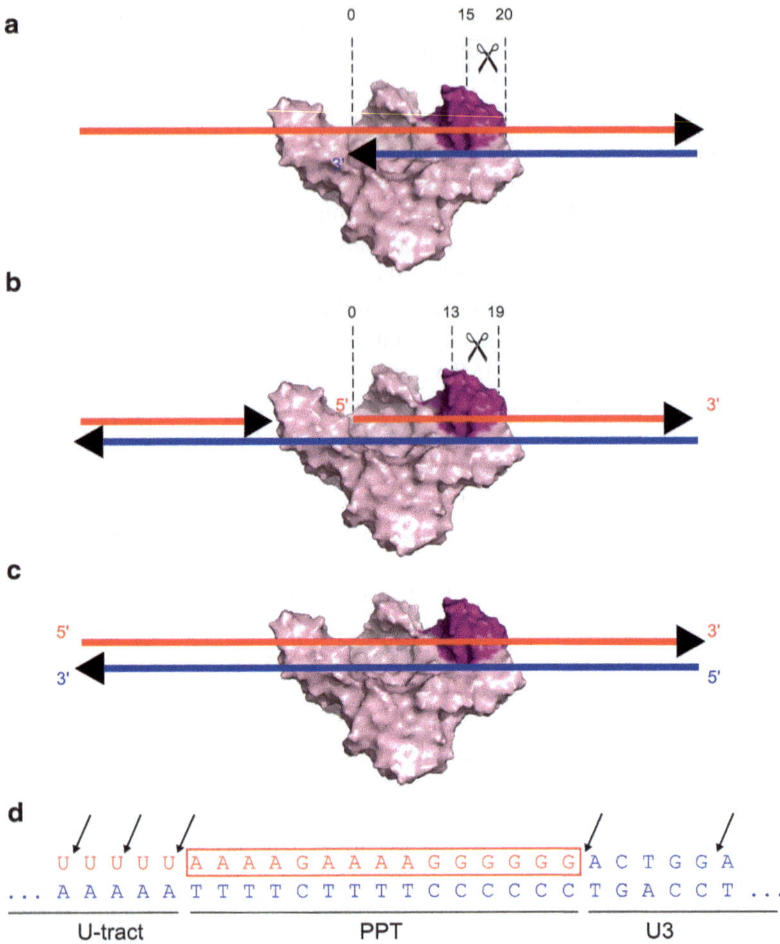


**Fig. 3.6** *RNase H activity in HIV-1 RT*. (**a–c**) Three modes of HIV-1 RT-mediated RNA cleavage. HIV-1 RT is represented in *pink*, the RNase H domain in *magenta*, RNA in *red*, and DNA in *blue*. Cleavage sites observed to the recessed end of the nucleic acid are indicated. (**a**) DNA 3′-directed cleavage. (**b**) RNA 5′-directed cleavage. (**c**) Internal cleavage. (**d**) Sequence of the PPT. The PPT primer is marked by the *red box*, and cleavage sites are indicated with *arrows*

the RNA remained in longer fragments stably associated with the DNA template. It has also been shown that RNase H hydrolysis events correlate with pause sites during DNA synthesis, which occur at secondary structures in the RNA (Suo and Johnson 1997; Purohit et al. 2007; Volkmann et al. 1993; Dudding et al. 1991). Cumulatively, these results show that the DNA polymerase and RNase H activities are not simultaneous, further implying that proper substrate engagement at one active site may preclude its productive interaction with the other (Nowotny et al. 2007; Lapkouski et al. 2013). An opposite view has also been presented, based on experiments in which HIV-1 RT was trapped in pre- and posttranslocation step by pyrophosphate analogue PFA. This study suggested that the substrate productively simultaneously interacts with both active sites (Beilhartz et al. 2009).

The second mode of RNase H cleavage is designated 5′-end-directed (Fig. 3.6b) and occurs when the DNA polymerase domain binds the recessed 5′ end of the RNA template (Wisniewski et al. 2000a; DeStefano et al. 2001; Palaniappan et al. 1996; Wisniewski et al. 2000b). In substrates with many RNA segments, the 5′ terminus has to be accessible and preceded by a gap of at least 2 nt (Schultz et al. 2004, 2006). It has been shown that the primary 5′-end-directed cleavages by HIV-1 RT occur at a distance of 13–19 nucleotides from the 5′ terminus and are followed by secondary cleavages 7–10 nts from the 5′ terminus and 5 nt from the 3′ RNA terminus (Wisniewski et al. 2000a, b, 2002). Detailed analysis of the primary cuts showed that the substrate sequence determines the actual sites of cleavage (Schultz et al. 2006). A or C (but not G) is preferred at position +1 from the scissile phosphate, C or G (not U) is preferred at position −2, and A is disfavored at −4. Hydrolysis occurs when the preferred sequence is located in the distance window from the 5′ end allowing for hydrolysis. The basis of the sequence preference of HIV-1 RNase H is not clear but may reflect contacts between the minor groove of the RNA/DNA hybrid and the RNase H domain, preferentially formed for particular bases.

The third RNase H activity is internal cleavage that does not depend on the primer or template termini (Fig. 3.6c) (Schultz et al. 2004), exhibiting similar sequence preference to 5′-end-directed cleavages (Schultz et al. 2004) – A or U at position +1 and C or G at position −2 with additional preferences observed for positions −4, −7, −12, and −14. The necessity for internal cleavages comes from the fact that polymerization-dependent 3′-end-directed cleavages are insufficient to completely remove genomic RNA from the RNA/DNA replication intermediate (DeStefano et al. 1994). Therefore, following DNA synthesis, HIV-1 RT executes polymerization-independent cleavage to complete hydrolysis. Internal cleavages occur first, and once gaps in the RNA strand are created, additional 5′-end-directed cleavage is supported.

Steps of proviral DNA synthesis which specifically require RNase H activity are the two strand transfers (Smith et al. 1999; Peliska and Benkovic 1992). $tRNA^{Lys3}$ hybridized to the primer-binding site (PBS) is used to initiate synthesis of the (−) stand of the DNA, and RNase H activity is required to remove RNA of the resulting RNA/DNA hybrid. This liberates newly synthesized DNA, allowing it to hybridize with the repeat (R) element at the other end of the genomic RNA to facilitate (−) strand DNA transfer. During (+) strand DNA synthesis, the tRNA primer has to be removed by RNase H to allow the second strand transfer event. The rate-limiting step in strand transfer is RNA degradation, reduction of which significantly hinders this event (Rausch et al. 2002). On the other hand, reducing DNA polymerase activity increases the rate of strand transfer, possibly by promoting more efficient RNase H activity (Gao et al. 2008).

In addition to nonspecific RNA hydrolysis, RNase H also performs specific and very precise cuts which are essential for proper completion of particular steps of reverse transcription. These cuts occur at special structures (RNA–DNA junctions) or special sequences, such as the PPT. tRNA primer removal, for example, must be precise because it defines 3′ LTR sequences required for integration. $tRNA^{Lys,3}$ extension produces a 5′-RNA–DNA-3′ junction and RNase H cleaves this junction 5′ to the last ribonucleotide. Consequently, a ribo-A is left at the 5′ end of (–) DNA strand (Furfine and Reardon 1991b; Pullen et al. 1992), likely reflecting the fact that the RNA strand interacts with RNase H through 2′-OH groups of two ribonucleotides on each side of the scissile phosphate. While the loss of one such interaction (as in the observed cleavage on the 5′ side of RNA–DNA junction) can be tolerated, lack of both 2′-OH interactions on one side of the scissile phosphate (as in the cleavage between RNA and DNA) probably abolishes activity. However, as an exception to this, the extended PPT is cleaved exactly at the RNA–DNA junction (Huber and Richardson 1990; Fuentes et al. 1995; Powell and Levin 1996; Gotte et al. 1999), suggesting the PPT sequence confers additional constraints for substrate positioning, overriding the cleavage preference of the RNase H domain.

An important step in proviral DNA synthesis is its initiation from RNA primers. In fact, retroviral RTs can efficiently extend only two types of primers, namely, the tRNA and PPT primers of (–) and (+) strand DNA synthesis, respectively, (Huber and Richardson 1990; Fuentes et al. 1995; Powell and Levin 1996; Palaniappan et al. 1998). The PPT comprises 15 ribonucleotides organized into short homopolymeric tracts (Fig. 3.6d) (Rausch and Le Grice 2004; Wohrl and Moelling 1990; Huber and Richardson 1990). Two such sequences – 3′ and central PPTs – are present in HIV-1 genome, with the former defining the 5′ LTR terminus of the double-stranded DNA provirus. HIV-1 RNase H fails to cleave in the middle of the PPT (Huber and Richardson 1990) but instead specifically cleaves at the upstream U-tract junction and at the PPT–U3 junction. Executing these cuts and protecting the PPT from degradation is a specific feature of RTs, since PPT sequences are efficiently cleaved by cellular RNases H1 (Huber and Richardson 1990) or the isolated MMLV RNase H domain (Zhan and Crouch 1997). Interestingly, mutating residues of the RNase H active site has a greater effect on specific PPT–U3 cleavage than on the hydrolysis of regular RNA/DNA hybrids (Rausch and Le Grice 1997), which might suggest that PPT cleavage requires more precise substrate positioning at the active site.

The exact mechanism of PPT cleavage specificity remains elusive, despite intense research in this area (Rausch and Le Grice 2004). The properties of the HIV-1 PPT clearly result from its specific sequence. For example, exchanging guanines in the 3′ region to cytosines abolished cleavage at the PPT–U3 junction (Powell and Levin 1996; Kvaratskhelia et al. 2002). The upstream $(rA)_4$ tracts are also required for accurate processing (Kvaratskhelia et al. 2002; McWilliams et al. 2003), while further studies showed that PPT rigidity may also contribute. Rausch et al. exploited substrates containing non-hydrogen bonding pyrimidine isosteres, which are known to form weaker base pairs and locally increase the flexibility of the duplex. Introducing such fluorinated bases into the DNA strand in the middle of PPT relocated the cleavage site from PPT–U3 junction to a position 3–4 bp downstream from the modification site (Rausch et al. 2003), suggesting that it is the flexible region of the PPT that directs its cleavages and rigidity of the middle portion may account for its resistance to degradation. Another possible mechanism by which the PPT avoids degradation is by adopting a unique sequence-dependent structure. Distortions in PPT structure have been detected by chemical footprinting (Kvaratskhelia et al. 2002). Also, in the reported crystal structure of HIV-1 RT in complex with a PPT fragment, a region of the duplex located in the vicinity of the RNase H domain contains deformations; in a stretch of five residues, the base pairing is very irregular (Sarafianos et al. 2001). It has been proposed that these deformations are PPT specific and preclude RNase H cleavage (Sarafianos et al. 2001). Not only generation of the PPT primer but also its removal from nascent (+) DNA relies on the RNase H activity (Huber and Richardson 1990). The PPT is first extended by 12 nucleotides after which RT switches orientation to place the template strand for cleavage specifically at the PPT–U3 junction (Gotte et al. 1999), defining the 5′ LTR terminus.

Several important regions and residues in HIV-1 RT have been demonstrated to influence the RNase H function. In agreement with the DNA polymerase domain directing RNase H cuts, mutations of several residues severely disturb RNase H activity, with some also affecting accurate PPT generation (Powell et al. 1999; Mandal et al. 2006; Boyer et al. 1992a, b, 1994). A particularly interesting residue is Trp266, mutation of which in either p66 or p51 inhibited RNase H activity (Gao et al. 1998). In the p66 subunit, Trp266 of the thumb subdomain positions the substrate and it thus can affect RNase H activity. Since the p51 thumb forms a platform for the p66 RNase H domain, mutating its Trp266 presumably affects the stability and positioning of the RNase H domain.

Another region shown to play an important role in RNase H function is the RNase H primer grip (Fig. 3.5). In the HIV-1 RT–RNA/DNA complex structure, this motif binds the backbone of the DNA strand between four and nine nucleotides from the RNase H active site (Sarafianos et al. 2001). The RNase H primer grip comprises Gly359, Ala360, and His361 of the p66 connection subdomain; Thr473, Asn474, Gln475, Lys476, Tyr501, and Ile505 of the RNase H domain; and p51 residues Lys395 and Glu396. This region has been extensively studied and found important for both DNA synthesis and RNA degradation (Julias et al. 2002; Rausch et al. 2002; Arion et al. 2002; Julias et al. 2003; McWilliams et al. 2006). For example, the importance of Gln475 was demonstrated by the fact that its substitution with Glu leads to defects in RNase H activity (Volkmann et al. 1993). Alanine-scanning mutagenesis illustrated that Thr473, Ans474, Gln475, and Tyr501 substitutions decreased RNase H activity and inhibited strand transfer (Rausch et al. 2002). In addition, a Y501A substitution essentially abolished PPT–U3 cleavage (Rausch et al. 2002). In vivo analysis (Julias et al. 2002) also demonstrated that Q475A and Y501A substitutions reduced the viral titer five- to tenfold and a T473A substitution abolished viral replication. Y501A variant and N474A+Q475A double mutant exhibited reduced synthesis of viral DNA and drastically increased the amount of aberrant PPT cleavages at the junctions with U3 region and U-tract.

Residues of the RNase H primer grip shown to be essential for proper synthesis of viral DNA are likely to form important contacts with the substrate not only when the RNase H domain binds the hybrid in the noncatalytic mode but also in the catalytic configuration. In the latter, the roles of these residues can be predicted based on the structures of Bh- and Hs-RNases H1 in complex with the substrate. As previously discussed, Thr473 is the strongly conserved residue at the heart of the phosphate-binding pocket. Asn474 and Gln475 are conserved residues located nearby and their counterparts in Bh- and Hs-RNases H1 (Asn105/Asn106 and Asn182/Gln183, respectively) are located in the minor groove of the substrate, forming a network of interactions with bases and deoxyriboses of the DNA. The equivalent of Y501 in human RNase H1 is Phe213 which is also located in the vicinity of the phosphate-binding pocket, forming a flat interface for binding of the backbone of the DNA strand. Therefore, the primer grip region can stabilize the two conformations of the substrate altering somewhat the network of interactions for each of them. This would explain why substitutions in this region have such a dramatic effect on RT function.

## 3.7 Perspectives

HIV-1 RT is arguably one of the most extensively studied retroviral enzymes, motivated mostly by its importance as a drug target for anti-AIDS therapy but also by the fundamental interest in understanding the intricate mechanism of reverse transcription. RNase H activity plays a pivotal role in this process. It is also the only enzymatic activity encoded by HIV-1 for which no inhibitors are clinically available. In light of the effort invested to study the mode of action of RNase H domain of HIV-1 RT, our understanding of its mechanism has been significantly advanced. Hopefully, this will lead to new and creative strategies to specifically inhibit this enzymatic activity.

## References

Arion D et al (2002) Mutational analysis of Tyr-501 of HIV-1 reverse transcriptase. Effects on ribonuclease H activity and inhibition of this activity by N-acylhydrazones. J Biol Chem 277(2):1370–1374

Arudchandran A et al (2000) The absence of ribonuclease H1 or H2 alters the sensitivity of Saccharomyces cerevisiae to hydroxyurea, caffeine and ethyl methanesulphonate: implications for roles of RNases H in DNA replication and repair. Genes Cells 5(10):789–802

Beilhartz GL et al (2009) HIV-1 reverse transcriptase can simultaneously engage its DNA/RNA substrate at both DNA polymerase and RNase H active sites: implications for RNase H inhibition. J Mol Biol 388(3):462–474

Ben-Artzi H et al (1992) Characterization of the double stranded RNA dependent RNase activity associated with recombinant reverse transcriptases. Nucleic Acids Res 20(19): 5115–5118

Boyer PL, Ferris AL, Hughes SH (1992a) Cassette mutagenesis of the reverse transcriptase of human immunodeficiency virus type 1. J Virol 66(2):1031–1039

Boyer PL, Ferris AL, Hughes SH (1992b) Mutational analysis of the fingers domain of human immunodeficiency virus type 1 reverse transcriptase. J Virol 66(12):7533–7537

Boyer PL et al (1994) Mutational analysis of the fingers and palm subdomains of human immunodeficiency virus type-1 (HIV-1) reverse transcriptase. J Mol Biol 243(3):472–483
Broccoli S et al (2004) Effects of RNA polymerase modifications on transcription-induced negative supercoiling and associated R-loop formation. Mol Microbiol 52(6):1769–1779
Cerritelli SM et al (1998) A common 40 amino acid motif in eukaryotic RNases H1 and caulimovirus ORF VI proteins binds to duplex RNAs. Nucleic Acids Res 26(7):1834–1840
Cerritelli SM et al (2003) Failure to produce mitochondrial DNA results in embryonic lethality in Rnaseh1 null mice. Mol Cell 11(3):807–815
Chung S et al (2011) Synthesis, activity, and structural analysis of novel alpha-hydroxytropolone inhibitors of human immunodeficiency virus reverse transcriptase-associated ribonuclease H. J Med Chem 54(13):4462–4473
Cirino NM et al (1995) Divalent cation modulation of the ribonuclease functions of human immunodeficiency virus reverse transcriptase. Biochemistry 34(31):9936–9943
Cote ML, Roth MJ (2008) Murine leukemia virus reverse transcriptase: structural comparison with HIV-1 reverse transcriptase. Virus Res 134(1–2):186–202
Davies JF 2nd et al (1991) Crystal structure of the ribonuclease H domain of HIV-1 reverse transcriptase. Science 252(5002):88–95
De Vivo M, Dal Peraro M, Klein ML (2008) Phosphodiester cleavage in ribonuclease H occurs via an associative two-metal-aided catalytic mechanism. J Am Chem Soc 130(33):10955–10962
DeStefano JJ et al (1991) Polymerization and RNase H activities of the reverse transcriptases from avian myeloblastosis, human immunodeficiency, and Moloney murine leukemia viruses are functionally uncoupled. J Biol Chem 266(12):7423–7431
DeStefano JJ et al (1994) Quantitative analysis of RNA cleavage during RNA-directed DNA synthesis by human immunodeficiency and avian myeloblastosis virus reverse transcriptases. Nucleic Acids Res 22(18):3793–3800
DeStefano JJ et al (2001) Physical mapping of HIV reverse transcriptase to the 5′ end of RNA primers. J Biol Chem 276(35):32515–32521
Drolet M et al (1995) Overexpression of RNase H partially complements the growth defect of an *Escherichia coli* delta topA mutant: R-loop formation is a major problem in the absence of DNA topoisomerase I. Proc Natl Acad Sci USA 92(8):3526–3530
Dudding LR, Nkabinde NC, Mizrahi V (1991) Analysis of the RNA- and DNA-dependent DNA polymerase activities of point mutants of HIV-1 reverse transcriptase lacking ribonuclease H activity. Biochemistry 30(43):10498–10506
Elsasser B, Fels G (2010) Atomistic details of the associative phosphodiester cleavage in human ribonuclease H. Phys Chem Chem Phys 12(36):11081–11088
Evans DB et al (1991) A recombinant ribonuclease H domain of HIV-1 reverse transcriptase that is enzymatically active. J Biol Chem 266(31):20583–20585
Fuentes GM et al (1995) Use of an oligoribonucleotide containing the polypurine tract sequence as a primer by HIV reverse transcriptase. J Biol Chem 270(47):28169–28176
Furfine ES, Reardon JE (1991a) Reverse transcriptase. RNase H from the human immunodeficiency virus. Relationship of the DNA polymerase and RNA hydrolysis activities. J Biol Chem 266(1):406–412
Furfine ES, Reardon JE (1991b) Human immunodeficiency virus reverse transcriptase ribonuclease H: specificity of tRNA(Lys3)-primer excision. Biochemistry 30(29):7041–7046
Gaidamakov SA et al (2005) Eukaryotic RNases H1 act processively by interactions through the duplex RNA-binding domain. Nucleic Acids Res 33(7):2166–2175
Gao HQ et al (1998) Effects of mutations in the polymerase domain on the polymerase, RNase H and strand transfer activities of human immunodeficiency virus type 1 reverse transcriptase. J Mol Biol 277(3):559–572
Gao L et al (2008) Apparent defects in processive DNA synthesis, strand transfer, and primer elongation of Met-184 mutants of HIV-1 reverse transcriptase derive solely from a dNTP utilization defect. J Biol Chem 283(14):9196–9205
Gilboa E et al (1979) A detailed model of reverse transcription and tests of crucial aspects. Cell 18(1):93–100

Goedken ER, Marqusee S (2001) Co-crystal of *Escherichia coli* RNase HI with Mn2+ ions reveals two divalent metals bound in the active site. J Biol Chem 276(10):7266–7271

Gopalakrishnan V, Peliska JA, Benkovic SJ (1992) Human immunodeficiency virus type 1 reverse transcriptase: spatial and temporal relationship between the polymerase and RNase H activities. Proc Natl Acad Sci USA 89(22):10763–10767

Gotte M et al (1998) Localization of the active site of HIV-1 reverse transcriptase-associated RNase H domain on a DNA template using site-specific generated hydroxyl radicals. J Biol Chem 273(17):10139–10146

Gotte M et al (1999) Temporal coordination between initiation of HIV (+)-strand DNA synthesis and primer removal. J Biol Chem 274(16):11159–11169

Hansen J, Schulze T, Moelling K (1987) RNase H activity associated with bacterially expressed reverse transcriptase of human T-cell lymphotropic virus III/lymphadenopathy-associated virus. J Biol Chem 262(26):12393–12396

Hansen J et al (1988) Identification and characterization of HIV-specific RNase H by monoclonal antibody. EMBO J 7(1):239–243

Haruki M et al (1994) Investigating the role of conserved residue Asp134 in *Escherichia coli* ribonuclease HI by site-directed random mutagenesis. Eur J Biochem 220(2):623–631

Haruki M et al (1997) Kinetic and stoichiometric analysis for the binding of *Escherichia coli* ribonuclease HI to RNA-DNA hybrids using surface plasmon resonance. J Biol Chem 272(35):22015–22022

Haruki M et al (2000) Catalysis by *Escherichia coli* ribonuclease HI is facilitated by a phosphate group of the substrate. Biochemistry 39(45):13939–13944

Himmel DM et al (2009) Structure of HIV-1 reverse transcriptase with the inhibitor beta-Thujaplicinol bound at the RNase H active site. Structure 17(12):1625–1635

Hostomsky Z et al (1991) Reconstitution in vitro of RNase H activity by using purified N-terminal and C-terminal domains of human immunodeficiency virus type 1 reverse transcriptase. Proc Natl Acad Sci USA 88(4):1148–1152

Hostomsky Z et al (1994) Redesignation of the RNase D activity associated with retroviral reverse transcriptase as RNase H. J Virol 68(3):1970–1971

Huang H et al (1998) Structure of a covalently trapped catalytic complex of HIV-1 reverse transcriptase: implications for drug resistance. Science 282(5394):1669–1675

Huber HE, Richardson CC (1990) Processing of the primer for plus strand DNA synthesis by human immunodeficiency virus 1 reverse transcriptase. J Biol Chem 265(18):10565–10573

Jacobo-Molina A et al (1993) Crystal structure of human immunodeficiency virus type 1 reverse transcriptase complexed with double-stranded DNA at 3.0 A resolution shows bent DNA. Proc Natl Acad Sci USA 90(13):6320–6324

Julias JG et al (2002) Mutations in the RNase H domain of HIV-1 reverse transcriptase affect the initiation of DNA synthesis and the specificity of RNase H cleavage in vivo. Proc Natl Acad Sci USA 99(14):9515–9520

Julias JG et al (2003) Mutation of amino acids in the connection domain of human immunodeficiency virus type 1 reverse transcriptase that contact the template-primer affects RNase H activity. J Virol 77(15):8548–8554

Kanaya S (1998) Enzymatic activity and protein stability of *E. coli* ribonuclease HI. In: Toulme JJ, Crouch HRJ (eds) Ribonucleases. INSERM, Paris, pp.1–38

Kanaya S et al (1990) Identification of the amino acid residues involved in an active site of *Escherichia coli* ribonuclease H by site-directed mutagenesis. J Biol Chem 265(8):4615–4621

Kanaya S, Katsuda-Nakai C, Ikehara M (1991) Importance of the positive charge cluster in *Escherichia coli* ribonuclease HI for the effective binding of the substrate. J Biol Chem 266(18):11621–11627

Katayanagi K et al (1990) Three-dimensional structure of ribonuclease H from *E. coli*. Nature 347(6290):306–309

Kati WM et al (1992) Mechanism and fidelity of HIV reverse transcriptase. J Biol Chem 267(36):25988–25997

Keck JL, Marqusee S (1995) Substitution of a highly basic helix/loop sequence into the RNase H domain of human immunodeficiency virus reverse transcriptase restores its Mn(2+)-dependent RNase H activity. Proc Natl Acad Sci USA 92(7):2740–2744

Keck JL, Marqusee S (1996) The putative substrate recognition loop of *Escherichia coli* ribonuclease H is not essential for activity. J Biol Chem 271(33):19883–19887

Keck JL, Goedken ER, Marqusee S (1998) Activation/attenuation model for RNase H. A one-metal mechanism with second-metal inhibition. J Biol Chem 273(51):34128–34133

Krakowiak A et al (2002) Stereochemical course of *Escherichia coli* RNase H. Chembiochem 3(12):1242–1250

Krug MS, Berger SL (1989) Ribonuclease H activities associated with viral reverse transcriptases are endonucleases. Proc Natl Acad Sci USA 86(10):3539–3543

Kvaratskhelia M, Budihas SR, Le Grice SF (2002) Pre-existing distortions in nucleic acid structure aid polypurine tract selection by HIV-1 reverse transcriptase. J Biol Chem 277(19): 16689–16696

Lapkouski et al (2013) Complexes of HIV-1 RT, NNRTI and RNA/DNA hybrid reveal a structure compatible with RNA degradation. Nat Struct Mol Biol 20(2):230–236

Lazzaro F et al (2012) RNase H and postreplication repair protect cells from Ribonucleotides incorporated in DNA. Mol Cell 45(1):99–110

Lim D, Orlova M, Goff SP (2002) Mutations of the RNase H C helix of the Moloney murine leukemia virus reverse transcriptase reveal defects in polypurine tract recognition. J Virol 76(16):8360–8373

Lima WF et al (2007a) Human RNase H1 discriminates between subtle variations in the structure of the heteroduplex substrate. Mol Pharmacol 71(1):83–91

Lima WF et al (2007b) The positional influence of the helical geometry of the heteroduplex substrate on human RNase H1 catalysis. Mol Pharmacol 71(1):73–82

Mandal D et al (2006) Analysis of HIV-1 replication block due to substitutions at F61 residue of reverse transcriptase reveals additional defects involving the RNase H function. Nucleic Acids Res 34(10):2853–2863

McWilliams MJ et al (2003) Mutations in the 5′ end of the human immunodeficiency virus type 1 polypurine tract affect RNase H cleavage specificity and virus titer. J Virol 77(20): 11150–11157

McWilliams MJ et al (2006) Combining mutations in HIV-1 reverse transcriptase with mutations in the HIV-1 polypurine tract affects RNase H cleavages involved in PPT utilization. Virology 348(2):378–388

Miller HI, Riggs AD, Gill GN (1973) Ribonuclease H (hybrid) in *Escherichia coli.* Identification and characterization. J Biol Chem 248(7):2621–2624

Mizrahi V et al (1990) Site-directed mutagenesis of the conserved Asp-443 and Asp-498 carboxy-terminal residues of HIV-1 reverse transcriptase. Nucleic Acids Res 18(18): 5359–5363

Mizrahi V, Brooksbank RL, Nkabinde NC (1994) Mutagenesis of the conserved aspartic acid 443, glutamic acid 478, asparagine 494, and aspartic acid 498 residues in the ribonuclease H domain of p66/p51 human immunodeficiency virus type I reverse transcriptase. Expression and biochemical analysis. J Biol Chem 269(30):19245–19249

Nowak E et al (2013) Structural analysis of monomeric retroviral reverse transcriptase in complex with an RNA/DNA hybrid. Nucleic Acids Res 41(6):3874–3887

Nowotny M (2009) Retroviral integrase superfamily: the structural perspective. EMBO Rep 10(2):144–151

Nowotny M, Yang W (2006) Stepwise analyses of metal ions in RNase H catalysis from substrate destabilization to product release. EMBO J 25(9):1924–1933

Nowotny M et al (2005) Crystal structures of RNase H bound to an RNA/DNA hybrid: substrate specificity and metal-dependent catalysis. Cell 121(7):1005–1016

Nowotny M et al (2007) Structure of human RNase H1 complexed with an RNA/DNA hybrid: insight into HIV reverse transcription. Mol Cell 28(2):264–276

Nowotny M et al (2008) Specific recognition of RNA/DNA hybrid and enhancement of human RNase H1 activity by HBD. EMBO J 27(7):1172–1181
Oda Y, Yoshida M, Kanaya S (1993) Role of histidine 124 in the catalytic function of ribonuclease HI from *Escherichia coli*. J Biol Chem 268(1):88–92
Ohtani N et al (1999a) Molecular diversities of RNases H. J Biosci Bioeng 88(1):12–19
Ohtani N et al (1999b) Identification of the genes encoding Mn2+–dependent RNase HII and Mg2+–dependent RNase HIII from Bacillus subtilis: classification of RNases H into three families. Biochemistry 38(2):605–618
Palaniappan C et al (1996) Helix structure and ends of RNA/DNA hybrids direct the cleavage specificity of HIV-1 reverse transcriptase RNase H. J Biol Chem 271(4):2063–2070
Palaniappan C et al (1998) Control of initiation of viral plus strand DNA synthesis by HIV reverse transcriptase. J Biol Chem 273(7):3808–3816
Pallan PS, Egli M (2008) Insights into RNA/DNA hybrid recognition and processing by RNase H from the crystal structure of a non-specific enzyme-dsDNA complex. Cell Cycle 7(16): 2562–2569
Peliska JA, Benkovic SJ (1992) Mechanism of DNA strand transfer reactions catalyzed by HIV-1 reverse transcriptase. Science 258(5085):1112–1118
Powell MD, Levin JG (1996) Sequence and structural determinants required for priming of plus-strand DNA synthesis by the human immunodeficiency virus type 1 polypurine tract. J Virol 70(8):5288–5296
Powell MD et al (1999) Residues in the alphaH and alphaI helices of the HIV-1 reverse transcriptase thumb subdomain required for the specificity of RNase H-catalyzed removal of the polypurine tract primer. J Biol Chem 274(28):19885–19893
Pullen KA, Ishimoto LK, Champoux JJ (1992) Incomplete removal of the RNA primer for minus-strand DNA synthesis by human immunodeficiency virus type 1 reverse transcriptase. J Virol 66(1):367–373
Purohit V et al (2007) Mechanisms that prevent template inactivation by HIV-1 reverse transcriptase RNase H cleavages. J Biol Chem 282(17):12598–12609
Rausch JW, Le Grice SF (1997) Substituting a conserved residue of the ribonuclease H domain alters substrate hydrolysis by retroviral reverse transcriptase. J Biol Chem 272(13):8602–8610
Rausch JW, Le Grice SF (2004) 'Binding, bending and bonding': polypurine tract-primed initiation of plus-strand DNA synthesis in human immunodeficiency virus. Int J Biochem Cell Biol 36(9):1752–1766
Rausch JW et al (2002) Altering the RNase H primer grip of human immunodeficiency virus reverse transcriptase modifies cleavage specificity. Biochemistry 41(15):4856–4865
Rausch JW et al (2003) Hydrolysis of RNA/DNA hybrids containing nonpolar pyrimidine isosteres defines regions essential for HIV type 1 polypurine tract selection. Proc Natl Acad Sci USA 100(20):11279–11284
Rosta E et al (2011) Catalytic mechanism of RNA backbone cleavage by ribonuclease H from quantum mechanics/molecular mechanics simulations. J Am Chem Soc 133(23):8934–8941
Sarafianos SG et al (2001) Crystal structure of HIV-1 reverse transcriptase in complex with a polypurine tract RNA:DNA. EMBO J 20(6):1449–1461
Schatz O et al (1989) Point mutations in conserved amino acid residues within the C-terminal domain of HIV-1 reverse transcriptase specifically repress RNase H function. FEBS Lett 257(2):311–314
Schultz SJ, Champoux JJ (1996) RNase H domain of Moloney murine leukemia virus reverse transcriptase retains activity but requires the polymerase domain for specificity. J Virol 70(12):8630–8638
Schultz SJ, Champoux JJ (2008) RNase H activity: structure, specificity, and function in reverse transcription. Virus Res 134(1–2):86–103
Schultz SJ, Zhang M, Champoux JJ (2004) Recognition of internal cleavage sites by retroviral RNases H. J Mol Biol 344(3):635–652
Schultz SJ, Zhang M, Champoux JJ (2006) Sequence, distance, and accessibility are determinants of 5′-end-directed cleavages by retroviral RNases H. J Biol Chem 281(4):1943–1955

Smith JS, Roth MJ (1993) Purification and characterization of an active human immunodeficiency virus type 1 RNase H domain. J Virol 67(7):4037–4049

Smith JS, Gritsman K, Roth MJ (1994) Contributions of DNA polymerase subdomains to the RNase H activity of human immunodeficiency virus type 1 reverse transcriptase. J Virol 68(9):5721–5729

Smith CM, Smith JS, Roth MJ (1999) RNase H requirements for the second strand transfer reaction of human immunodeficiency virus type 1 reverse transcription. J Virol 73(8):6573–6581

Stahl SJ et al (1994) Construction of an enzymatically active ribonuclease H domain of human immunodeficiency virus type 1 reverse transcriptase. Protein Eng 7(9):1103–1108

Starnes MC, Cheng YC (1989) Human immunodeficiency virus reverse transcriptase-associated RNase H activity. J Biol Chem 264(12):7073–7077

Steitz TA, Steitz JA (1993) A general two-metal-ion mechanism for catalytic RNA. Proc Natl Acad Sci USA 90(14):6498–6502

Suo Z, Johnson KA (1997) Effect of RNA secondary structure on RNA cleavage catalyzed by HIV-1 reverse transcriptase. Biochemistry 36(41):12468–12476

Tanese N, Goff SP (1988) Domain structure of the Moloney murine leukemia virus reverse transcriptase: mutational analysis and separate expression of the DNA polymerase and RNase H activities. Proc Natl Acad Sci USA 85(6):1777–1781

Telesnitsky A, Blain SW, Goff SP (1992) Defects in Moloney murine leukemia virus replication caused by a reverse transcriptase mutation modeled on the structure of *Escherichia coli* RNase H. J Virol 66(2):615–622

Tisdale M et al (1991) Mutations within the RNase H domain of human immunodeficiency virus type 1 reverse transcriptase abolish virus infectivity. J Gen Virol 72(Pt 1):59–66

Volkmann S et al (1993) Enzymatic analysis of two HIV-1 reverse transcriptase mutants with mutations in carboxyl-terminal amino acid residues conserved among retroviral ribonucleases H. J Biol Chem 268(4):2674–2683

Wahba L et al (2011) RNase H and multiple RNA biogenesis factors cooperate to prevent RNA:DNA hybrids from generating genome instability. Mol Cell 44(6):978–988

Wang J et al (1994) Structural basis of asymmetry in the human immunodeficiency virus type 1 reverse transcriptase heterodimer. Proc Natl Acad Sci USA 91(15):7242–7246

Wisniewski M et al (2000a) Unique progressive cleavage mechanism of HIV reverse transcriptase RNase H. Proc Natl Acad Sci USA 97(22):11978–11983

Wisniewski M et al (2000b) The sequential mechanism of HIV reverse transcriptase RNase H. J Biol Chem 275(48):37664–37671

Wisniewski M et al (2002) Substrate requirements for secondary cleavage by HIV-1 reverse transcriptase RNase H. J Biol Chem 277(32):28400–28410

Wohrl BM, Moelling K (1990) Interaction of HIV-1 ribonuclease H with polypurine tract containing RNA-DNA hybrids. Biochemistry 29(44):10141–10147

Yang W et al (1990) Structure of ribonuclease H phased at 2 A resolution by MAD analysis of the selenomethionyl protein. Science 249(4975):1398–1405

Yang W, Lee JY, Nowotny M (2006) Making and breaking nucleic acids: two-Mg2+–ion catalysis and substrate specificity. Mol Cell 22(1):5–13

Zhan X, Crouch RJ (1997) The isolated RNase H domain of murine leukemia virus reverse transcriptase. Retention of activity with concomitant loss of specificity. J Biol Chem 272(35): 22023–22029

# Chapter 4
# Conformational Dynamics of Reverse Transcription

**Stuart F.J. LeGrice**

## Abbreviations

| | |
|---|---|
| Cy3 | Cyanine 3 |
| Cy5 | Cyanine 5 |
| ddNTP | Dideoxynucleoside triphosphate |
| dNTP | Deoxynucleoside triphosphate |
| FRET | Fluorescence resonance energy transfer |
| HIV | Human immunodeficiency virus |
| LTR | Long terminal repeat |
| MLV | Murine leukemia virus |
| NC | Nucleocapsid protein |
| NNRTI | Nonnucleoside RT inhibitor |
| PPT | Polypurine tract |
| RSV | Rous sarcoma virus |
| RT | Reverse transcriptase |
| UTR | Untranslated region |

*Conflict of interest statement*: None declared

S.F.J. LeGrice (✉)
HIV Drug Resistance Program, National Laboratory for Cancer Research,
Building 535, Room 312, Frederick, MD 21702, USA
e-mail: legrices@mail.nih.gov

S.F.J. LeGrice and M. Götte (eds.), *Human Immunodeficiency Virus Reverse Transcriptase: A Bench-to-Bedside Success*, DOI 10.1007/978-1-4614-7291-9_4,

## 4.1 Introduction

Synthesis of double-stranded DNA from the retrovirus (+)-strand RNA genome is a multi-step process catalyzed by the virus-coded reverse transcriptase (RT), using a combination of synthetic (RNA- and DNA-dependent DNA synthesis) and degradative activities (ribonuclease H, or RNase H). For human immunodeficiency virus type 1 (HIV-1) RT, the two-subunit p66/p51 heterodimer must accommodate a variety of conformationally distinct nucleic acid substrates, including duplex RNA, RNA/DNA hybrids duplex DNA, and RNA-DNA chimeras to complete this process (Telesnitsky and Goff 1997). Despite a wealth of structural and biochemical studies, the mechanism whereby RT adopts different orientations to coordinate its DNA polymerase and RNase H activities has remained elusive. The lower processivity of HIV-1 RT (Bibillo and Eickbush 2002) also raises the issue of polymerization site targeting when the enzyme re-engages its nucleic acid substrate distal from the primer terminus. Although X-ray crystallography has contributed significantly to our understanding of reverse transcription complexes and resistance to antiviral drugs, it provides a static picture, revealing few details regarding motion of the enzyme on its substrate. In contrast, single-molecule spectroscopy has proven an invaluable tool capable of resolving static and dynamic heterogeneity of an ensemble (Lu et al. 1998; Moerner and Orrit 1999; Weiss 2000). An earlier study (Rothwell et al. 2003) used a fluorescence resonance energy transfer (FRET) approach to distinguish three structurally distinct HIV-1 RT species, including one form where nucleic acid was positioned to permit nucleotide incorporation and a second corresponding to a nucleoprotein complex formed following nucleotide addition but prior to translocation. However, since this seminal study, little progress has been made in dissection of the conformational dynamics of reverse transcription. Progress in these areas and the implications for investigational and established inhibitors that interfere with HIV-1 RT function will be reviewed here.

## 4.2 Single-Molecule Assay System

Our application FRET to HIV-1 reverse transcription complexes took advantage of previous observations that (i) an active p66/p51 RT heterodimer could be reconstituted in vitro from its p66 and p51 subunits, which allowed their independent manipulation (Le Grice et al. 1991a), and (ii) a "Cys-free" p66 subunit (i.e., containing Cys38Ala and Cys280Ala substitutions) could be reconstituted with a wild-type p51 subunit and likewise displayed minimal loss of activity (Rausch et al. 2000b). The FRET donor dye Cy3 could therefore be disulfide linked to either the p66 N-terminal DNA polymerase or C-terminal RNase H domain of such modified and reconstituted enzymes. The FRET acceptor dye Cy5 was in turn attached to a biotinylated nucleic acid substrate, which was subsequently surface immobilized on a streptavidin-coated slide. As indicated in the scheme of Fig. 4.1a, b surface-immobilized

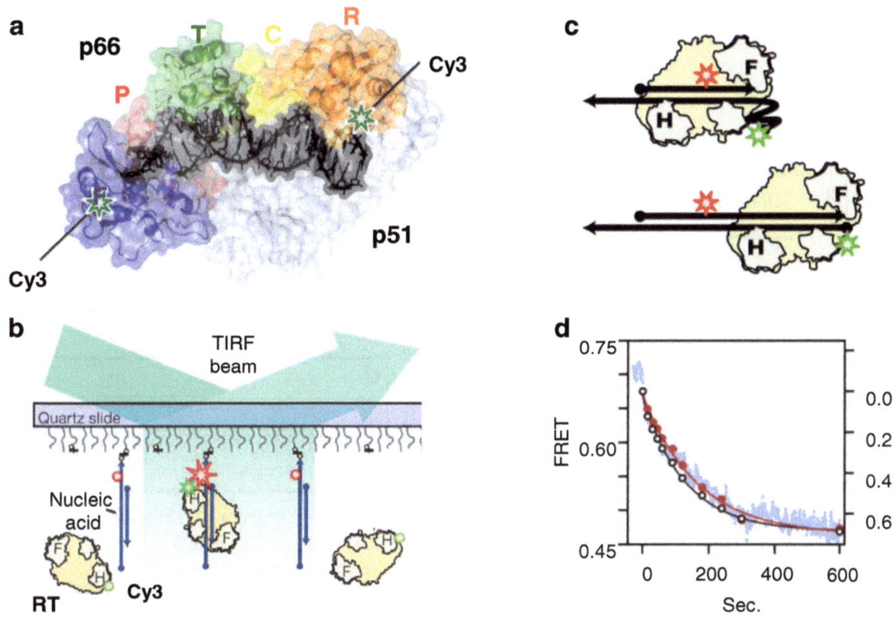


**Fig. 4.1** Experimental system for probing HIV-1 RT conformational dynamics by single-molecule spectroscopy. (**a**) Structure of the p66/p51 RT heterodimer containing duplex DNA. p66fingers (**F**), palm (**P**), thumb (**T**), and connection (**C**) subdomains are color-coded blue, red, green, and yellow, respectively, while the RNase H domain (**R**) is depicted in gold and the p51 subunit in gray. Sites of Cy3 introduction at the N-terminal, DNA polymerase, or C-terminal RNase H domain of the p66 subunit are indicated by green stars. (**b**) Cy5-containing nucleic acid substrates are immobilized on a quartz plate, and binding/dissociation of Cy3-labeled RT is determined by FRET. Stars and spheres represent dyes that do or do not emit fluorescence, respectively. (**c**) In situ examination of HIV-1 RT activity. Immobilized duplex DNA (*black arrows*) labeled with Cy3 (*green*) and Cy5 (*red*) is incubated with HIV-1 RT and dNTPs. Primer-extension converts the ssDNA template overhang to dsDNA, separating the dyes and decreasing the FRET signal. (**d**) Comparison of primer-extension kinetics from the in situ FRET assay with those obtained by gel electrophoresis. Light blue dots represent an average of ~450 FRET traces from a single field of view. Overlaid are the extension kinetics determined for non-immobilized primer-template complexes by gel electrophoresis using Cy3-labeled RT (*red circles*) or unlabeled RT (*black circles*). Equivalent primer-extension kinetics obtained for unimmobilized and immobilized substrates and for unlabelled and dye-labeled RT shows that dye labeling and surface immobilization do not significantly perturb DNA polymerase activity

substrates are immersed in a solution of Cy3-labeled RT and fluorescence of individual substrates thereafter monitored via total-internal-reflection fluorescence (TIRF) microscopy.

When considering the modifications introduced into both the nucleic acid duplexes and the HIV-1 RT heterodimer, an important aspect was whether single-molecule data these complexes would provide was representative of the biological system, i.e., was DNA synthesis possible on an immobilized, fluorescently labeled substrate? This issue could be addressed by immobilizing duplex DNA containing

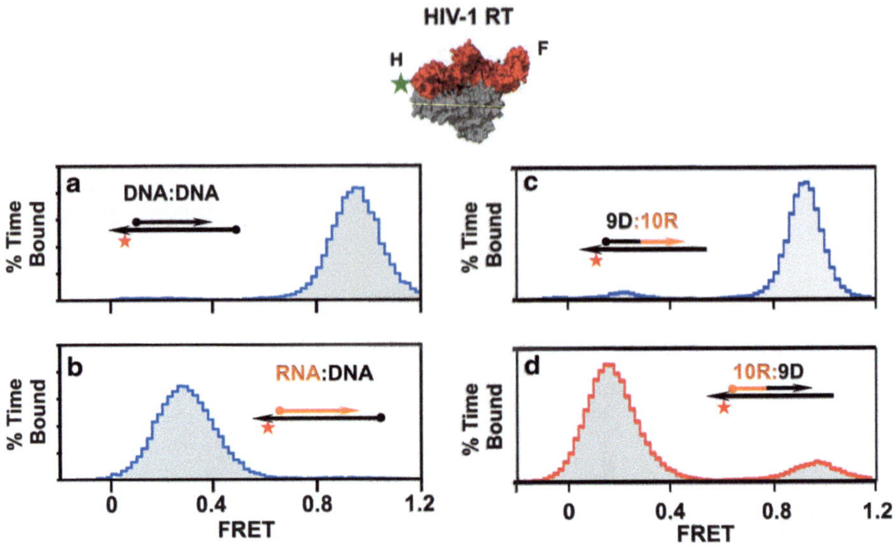


**Fig. 4.2** FRET distributions of Cy3, RNase H-labeled HIV-1 RT on nucleic acid duplexes reveal distinct binding orientations that are dependent on the nature of the primer. In the cartoon of the p66/p51 RT heterodimer, Cy3 is depicted in green. In (**a**)–(**d**), DNA and RNA strands are shown in *black* and *orange*, respectively. (**a**) DNA template/DNA primer. 5′ and 3′ nucleic acid termini are shown as spheres and arrows, respectively, and the DNA template is labeled near the 5′ terminus with Cy5 (*red*). (**b**) DNA template/RNA primer of equivalent length and sequence. (**c**), (**d**) Binding orientations on chimeric substrates. DNA template is as in (**a**) and (**b**), while the primer is either 5′-RNA/3′-DNA (**c**) or 5′-DNA/3′-RNA (**d**)

FRET dyes internal to the primer and on the single-stranded template overhang (Fig. 4.1c) and incubating this in the presence of HIV-1 RT and a dNTP cocktail. Successful synthesis of a DNA duplex would induce gradual separation of the Cy3 and Cy5 dyes, resulting in a gradual loss of FRET signal with incubation time. As shown in Fig. 4.1d, this was observed experimentally, and primer-extension kinetics on an immobilized substrate were in good agreement with those obtained through an equivalent gel-based analysis. Thus, at the outset of our studies, we could demonstrate that both dye labeling and surface immobilization of the substrate did not significantly affect the DNA polymerase activity of HIV-1 RT.

In keeping with biochemical data, the high-FRET signal obtained when RT labeled with Cy3 on the RNase H domain was incubated with duplex DNA whose template was Cy5 modified was consistent with an orientation positioning the DNA polymerase catalytic center over the primer 3′ terminus (Fig. 4.2a). To our surprise, Fig. 4.2b indicates that substituting an RNA primer of identical sequence and length supported enzyme binding in the opposite orientation, i.e., with the RNase H domain positioned over the primer terminus. Figure 4.2c, d show data obtained from RNA/DNA primer chimeras of equivalent length, clearly demonstrating that their 5′ region was the critical determinant of enzyme orientation. In an extension to the experiment of Fig. 4.2d, increasing the RNA:DNA composition of the primer in single-nucleotide increments indicated that as few as two 5′ ribonucleotides (i.e., 2RNA/17DNA) sufficed to "initiate" inversion of enzyme orientation on the primer terminus from a DNA polymerase to an RNase H mode and that exclusive binding in this mode could be achieved when the primer contained five 5′ ribonucleotides (5RNA/14DNA). Thus, while the structural basis for the observations of Fig. 4.2 remains to be fully established, our preliminary data (Abbondanzieri et al. 2008) indicated a possible allosteric effect, where contacts involving the primer 5′ terminus and structural motifs of the HIV-1 enzyme (possibly the RNase H primer grip, Sarafianos et al. 2001) regulate DNA polymerase activity affecting enzyme orientation. In the following sections, specific steps in the reverse transcription process will be discussed with respect to such substrate-induced alterations in HIV-1 RT orientation.

## 4.3 "Programmed" Initiation of (−)-Strand DNA Synthesis

Retroviruses initiate first (or (−))-strand DNA synthesis from the 3′ terminus of a host-coded tRNA primer hybridized to a complementary sequence on their genome designated the primer binding site (PBS) (Telesnitsky and Goff 1997). tRNA primer usage is surprisingly varied, including $tRNA^{Pro}$ in murine leukemia virus (MLV) (Levin and Seidman 1979), $tRNA^{Trp}$ in Rous sarcoma virus (RSV) (Taylor and Illmensee 1975), $tRNA^{iMet}$ in the long terminal repeat (LTR)-retrotransposon Ty1 (Chapman et al. 1992), and $tRNA^{Lys3}$ in HIV-1 and HIV-2 (Leis 1993). Additional intermolecular interactions between the tRNA primer and the 5′ untranslated region (UTR) of the (+) RNA genome have been demonstrated, and while alternative structural models have been proposed (Beerens and Berkhout 2002; Beerens et al. 2001; Isel et al. 1995, 1999), both have been suggested to control (−)-strand DNA synthesis at the level of initiation. This notion was further strengthened by kinetic analysis that indicated a slow and distributive initiation phase was followed by a transition into rapid and processive DNA synthesis (Lanchy et al. 1996a, b.) This slow initiation phase, followed by a rapid elongation phase, is many respects reminiscent of "cycling" of *Escherichia coli* RNA polymerase at several promoters, where productive transcription involves escape from abortive initiation events and synthesis of short RNAs (Carpousis and Gralla 1980).

Using in vitro systems designed to mimic $tRNA^{Lys,3}$-primed (−)-strand DNA synthesis, Fig. 4.3 illustrates significant pausing of the replication complex on the genomes of both HIV-1 and equine infectious anemia virus (EIAV, a related lentivirus which also exploits $tRNA^{Lys,3}$ as its replication primer, Le Grice et al. 1991b; Leis 1993) during addition of the first 3–5 deoxynucleotides. Since HIV-1 (−)-strand DNA synthesis from a tRNA primer prepared by in vitro transcription exhibited similar pausing (Miller et al. 2004), this suggested that modified bases of the natural tRNA were not responsible for initiation control. Single-molecule spectroscopy has subsequently been successful in dissecting individual events of the initiation phase with respect to multiple inversions of HIV-1 RT orientation (Liu et al. 2010).

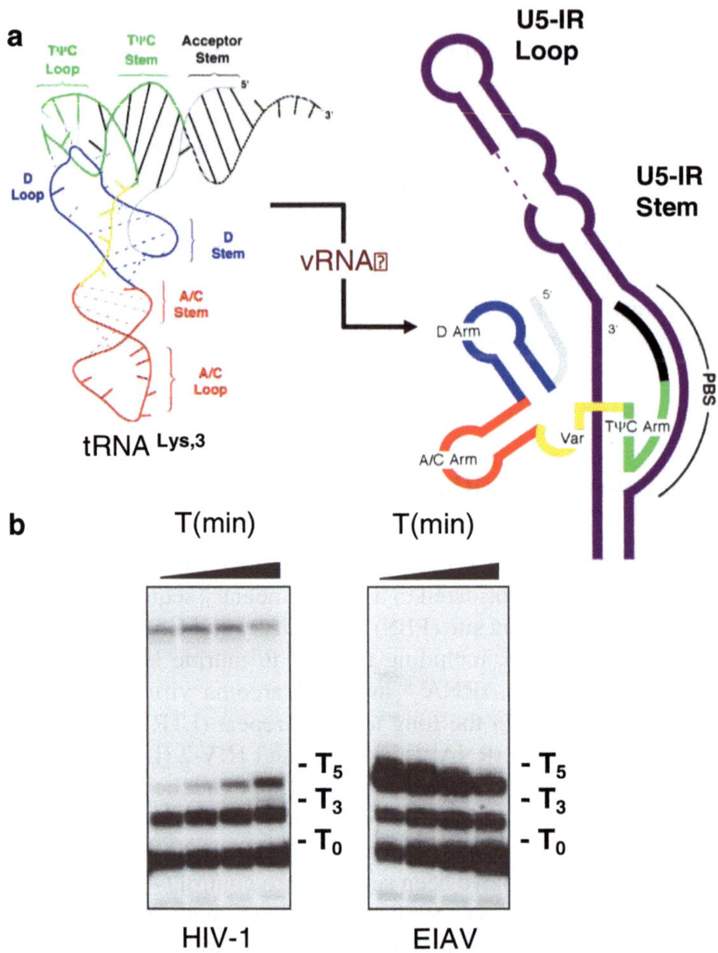


**Fig. 4.3** tRNA$^{Lys,3}$-primed initiation of HIV-1 reverse transcription in retroviruses. Panel (**a**), *Left* L-shaped tRNA structure. Anticodon, D, TΨC, and variable loops are depicted in *red*, *blue*, *green* and *yellow*, respectively, and the 3′ acceptor stem in black. *Right*, interaction of the tRNA primer with the viral RNA genome (magenta) in and around the PBS. Stem-loop structures comprising the U5-IR hairpin are illustrated. (**b**) RT-catalyzed (–)-strand DNA synthesis from tRNA$^{Lys,3}$ hybridized to the PBS of the HIV-1 (*left*) and EIAV genomes (*right*) is characterized by stalling of the replication complex in the immediate vicinity of the PBS. For simplicity, only a portion of an autoradiogram following denaturing polyacrylamide gel electrophoresis of the primer-extension products is presented *T0*, unextended tRNA$^{Lys,3}$; *T+3*, *T+5*, tRNA$^{Lys,3}$ extended by 3 and 5 deoxynucleotides, respectively

Examining RT orientation on complexes mirroring initiation of (–)-strand DNA synthesis in single-nucleotide increments required constructing tRNA-DNA chimeras, which was achieved by enzymatically removing several 3′ tRNA nucleotides from the natural tRNA and splint ligation of RNA-DNA chimeras to regenerate

76-nttRNA molecules extended by a predetermined number of deoxynucleotides (T+1, T+2, etc., where T designates the 76-nt RNA). These chimeras were hybridized to a Cy5-labeled viral RNA containing the PBS and the immediately adjacent U5-IR hairpin of the 5′ UTR and incubated in the presence of Cy3, RNase H-labeled HIV-1 RT. Nucleoprotein complexes with RT bound to the unextended tRNA produce both high- and low-FRET peaks, with a predominant orientation positioning the RNase H domain over the primer terminus. As might be predicted, inclusion of the incoming dNTP, thereby initiating DNA synthesis (dCTP), shifts enzyme distribution in favor of the polymerization-competent orientation (Fig. 4.4a). However, hybridization of T+3 and T+5 tRNA/DNA chimeras favors enzyme binding in the RNase H mode. While surprising, these observations are in keeping with gel electrophoresis data of Fig. 4.3, which indicates T+3 and T+5 as major pause sites ahead of the PBS. FRET distribution is, however, abruptly reversed in the presence of a T+6 chimera, with the polymerization-competent binding mode favored, in conjunction with which Fig. 4.3 indicates that the impediment to reverse transcription is thereafter relieved.

Since early events of HIV-1 (−)-strand DNA synthesis predict that the U5-IR hairpin upstream of the PBS must be disrupted (Fig. 4.3), disassembly of this structure (via RNA strand displacement synthesis) seemed the logical event that controlled HIV-1 RT orientation and the associated transition from initiation to elongation of DNA synthesis. In order to test this hypothesis experimentally, tRNA-DNA chimeras were annealed to a viral RNA template containing both Cy3 and Cy5, after which the duplex was incubated with HIV-1 RT (Fig. 4.4b). FRET dyes were introduced such that, if the integrity of the U5-IR hairpin was maintained, this would generate a high-FRET signal, whereas its disruption would be reflected by a low-FRET signal. As demonstrated in Fig. 4.4b, the high-FRET signal persists on nucleoprotein complexes whose tRNA primer is extended by as many as five dNTPs, indicating the integrity of the U5-IR hairpin is largely maintained. This situation is dramatically reversed in the presence of a T+6 tRNA/DNA chimera, i.e., the low-FRET signal indicates disruption of the hairpin, allowing the retroviral polymerase to thereafter assume a polymerization-competent orientation. Our single-molecule spectroscopy studies thus allowed us to propose the following model with respect to the steps of tRNA-primed initiation of HIV-1 (−)-strand DNA synthesis:

- It the absence of DNA synthesis, RT interacts with the initiation complex in two binding modes but spends considerable time in the "flipped," polymerase-incompetent RNase H orientation.
- Following addition of two dNTPs to $tRNA^{Lys,3}$, the binding equilibrium shifts to favor the polymerization-competent orientation, increasing the DNA synthesis rate.
- At position T+3, the replication complex encounters the U5-IR hairpin, which induces reorientation of RT into the RNase H mode, increasing the probability of pausing and dissociation.
- RNA strand displacement synthesis to position T+6 destabilizes the base of the U5-IR hairpin, which redistributes RT into the orientation competent for catalyzing processive DNA synthesis.

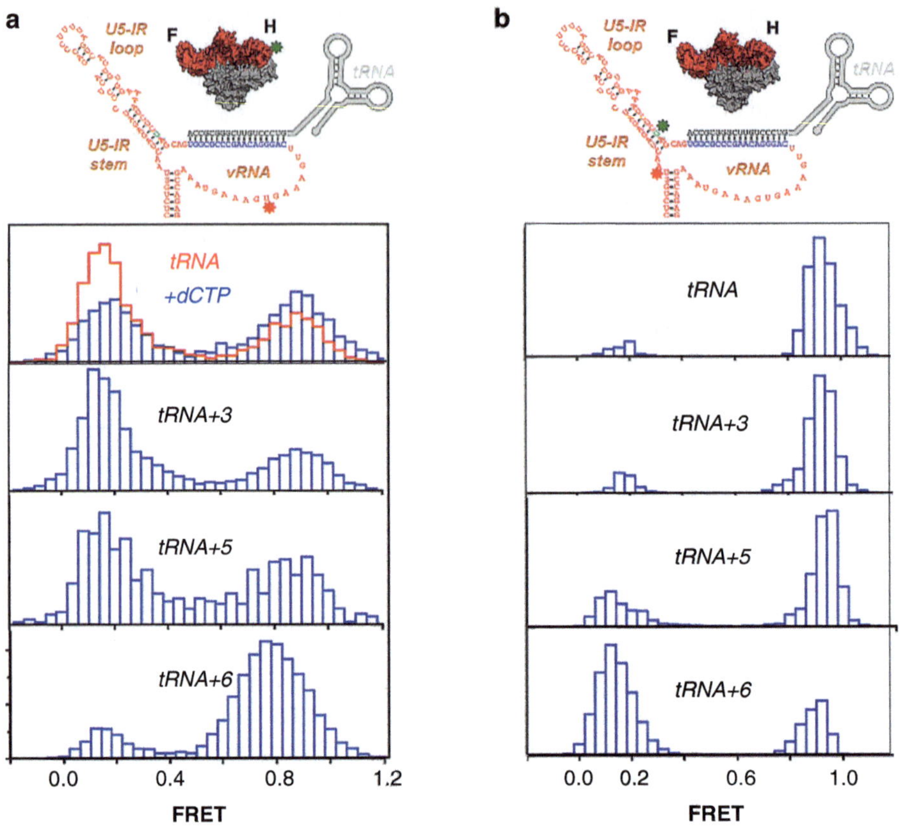


**Fig. 4.4** (**a**) Alterations in HIV-1 RT binding orientation during tRNA$^{Lys,3}$-primed initiation of (–)-strand DNA synthesis. The cartoon in describes the experimental system, where Cy3, RNase H-labeled RT is incubated with a tRNA/vRNA duplex whose viral RNA is internally labeled with Cy5. The lower portion of the figure indicates FRET distribution obtained when Cy3-labeled RT binds to Cy5-labeled tRNA + $n$–vRNA complexes in the presence of 200 μM cognate dNTP. In the case of tRNA ($n$=0), the FRET distribution in the absence of dNTP (*red*) is also shown. (**b**) Disruption of the U5-IR hairpin upstream of the PBS occurs upon addition of the sixth deoxynucleotide to tRNA$^{Lys,3}$. In this experiment, the cartoon depicts a doubly labeled tRNA/vRNA construct used to monitor the folding of the stem-loop structure, while RT is unlabelled. Cy3 (*green star*) and Cy5 (*red star*) were attached to vRNA at positions U132 and U177. The lower portion of the figure shows FRET distributions obtained when 100 nM RT was added to the doubly labeled template annealed to the tRNA, tRNA + 3, tRNA + 5, and tRNA + 6 primers

Although not shown here, addition of the HIV-1 nucleocapsid protein, NCp7, to the double-labeled tRNA-vRNA duplex resulted in considerable broadening of the FRET distribution, suggesting this RNA chaperone was capable of disrupting the U5-IR hairpin. These observations are consistent with reports suggesting a role for NC in disrupting RNA secondary structures that present a barrier to the replication complex (Ji et al. 1996; Rodriguez-Rodriguez et al. 1995; Wu et al. 1996). However, since NCp7 is a maturation product of the HIV-1 Pr55 gag precursor polyprotein, and only proteolytically released following virus maturation, the initiation program of (–)-strand DNA synthesis we and others have characterized (Isel et al. 1995, 1999; Lanchy et al. 1996a, b) may represent a control mechanism that the virus has evolved to avoid premature reverse transcription prior to budding. In support of this model, HIV-1 virion-associated $tRNA^{Lys,3}$ has been demonstrated to contain a 2-deoxynucleotide extension at its 3′ terminus (Y. Huang et al. 1997), showing that DNA synthesis can be initiated prior to budding but is immediately shut down until a later stage in the virus life cycle.

## 4.4 Dynamics of Polypurine Tract (PPT)-Primed (+)-Strand DNA Synthesis

Following tRNA-primed initiation and transfer of (–)-strand strong-stop DNA to the genome 3′ terminus (Basu et al. 2008), RNA-templated DNA synthesis continues, concomitant with which RT-associated RNase H activity nonspecifically hydrolyzes RNA of the RNA/DNA replication intermediate. An exception to this is the polypurine tract (PPT) located at the 3′ terminus of the (+) RNA genome, which is refractory to hydrolysis and provides the RNA primer for initiation of (+)-strand DNA-dependent DNA synthesis (although HIV-1 contains a second, central PPT, its role remains controversial and is not discussed here). Subsequent to initiation of (+)-strand DNA synthesis, the PPT RNA primer must be accurately excised from nascent DNA in order to liberate terminal sequences of the 5′ LTR necessary for integration of the DNA duplex into the host chromosome. Using an in vitro model comprising an RNA/DNA hybrid (within which the PPT was positioned), HIV-1 RT, and a dNTP cocktail (one of which was radiolabeled), we could demonstrate (i) accurate selection of the PPT 3′ terminus, (ii) extension of the PPT into (+)-strand DNA, and (iii) precise removal from the resulting RNA-DNA chimera (Rausch and Le Grice 1997).To successfully accomplish these steps, the PPT 3′ terminus must be accurately recognized by both the N-terminal DNA polymerase and C-terminal RNase H catalytic centers of HIV-1 RT, and while numerous mutagenesis studies have been undertaken to understand the structural basis for PPT recognition (Rausch and Le Grice 2004, 2007; Turner et al. 2008), the contribution from enzyme conformational dynamics had not been not examined. In Fig. 4.5, the interaction of Cy3, RNase H-labeled HIV-1 RT (Fig. 4.5a) with several Cy5-containing PPT substrates mimicking different steps in (+)-strand DNA synthesis was examined by FRET (Fig. 4.5b, d). In Fig. 4.5b, the PPT primer was extended at its 3′ terminus by two ribonucleotides (PPT:r2) to simulate the scenario prior to cleavage at the PPT/U3 junction, while the substrate of Fig. 4.5c simulates RNase H-mediated removal of these ribonucleotides, and the PPT:d2 substrate (Fig. 4.5d) mimics initiation of (+)-strand DNA synthesis. The low-FRET signal obtained with the PPT:r2duplex was similar to that observed with a nonspecific RNA primer (Fig. 4.2b), indicating that RT was oriented with its C-terminal RNase H domain in the vicinity of the PPT-U3 junction. In contrast, when these two ribonucleotides are removed (Fig. 4.5c) or

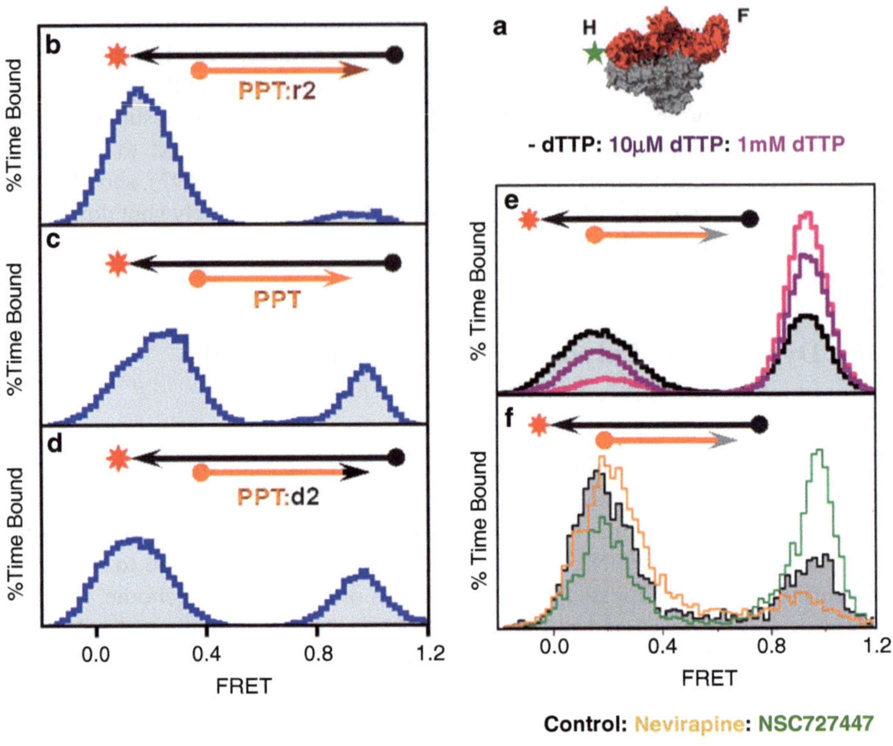


**Fig. 4.5** Dynamic orientations of HIV-1 RT on the PPT. (**a**) cartoon of Cy3, RNase H-labeled RT. (**b**)–(**d**) FRET histograms of RT bound to Cy5-labeled DNA templates hybridized to PPT:r2 (**b**), PPT (**c**), or PPT:d2 primers (**d**). DNA and RNA strands are represented in black and orange, respectively, the r2 PPT extension in magenta and the d2 primer extension in black. Nucleic acid 5′ and 3′ termini are represented by filled circles and arrows, respectively. (**e**) FRET histograms of HIV-1 RT bound to PPT:dd2 substrates (*black*) and in the presence of 10 μM (*purple*) or 1 mM dTTP (*magenta*). (**f**) FRET histograms of RT bound to PPT:dd2 substrates in the presence of the NNRTI nevirapine (*yellow*) and the RNase H inhibitor NSC727447 (*green*)

replaced with 2 deoxynucleotides (Fig. 4.5d), a considerable fraction of enzyme molecules are bound in the polymerization-competent mode. This was particularly striking for the PPT substrate since, like PPT:r2, this is all-RNA, suggesting that PPT geometry (such as its "modular nature" of adjacent r(A)4:d$(T)_4$ and $r(G)_6$:d$(C)_6$ tracts) may also influence RT orientation. The ability of the PPT RNA to direct HIV-1 RT binding in a polymerization-competent mode may also illustrate at property that distinguishes it from random RNA primers generated during (–)-strand RNA-dependent DNA synthesis.

The ability of RT to adopt different orientations on the PPT prompted us to investigate conditions that might favor binding in one conformation. In Fig. 4.5e, the PPT:d2 primer was replaced with a counterpart containing a chain-terminating dideoxynucleotide (PPT:dd2). Crystallographic data (Huang et al. 1998) has indicated that, with a chain-terminated primer, including the incoming dNTP (dTTP, which could not be incorporated) establishes a stable ternary complex. In support of this, Fig. 4.5e shows that as the dTTP concentration increases, the high-FRET, polymerase-competent enzyme orientation is stabilized.

The nonnucleoside RT inhibitor (NNRTI) nevirapine, a critical component of combination antiretroviral therapy (Croxtall 2012), has been shown by X-ray crystallography to bind near the base of the p66 thumb subdomain and allosterically affect DNA synthesis by displacing the 3′-end of DNA primer from its polymerase active site position (Das et al. 2012). At the same time, Grobler and colleagues (Grobler et al. 2007) demonstrated that PPT-primed (+)-strand DNA synthesis was preferentially inhibited by NNRTIs. Figure 4.5f examines the effect of small-molecule DNA polymerase and RNase H inhibitors on HIV-1 RT conformational dynamics. Surprisingly, addition of nevirapine resulted in destabilization of the high-FRET, polymerization-competent orientation. Although not shown, a similar effect was noted for the NNRTI efavirenz (Abbondanzieri et al. 2008). Such observations would be in line with those reported by Grobler and colleagues (2007), demonstrating that an additional and unexpected property of NNRTIs is their ability to promote an alteration in enzyme conformation that is inconsistent with DNA synthesis. In contrast, the vinylogous urea RNase H inhibitor NSC727447, which binds at the interface between the p51 thumb and the p66 RNase H domain (Wendeler et al. 2008), may have the opposite effect, namely, stabilizing RT on the PPT in a polymerization-competent mode, which would have consequences for PPT cleavage. Although the application of single-molecule spectroscopy to drug screening is impractical, data of Fig. 4.5 clearly demonstrates its value in unveiling novel and unexpected properties of antiviral drugs currently in clinical use.

## 4.5 Shuttling of HIV-1 RT Between Termini of Nucleic Acid Duplexes

Processivity of nucleic acid polymerizing enzymes is defined by the number of nucleotides added to the primer 3′ terminus in a single binding event. Distributive enzymes (such as *E. coli* DNA polymerase IV) in general dissociate from their substrate after addition of only a few nucleotides (Wagner and Nohmi 2000), whereas processive enzymes (e.g., *E. coli* DNA polymerase III) can polymerize hundreds to several thousand nucleotides in a single binding event (Rothwell et al. 2003). For a low-processivity enzyme such as HIV-1 RT (Huber et al. 1989; Williams et al. 1990), this raises the important issue of how the enzyme re-engages the primer terminus in the event of dissociation and rebinding. Secondly, assuming that, similar to RNA polymerases, RT can access the primer by "facilitated diffusion" (Kim and Larson 2007), since the nucleic acid duplex would seem incapable of providing an orientational bias, what are the consequences of enzyme encountering the primer terminus in a polymerase-incompetent orientation? Lastly, how, following dissociation, is the primer terminus relocated during strand displacement synthesis, considering this might be displaced from the template by the competing non-template strand?

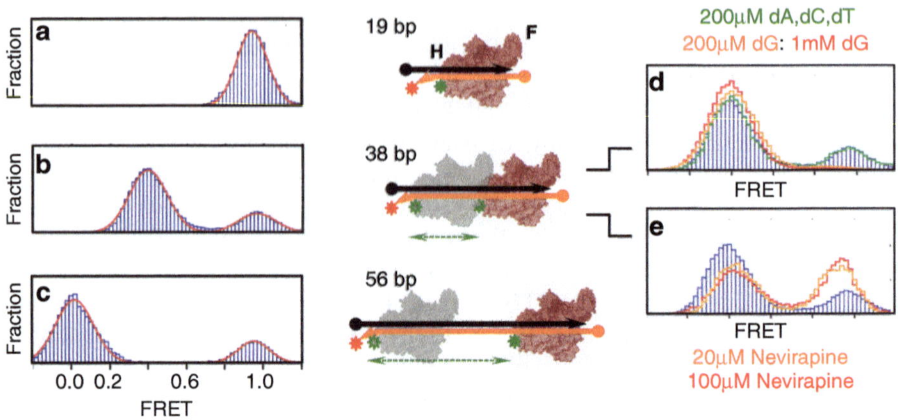


**Fig. 4.6** Sliding of HIV-1 RT on nucleic acid duplexes. The cartoon in the center of the figure depicts the location of RT on RNA/DNA hybrids of 19 (*upper*), 38 (*middle*), and 56 bp (*lower*), with the 3′, Cy5-labeled RNA primarin orange and the DNA template in black. RNA and DNA 5′ and 3′ termini are depicted by filled spheres and circles, respectively. (**a**)–(**c**) FRET histograms obtained from Cy3, RNase H-labeled RT bound to the 19 bp. 38 bp, and 56 bp RNA/DNA hybrids, respectively. Sliding of RT in (**b**) and (**c**) is depicted in the cartoon by the gray molecule. (**d**) Effect of dNTPs on RT sliding on the 38 bp RNA/DNA hybrid. Histograms were obtained in the absence of dNTPS (*green*), the cognate dNTP (dGTP, *orange*), and the non-cognate dNTP (*green*). (**e**) FRET histograms from RT bound to the 38 bp RNA/DNA hybrid in the absence (*blue*) or presence of 20 μM (*orange*) and 100 μM nevirapine (*red*)

In order to gain a more comprehensive understanding of movement of HIV-1 RT along its nucleic acid duplexes, we exploited single-molecule spectroscopy to examine interactions with RNA template/DNA primer combinations of increasing length (Fig. 4.6), making use of Cy3-RNase H-labeled RT containing the RNase H-inactivating mutation Glu478Gln (Rausch et al. 2000b). In these experiments, a 3-nt "flap" was also introduced at the 3′ terminus of the RNA template. In Fig. 4.6a, a 19 bp duplex was chosen to mimic the RT "footprint" demonstrated by X-ray crystallography (Jacobo-Molina et al. 1993) and chemical footprinting (Metzger et al. 1993). On this substrate, the single, high-FRET signal obtained indicated a unique conformation of RT with its DNA polymerase active site in the immediate vicinity of the DNA primer 3′ terminus. In sharp contrast, increasing the length of the template/primer duplex from 19 to 38 bp yielded both high- and low-FRET peaks, with the latter predominating. As illustrated in the cartoon of Fig. 4.6, the low-FRET signal is consistent with the DNA polymerase active site still located over the DNA primer 3′ terminus, which would separate the Cy3 and Cy5 dyes by ~19 bp. The high-FRET signal indicated a second binding mode where RT presumably "slides" to the other end of the duplex, while maintaining its orientation (indicated by the gray molecule). If this notion was correct, it predicted that increasing the length of the RNA/DNA hybrid would further increase separation of the FRET peaks, as was observed experimentally in the data of Fig. 4.6c. Equivalent observations were made when the RNA/DNA hybrids were substituted with duplex DNA (mimicking DNA-dependent DNA synthesis), indicating that the ability of RT to slide on nucleic acid was a general property (Liu et al. 2008). Finally, the observation that shuttling between duplex termini was enhanced as with increasing temperature and independent of dNTP hydrolysis suggested that movement was a thermally driven diffusion process.

Conformational alterations that might allow RT to leave the primer terminus and slide along its substrate were examined with the 38 bp RNA/DNA hybrid in the presence of small-molecule ligands. Simulating conditions that establish a stable ternary complex (Huang et al. 1998), where the fingers and palm subdomains of RT are clamped around the substrate, would be predicted to favor an interaction with the primer terminus, thereby reducing sliding. Indeed, data of Fig. 4.6d shows that inclusion of the incoming dNTP (dGTP) stabilized the low-FRET orientation indicative of RT bound at the DNA primer 3′ terminus in a polymerization-competent mode (the same figure shows that the presence of non-cognate dNTPs, but withholding dGTP, induced equivalent sliding kinetics). These effects were reversed in the presence of the NNRTI nevirapine, i.e., shuttling between the high- and low-FRET states was observed. Since, mechanistically, NNRTI binding has been shown to relax the grip that the p66 fingers and thumb impose on nucleic acid (Das et al. 2012), this might induce the opposite effect, i.e., enhancing "dislocation" from the primer terminus and increasing sliding. This outcome was realized experimentally and illustrated in Fig. 4.6e, where RT is located with almost equal probability at either end of the RNA/DNA hybrid in the presence of nevirapine. Such a property of nevirapine may be another unexpected and advantageous feature of this inhibitor class, possibly explaining previous observations that long DNA synthesis is preferentially inhibited by NNRTIs (Quan et al. 1998).

Finally, the manner whereby RT might perform a one-dimensional search to locate the primer terminus on longer DNAs was examined on a Cy5-labeled 550 bp DNA duplex. Contrary to what had been demonstrated on short RNA/DNA hybrids, FRET traces indicated that the enzyme bound the DNA outside the polymerization target site and only thereafter accessed the primer terminus via sliding. However, as indicated earlier, the lack of an orientational bias would have the consequence that RT accesses the primer terminus in either a polymerization-competent orientation or in a polymerization-incompetent RNase H orientation. As predicted, FRET traces did indeed provide evidence of enzyme accessing the primer terminus in either of these orientations. Remarkably, and for reasons that remain to be fully elucidated, incorrectly oriented molecules displayed a capacity to "flip" into the polymerization-competent orientation, and such an inversion of orientation was accomplished without enzyme dissociation. Conceivably, these unusual conformational "gymnastics" (Liu et al. 2008) may contribute towards increasing the efficiency with which the polymerization target site is accessed.

*Dynamics of Strand Displacement Synthesis on RNA and DNA Templates*. As outlined earlier, the HIV-1 5′ UTR comprises a number of RNA secondary structure elements that play critical roles in the virus replication cycle, including the TAR loop, the U5-IR hairpin, the dimer linkage sequence (DLS), and the encapsidation site (Ψ) (Jakobsen et al. 2007). Disrupting these secondary structures during (−)-strand DNA synthesis presents a formidable challenge to the replication machinery since, if they induce enzyme dissociation, the greater stability of duplex RNA compared to the DNA/RNA product of (−)-strand synthesis might promote refolding and displacement of the nascent DNA primer terminus. Although the problem might not be as acute, (+)-strand, DNA-dependent DNA synthesis likewise mandates strand displacement activity, e.g., termination of DNA synthesis downstream of the HIV-1 central PPT (Charneau et al. 1992). To examine the dynamics of RNA, strand displacement synthesis required incubating HIV-1 RT with doubly Cy3/Cy5-labeled, three-stranded substrates containing an RNA template (T) to which a DNA primer (p) and a non-template (NT) RNA strand were hybridized, the latter immediately adjacent to the primer 3′ terminus (Liu et al. 2008). A single FRET peak with a value of 0.35 was observed in the absence of DNA synthesis, consistent with enzyme positioning in a polymerization-competent orientation over the DNA primer 3′ terminus. Subsequent inclusion of a chain-terminating dNTP/ddNTP cocktail supporting a two-nucleotide primer extension gave rise to a FRET peak with a value of 0.45, while a cocktail which supported a four nucleotide extension resulted in a FRET peak of 0.6. While such observations were consistent with unwinding of the T/NT duplex in response to translocation of the polymerization machinery, repetitive transitions between the 0.6 and 0.3 FRET states suggested reannealing of the template and non-template strands and displacement of the prime terminus.

Such repetitive transitions in FRET states were further investigated on p/T/NT three-stranded duplexes whose primers had been chain terminated with a ddNTP to prevent their extension. Under these conditions, transitions between high- and low-FRET states were observed, indicating that even for strand displacement synthesis, HIV-1 RT displayed a capacity to disengage from the primer terminus and slide to the other end of the duplex. Further analysis of FRET traces indicated that most binding events initiated downstream of the primer terminus and were followed by sliding of RT to the polymerization site. Such a mechanism would allow the HIV enzyme to access the "disrupted" polymerization site, promote reannealing of the primer, and facilitate RNA strand displacement synthesis. However, one intriguing observation was that RT shuttled between *equivalent* FRET states on three-stranded substrates whose primer had been extended by both 4 (p4/T/NT) and 10 deoxynucleotides (p10/T/NT). Our interpretation of this result was that, despite having the ability to shuttle, RT could not displace more than four template ribonucleotides, a notion demonstrated experimentally in an ensemble-based primer-extension experiment, where RNA strand displacement synthesis was severely impaired following addition of five nucleotides to the DNA primer.

During DNA strand displacement synthesis, the primer-extension product and displaced non-template DNA are energetically similar. Equivalent single-molecule experiments with three-stranded DNA substrates mimicking discrete primer-extension events indicate that HIV-1 RT readily accesses the primer terminus and unwinds the adjacent template/non-template duplex, regardless of the length of primer extension. A related single-molecule study (Kim et al. 2010) showed that on long DNA strands, the primer-extension rate decreased almost eightfold at hairpins. The rate of DNA strand displacement was also dependent on the G:C content of the duplex, from which these authors proposed that hairpin melting was driven by a combination of free energy released during dNTP hydrolysis and thermal fraying of base pairs. In agreement with these single-molecule studies, biochemical studies of Whiting and Champoux have shown that HIV-1 RT catalyzes strand displacement synthesis over several hundred nucleotides on DNA duplexes (Whiting and Champoux 1994).

## 4.6 Conclusions and Perspectives

Single-molecule spectroscopy has demonstrated that the two-subunit, 117 kDa HIV-1 RT is a highly dynamic molecule, capable of not only sliding along its nucleic acid substrate to access the polymerization site but also altering its orientation on the substrate during tRNA-primed initiation of (−)-strand DNA synthesis as a possible means of regulating reverse transcription prior to virion budding. An unexpected finding of our work was demonstrating that anti-HIV RT drugs in clinical use can not only reverse the polymerization-competent enzyme orientation but also induce "dislocation" from the primer terminus, both of which likely contribute to the efficiency of NNRTIs as DNA polymerase inhibitors. Whether such observations will open new therapeutic avenues, such as small molecules targeting the initiation complex to stabilize a binding mode incompatible with DNA synthesis, remains to be seen. Nonetheless, single-molecule spectroscopy has shown how HIV-1 RT performs an elegant series of molecular gymnastics as it converts viral RNA into double-stranded DNA.

While illuminating several surprising features of HIV-1 RT, our studies have also raised a number of questions. Uppermost among these is the ability to spontaneously change orientation, or flip, exemplified by the FRET traces of Fig. 4.7. These experiments were performed on the PPT:d2 substrate of Fig. 4.5 (Fig. 4.7a, b) and a 550 bp DNA duplex was used to demonstrate RT accessing the polymerization target site (Liu et al. 2008). On the PPT:d2 substrate, a stable FRET trace is derived from a ternary complex, while in the presence of nevirapine, multiple transitions between the high- and low-FRET state are apparent. Surprisingly, control experiments indicated that these flipping events are supported without dissociation of RT from its substrate nor did it require multiple enzymes. Similarly, Fig. 4.7c, d illustrates the two scenarios observed when RT slides along a long DNA substrate to engage the polymerization target site. While the monophasic FRET trace of Fig. 4.7c demonstrates an enzyme accessing its target site in the polymerization-competent orientation, the rapid exchange between high- and low-FRET states of Fig. 4.7d is indicative of an enzyme flipping from the RNase H to polymerization-competent mode, which again occurs without enzyme dissociation. Further experimentation will be required to determine how HIV-1 RT relinquishes and renews contact with nucleic acid to support these changes in orientation. Another issue which requires

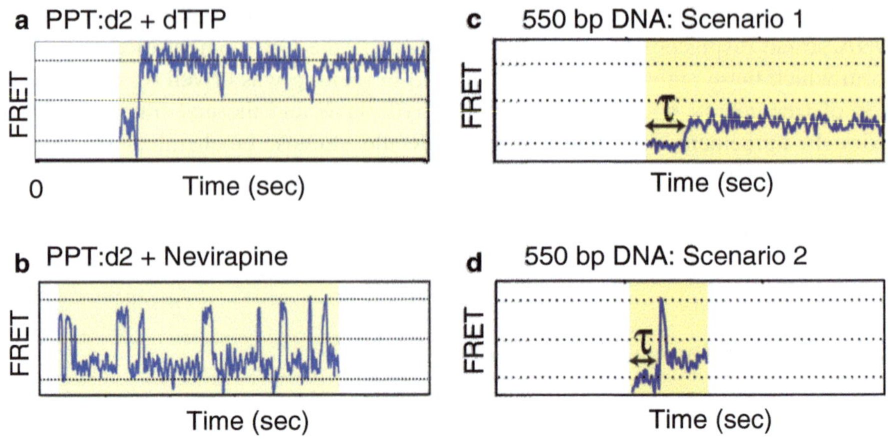


**Fig. 4.7** Dynamic flipping of HIV-1 RT. (**a**) FRET trace from Cy3, RNase H-labeled RT bound to a Cy5-labeled PPT substrate in the presence of the incoming dNTP, which stabilizes ternary complex formation. (**b**) FRET trace from Cy3-labeled RT bound to a Cy5-labeled PPT:d2 containing RNA/DNA hybrid in the presence of the NNRTI nevirapine(Abbondanzieri et al. 2008). (**c**), (**d**) FRET traces from Cy3, RNase H-labeled RT bound to a 550 bp Cy5-labeled DNA duplex. The trace of (**c**) represents an enzyme reaching the Cy5-labeled primer terminus in a polymerization-competent orientation, resulting in a low-FRET signal. In (**d**), RT encounters the primer terminus in the opposite orientation, giving rise to a transient high-FRET signal but thereafter flips into the low-FRET, polymerization-competent orientation (Liu et al. 2008)

further investigation is the mechanistic basis of pausing, which could be inferred from FRET traces of DNA strand displacement synthesis. Structural motifs of RT that might contribute to these events include the primer and template grips of the p66 subunit (Jacobo-Molina et al. 1993), the RNase H primer grip (Sarafianos et al. 2001), or the phosphate binding pocket that precedes the RNase H catalytic center (Dash et al. 2008). Finally, it is important to examine whether conformational dynamics displayed by HIV-1 RT are also a feature of monomeric enzymes, such as those from Moloney murine leukemia virus (Tanese and Goff 1988) or the *Saccharomyces cerevisiae* LTR retrotransposon Ty1 (Rausch et al. 2000a).

## References

Abbondanzieri EA, Bokinsky G, Rausch JW, Zhang JX, Le Grice SF, Zhuang X (2008) Dynamic binding orientations direct activity of HIV reverse transcriptase. Nature 453(7192):184–189

Basu VP, Song M, Gao L, Rigby ST, Hanson MN, Bambara RA (2008) Strand transfer events during HIV-1 reverse transcription. Virus Res 134(1–2):19–38

Beerens N, Berkhout B (2002) The tRNA primer activation signal in the human immunodeficiency virus type 1 genome is important for initiation and processive elongation of reverse transcription. J Virol 76(5):2329–2339

Beerens N, Groot F, Berkhout B (2001) Initiation of HIV-1 reverse transcription is regulated by a primer activation signal. J Biol Chem 276(33):31247–31256

Bibillo A, Eickbush TH (2002) High processivity of the reverse transcriptase from a non-long terminal repeat retrotransposon. J Biol Chem 277(38):34836–34845 [Research Support, U.S. Gov't, P.H.S.]

Carpousis AJ, Gralla JD (1980) Cycling of ribonucleic acid polymerase to produce oligonucleotides during initiation in vitro at the lac UV5 promoter. Biochemistry 19(14):3245–3253

Chapman KB, Bystrom AS, Boeke JD (1992) Initiator methionine tRNA is essential for Ty1 transposition in yeast. Proc Natl Acad Sci USA 89(8):3236–3240

Charneau P, Alizon M, Clavel F (1992) A second origin of DNA plus-strand synthesis is required for optimal human immunodeficiency virus replication. J Virol 66(5):2814–2820 [Comparative Study Research Support, Non-U.S. Gov't]

Croxtall JD (2012) Etravirine: a review of its use in the management of treatment-experienced patients with HIV-1 infection. Drugs 72(6):847–869 [Review]

Das K, Martinez SE, Bauman JD, Arnold E (2012) HIV-1 reverse transcriptase complex with DNA and nevirapine reveals non-nucleoside inhibition mechanism. Nat Struct Mol Biol 19(2):253–259 [Research Support, N.I.H., Extramural]

Dash C, Scarth BJ, Badorrek C, Gotte M, Le Grice SF (2008) Examining the ribonuclease H primer grip of HIV-1 reverse transcriptase by charge neutralization of RNA/DNA hybrids. Nucleic Acids Res 36(20):6363–6371 [Research Support, N.I.H., Intramural Research Support, Non-U.S. Gov't]

Grobler JA, Dornadula G, Rice MR, Simcoe AL, Hazuda DJ, Miller MD (2007) HIV-1 reverse transcriptase plus-strand initiation exhibits preferential sensitivity to non-nucleoside reverse transcriptase inhibitors in vitro. J Biol Chem 282(11):8005–8010 [In Vitro]

Huang Y, Wang J, Shalom A, Li Z, Khorchid A, Wainberg MA et al (1997) Primer tRNA3Lys on the viral genome exists in unextended and two-base extended forms within mature human immunodeficiency virus type 1. J Virol 71(1):726–728 [Research Support, Non-U.S. Gov't]

Huang H, Chopra R, Verdine GL, Harrison SC (1998) Structure of a covalently trapped catalytic complex of HIV-1 reverse transcriptase: implications for drug resistance. Science 282(5394): 1669–1675

Huber HE, McCoy JM, Seehra JS, Richardson CC (1989) Human immunodeficiency virus 1 reverse transcriptase. Template binding, processivity, strand displacement synthesis, and template switching. J Biol Chem 264(8):4669–4678 [Comparative Study Research Support, Non-U.S. Gov't Research Support, U.S. Gov't, P.H.S.]

Isel C, Ehresmann C, Keith G, Ehresmann B, Marquet R (1995) Initiation of reverse transcription of HIV-1: secondary structure of the HIV-1 RNA/tRNA(3Lys) (template/primer). J Mol Biol 247(2):236–250

Isel C, Westhof E, Massire C, Le Grice SF, Ehresmann B, Ehresmann C et al (1999) Structural basis for the specificity of the initiation of HIV-1 reverse transcription. EMBO J 18(4):1038–1048

Jacobo-Molina A, Ding J, Nanni RG, Clark AD Jr, Lu X, Tantillo C et al (1993) Crystal structure of human immunodeficiency virus type 1 reverse transcriptase complexed with double-stranded DNA at 3.0 a resolution shows bent DNA. Proc Natl Acad Sci USA 90(13):6320–6324

Jakobsen MR, Haasnoot J, Wengel J, Berkhout B, Kjems J (2007) Efficient inhibition of HIV-1 expression by LNA modified antisense oligonucleotides and DNAzymes targeted to functionally selected binding sites. Retrovirology 4:29 [Research Support, Non-U.S. Gov't]

Ji X, Klarmann GJ, Preston BD (1996) Effect of human immunodeficiency virus type 1 (HIV-1) nucleocapsid protein on HIV-1 reverse transcriptase activity in vitro. Biochemistry 35(1):132–143 [Research Support, Non-U.S. Gov't Research Support, U.S. Gov't, P.H.S.]

Kim JH, Larson RG (2007) Single-molecule analysis of 1D diffusion and transcription elongation of T7 RNA polymerase along individual stretched DNA molecules. Nucleic Acids Res 35(11):3848–3858 [Research Support, U.S. Gov't, Non-P.H.S.]

Kim S, Schroeder CM, Xie XS (2010) Single-molecule study of DNA polymerization activity of HIV-1 reverse transcriptase on DNA templates. J Mol Biol 395(5):995–1006 [In Vitro Research Support, N.I.H., Extramural Research Support, Non-U.S. Gov't]

Lanchy JM, Ehresmann C, Le Grice SF, Ehresmann B, Marquet R (1996a) Binding and kinetic properties of HIV-1 reverse transcriptase markedly differ during initiation and elongation of reverse transcription. EMBO J 15(24):7178–7187

Lanchy JM, Isel C, Ehresmann C, Marquet R, Ehresmann B (1996b) Structural and functional evidence that initiation and elongation of HIV-1 reverse transcription are distinct processes. Biochimie 78(11–12):1087–1096

Le Grice SF, Naas T, Wohlgensinger B, Schatz O (1991a) Subunit-selective mutagenesis indicates minimal polymerase activity in heterodimer-associated p51 HIV-1 reverse transcriptase. EMBO J 10(12):3905–3911

Le Grice SF, Panin M, Kalayjian RC, Richter NJ, Keith G, Darlix JL et al (1991b) Purification and characterization of recombinant equine infectious anemia virus reverse transcriptase. J Virol 65(12):7004–7007 [Research Support, Non-U.S. Gov't Research Support, U.S. Gov't, P.H.S.]

Leis J (1993) Regulation of initiation of reverse transcription of retroviruses. In: Skalka AM, Goff SP (eds) Reverse transcriptase. Cold Spring Harbor Laboratory Press, Plainview, pp. 33–48

Levin JG, Seidman JG (1979) Selective packaging of host tRNA's by murine leukemia virus particles does not require genomic RNA. J Virol 29(1):328–335

Liu S, Abbondanzieri EA, Rausch JW, Le Grice SF, Zhuang X (2008) Slide into action: dynamic shuttling of HIV reverse transcriptase on nucleic acid substrates. Science 322(5904):1092–1097 [Research Support, N.I.H., Extramural Research Support, N.I.H., Intramural Research Support, Non-U.S. Gov't]

Liu S, Harada BT, Miller JT, Le Grice SF, Zhuang X (2010) Initiation complex dynamics direct the transitions between distinct phases of early HIV reverse transcription. Nat Struct Mol Biol 17(12):1453–1460 [Research Support, N.I.H., Extramural Research Support, N.I.H., Intramural]

Lu HP, Xun L, Xie XS (1998) Single-molecule enzymatic dynamics. Science 282(5395):1877–1882 [Research Support, Non-U.S. Gov't Research Support, U.S. Gov't, Non-P.H.S.]

Metzger W, Hermann T, Schatz O, Le Grice SF, Heumann H (1993) Hydroxyl radical footprint analysis of human immunodeficiency virus reverse transcriptase-template.primer complexes. Proc Natl Acad Sci U S A 90(13):5909–5913 [Research Support, Non-U.S. Gov't Research Support, U.S. Gov't, P.H.S.]

Miller JT, Khvorova A, Scaringe SA, Le Grice SF (2004) Synthetic tRNALys, 3 as the replication primer for the HIV-1HXB2 and HIV-1Mal genomes. Nucleic Acids Res 32(15):4687–4695 [Comparative Study]

Moerner WE, Orrit M (1999) Illuminating single molecules in condensed matter. Science 283(5408):1670–1676 [Research Support, U.S. Gov't, Non-P.H.S. Review]

Quan Y, Liang C, Inouye P, Wainberg MA (1998) Enhanced impairment of chain elongation by inhibitors of HIV reverse transcriptase in cell-free reactions yielding longer DNA products. Nucleic Acids Res 26(24):5692–5698

Rausch JW, Le Grice SF (1997) Substituting a conserved residue of the ribonuclease H domain alters substrate hydrolysis by retroviral reverse transcriptase. J Biol Chem 272(13):8602–8610 [Research Support, U.S. Gov't, P.H.S.]

Rausch JW, Le Grice SF (2004) 'Binding, bending and bonding': polypurine tract-primed initiation of plus-strand DNA synthesis in human immunodeficiency virus. Int J Biochem Cell Biol 36(9):1752–1766 [Review]

Rausch JW, Le Grice SF (2007) Purine analog substitution of the HIV-1 polypurine tract primer defines regions controlling initiation of plus-strand DNA synthesis. Nucleic Acids Res 35(1):256–268 [Research Support, N.I.H., Intramural]

Rausch JW, Grice MK, Henrietta M, Nymark M, Miller JT, Le Grice SF (2000a) Interaction of p55 reverse transcriptase from the *Saccharomyces cerevisiae* retrotransposon Ty3 with conformationally distinct nucleic acid duplexes. J Biol Chem 275(18):13879–13887 [Research Support, U.S. Gov't, P.H.S.]

Rausch JW, Sathyanarayana BK, Bona MK, Le Grice SF (2000b) Probing contacts between the ribonuclease H domain of HIV-1 reverse transcriptase and nucleic acid by site-specific photocross-linking. J Biol Chem 275(21):16015–16022 [Research Support, U.S. Gov't, P.H.S.]

Rodriguez-Rodriguez L, Tsuchihashi Z, Fuentes GM, Bambara RA, Fay PJ (1995) Influence of human immunodeficiency virus nucleocapsid protein on synthesis and strand transfer by the reverse transcriptase in vitro. J Biol Chem 270(25):15005–15011 [Research Support, Non-U.S. Gov't Research Support, U.S. Gov't, P.H.S.]

Rothwell PJ, Berger S, Kensch O, Felekyan S, Antonik M, Wohrl BM et al (2003) Multiparameter single-molecule fluorescence spectroscopy reveals heterogeneity of HIV-1 reverse transcriptase: primer/template complexes. Proc Natl Acad Sci U S A 100(4):1655–1660 [Research Support, Non-U.S. Gov't]

Sarafianos SG, Das K, Tantillo C, Clark AD Jr, Ding J, Whitcomb JM et al (2001) Crystal structure of HIV-1 reverse transcriptase in complex with a polypurine tract RNA: DNA. EMBO J 20(6):1449–1461

Tanese N, Goff SP (1988) Domain structure of the Moloney murine leukemia virus reverse transcriptase: mutational analysis and separate expression of the DNA polymerase and RNase H activities. Proc Natl Acad Sci U S A 85(6):1777–1781 [Research Support, U.S. Gov't, P.H.S.]

Taylor JM, Illmensee R (1975) Site on the RNA of an avian sarcoma virus at which primer is bound. J Virol 16(3):553–558

Telesnitsky A, Goff SP (1997) Reverse transcriptase and the generation of retroviral DNA. In: Coffin JM, Hughes SH, Varmus HE (eds) Retroviruses. Cold Spring Harbor Laboratory Press, Plainview

Turner KB, Brinson RG, Yi-Brunozzi HY, Rausch JW, Miller JT, Le Grice SF et al (2008) Structural probing of the HIV-1 polypurine tract RNA: DNA hybrid using classic nucleic acid ligands. Nucleic Acids Res 36(8):2799–2810 [Research Support, N.I.H., Extramural Research Support, N.I.H., Intramural Research Support, U.S. Gov't, Non-P.H.S.]

Wagner J, Nohmi T (2000) Escherichia coli DNA polymerase IV mutator activity: genetic requirements and mutational specificity. J Bacteriol 182(16):4587–4595 [Research Support, Non-U.S. Gov't]

Weiss S (2000) Measuring conformational dynamics of biomolecules by single molecule fluorescence spectroscopy. Nat Struct Biol 7(9):724–729 [Research Support, Non-U.S. Gov't Research Support, U.S. Gov't, Non-P.H.S. Review]

Wendeler M, Lee HF, Bermingham A, Miller JT, Chertov O, Bona MK et al (2008) Vinylogous ureas as a novel class of inhibitors of reverse transcriptase-associated ribonuclease H activity. ACS Chem Biol 3(10):635–644 [Research Support, N.I.H., Extramural Research Support, N.I.H., Intramural]

Whiting SH, Champoux JJ (1994) Strand displacement synthesis capability of Moloney murine leukemia virus reverse transcriptase. J Virol 68(8):4747–4758 [Research Support, Non-U.S. Gov't Research Support, U.S. Gov't, P.H.S.]

Williams KJ, Loeb LA, Fry M (1990) Synthesis of DNA by human immunodeficiency virus reverse transcriptase is preferentially blocked at template oligo(deoxyadenosine) tracts. J Biol Chem 265(30):18682–18689 [Research Support, Non-U.S. Gov't Research Support, U.S. Gov't, Non-P.H.S. Research Support, U.S. Gov't, P.H.S.]

Wu W, Henderson LE, Copeland TD, Gorelick RJ, Bosche WJ, Rein A et al (1996) Human immunodeficiency virus type 1 nucleocapsid protein reduces reverse transcriptase pausing at a secondary structure near the murine leukemia virus polypurine tract. J Virol 70(10):7132–7142 [Research Support, U.S. Gov't, P.H.S.]

# Part II
# Mechanism of Action of Approved RT Inhibitors

# Chapter 5
# Nucleoside RT Inhibitors: Structural and Molecular Biology

**Gaofei Lu, Antonio J. Acosta-Hoyos, and Walter A. Scott**†

## 5.1 Introduction

The ability of HIV-1 reverse transcriptase (HIV-1 RT) to incorporate nucleotide analogs with structural diversity has been exploited to develop a wide variety of antiretroviral drugs that mimic natural dNTPs. These compounds are typically administered as nucleosides (which are taken up by cells and converted to the triphosphate form by the action of cellular kinases) or as nucleoside phosphonates (which are also taken up by cells after they have been decorated with hydrophobic groups and are converted to the active form by cellular enzymes that remove the hydrophobic groups and add two phosphate groups to the phosphonate residue). The former class of compounds is referred to as nucleoside RT inhibitors (NRTIs) and the latter as nucleotide RT inhibitors (NtRTIs). Seven FDA-approved NRTIs or NtRTIs are shown in Fig. 5.1. All of these compounds are incorporated by HIV-1 RT at the 3′ end of a growing DNA chain and are chain terminators – i.e., they block further chain elongation because they lack a 3′-0H group that is required for incorporation of additional nucleotides.

The antiviral activities of NRTIs and NtRTIs are determined by a large number of factors including the ability of the compound to enter the blood stream, to be taken up by target cells, and to serve as substrate for cellular enzymes needed to produce the active metabolite. In addition, the active compound must (i) bind to HIV-1 RT with an affinity sufficient to compete with intracellular dNTP pools, (ii) be able to take part in phosphodiester bond synthesis, and (iii) inhibit further DNA chain elongation after it has been incorporated. Additional factors that determine

G. Lu (✉) • W.A. Scott (deceased)
Department of Biochemistry & Molecular Biology, University of Miami Miller School of Medicine, Miami, FL 33101, USA
e-mail: GLu@med.miami.edu

A.J. Acosta-Hoyos
Medicina, Universidad Simón Bolívar, Barranquilla, Atlantico, Colombia

S.F.J. LeGrice and M. Götte (eds.), *Human Immunodeficiency Virus Reverse Transcriptase: A Bench-to-Bedside Success*, DOI 10.1007/978-1-4614-7291-9_5,

| NRTI | Name | Structure | IC50 (M)[a] | IC50 (μM)[b] |
|---|---|---|---|---|
| **AZT** | Zidovudine | 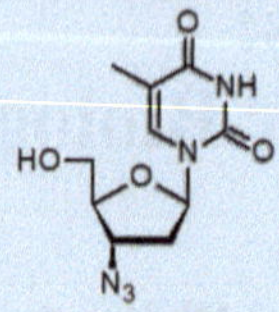 | 0.02 | 0.031 |
| **ddI** | Didanosine | 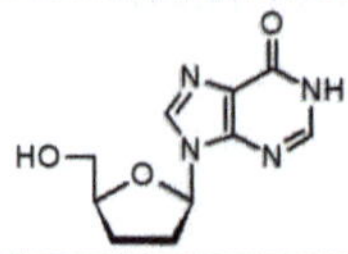 | 5.0 | 0.49 |
| **d4T** | Stavudine | 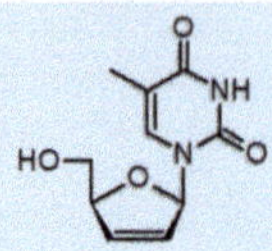 | 0.5 | 1.9 |
| **3TC** | Lamivudine | 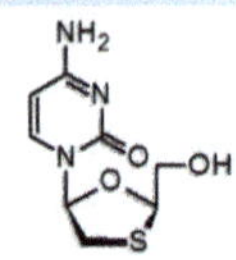 | 2.4 | 0.60 |
| **ABC** | Abacavir | 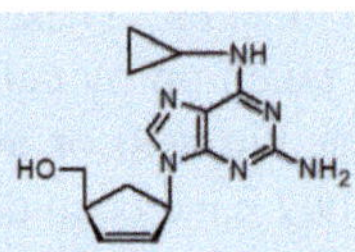 | 1.6 | 4.8 |
| **TFV** | Tenofovir | 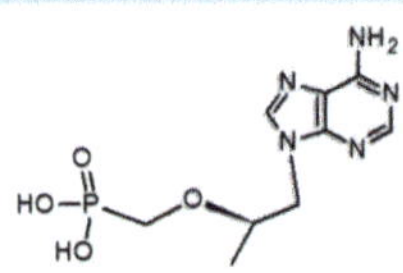 | 0.76[c] | 0.54[c] |
| **FTC** | Emtricitabine | 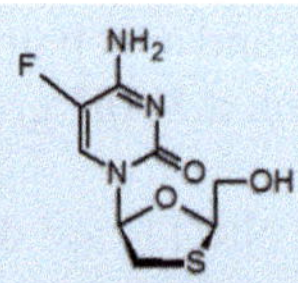 | ND | 0.044 |

**Fig. 5.1** Structure and anti-HIV activities of FDA-approved NRTIs and NtRTIs. (**a**) IC50 is the drug concentration that gives 50 % inhibition of infection in HEK 293 cells. Drug susceptibility was assayed with recombinant viruses containing the protease and RT sequences for wild-type HIV-1 isolates derived from patient plasma samples inserted into a wild-type HIV-1 NL4-3 background (Parkin et al. 2004; Petropoulos et al. 2000). (**b**) IC50 determined for wild-type HIV-1 LAI in MT-2 cells (Sluis-Cremer et al. 2009). (**c**) IC50 determined for the prodrug TDF (see text)

the activities of these compounds include vulnerability to degradation in the gut, in the blood, or in cells; susceptibility of the nucleotide analog to be removed by the excision activity of RT after it has been incorporated; and the ease with which drug-resistant mutations are selected. All of these factors have been intensely studied over the past 25 years, leading to development of drugs that are presently used to treat HIV-infected patients. Additional promising compounds are currently under study as possible antiviral drugs.

In addition to chain-terminating nucleosides, other drugs have been developed that target HIV-1 RT. The most significant, so far, have been non-nucleoside RT inhibitors (NNRTI) that act allosterically at a site near the polymerase active site to inhibit DNA synthesis. NNRTIs are the subject of a later chapter and will not be discussed here. In addition, a class of non-nucleoside compounds have been identified that inhibit by competing for dNTP binding (NcRTIs) (Maga et al. 2010), and a class of nucleosides that contain a 3′OH but stall chain elongation due to a steric clash that occurs after a few additional nucleotides have been incorporated (Boyer et al. 2009). This chapter will focus on classical chain-terminating drugs that have played a major role in chemotherapeutic efforts to control the HIV/AIDS epidemic.

## 5.2 HIV-1 RT Interaction with Substrates for DNA Synthesis

### *5.2.1 Interaction Between HIV-1 RT and Nucleoside Triphosphates*

Most of what we know about interaction between HIV-1 RT and its substrates is derived from studies with purified recombinant enzyme through enzyme kinetics, in vitro binding studies, and X-ray crystallography. These studies have shown that RT has much in common with other DNA polymerases. To catalyze DNA synthesis, the enzyme must bind at the 3′ end of the growing DNA chain. Selection of the incoming nucleotide is determined by base-pairing with the next nucleotide residue on the template strand. Accurate incorporation is enhanced by a two-step binding process – initial binding in the dNTP binding site on the RT-primer/template (P/T) complex followed by a conformational change that produces a stable complex in which the phosphodiester bond is formed. When the 3′ end of the primer is occupied by a chain- terminating nucleotide, it cannot be elongated, and dNTP binding forms a dead-end complex (RT•Pterm/T•dNTP) that dissociates very slowly (Kati et al. 1992; Müller et al. 1991; Tong et al. 1997). Induction of a major conformational change upon dNTP binding is widely observed among DNA polymerases. Detailed studies of conformational changes in the RT•P/T complex that accompany DNA synthesis have recently been carried out by introducing fluorescent tags (Abbondanzieri et al. 2008; Kellinger and Johnson 2010, 2011).

The dead-end complex observed with chain-terminated P/T, RT, and the next complementary dNTP (RT•Pterm/T•dNTP complex) can be detected by electrophoretic mobility shift assay (EMSA) experiments (Sharma et al. 2002; Tong et al. 1997) and by changes induced in the DNase I (Meyer et al. 2007) and chemical footprints (Marchand and Götte 2003). Conformational changes associated with dead-end complex formation have also been characterized by exonuclease protection experiments and by increased susceptibility to chemical cross-linking (Rutvisuttinunt et al. 2008). The RT•Pterm/T•dNTP complex is also called a "ternary complex" by comparison with the binary complex (RT•P/T), which is formed by RT binding to P/T in the absence of dNTP. Both binary and ternary complexes have similar electrophoretic mobilities in EMSA experiments but are distinguishable due to a large difference in complex stability. The binary complex is readily dissociated by poly(rA)/oligo(dT) or heparin, whereas the ternary complex is resistant to these treatments (Meyer et al. 1999; Tong et al. 1997). These results suggest that dNTP binding induces a conformational change in which the protein chain folds down around the P/T, increasing the protection from nucleases and decreasing the distances between the P/T and reactive groups in the enzyme. This interpretation is now supported by several crystal structures of both binary and ternary complexes.

### *5.2.2 Crystallographic Structural Studies*

Numerous crystal structures of HIV-1 RT support the conclusion that the enzyme can be envisioned as a right hand grasping the primer-template, which is similar in general terms to many other DNA polymerases. The active form of HIV-1 RT is a heterodimer of two subunits of 66 kDa (p66) and 51 kDa (p51) with the catalytic activities restricted to the p66 subunit (Hostomsky et al. 1992; Le Grice et al. 1991; Le Grice 1993). The N-terminal portions of the two subunits have the same sequence, but p51 lacks 120 C-terminal residues that are present in p66. Early crystal structures of HIV-1 RT were determined with no ligand, or in complex with nevirapine (a non-nucleoside RT inhibitor), or with DNA duplex (Ding et al. 1998; Jacobo-Molina et al. 1993; Kohlstaedt et al. 1992; Rodgers et al. 1995). The development of our understanding of RT through these structures is discussed in detail in recent review articles (Sarafianos et al. 2009; Singh et al. 2010). The first ternary complex structure of HIV-1 RT was reported in 1998 (Huang et al. 1998) and showed that binding of the incoming dNTP was accompanied by a major conformational change in which the "fingers" subdomain of the p66 RT subunit closed around the nucleic acid substrate and the bound dNTP, placing the active site residues in position for catalyzing phosphodiester bond synthesis. After phosphodiester bond formation, the enzyme translocates forward into position to accept the next dNTP. Several high-resolution structures of HIV-1 RT are now available, including binary complexes with the primer strand terminated with AZTMP (Das et al. 2012; Sarafianos et al. 2002) or TFV-MP (Tuske et al. 2004) and ternary complexes in which TFV-DP (Das et al. 2009; Tuske et al. 2004), GS-9148 (a dATP analog)(Lansdon et al. 2010), or AZTTP (Das et al. 2012) is bound as the incoming dNTP. Ternary structures in which the excision product Ap4AZT is bound (Tu et al. 2010) will be discussed in more detail below, as well as crystal structures for several NRTI-resistant mutants of HIV-1 RT (Chamberlain et al. 2002; Das et al. 2009; Sarafianos et al. 1999; Tu et al. 2010). These structural studies represent major advances in our detailed understanding of the mechanisms of resistance to NRTIs and NtRTIs.

## 5.3 Mechanisms of Drug Resistance

NRTIs and NtRTIs are prodrugs. Once inside the cell, two or three enzyme-catalyzed phosphorylation events must occur before the active chain-terminating nucleoside triphosphate analog is formed. There are approximately 20,000 incorporation events during the complete synthesis of double-stranded HIV-1 DNA. For each activated NRTI or NtRTI, this corresponds to about 5,000 sites where RT could incorporate either the natural nucleotide or the chain- terminating analog. During prolonged exposure to NRTIs or NtRTIs, HIV-1 develops resistance through the selection of specific mutations, rendering the drugs less effective. Reconstitution of RT with mutations in one of the two subunits indicates that resistance is usually determined by the p66 subunit. This resistance, as well as the toxic side effects of long-term use of these drugs, is the main limitation on the use of NRTIs and NtRTIs as antiviral therapy. The primary molecular mechanisms of drug resistance are (i) discrimination against the active form of the drug in favor of the natural dNTP and (ii) the ability of RT to remove the chain-terminating residue after it has been incorporated. Removal occurs through an excision reaction by a mechanism related to pyrophosphorolysis.

### *5.3.1 Resistance Through Increased Discrimination Against the Active Form of the Drug*

Drug resistance may arise through mutations that selectively bind the natural nucleotide and exclude the nucleoside triphosphate analog. This is the major mechanism for resistance to TDF, 3TC, FTC, ABC, and ddI. Therapy with these drugs results in selection of mutations in RT including L74V (ddI and ABC), M184V (3TC, FTC, and ABC), and K65R (TDF, ddI, 3TC, FTC, and ABC), as well as other mutations such as the "Q151M complex" (A62V, V75I, F77L, F116Y, and Q151M), which confers multidrug resistance against AZT, d4T, ddI, ABC, and TDF (Menéndez-Arias 2008; Shirasaka et al. 1995).

Residue M184 lies close to the primer terminus and near to the binding site for the incoming dNTP. 3TC initially selects an isoleucine at this position, but it is rapidly replaced by valine, which supports more efficient replication of the virus (Schinazi et al. 1993; Schuurman et al. 1995). These mutations alter the environment around the polymerase active site and the dNTP binding site, preventing binding and optimal positioning of the analog. The mutant enzyme retains binding activity for the natural dNTP, although with reduced affinity. Structural studies focused on M184I since that mutation is selected first. The isoleucine at this position sterically interacts with the oxathiolane ring of 3TC-TP or FTC-TP, but not with the natural deoxynucleotide, dCTP (Sarafianos et al. 1999). At the molecular level, binding of the incoming nucleotide analog is altered, and the primer/template is repositioned so that the catalytic complex is destabilized for both M184V and M184I (Gao et al. 2000).

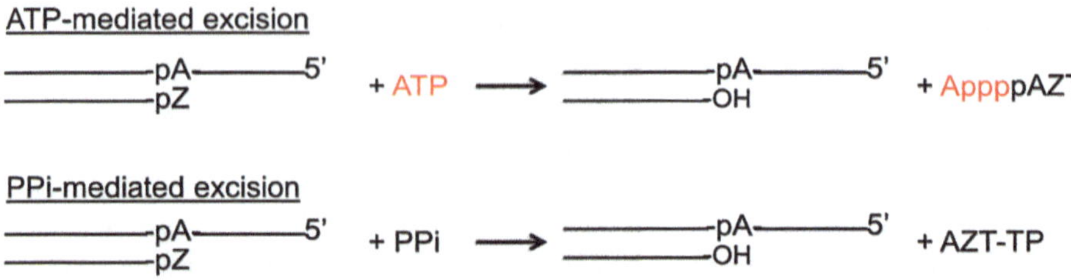


**Fig. 5.2** Excision reaction by HIV-1 RT with either PPi or ATP as acceptor substrate. Z denotes AZTMP as the 3′ terminus of the primer DNA

The K65R mutation is frequently selected during TDF therapy. A K65R substitution creates a structure with reduced mobility of the amino acid side chains around the polymerase active site (Das et al. 2009). The more rigid structure allows the natural nucleotide to bind but discriminates against TNF-DP binding. The structural effects of the K65R mutation are discussed in more detail below.

### 5.3.2 *Resistance Through ATP-Mediated Excision of Chain-Terminating Nucleotides*

During treatment with thymidine analogs, specific mutations are selected at positions 41, 67, 70, 210, 215, and 219 that confer high levels of resistance to AZT and lower levels of resistance to other NRTIs. The laboratories of Parniak (Arion et al. 1998) and Scott (Meyer et al. 1998, 1999) described a biochemical mechanism of resistance that involves unblocking of the DNA chain rather than substrate discrimination. RT bearing the mutations D67N, K70R, T215F, and K219Q showed an elevated ability to remove AZT-monophosphate (AZTMP) from the 3′ end of the DNA primer. This reaction (usually referred to as "excision") involves an attack by compounds containing an exposed di- or triphosphate moiety (e.g., ATP, PPi) on the phosphodiester bond linking AZTMP to the primer strand, producing an unblocked primer end with the capacity to be extended (Fig. 5.2). Transfer of the AZTMP to PPi regenerates AZTTP. Transfer to a nucleoside triphosphate acceptor produces a dinucleoside polyphosphate (e.g., transfer of AZTMP to ATP produces 3′-azido-3′ deoxythymidine-(5′)-tetraphospho-(5′)-adenosine, Ap4AZT). Mutations that enhance excision activity are usually referred to as TAMs to indicate their initial discovery through resistance to thymidine analog NRTIs; however, TAMs are associated with decreased drug susceptibility for most NRTIs and NtRTIs (Whitcomb et al. 2003). Structural and metabolic considerations have suggested that ATP is usually the intracellular substrate for the excision reaction, at least in unstimulated primary lymphocytes or in monocyte-derived macrophages (Acosta-Hoyos and Scott 2010; Boyer et al. 2001; Smith et al. 2005; Smith and Scott 2006; Tu et al. 2010).

The RNase H activity of HIV-1 RT, an essential component in the viral replication process, plays an indirect role in excision-mediated resistance (reviewed by Delviks-Frankenberry et al. 2010, and Ehteshami and Götte 2008). Nikolenko et al. (2005) proposed that RNase H cleavages near the 3′ end of the primer could enhance the inhibitory ability of NRTIs by causing the dissociation of the DNA/RNA duplex before excision could take place. RT mutations with reduced RNase H activity were proposed to confer resistance to a chain-terminating nucleoside analog by increasing the time available for excision to occur before the primer dissociates from the cleaved template. Several RNase H and connection-domain mutations isolated from treatment-experienced HIV-1 infected patients increased AZT resistance in the context of TAMs (Brehm et al. 2012; Nikolenko et al. 2007; Yap et al. 2007). RNase H cleavages occur approximately 18 nt upstream from the primer terminus when RT binds with the 3′ end of the primer in the polymerase active site (polymerase- dependent cleavage). The resulting 18 bp duplex is stable, and these cleavages do not lead to P/T dissociation; however, secondary RNase H cleavages can occur within 10 nt of the primer terminus when RT binds in a polymerase-independent fashion and these cleavages result in P/T dissociation (Acosta-Hoyos et al. 2012; Brehm et al. 2008; Ehteshami et al. 2008; Radzio and Sluis-Cremer 2008). The roles of RNase H in viral replication and drug resistance are discussed in later chapters.

### *5.3.3 Mutations in RT that Antagonize the Effect of TAMs*

Shortly after introduction of 3TC as a new antiretroviral therapy in the early 1990s, it was evident that the selection of M184V as the dominant virus population was a potential benefit of this therapy because the mutation was associated with reduced RT processivity and impaired viral fitness and because this mutation restored a large part of the sensitivity to AZT in virus that was AZT resistant due to TAMs (Boucher et al. 1993; Larder et al. 1995; Petrella and Wainberg 2002; Tisdale et al. 1993). Several laboratories have shown that RTs containing the M184V mutation have reduced excision activity in vitro (cited in Acosta-Hoyos et al. 2012). Recent experiments suggest that the mechanism of reduced excision is at least partly indirect. M184V reduces excision to a greater extent on an RNA template than on a DNA template, and RNase H cleavage at secondary sites is enhanced when this mutation is present. M184V RT has reduced affinity for the 3′ end of the primer allowing the enzyme to dissociate more readily and rebind at secondary sites on the P/T. This leads to increased RNase H cleavages near the primer terminus, increased dissociation of the RNA/DNA duplex, and reduced excision activity (Acosta-Hoyos et al. 2012).

Several other mutations selected in the RT-coding region during NRTI or NNRTI therapy, including K65R, L74V, V75I, L100I, and Y181C, have also been shown to antagonize the resistance mediated by TAMs (reviewed by Acosta-Hoyos and Scott 2010). In vitro studies show that the excision is substantially decreased when K65R is present, which may explain why TAMs and K65R are rarely observed together (Frankel et al. 2007; Parikh et al. 2006; White et al. 2006). In contrast to M184V, the

K65R mutation reduces excision to approximately the same extent on both RNA and DNA templates, suggesting that RNase H activity does not play a significant role in the ability of this mutation to suppress TAMs (Acosta-Hoyos et al. 2012). Structural studies on the mutant enzyme provide insight into the mechanism of reduced excision by K65R (Das et al. 2009).

## 5.4 Structural Dissection of Drug-Resistant RT

### *5.4.1 TAMs and Excision of AZTMP*

Recently, the structure has been solved of a ternary complex of RT bound to AZTMP- terminated primer/template, with AZTTP as the incoming nucleotide (Das et al. 2012). The structure is very similar to the ternary complex containing TTP that was previously reported by Huang et al. (1998) and to a ternary complex containing GS-9148, a dATP analog (Lansdon et al. 2010), indicating that no major distortions are required to bind either AZTTP or the AZTMP- terminated P/T. These results are in agreement with numerous reports showing that both WT and TAM-containing RTs can efficiently use either AZTTP or TTP as a substrate for nucleotide incorporation.

Excision occurs when the bond connecting the nucleotide on the 3′ end of the primer strand is positioned in the polymerase active site. As shown in Fig. 5.2, excision is the reversal of polymerization and utilizes the polymerase active site of the enzyme to transfer the chain-terminating residue to the acceptor ATP or PPi. When RT occupies this position on the P/T, the 3′-terminal residue of the primer is placed in the dNTP binding site (termed the N-site) on the enzyme. This is the pretranslocation complex, and RT must bind in this position to perform the excision reaction. To accept the next complementary dNTP for DNA elongation, RT must translocate forward so that the N-site is open (Götte 2006). When RT occupies the forward position on the P/T, the primer terminus occupies a site defined as the P-site on the enzyme. AZTMP-terminated P/T favors binding to RT in the pretranslocation position (Marchand and Götte 2003; Meyer et al. 2007). RT can be forced to move to the postranslocation by adding the next complementary dNTP, but much higher dNTP concentrations are required than when the primer is terminated with ddTMP or d4T-MP, suggesting either that translocation is inhibited by the structure of the terminal AZTMP or that even after translocation the AZTMP interferes with dNTP binding in the N-site (Boyer et al. 2001; Marchand and Götte 2003; Meyer et al. 1999, 2000; Tong et al. 1997). In fact, recent structures of complex between RT and AZTMP- terminated P/T show that the azido group on the AZTMP-terminated primer occupies the space that is otherwise occupied by $Mg^{++}$ ion "A" in the polymerase active site (Das et al. 2012). The loss of this $Mg^{++}$ ion may result in altered binding affinity for the incoming dNTP. However, Das et al. (2012) were able to overcome the inhibition of next nucleotide binding to form the ternary complex with AZTTP and AZTMP-terminated P/T by soaking the binary complex crystals in a solution containing AZTTP.

Crystal structures of binary complexes of WT RT with AZTMP-terminated P/T were reported in 2003 (Sarafianos et al. 2003). By chemical cross-linking between the template strand and a cysteine residue on the enzyme, complexes were captured in which RT was bound in the pretranslocation position with the primer terminus in the N-site (N-site complex) or in the post-translocational position with the primer terminus in the P-site (P-site complex). Both of these structures were open complexes; however, a recent structure of an N-site binary complex containing mutant RT and AZTMP-terminated P/T was crystallized in a closed configuration (Tu et al. 2010). This may indicate a unique interaction between the terminal AZTMP and the N-site residues of RT. Tu et al. (2010) have also reported a ternary complex, which is a closed structure containing AZT-resistant RT bound to ddC-terminated P/T and the excision product, Ap4AZT (Fig. 5.3). By comparing the unliganded mutant enzyme, the binary complex containing RT in the post-translocation position, and the ternary complex with Ap4AZT, these authors provide a rationale for enhanced ATP-dependent excision, showing that an AMP- binding cleft is formed in the mutant enzyme complex that accommodates the AMP moiety of Ap4AZT. Other TAMs (except M41L) appear to interact directly with the ATP portion of the excision product in the mutant ternary complex. The primary resistance mutations T215Y and K70R form interactions that hold the AMP moiety in the newly created binding cleft. Y215 forms π–π stacking interactions with the adenine ring, and the guanidinium group of R70 forms hydrogen bonds with the 3′ OH and 5′ oxygen atom of the ribose ring and the α−phosphate. The position of the AMP moiety on the surface of the enzyme is very different between the mutant and wild-type enzymes, and the position is more precisely defined on the mutant enzyme, indicating that a specific binding site has been created by the amino acid substitutions. Enhanced excision can be explained by precise positioning of the attacking phosphates of the excision substrate and by tighter interactions between the mutant enzyme and the AMP component of the reaction intermediate, which facilitates the transfer reaction.

### *5.4.2 The K65R Mutation: Resistance to Tenofovir Disoproxil Fumarate (TDF) and Suppression of TAMs*

The K65R mutation is selected during TDF treatment and gives rise to resistance to this NtRTI. Ternary complex structures described by Tuske et al. (2004) and Das et al. (2009) help to understand the mechanism of resistance to this drug and the effects of K65R on excision. The ternary complexes containing wild-type RT/dsDNA with TFV-DP or dATP show that K65 interacts with the γ-phosphate of the incoming nucleotide (Tuske et al. 2004). When K65 is mutated to R65, the planar guanidinium moiety stacks with the guanidinium of R72, which interacts with the adenine base of the incoming nucleotide, restricting the conformational change that precedes phosphodiester bond synthesis (Das et al. 2009) (Fig. 5.3). R65-R72 stacking interactions result in a rigid platform formation that is not observed between K65 and R72 in wild-type RT (Tuske et al. 2004). Such loss of flexibility provides

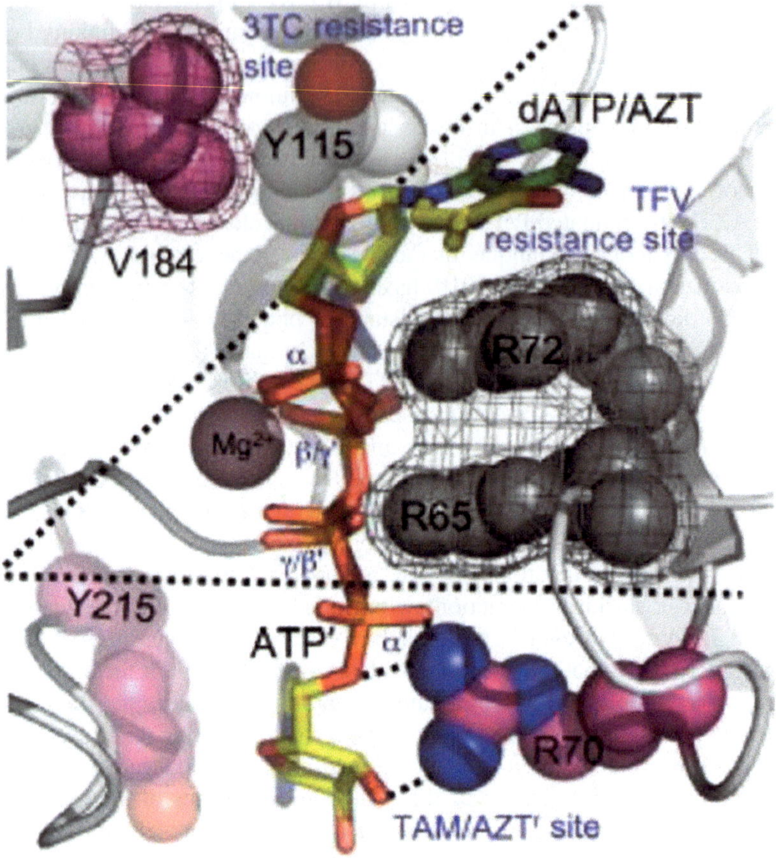


**Fig. 5.3** Summary of structures showing mutations in RT that confer NRTI and NtRTI resistance. Mutations in distinct regions around the polymerase active site of RT confer drug resistance by distinct mechanisms. 3TC resistance site: M184V (modeled from the binary structure, Sarafianos et al. 1999) interacts with the ribose ring and restricts the dNTP binding site. TFV resistance site: K65R participates in guanidinium stacking with R72, which restricts the flexibility of the dNTP binding site and also stacks with the adenine base of TFV-DP when it binds in the dNTP site (Das et al. 2009). TAM/AZTr site: T215Y and K70R interact with the adenosine portion of the ATP excision acceptor substrate creating a stable complex with the AZT-excision product Ap4AZT (Tu et al. 2010) (Reprinted with permission from Das et al. 2009. © 2009 The American Society for Biochemistry and Molecular Biology. All rights reserved)

a structural rationale for both the drug resistance and reduced excision activity conferred by this mutation. The ability of the stacked guanidinium groups to interact with the adenine base is more restricted in the TFV-DP ternary complex than in the dATP ternary complex, allowing the enzyme to distinguish between these two substrates and preferentially avoid the TFV-DP substrate. Reduced flexibility conferred by the stacked guanidinium structure in the K65R mutant enzyme can also be invoked to explain reduced excision activity, since structural flexibility is also reduced in the finger-loop region of the enzyme where the TAMs are located. The R65-R72 platform can interact with the side chains of mutated residues associated with AZT resistance, i.e., R70 and Y215. These interactions restrict movements on the surface of the enzyme that are needed to accommodate the AMP moiety of ATP during excision and interfere with the ability of Y215 to form π–π interactions with the adenine of ATP. This interaction is a major contributor to enhanced excision when TAMs are present.

## 5.5 Uptake and Phosphorylation of NRTIs and NtRTIs

The antiviral effect of orally administered NRTIs and NtRTIs is determined by the efficiency with which they are taken up in the gut, their stability in the blood stream, and their ability to enter into cells that are targets for HIV-1 infection. Most cells have transport systems that allow them to salvage nucleosides and nucleobases from the blood or lymph. It is generally believed that NRTIs enter cells using the transporters responsible for natural nucleoside uptake (Pastor-Anglada et al. 2005). While transport may be active or passive, intracellular phosphorylation can drive the concentration gradient in favor of intracellular accumulation of the active metabolite, even if the transport system is not active. The affinity for each transport system differs for different NRTIs, which could have some impact on uptake of NRTIs in different tissues.

NtRTIs are more problematic, largely because the highly charged phosphonates are not readily taken up by cells. Tenofovir, an acyclic nucleotide analog of dAMP, is a dianion at physiological pH. The uptake problem has been addressed by decoration of the phosphonate group with hydrophobic structures (Ray and Hostetler 2011; Robbins et al. 1998). The FDA-approved form of the drug (TDF) is a bis-isopropoxycarbonyloxymethyl ester prodrug of tenofovir (Fig. 5.4). The aliphatic structures are attached as phosphoesters and later removed by cellular hydrolases after the compounds have entered the cell. The aliphatic groups are also susceptible to hydrolysis by plasma esterases (Naesens et al. 1998). TDF is associated with renal toxicity in a significant fraction of treated patients (Gallant et al. 2005), and this might be avoided by developing prodrugs that deliver tenofovir more efficiently in lymphoid tissue, so that lower doses could be used. Two candidate prodrugs, GS-7340 and CMX157, that might serve as alternatives to TDF are also shown in Fig. 5.4. GS-7340 (isopropylalaninyl monoamidate phenyl monoester of TNF) has enhanced ability to deliver tenofovir to lymphatic tissues (Lee et al. 2005; Markowitz et al. 2011). The potency of GS-7340 was fivefold greater than TDF in peripheral blood mononuclear cells (PBMCs) and 400-fold greater than unmodified tenofovir, presumably due to improved cellular uptake. The compound CMX157 (hexadecyloxypropyl-TNF) was designed to mimic lysophosphatidylcholine, which enters cells through natural lipid uptake pathways (Lanier et al. 2010; Painter et al. 2007). CMX157 is 270-fold more active than tenofovir in PBMCs. Both compounds are currently in the early stages of clinical testing for therapy against HIV infection.

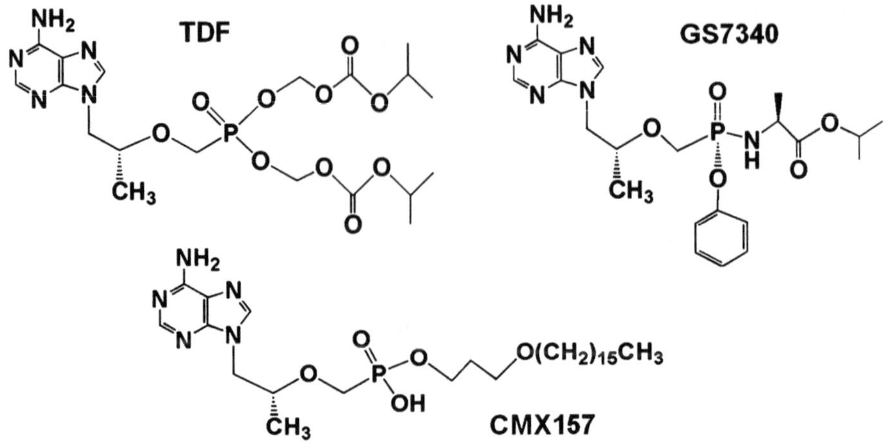


**Fig. 5.4** Structures of three prodrugs for TFV showing structural modifications that facilitate uptake and delivery into cells that are targets for HIV-1 infection. TDF, tenofovir-disoproxil fumarate; GS7340, isopropylalaninyl monoamidate phenyl monoester of TFV; CMX157, hexadecyloxypropyl monoester of TFV

A more comprehensive discussion of prodrug strategies used with nucleoside phosphonates and similar structures can be found in recent reviews (De Clercq 2011; Ray and Hostetler 2011).

Formation of the active metabolites of NRTIs and NtRTIs requires the addition of phosphate residues by cellular kinases (Cihlar and Ray 2010; Van Rompay et al. 2000, 2003). The first phosphate (α-position) is added by nucleoside-specific kinases. For the thymidine analog AZT, cytoplasmic thymidine kinase (TK1) is most important, and this analog is a poor substrate for the mitochondrial thymidine kinase (TK2). However, TK2 does carry out the phosphorylation when TK2 activity is high as in muscle tissue, which may play a major role in the muscle-specific drug toxicity of AZT. The thymidine analog d4T is very poorly phosphorylated by either TK1 or TK2, and d4T phosphorylation is only slightly diminished in a TK1-deficient cell line (Balzarini et al. 1989). This suggests an alternative pathway for d4T phosphorylation, but none has been identified to date. For cytidine analogs (3TC, FTC, ddC), the α-phosphate is added by deoxycytidine kinase (dCK). For the adenine analog (ddI), initial phosphorylation is carried out by cytosolic 5′-nucleotidase (5′NT) acting as an inosine phosphotransferase (Johnson and Fridland 1989).

The activation pathway for abacavir differs from that of other nucleoside analogs. Addition of the first phosphate is carried out by adenosine phosphotransferase. This step is followed by deamination by a cytosolic enzyme to form carbovir monophosphate which is a dGMP analog (Faletto et al. 1997). For each NRTI, the β-phosphate is added by a nucleotide-specific enzyme (thymidylate kinase, uridylate-cytidylate kinase, adenylate kinase, or guanylate kinase), and the γ-phosphate is added by nucleoside diphosphate kinase (NDK). Tenofovir resembles dAMP and is converted to the triphosphate analog by cellular adenylate kinase and NDK. It has been reported that TFV-DP can reach intracellular concentrations greater than 2 mM in PBMCs (Lee et al. 2005; Robbins et al. 1998). Expression and activity of the cellular kinases are regulated by the cell cycle, and the state of activation of the cells at the time of infection plays an important role in the levels of the active metabolites of these drugs (Van Rompay et al. 2000).

## 5.6 Intracellular dNTP Concentrations

### *5.6.1 Variations in dNTP Levels in Different Tissue Compartments*

The antiviral activity of NRTIs and NtRTIs is influenced by the intracellular concentrations of natural nucleosides and nucleotides at several levels. Once inside the cell, NRTIs and NtRTIs must compete with natural nucleosides for activation by kinases. In addition, the kinase levels may be regulated by the end-product dNTPs, and this regulation may be altered by the drugs. For example, feedback inhibition of thymidine kinase is an important factor limiting the activation of thymidine analog fluorodeoxyuridine in human bladder carcinoma cells (Vazquez-Padua et al. 1989) and may also limit activation of other thymidine analogs such as AZT and d4T. Activated NRTIs and NtRTIs also compete with the natural dNTPs for incorporation into proviral DNA. Intracellular dNTP levels are also subject to cellular control mechanisms related to the cell cycle (Gao et al. 1993; Van Rompay et al. 2003). The dNTP levels in unstimulated PBMC are about 1.7 μM (Diamond et al. 2004). After stimulation with phytohemagglutinin and interleukin-2, levels of 16 μM can be attained (Kennedy et al. 2010). Levels of dNTPs are very low in non-dividing cells such as macrophages (20–70 nM) (Diamond et al. 2004; Kennedy et al. 2010; Perez-Bercoff et al. 2007). HIV-1 RT has high affinity for dNTPs, which is required for the virus to replicate under the conditions found in macrophages. Cell lines frequently have dNTP levels as high as 15 to 170 μM (Perez-Bercoff et al. 2007; Roy et al. 1999) and correspondingly higher IC50s for NRTIs and NtRTIs. Primary rectal lymphocytes in macaques have recently been shown to have 100-fold higher dATP concentrations than circulating lymphocytes and lymphoid tissues in these animals (García-Lerma et al. 2011). Pretreatment with TFV prodrug GS-7340 was unable to protect the animals against rectal challenge with simian-human immunodeficiency virus (SHIV) despite accumulation of relatively high levels of TFV-DP in the lymphoid and rectal tissues. Apparently, the high intracellular levels of TFV-DP that can be achieved are no match for the very high endogenous levels of dATP in the rectal tissues.

Resistance mutations in RT frequently have large effects on the affinity of the enzyme for dNTPs, resulting in poor viral replication in cells that contain low levels of dNTPs (Jamburuthugoda et al. 2006, 2008; Van Cor-Hosmer et al. 2010, 2012). Since drug sensitivity is often measured using cell lines, the effect of reduced affinity for dNTPs may be masked by the high intracellular dNTP concentrations. In addition, mutant viruses are often compared for differences in replication capacity in experiments that are also performed in cell lines. To the extent that mutant effects on fitness may be due to effects on affinity of RT for dNTPs, the replication capacity measurements may be highly dependent on the type and metabolic state of the cells used for these measurements (Perez-Bercoff et al. 2007).

### *5.6.2 Control of dNTP Intracellular Pools in Monocyte-Derived Macrophages (MDMs) and Dendritic Cells (DCs)*

HIV-1 replicates very poorly in macrophages, monocytes, and DCs; however, introducing the Vpx protein from SIV or HIV-2 into these cells repairs the defect and stimulates HIV-1 replication (Goujon et al. 2007; Kaushik et al. 2009). It has recently been reported that a cellular antiviral restriction factor, SAMHD1, is responsible for the HIV-1 replication defect (Hrecka et al. 2011; Laguette et al. 2011). Vpx counteracts the activity of SAMHD1 by targeting it for proteasomal degradation. SAMHD1 is a potent dGTP-stimulated triphosphohydrolase that converts dNTPs to deoxynucleosides and inorganic triphosphate (Goldstone et al. 2011; Hrecka et al. 2011; Laguette et al. 2011; Lahouassa et al. 2012; Powell et al. 2011). SAMHD1 is expressed in most cell types, but the expression level is high in myeloid-lineage cells, and its presence in terminally differentiated, nonreplicating macrophages can account for their very low intracellular dNTP concentrations. When SIV/HIV-2 protein Vpx was introduced into MDM, the level of dNTPs in these cells increased 5- to 33-fold (Lahouassa et al. 2012). Expression of SAMHD1 does not restrict HIV-1 replication in activated PBMCs where the dNTP levels are high enough that SAMHD1 activity has little effect. These results show that HIV-1 replication is naturally regulated by depletion of dNTP pools, at least in nondividing cells where dNTPs can be lowered without interfering with host cell functions, and suggest that dNTP pool size is another possible target for antiviral therapy.

### *5.6.3 Drugs that Alter Intracellular dNTP Pools*

Hydroxyurea (HU) inhibits cellular ribonucleotide reductase, which is responsible for the formation of dNTPs from ribonucleotides. Treatment with HU causes depletion of the cellular dNTP pools, which can increase the incorporation of nucleotide analogs. The use of HU has been explored in the treatment of HIV-AIDS to enhance the activity of NRTIs and NtRTIs (Bakshi et al. 2007; Lisziewicz et al. 2003). Another example is mycophenolic acid (MPA), an inosine 5′-monophosphate dehydrogenase inhibitor that blocks the conversion of inosine monophosphate to guanosine monophosphate leading to decreased intracellular dGTP concentration.

Carbovir triphosphate, the active metabolite of abacavir, is a dGTP analog, and the presence of MPA increases the antiviral activity of abacavir in primary PBMC culture (Borroto-Esoda et al. 2004; Margolis et al. 2002). On the other hand, aphidicolin, a DNA polymerase alpha inhibitor that arrests cells in the S/G2 phase, causes a significant increase in intracellular TTP (Wurtzer et al. 2005), and the IC50 for AZT is increased 23-fold as a result of aphidicolin treatment. So far, pharmacological alteration of intracellular dNTP pools has not been successfully incorporated into antiviral therapies; however, as mentioned above, decrease of dNTP levels in monocytes and DCs, due to cellular SAMHD1, effectively blocks HIV-1 infection, suggesting that pharmacological manipulation of dNTP pools may yet have a place in anti-HIV therapies.

## 5.7 dUTP Incorporation and Removal, Deoxycytidine Deamination, and Ribonucleotide Incorporation

Deoxyuridine may be introduced into DNA by misincorporation of dUMP during DNA synthesis or by deamination of deoxycytidine after it has been incorporated (Chen et al. 2002; Krokan et al. 2002; Sire et al. 2008). HIV-RT shows little ability to discriminate between dUTP and TTP as substrates for incorporation. Many viruses encode dUTPase which limits dUMP incorporation by depleting the dUTP pool, but HIV-1 does not encode such an enzyme. During HIV-1 assembly, the nuclear isoform of uracil-DNA glycosylase (UNG-2) is incorporated into virus particles and may play a role in removing uracil residues that are incorporated or are formed by deamination (Chen et al. 2004; Mansky et al. 2000; Priet et al. 2005; Sire et al. 2008; Willetts et al. 1999). Cytosine deamination is highly mutagenic if not repaired, since the product dU base-pairs with dA resulting in CG to TA transition mutations. The deleterious effects of dUMP misincorporation in place of TMP are less obvious since both nucleotides base-pair with dA. However, dUMP misincorporation has been associated with decreased initiation of plus-strand DNA synthesis in vitro, suggesting that substituting dU for T has structural consequences that affect the P/T interaction with HIV-1 RT (Klarmann et al. 2003). Roles for UNG-2 in virus replication have been called into question because viruses produced from human $UNG^{-/-}$ cells infect as efficiently as virus from wild-type cells and cells that express an inhibitor of UNG are also susceptible to infection (Kaiser and Emerman 2006).

A potent antiviral defense mechanism has recently been discovered that is mediated by a cytidine deaminase (APOBEC3G) that converts dC to dU after dC has been incorporated into DNA (Harris et al. 2003; Sheehy et al. 2002; for recent reviews see Malim 2009; Navarro and Landau 2004). APOBEC3G is incorporated into virus particles during assembly and introduces dC to dU mutations in the newly synthesized single-stranded DNA that is formed after the virus enters a target cell. Cytidine deamination by APOBEC3G is so extensive that the proviral DNA is catastrophically mutated and viable virus cannot be made. Attempts to repair the resulting dUs in the newly synthesized viral DNA could also lead to fragmentation of the

DNA viral genome. Since the product of APOBEC3G is the substrate for UNG-2 activity, it might be expected that the presence of UNG-2 in HIV-1 virions would enhance the antiviral effect of APOBEC3G; however, enhanced APOBEC3G-mediated inhibition of HIV-1 infection has not been consistently observed (Kaiser and Emerman 2006; Sire et al. 2008; Yang et al. 2007). Wild-type HIV-1 avoids the antiviral effects of APOBEC3G by encoding the Vif protein, which binds to APOBEC3G, prevents its incorporation into virus particles, and redirects the protein to cellular proteasomes for degradation.

Yan et al. (2011) have developed a method for detecting uracil in HIV DNA using the differential ability of Taq and Pfu DNA polymerases to amplify DNA containing uracil residues. These authors concluded that cytoplasmic HIV DNA formed in primary CD4+ T cells and macrophages contains up to 500 uracils per 10 kb genome. A higher level of dUMP incorporation would be expected in terminally differentiated macrophages where dUTP concentration is reported to be about 60 times higher than TTP, while dUTP and TTP concentrations are similar in activated PBMCs (Kennedy et al. 2011). Yan et al. (2011) did not detect a difference in the amount of uracil in cytoplasmic HIV DNA between CD4+ T cells and macrophages; however, the total uracil content of the viral DNA depends on multiple factors in addition to dUTP concentration including of the rate of cytidine deamination, the rate of repair of uracil lesions after they are formed, and the relative susceptibility of uracilated DNA to intracellular nucleases. Nonetheless, it appears that replicating HIV-1 can produce viral DNA containing a large number of uracils without losing infectivity. Yan et al. (2011) have suggested that the uracil in cytosolic HIV DNA may play a beneficial role for the virus by inhibiting autointegration. These authors have also suggested that heavily uracilated DNA is a poor inducer of interferon and this may provide a camouflage that allows the virus to circumvent antiviral responses in the target cell.

Finally, in in vitro experiments, a very large excess of rNTPs over dNTPs causes HIV-1 RT to misincorporate ribonucleotides into DNA. This situation occurs in terminally differentiated macrophages where ribonucleotide concentrations are in excess over dNTPs by 360- to 28000-fold (Kennedy et al. 2010). HIV-1 RT incorporated one ribonucleotide for every 146 nucleotides incorporated during macrophage infection (Kennedy et al. 2012), whereas little or no rNTP incorporation was detected in CD4+ T cells. Repair of incorporated rNMPs by cellular repair enzymes was also reduced in macrophages. These findings indicate that ribonucleotide incorporation provides an additional factor at play in macrophage infection and suggests that a potentially new class of RT inhibitors could uniquely target macrophages.

## 5.8 Conclusions

HIV-1 RT, in common with other DNA polymerases, usually binds the correct incoming dNTP, which induces conformational changes in the enzyme leading to nucleotide incorporation and chain elongation. NRTIs and NtRTIs can also be incorporated through this process leading to chain termination. The ability of HIV-1

RT to avoid NRTI incorporation is less than for the cellular enzymes with the result that these compounds are effective inhibitors of HIV replication. HIV can become resistant to NRTIs through mutations that increase the ability of RT to discriminate between NRTIs and natural nucleotides or through mutations that increase the ability of RT to remove NRTIs after they have been incorporated by transfer to a nucleotide acceptor in a reaction related to pyrophosphorolysis.

The role of intracellular dNTP concentrations in controlling HIV-1 DNA synthesis is coming into focus. The discovery that the antiviral activity of SAMHD1 in MDMs and DCs can be explained by its ability to reduce dNTP concentrations below a level that will support viral DNA synthesis demonstrates that dNTP pools may potentially be manipulated by antiviral drugs. Intracellular concentrations of the activated form of NRTIs are often limited at the level of the first phosphorylation step, and the ability to deliver nucleoside phosphonates to the intracellular compartment allows us to bypass the initial kinase step and obtain much higher intracellular concentrations of the triphosphate analog.

Despite years of research on HIV-1 RT and the development of many important drugs in this class of antiviral inhibitors, our knowledge is far from complete. When HIV-1 infects a target cell, it enters in a cage that is created by the viral capsid proteins, and there is evidence that this protected environment persists for some time after infection (Arhel et al. 2007; Dismuke and Aiken 2006; Gao et al. 2008; Hulme et al. 2011; Schaller et al. 2011). Natural dNTPs and activated forms of NRTIs and NtRTIs penetrate into this structure, but we do not know their concentrations at the site of nucleotide incorporation. In addition, excision rescue of chain-terminated DNA chains likely occurs within this structure, and we do not know the concentrations of excision acceptor substrates in this environment. Repair of dUMP that is created by deamination of newly synthesized DNA may be initiated by uracil-DNA glycosidase that is packaged into the virus particles during virus assembly. Alternatively, repair of dUMP and ribonucleotides in the viral DNA may have to wait until uncoating is complete and possibly until the preintegration complex has entered into the nucleus. Resolving these steps in viral replication will be technically challenging, but our better understanding of the early stages of viral infection may be key to further development of RT as a drug target.

**Acknowledgments** G.L. and A. J. A-H were supported by predoctoral fellowships from the American Heart Association (2280106 and 0615079B). This work was supported by the US Public Health Service grant AI-39973 to W.A.S. and the University of Miami Developmental Center for AIDS Research (P30-AI-073961).

## References

Abbondanzieri EA, Bokinsky G, Rausch JW, Zhang JX, Le Grice SFJ, Zhuang X (2008) Dynamic binding orientations direct activity of HIV reverse transcriptase. Nature 453(7192):184–189. doi:10.1038/nature06941

Acosta-Hoyos AJ, Scott WA (2010) The role of nucleotide excision by reverse transcriptase in HIV drug resistance. Viruses 2(2):372–394

Acosta-Hoyos AJ, Matsuura SE, Meyer PR, Scott WA (2012) A role of template cleavage in reduced excision of chain-terminating nucleotides by human immunodeficiency virus type 1 reverse transcriptase containing the M184V mutation. J Virol 86(9):5122–5133. doi:10.1128/JVI.05767-11

Arhel NJ, Souquere-Besse S, Munier S, Souque P, Guadagnini S, Rutherford S et al (2007) HIV-1 DNA flap formation promotes uncoating of the pre- integration complex at the nuclear pore. EMBO J 26(12):3025–3037. doi:10.1038/sj.emboj.7601740

Arion D, Kaushik N, McCormick S, Borkow G, Parniak MA (1998) Phenotypic mechanism of HIV-1 resistance to 3′-azido-3′-deoxythymidine (AZT): increased polymerization processivity and enhanced sensitivity to pyrophosphate of the mutant viral reverse transcriptase. Biochemistry 37(45):15908–15917

Bakshi RP, Hamzeh F, Frank I, Eron JJ Jr, Bosch RJ, Rosenkranz SL et al (2007) Effect of hydroxyurea and dideoxyinosine on intracellular 3′- deoxyadenosine-5′-triphosphate concentrations in HIV-infected patients. AIDS Res Hum Retroviruses 23(11):1360–1365. doi:10.1089/aid.2007.0078

Balzarini J, Herdewijn P, De Clercq E (1989) Differential patterns of intracellular metabolism of 2′,3′-didehydro-2′3′-dideoxythymidine and 3′-azido-2′,3′-dideoxythymidine, two potent anti-human immunodeficiency virus compounds. J Biol Chem 264(11):6127–6133

Borroto-Esoda K, Myrick F, Feng J, Jeffrey J, Furman P (2004) In vitro combination of amdoxovir and the inosine monophosphate dehydrogenase inhibitors mycophenolic acid and ribavirin demonstrates potent activity against wild-type and drug-resistant variants of human immunodeficiency virus type 1. Antimicrob Agents Chemother 48(11):4387–4394. doi:10.1128/AAC.48.11.4387-4394.2004

Boucher CAB, Cammack N, Schipper P, Schuurman R, Rouse P, Wainberg MA, Cameron JM (1993) High-level resistance to (-) enantiomeric 2′-deoxy-3′-thiacytidine in vitro is due to one amino acid substitution in the catalytic site of human immunodeficiency virus type 1 reverse transcriptase. Antimicrob Agents Chemother 37(10):2231–2234

Boyer PL, Sarafianos SG, Arnold E, Hughes SH (2001) Selective excision of AZTMP by drug-resistant human immunodeficiency virus reverse transcriptase. J Virol 75(10):4832–4842

Boyer PL, Vu BC, Ambrose Z, Julias JG, Warnecke S, Liao C et al (2009) The nucleoside analogue D-carba T blocks HIV-1 reverse transcription. J Med Chem 52(17):5356–5364

Brehm JH, Mellors JW, Sluis-Cremer N (2008) Mechanism by which a glutamine to leucine substitution at residue 509 in the ribonuclease H domain of HIV-1 reverse transcriptase confers zidovudine resistance. Biochemistry 47(52):14020–14027

Brehm JH, Scott Y, Koontz DL, Perry S, Hammer S, Katzenstein D, AIDS Clinical Trials Group Study 175 Protocol Team et al (2012) Zidovudine (AZT) monotherapy selects for the A360V mutation in the connection domain of HIV-1 reverse transcriptase. PLoS One 7(2):e31558. doi:10.1371/journal.pone.0031558

Chamberlain PP, Ren J, Nichols CE, Douglas L, Lennerstrand J, Larder BA et al (2002) Crystal structures of zidovudine- or lamivudine-resistant human immunodeficiency virus type 1 reverse transcriptases containing mutations at codons 41, 184, and 215. J Virol 76(19):10015–10019

Chen R, Wang H, Mansky LM (2002) Roles of uracil-DNA glycosylase and dUTPase in virus replication. J Gen Virol 83(10):2339–2345

Chen R, Le Rouzic E, Kearney JA, Mansky LM, Benichou S (2004) Vpr-mediated incorporation of UNG2 into HIV-1 particles is required to modulate the virus mutation rate and for replication in macrophages. J Biol Chem 279(27):28419–28425. doi:10.1074/jbc.M403875200

Cihlar T, Ray AS (2010) Nucleoside and nucleotide HIV reverse transcriptase inhibitors: 25 years after zidovudine. Antiviral Res 85(1):39–58. doi:10.1016/j.antiviral.2009.09.014

Das K, Bandwar RP, White KL, Feng JY, Sarafianos SG, Tuske S et al (2009) Structural basis for the role of the K65R mutation in HIV-1 reverse transcriptase polymerization, excision antagonism, and tenofovir resistance. J Biol Chem 284(50):35092–35100

Das K, Martinez SE, Bauman JD, Arnold E (2012) HIV-1 reverse transcriptase complex with DNA and nevirapine reveals non-nucleoside inhibition mechanism. Nat Struct Mol Biol 19(2):253–259. doi:10.1038/nsmb.2223

De Clercq E (2011) The clinical potential of the acyclic (and cyclic) nucleoside phosphonates. The magic of the phosphonate bond. Biochem Pharmacol 82(2):99–109. doi:10.1016/j.bcp.2011.03.027

Delviks-Frankenberry KA, Nikolenko GN, Pathak VK (2010) The "connection" between HIV drug resistance and RNase H. Viruses 2(7):1476–1503. doi:10.3390/v2071476

Diamond TL, Roshal M, Jamburuthugoda VK, Reynolds HM, Merriam AR, Lee KY et al (2004) Macrophage tropism of HIV-1 depends on efficient cellular dNTP utilization by reverse transcriptase. J Biol Chem 279(49):51545–51553. doi:10.1074/jbc.M408573200

Ding J, Das K, Hsiou Y, Sarafianos SG, Clark AD Jr, Jacobo-Molina A et al (1998) Structure and functional implications of the polymerase active site region in a complex of HIV-1 RT with a double-stranded DNA template-primer and an antibody Fab fragment at 2.8 Å resolution. J Mol Biol 284(4):1095–1111. doi:10.1006/jmbi.1998.2208

Dismuke DJ, Aiken C (2006) Evidence for a functional link between uncoating of the human immunodeficiency virus type 1 core and nuclear import of the viral preintegration complex. J Virol 80(8):3712–3720. doi:10.1128/JVI.80.8.3712-3720.2006

Ehteshami M, Götte M (2008) Effects of mutations in the connection and RNase H domains of HIV-1 reverse transcriptase on drug susceptibility. AIDS Rev 10(4):224–235

Ehteshami M, Beilhartz GL, Scarth BJ, Tchesnokov EP, McCormick S, Wynhoven B et al (2008) Connection domain mutations N348I and A360V in HIV-1 reverse transcriptase enhance resistance to 3′-azido-3′-deoxythymidine through both RNase H- dependent and -independent mechanisms. J Biol Chem 283(32):22222–22232

Faletto MB, Miller WH, Garvey EP, St. Clair MH, Daluge SM, Good SS (1997) Unique intracellular activation of the potent anti-human immunodeficiency virus agent 1592U89. Antimicrob Agents Chemother 41(5):1099–1107

Frankel FA, Invernizzi CF, Oliveira M, Wainberg MA (2007) Diminished efficiency of HIV-1 reverse transcriptase containing the K65R and M184V drug resistance mutations. AIDS 21(6):665–675. doi:10.1097/QAD.0b013e3280187505

Gallant JE, Parish MA, Keruly JC, Moore RD (2005) Changes in renal function associated with tenofovir disoproxil fumarate treatment, compared with nucleoside reverse-transcriptase inhibitor treatment. Clin Infect Dis 40(8):1194–1198. doi:10.1086/428840

Gao W-Y, Shirasaka T, Johns DG, Broder S, Mitsuya H (1993) Differential phosphorylation of azidothymidine, dideoxycytidine, and dideoxyinosine in resting and activated peripheral blood mononuclear cells. J Clin Invest 91(5):2326–2333

Gao H-Q, Boyer PL, Sarafianos SG, Arnold E, Hughes SH (2000) The role of steric hindrance in 3TC resistance of human immunodeficiency virus type-1 reverse transcriptase. J Mol Biol 300(2):403–418

Gao Y, Lobritz MA, Roth J, Abreha M, Nelson KN, Nankya I et al (2008) Targets of small interfering RNA restriction during human immunodeficiency virus type 1 replication. J Virol 82(6):2938–2951. doi:10.1128/JVI.02126-07

García-Lerma JG, Aung W, Cong M-e, Zheng Q, Youngpairoj AS, Mitchell J, Heneine W (2011) Natural substrate concentrations can modulate the prophylactic efficacy of nucleotide HIV reverse transcriptase inhibitors. J Virol 85(13):6610–6617. doi:10.1128/JVI.00311-11

Goldstone DC, Ennis-Adeniran V, Hedden JJ, Groom HCT, Rice GI, Christodoulou E et al (2011) HIV-1 restriction factor SAMHD1 is a deoxynucleoside triphosphate triphosphohydrolase. Nature 480(7377):379–382. doi:10.1038/nature10623

Götte M (2006) Effects of nucleotides and nucleotide analogue inhibitors of HIV-1 reverse transcriptase in a ratchet model of polymerase translocation. Curr Pharm Des 12(15):1867–1877

Goujon C, Rivière L, Jarrosson-Wuilleme L, Bernaud J, Rigal D, Darlix J-L, Cimarelli A (2007) $SIV_{SM}$/HIV-2 Vpx proteins promote retroviral escape from a proteasome- dependent restriction pathway present in human dendritic cells. Retrovirology 4:2. doi:10.1186/1742-4690-4-2

Harris RS, Bishop KN, Sheehy AM, Craig HM, Petersen-Mahrt SK, Watt IN, Malim MH (2003) DNA deamination mediates innate immunity to retroviral infection. Cell 113(6):803--809

Hostomsky Z, Hostomska Z, Fu T-B, Taylor J (1992) Reverse transcriptase of human immunodeficiency virus type 1: functionality of subunits of the heterodimer in DNA synthesis. J Virol 66(5):3179–3182

Hrecka K, Hao C, Gierszewska M, Swanson SK, Kesik-Brodacka M, Srivastava S, Skowronski J (2011) Vpx relieves inhibition of HIV-1 infection of macrophages mediated by the SAMHD1 protein. Nature 474(7353):658–661. doi:10.1038/nature10195

Huang H, Chopra R, Verdine GL, Harrison SC (1998) Structure of a covalently trapped catalytic complex of HIV-1 reverse transcriptase: implications for drug resistance. Science 282(5394):1669–1675

Hulme AE, Perez O, Hope TJ (2011) Complementary assays reveal a relationship between HIV-1 uncoating and reverse transcription. Proc Natl Acad Sci USA 108(24):9975–9980. doi:10.1073/pnas.1014522108

Jacobo-Molina A, Ding J, Nanni RG, Clark AD Jr, Lu X, Tantillo C et al (1993) Crystal structure of human immunodeficiency virus type 1 reverse transcriptase complexed with double-stranded DNA at 3.0 Å resolution shows bent DNA. Proc Natl Acad Sci USA 90(13):6320–6324

Jamburuthugoda VK, Chugh P, Kim B (2006) Modification of human immunodeficiency virus type 1 reverse transcriptase to target cells with elevated cellular dNTP concentrations. J Biol Chem 281(19):13388–13395. doi:10.1074/jbc.M600291200

Jamburuthugoda VK, Santos-Velazquez JM, Skasko M, Operario DJ, Purohit V, Chugh P et al (2008) Reduced dNTP binding affinity of 3TC-resistant M184I HIV-1 reverse transcriptase variants responsible for viral infection failure in macrophage. J Biol Chem 283(14):9206–9216. doi:10.1074/jbc.M710149200

Johnson MA, Fridland A (1989) Phosphorylation of 2′,3′-dideoxyinosine by cytosolic 5′- nucleotidase of human lymphoid cells. Mol Pharmacol 36(2):291–295

Kaiser SM, Emerman M (2006) Uracil DNA glycosylase is dispensable for human immunodeficiency virus type 1 replication and does not contribute to the antiviral effects of the cytidine deaminase Apobec3G. J Virol 80(2):875–882. doi:10.1128/JVI.80.2.875-882.2006

Kati WM, Johnson KA, Jerva LF, Anderson KS (1992) Mechanism and fidelity of HIV reverse transcriptase. J Biol Chem 267(36):25988–25997

Kaushik R, Zhu X, Stranska R, Wu Y, Stevenson M (2009) A cellular restriction dictates the permissivity of nondividing monocytes/macrophages to lentivirus and gammaretrovirus infection. Cell Host Microbe 6(1):68–80. doi:10.1016/j.chom.2009.05.022

Kellinger MW, Johnson KA (2010) Nucleotide-dependent conformational change governs specificity and analog discrimination by HIV reverse transcriptase. Proc Natl Acad Sci USA 107(17):7734–7739. doi:10.1073/pnas.0913946107

Kellinger MW, Johnson KA (2011) Role of induced fit in limiting discrimination against AZT by HIV reverse transcriptase. Biochemistry 50(22):5008–5015. doi:10.1021/bi200204m

Kennedy EM, Gavegnano C, Nguyen L, Slater R, Lucas A, Fromentin E et al (2010) Ribonucleoside triphosphates as substrate of human immunodeficiency virus type 1 reverse transcriptase in human macrophages. J Biol Chem 285(50):39380–39391. doi:10.1074/jbc.M110.178582

Kennedy EM, Daddacha W, Slater R, Gavegnano C, Fromentin E, Schinazi RF, Kim B (2011) Abundant non-canonical dUTP found in primary human macrophages drives its frequent incorporation by HIV-1 reverse transcriptase. J Biol Chem 286(28):25047–25055. doi:10.1074/jbc.M111.234047

Kennedy EM, Amie SM, Bambara RA, Kim B (2012) Frequent incorporation of ribonucleotides during HIV-1 reverse transcription and their attenuated repair in macrophages. J Biol Chem 287(17):14280–14288. doi:10.1074/jbc.M112.348482

Klarmann GJ, Chen X, North TW, Preston BD (2003) Incorporation of uracil into minus strand DNA affects the specificity of plus strand synthesis initiation during lentiviral reverse transcription. J Biol Chem 278(10):7902–7909. doi:10.1074/jbc.M207223200

Kohlstaedt LA, Wang J, Friedman JM, Rice PA, Steitz TA (1992) Crystal structure at 3.5 Å resolution of HIV-1 reverse transcriptase complexed with an inhibitor. Science 256(5065):1783–1790

Krokan HE, Drabløs F, Slupphaug G (2002) Uracil in DNA–occurrence, consequences and repair. Oncogene 21(58):8935–8948. doi:10.1038/sj.onc.1205996

Laguette N, Sobhian B, Casartelli N, Ringeard M, Chable-Bessia C, Ségéral E et al (2011) SAMHD1 is the dendritic- and myeloid-cell-specific HIV-1 restriction factor counteracted by Vpx. Nature 474(7353):654–657. doi:10.1038/nature10117

Lahouassa H, Daddacha W, Hofmann H, Ayinde D, Logue EC, Dragin L et al (2012) SAMHD1 restricts the replication of human immunodeficiency virus type 1 by depleting the intracellular pool of deoxynucleoside triphosphates. Nature Immunol 13(3):223–228. doi:10.1038/ni.2236

Lanier ER, Ptak RG, Lampert BM, Keilholz L, Hartman T, Buckheit RW Jr et al (2010) Development of hexadecyloxypropyl tenofovir (CMX157) for treatment of infection caused by wild-type and nucleoside/nucleotide-resistant HIV. Antimicrob Agents Chemother 54(7):2901–2909

Lansdon EB, Samuel D, Lagpacan L, Brendza KM, White KL, Hung M et al (2010) Visualizing the molecular interactions of a nucleotide analog, GS-9148, with HIV-1 reverse transcriptase-DNA complex. J Mol Biol 397(4):967–978. doi:10.1016/j.jmb.2010.02.019

Larder BA, Kemp SD, Harrigan PR (1995) Potential mechanism for sustained antiretroviral efficacy of AZT-3TC combination therapy. Science 269(5224):696–699

Le Grice SFJ (1993) Human immunodeficiency virus reverse transcriptase. In: Skalka AM, Goff SP (eds) Reverse Transcriptase. Ch. 9. Cold Spring Harbor Laboratory Press, Cold Spring Harbor, pp. 163-–191

Le Grice SFJ, Naas T, Wohlgensinger B, Schatz O (1991) Subunit-selective mutagenesis indicates minimal polymerase activity in heterodimer-associated p51 HIV-1 reverse transcriptase. EMBO J 10(12):3905–3911

Lee WA, He G-X, Eisenberg E, Cihlar T, Swaminathan S, Mulato A, Cundy KC (2005) Selective intracellular activation of a novel prodrug of the human immunodeficiency virus reverse transcriptase inhibitor tenofovir leads to preferential distribution and accumulation in lymphatic tissue. Antimicrob Agents Chemother 49(5):1898–1906. doi:10.1128/AAC.49.5.1898-1906.2005

Lisziewicz J, Foli A, Wainberg M, Lori F (2003) Hydroxyurea in the treatment of HIV infection: clinical efficacy and safety concerns. Drug Saf 26(9):605–624

Maga G, Radi M, Gerard M-A, Botta M, Ennifar E (2010) HIV-1 RT inhibitors with a novel mechanism of action: NNRTIs that compete with the nucleotide substrate. Viruses 2(4):880–899

Malim MH (2009) APOBEC proteins and intrinsic resistance to HIV-1 infection. Philos Trans R Soc Lond B Biol Sci 364(1517):675–687. doi:10.1098/rstb.2008.0185

Mansky LM, Preveral S, Selig L, Benarous R, Benichou S (2000) The interaction of vpr with uracil DNA glycosylase modulates the human immunodeficiency virus type 1 In vivo mutation rate. J Virol 74(15):7039–7047

Marchand B, Götte M (2003) Site-specific footprinting reveals differences in the translocation status of HIV-1 reverse transcriptase. Implications for polymerase translocation and drug resistance. J Biol Chem 278(37):35362–35372

Margolis DM, Kewn S, Coull JJ, Ylisastigui L, Turner D, Wise H et al (2002) The addition of mycophenolate mofetil to antiretroviral therapy including abacavir is associated with depletion of intracellular deoxyguanosine triphosphate and a decrease in plasma HIV-1 RNA. J Acquir Immune Defic Syndr 31(1):45–49

Markowitz M, Zolopa A, Ruane P, Squires K, Zhong L, Kearney B, Lee W (2011) GS-7340 demonstrates greater declines in HIV-1 RNA than TDF during 14 days of monotherapy in HIV-1-infected subjects. Paper presented at the 18th conference on retroviruses and opportunistic infections. Abstract retrieved from http://www.retroconference.org/2011/Abstracts/42549.htm

Menéndez-Arias L (2008) Mechanisms of resistance to nucleoside analogue inhibitors of HIV-1 reverse transcriptase. Virus Res 134(1–2):124–146

Meyer PR, Matsuura SE, So AG, Scott WA (1998) Unblocking of chain-terminated primer by HIV-1 reverse transcriptase through a nucleotide-dependent mechanism. Proc Natl Acad Sci USA 95(23):13471–13476

Meyer PR, Matsuura SE, Mian AM, So AG, Scott WA (1999) A mechanism of AZT resistance: an increase in nucleotide-dependent primer unblocking by mutant HIV-1 reverse transcriptase. Mol Cell 4(1):35–43

Meyer PR, Matsuura SE, Schinazi RF, So AG, Scott WA (2000) Differential removal of thymidine nucleotide analogues from blocked DNA chains by human immunodeficiency virus reverse transcriptase in the presence of physiological concentrations of 2′-deoxynucleoside triphosphates. Antimicrob Agents Chemother 44(12):3465–3472

Meyer PR, Rutvisuttinunt W, Matsuura SE, So AG, Scott WA (2007) Stable complexes formed by HIV-1 reverse transcriptase at distinct positions on the primer- template controlled by binding deoxynucleoside triphosphates or foscarnet. J Mol Biol 369(1):41–54

Müller B, Restle T, Reinstein J, Goody RS (1991) Interaction of fluorescently labeled dideoxynucleotides with HIV-1 reverse transcriptase. Biochemistry 30(15):3709–3715

Naesens L, Bischofberger N, Augustijns P, Annaert P, Van den Mooter G, Arimilli MN et al (1998) Antiretroviral efficacy and pharmacokinetics of oral bis(isopropyloxycarbonyloxymethyl)-9-(2-phosphonylmethoxypropyl)adenine in mice. Antimicrob Agents Chemother 42(7):1568–1573

Navarro F, Landau NR (2004) Recent insights into HIV-1 Vif. Curr Opin Immunol 16(4):477–482. doi:10.1016/j.coi.2004.05.006

Nikolenko GN, Palmer S, Maldarelli F, Mellors JW, Coffin JM, Pathak VK (2005) Mechanism for nucleoside analog-mediated abrogation of HIV-1 replication: balance between RNase H activity and nucleotide excision. Proc Natl Acad Sci USA 102(6):2093–2098

Nikolenko GN, Delviks-Frankenberry KA, Palmer S, Maldarelli F, Fivash MJ Jr, Coffin JM, Pathak VK (2007) Mutations in the connection domain of HIV-1 reverse transcriptase increase 3′-azido-3′-deoxythymidine resistance. Proc Natl Acad Sci USA 104(1):317–322. doi:10.1073/pnas.0609642104

Painter GR, Almond MR, Trost LC, Lampert BM, Neyts J, De Clercq E et al (2007) Evaluation of hexadecyloxypropyl-9-R-[2- (Phosphonomethoxy)propyl]-adenine, CMX157, as a potential treatment for human immunodeficiency virus type 1 and hepatitis B virus infections. Antimicrob Agents Chemother 51(10):3505–3509

Parikh UM, Barnas DC, Faruki H, Mellors JW (2006) Antagonism between the HIV-1 reverse-transcriptase mutation K65R and thymidine-analogue mutations at the genomic level. J Infect Dis 194(5):651–660

Parkin NT, Hellmann NS, Whitcomb JM, Kiss L, Chappey C, Petropoulos CJ (2004) Natural variation of drug susceptibility in wild-type human immunodeficiency virus type 1. Antimicrob Agents Chemother 48(2):437–443

Pastor-Anglada M, Cano-Soldado P, Molina-Arcas M, Lostao MP, Larráyoz I, Martínez- Picado J, Casado FJ (2005) Cell entry and export of nucleoside analogues. Virus Res 107(2):151–164

Perez-Bercoff D, Wurtzer S, Compain S, Benech H, Clavel F (2007) Human immunodeficiency virus type 1: resistance to nucleoside analogues and replicative capacity in primary human macrophages. J Virol 81(9):4540–4550

Petrella M, Wainberg MA (2002) Might the M184V substitution in HIV-1 RT confer clinical benefit? AIDS Rev 4(4):224–232

Petropoulos CJ, Parkin NT, Limoli KL, Lie YS, Wrin T, Huang W et al (2000) A novel phenotypic drug susceptibility assay for human immunodeficiency virus type 1. Antimicrob Agents Chemother 44(4):920–928

Powell RD, Holland PJ, Hollis T, Perrino FW (2011) Aicardi-Goutières syndrome gene and HIV-1 restriction factor SAMHD1 is a dGTP-regulated deoxynucleotide triphosphohydrolase. J Biol Chem 286(51):43596–43600. doi:10.1074/jbc.C111.317628

Priet S, Gros N, Navarro J-M, Boretto J, Canard B, Quérat G, Sire J (2005) HIV-1- associated uracil DNA glycosylase activity controls dUTP misincorporation in viral DNA and is essential to the HIV-1 life cycle. Mol Cell 17(4):479–490. doi:10.1016/j.molcel.2005.01.016

Radzio J, Sluis-Cremer N (2008) Efavirenz accelerates HIV-1 reverse transcriptase ribonuclease H cleavage, leading to diminished zidovudine excision. Mol Pharmacol 73(2):601–606

Ray AS, Hostetler KY (2011) Application of kinase bypass strategies to nucleoside antivirals. Antiviral Res 92(2):277–291. doi:10.1016/j.antiviral.2011.08.015

Robbins BL, Srinivas RV, Kim C, Bischofberger N, Fridland A (1998) Anti-human immunodeficiency virus activity and cellular metabolism of a potential prodrug of the acyclic nucleoside phosphonate 9-R-(2-phosphonomethoxypropyl)adenine (PMPA), Bis (isopropyloxymethylcarbonyl)PMPA. Antimicrob Agents Chemother 42(3):612–617

Rodgers DW, Gamblin SJ, Harris BA, Ray S, Culp JS, Hellmig B et al (1995) The structure of unliganded reverse transcriptase from the human immunodeficiency virus type 1. Proc Natl Acad Sci USA 92(4):1222–1226

Roy B, Beuneu C, Roux P, Buc H, Lemaire G, Lepoivre M (1999) Simultaneous determination of pyrimidine or purine deoxyribonucleoside triphosphates using a polymerase assay. Anal Biochem 269(2):403–409. doi:10.1006/abio.1999.4051

Rutvisuttinunt W, Meyer PR, Scott WA (2008) Interactions between HIV-1 reverse transcriptase and the downstream template strand in stable complexes with primer- template. PLoS One 3(10):e3561

Sarafianos SG, Das K, Clark AD Jr, Ding J, Boyer PL, Hughes SH, Arnold E (1999) Lamivudine (3TC) resistance in HIV-1 reverse transcriptase involves steric hindrance with β-branched amino acids. Proc Natl Acad Sci USA 96(18):10027–10032

Sarafianos SG, Clark AD Jr, Das K, Tuske S, Birktoft JJ, Ilankumaran P et al (2002) Structures of HIV-1 reverse transcriptase with pre- and post- translocation AZTMP-terminated DNA. EMBO J 21(23):6614–6624

Sarafianos SG, Clark AD Jr, Tuske S, Squire CJ, Das K, Sheng D et al (2003) Trapping HIV-1 reverse transcriptase before and after translocation on DNA. J Biol Chem 278(18): 16280–16288

Sarafianos SG, Marchand B, Das K, Himmel DM, Parniak MA, Hughes SH, Arnold E (2009) Structure and function of HIV-1 reverse transcriptase: molecular mechanisms of polymerization and inhibition. J Mol Biol 385(3):693–713

Schaller T, Ocwieja KE, Rasaiyaah J, Price AJ, Brady TJ, Roth SL et al (2011) HIV-1 capsid-cyclophilin interactions determining nuclear import pathway, integration targeting and replication efficiency. PLoS Pathog 7(12):e1002439. doi:10.1371/journal.ppat.1002439

Schinazi RF, Lloyd RM Jr, Nguyen M-H, Cannon DL, McMillan A, Ilksoy N et al (1993) Characterization of human immunodeficiency viruses resistant to oxathiolane-cytosine nucleosides. Antimicrob Agents Chemother 37(4):875–881

Schuurman R, Nijhuis M, van Leeuwen R, Schipper P, de Jong D, Collis P et al (1995) Rapid changes in human immunodeficiency virus type 1 RNA load and appearance of drug-resistant virus populations in persons treated with lamivudine (3TC). J Infect Dis 171(6):1411–1419

Sharma B, Kaushik N, Singh K, Kumar S, Pandey VN (2002) Substitution of conserved hydrophobic residues in motifs B and C of HIV-1 RT alters the geometry of its catalytic pocket. Biochemistry 41(52):15685–15697

Sheehy AM, Gaddis NC, Choi JD, Malim MH (2002) Isolation of a human gene that inhibits HIV-1 infection and is suppressed by the viral Vif protein. Nature 418(6898):646–650. doi:10.1038/nature00939

Shirasaka T, Kavlick MF, Ueno T, Gao W-Y, Kojima E, Alcaide ML et al (1995) Emergence of human immunodeficiency virus type 1 variants with resistance to multiple dideoxynucleosides in patients receiving therapy with dideoxynucleosides. Proc Natl Acad Sci USA 92(6): 2398–2402

Singh K, Marchand B, Kirby KA, Michailidis E, Sarafianos SG (2010) Structural aspects of drug resistance and Inhibition of HIV-1 reverse transcriptase. Viruses 2(2):606–638

Sire J, Quérat G, Esnault C, Priet S (2008) Uracil within DNA: an actor of antiviral immunity. Retrovirology 5:45. doi:10.1186/1742-4690-5-45

Sluis-Cremer N, Koontz D, Bassit L, Hernandez-Santiago BI, Detorio M, Rapp KL et al (2009) Anti-human immunodeficiency virus activity, cross-resistance, cytotoxicity, and intracellular pharmacology of the 3′-azido-2′,3′-dideoxypurine nucleosides. Antimicrob Agents Chemother 53(9):3715–3719

Smith AJ, Scott WA (2006) The influence of natural substrates and inhibitors on the nucleotide-dependent excision activity of HIV-1 reverse transcriptase in the infected cell. Curr Pharm Des 12(15):1827–1841

Smith AJ, Meyer PR, Asthana D, Ashman MR, Scott WA (2005) Intracellular substrates for the primer-unblocking reaction by human immunodeficiency virus type 1 reverse transcriptase: detection and quantitation in extracts from quiescent- and activated-lymphocyte subpopulations. Antimicrob Agents Chemother 49(5):1761–1769

Tisdale M, Kemp SD, Parry NR, Larder BA (1993) Rapid in vitro selection of human immunodeficiency virus type 1 resistant to 3′-thiacytidine inhibitors due to a mutation in the YMDD region of reverse transcriptase. Proc Natl Acad Sci USA 90(12):5653–5656

Tong W, Lu C-D, Sharma SK, Matsuura S, So AG, Scott WA (1997) Nucleotide- induced stable complex formation by HIV-1 reverse transcriptase. Biochemistry 36(19):5749–5757

Tu X, Das K, Han Q, Bauman JD, Clark AD Jr, Hou X et al (2010) Structural basis of HIV-1 resistance to AZT by excision. Nat Struct Mol Biol 17(10):1202–1209

Tuske S, Sarafianos SG, Clark AD Jr, Ding J, Naeger LK, White KL et al (2004) Structures of HIV-1 RT-DNA complexes before and after incorporation of the anti-AIDS drug tenofovir. Nat Struct Mol Biol 11(5):469–474

Van Cor-Hosmer SK, Daddacha W, Kim B (2010) Mechanistic interplay among the M184I HIV-1 reverse transcriptase mutant, the central polypurine tract, cellular dNTP concentrations and drug sensitivity. Virology 406(2):253–260. doi:10.1016/j.virol.2010.07.028

Van Cor-Hosmer SK, Daddacha W, Kelly Z, Tsurumi A, Kennedy EM, Kim B (2012) The impact of molecular manipulation in residue 114 of human immunodeficiency virus type-1 reverse transcriptase on dNTP substrate binding and viral replication. Virology 422(2):393–401. doi:10.1016/j.virol.2011.11.004

Van Rompay AR, Johansson M, Karlsson A (2000) Phosphorylation of nucleosides and nucleoside analogs by mammalian nucleoside monophosphate kinases. Pharmacol Ther 87(2–3):189–198

Van Rompay AR, Johansson M, Karlsson A (2003) Substrate specificity and phosphorylation of antiviral and anticancer nucleoside analogues by human deoxyribonucleoside kinases and ribonucleoside kinases. Pharmacol Ther 100(2):119–139

Vazquez-Padua MA, Kunugi K, Risueno C, Fischer PH (1989) Modulation of the feedback regulation of thymidine kinase activity by pH in 647 V cells. Cancer Res 49(20):5644–5649

Whitcomb JM, Parkin NT, Chappey C, Hellmann NS, Petropoulos CJ (2003) Broad nucleoside reverse-transcriptase inhibitor cross-resistance in human immunodeficiency virus type 1 clinical isolates. J Infect Dis 188(7):992–1000. doi:10.1086/378281

White KL, Chen JM, Feng JY, Margot NA, Ly JK, Ray AS et al (2006) The K65R reverse transcriptase mutation in HIV-1 reverses the excision phenotype of zidovudine resistance mutations. Antivir Ther 11(2):155–163

Willetts KE, Rey F, Agostini I, Navarro J-M, Baudat Y, Vigne R, Sire J (1999) DNA repair enzyme uracil DNA glycosylase is specifically incorporated into human immunodeficiency virus type 1 viral particles through a Vpr-independent mechanism. J Virol 73(2):1682–1688

Wurtzer S, Compain S, Benech H, Hance AJ, Clavel F (2005) Effect of cell cycle arrest on the activity of nucleoside analogues against human immunodeficiency virus type 1. J Virol 79(23):14815–14821. doi:10.1128/JVI.79.23.14815-14821.2005

Yan N, O'Day E, Wheeler LA, Engelman A, Lieberman J (2011) HIV DNA is heavily uracilated, which protects it from autointegration. Proc Natl Acad Sci USA 108(22):9244–9249. doi:10.1073/pnas.1102943108

Yang B, Chen K, Zhang C, Huang S, Zhang H (2007) Virion-associated uracil DNA glycosylase-2 and apurinic/apyrimidinic endonuclease are involved in the degradation of APOBEC3G-edited nascent HIV-1 DNA. J Biol Chem 282(16):11667–11675. doi:10.1074/jbc.M606864200

Yap S-H, Sheen C-W, Fahey J, Zanin M, Tyssen D, Lima VD et al (2007) N348I in the connection domain of HIV-1 reverse transcriptase confers zidovudine and nevirapine resistance. PLoS Med 4(12):e335. doi:10.1371/journal.pmed.0040335

# Chapter 6
# Nonnucleoside Reverse Transcriptase Inhibitors (NNRTIs)

**Kalyan Das, Eddy Arnold, and Stephen H. Hughes**

## 6.1 Introduction

HIV is a retrovirus; the retroviral life cycle is characterized by two specific steps: (1) the conversion of the single-stranded RNA genome found in the virion into double-stranded DNA by the viral enzyme reverse transcriptase (RT) and (2) the subsequent insertion of this DNA copy into the host genome by the viral enzyme integrase. Both of these steps are essential for the viral life cycle; both are targets for important anti-HIV drugs. There are two broad classes of drugs that block reverse transcription: nucleoside/nucleotide analogs (NRTIs) and nonnucleoside inhibitors (NNRTIs). The conversion of the RNA genome into DNA involves the collaboration of the two enzymatic activities of RT, a DNA polymerase that can copy either an RNA or a DNA substrate, and an RNase H that cleaves RNA if, and only if, it is part of an RNA/DNA duplex (see Chap. 1). The first anti-HIV drugs to be identified and used in clinic were NRTIs (see Chap. 2); the second class were the NNRTIs. NNRTIs bind to RT and allosterically block its polymerase activity. NRTIs and NNRTIs are both used extensively to treat HIV-1 infected patients in the highly active antiviral therapies (HAART); both classes of drugs hold the promise that they can be used to reduce the transmission of the virus (Chap. 15). Because HIV

K. Das, Ph.D. • E. Arnold, Ph.D.
Department of Chemistry & Chemical Biology, CABM, Rutgers University, Room 016, 679 Hoes Lane, Piscataway, NJ 08854, USA
e-mail: kalyan@cabm.rutgers.edu; arnold@cabm.rutgers.edu

S.H. Hughes, Ph.D. (✉)
HIV Drug Resistance Program, National Cancer Institute, P.O. Box B, Bldg. 539, Rm. 130A, Frederick, MD 21702-1201, USA
e-mail: hughesst@mail.nih.gov

S.F.J. LeGrice and M. Götte (eds.), *Human Immunodeficiency Virus Reverse Transcriptase: A Bench-to-Bedside Success*, DOI 10.1007/978-1-4614-7291-9_6, 

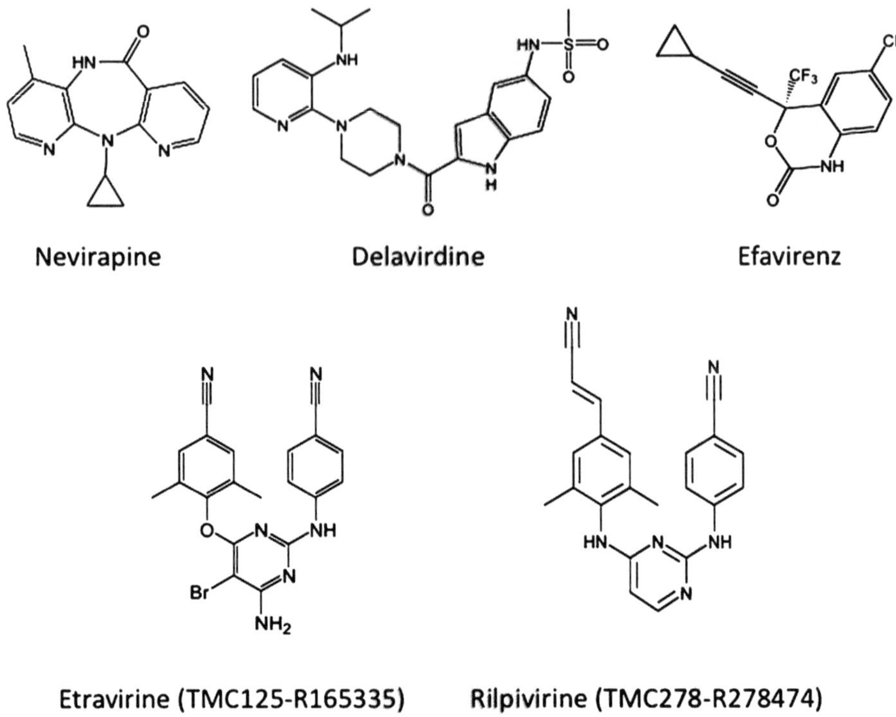


**Fig. 6.1** The five NNRTIs approved for treating HIV-1 infections

infections cannot be cured by drug treatment, patients who begin drug therapy are usually on therapy for the rest of their lives. This makes the problem of developing anti-HIV drugs particularly challenging and places special emphasis on two fundamental considerations: the toxicity of long-term therapy and the development of resistance. As will be discussed in more detail later, for NNRTIs, the development of resistance has been the larger issue; in contrast to NRTIs, the toxicity of most NNRTIs has been a lesser concern. This chapter will discuss NNRTIs and will be divided into three parts: (1) mechanism of action, (2) the problem of resistance, and (3) a brief look forward at what the future might hold, both in terms of the development of new NNRTIs and in their uses.

### *6.1.1 Mechanism of Action*

Although NNRTIs are chemically diverse, they share certain characteristics (being predominantly hydrophobic and often being comprised of two or three linked aromatic rings) (Fig. 6.1 shows five approved NNRTI drugs, nevirapine, efavirenz, delavirdine, etravirine, and rilpivirine). All NNRTIs bind to a flexible hydrophobic pocket in the palm subdomain of the p66 subunit, located about 10 Å away from the polymerase active site. NNRTIs are much more selective for HIV-1 than are NRTIs; most NNRTIs have little or no activity against HIV-2, and none are active against MLV. This is because the amino acids that form the hydrophobic pocket to which NNRTIs bind are not evolutionarily conserved (Barber et al. 1990). This lack of conservation is both a blessing and a curse. NNRTIs do not inhibit cellular DNA polymerases; this explains why NNRTIs are, in general, nontoxic. However, the fact that the binding pocket is not evolutionarily conserved is the root cause of the problems associated with the development of resistance, a topic that will be considered in the next section.

Crystal structures of HIV-1 RT have been determined in multiple conformational states/complexes including (1) unliganded RT (Hsiou et al. 1996); (2) RT/NNRTI complexes (Das et al. 2008, 1996; Ding et al. 1995; Hopkins et al. 1996; Hsiou et al. 1998; Kohlstaedt et al. 1992; Ren et al. 2002, 1995, 1999); (3) RT/nucleic acid binary complex (Jacobo-Molina et al. 1993; Sarafianos et al. 2002, 2001) and RT/DNA/dNTP (or dNTP analog) ternary complexes (Das et al. 2012; Huang et al. 1998; Lansdon et al. 2010; Tu et al. 2010; Tuske et al. 2004); and (4) an RT/DNA/NNRTI complex (Das et al. 2012). Comparing structures that do and do not have a bound NNRTI has been critical for our understanding of how NNRTIs bind to and inhibit RT; structures in which NNRTIs are bound to mutant forms of RT have been critical for understanding the mechanisms of resistance. RT structures that do not contain a bound NNRTI lack the NNRTI-binding pocket. A recent study showed that soaking nevirapine into preformed crystals of an RT/DNA complex induced the conformational changes in the pocket region required to accommodate nevirapine (Das et al. 2012) (Fig. 6.2). The major structural rearrangements include shifting of the β12–β13–β14 sheet that contains the DNA primer grip and flipping the side chains of the aromatic pocket residues Tyr181 and Tyr188. Binding of an NNRTI also shifts the position of the thumb subdomain of the p66 subunit, primarily by lifting the β12–β13–β14 sheet at the hinge between p66 palm and thumb; as will be discussed below, this is important for the mechanism of inhibition of the polymerase activity of RT. NNRTIs differ in both size and shape, and because the pocket region is flexible; the binding of different NNRTIs affects the size and shape of the binding pocket in different ways, an effect that has been referred to as a "shrink wrap" fit, because the shape of the pocket is, to some degree, dictated by the shape of the bound NNRTI. Although all of the first-generation NNRTIs bound to RT in a similar way, the so-called butterfly mode (Ding et al. 1995) with one wing of the butterfly interacting with two tyrosines (Y181 and Y188, Fig. 6.3), second-generation NNRTIs can bind in other conformations. Recently approved DAPY NNRTI drugs have a U shape; the rotatable bonds of these NNRTIs permit conformational flexibility, which allows them to bind in additional conformations (Fig. 6.4) (Das et al. 2004). The additional binding modes and intrinsic flexibility were important for the development of NNRTIs that are more broadly effective against some of the common NNRTI-resistant RTs; this will be discussed in more detail later, in the section on resistance.

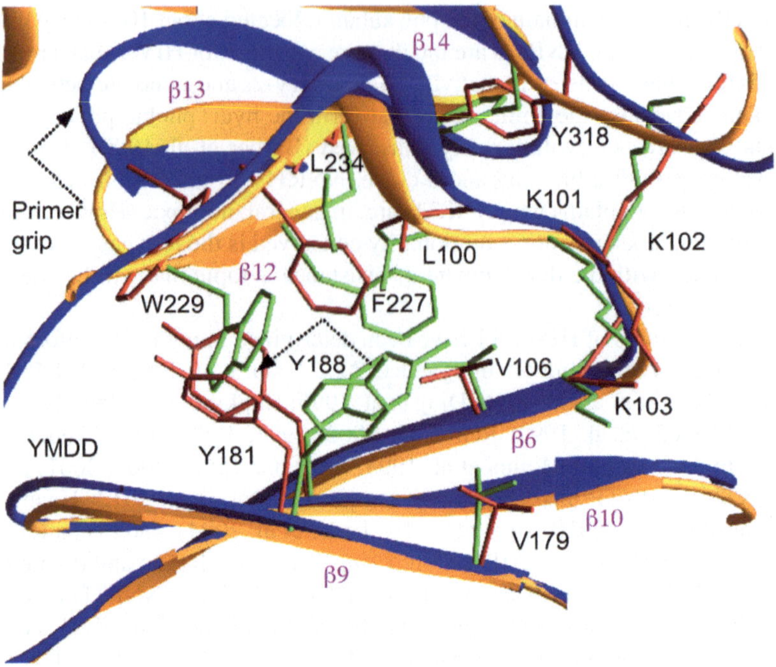


**Fig. 6.2** Changes in the vicinity of the NNRTI-binding pocket of HIV-1 RT that accompany NNRTI binding. The figure shows a *ribbon diagram* of the portion of RT near the NNRTI-binding pocket, and the positions of the side chains of some of the amino acids near the pocket are also shown. The position of the amino acid backbone (*ribbon*) in the absence of a bound NNRTI is shown in *gold*, and the position in the presence of a bound NNRTI is shown in *blue*. The positions of the amino acid side chains in the absence of an NNRTI are shown in *green* and in its presence are shown in *red* (Figure from Hsiou et al. 1996)

Although it has been known for some time, from biochemical assays, that binding of an NNRTI blocks the chemical step of polymerization, without interfering with either the binding of the nucleic acid or the incoming dNTP (Rittinger et al. 1995; Spence et al. 1995), the exact mechanism of inhibition was unclear. The first structure of HIV-1 RT with a bound NNRTI was solved approximately 20 years ago (Kohlstaedt et al. 1992), and there are now a number of structures of RT/NNRTI complexes; however, until very recently, there were no structures of HIV-1 RT bound to both an NNRTI and a nucleic acid substrate. This problem has recently been solved (Das et al. 2012), and the structure provides a clear answer to this long-standing question. The β12–β13–β14 sheet of the p66 subunit of RT lies adjacent to the double-stranded nucleic acid substrate. When the binding of the NNRTI displaces the sheet, this, in turn, pushes the end of the primer strand away from the polymerase active site (by 5.5 Å). Thus, the nucleic acid and the incoming dNTP may still bind to RT, but, in the presence of an NNRTI, neither the primer 3'-end nor a dNTP is appropriately positioned at the active site for effective catalysis (Fig. 6.5). This structure also suggests an answer to another long-standing puzzle: Although

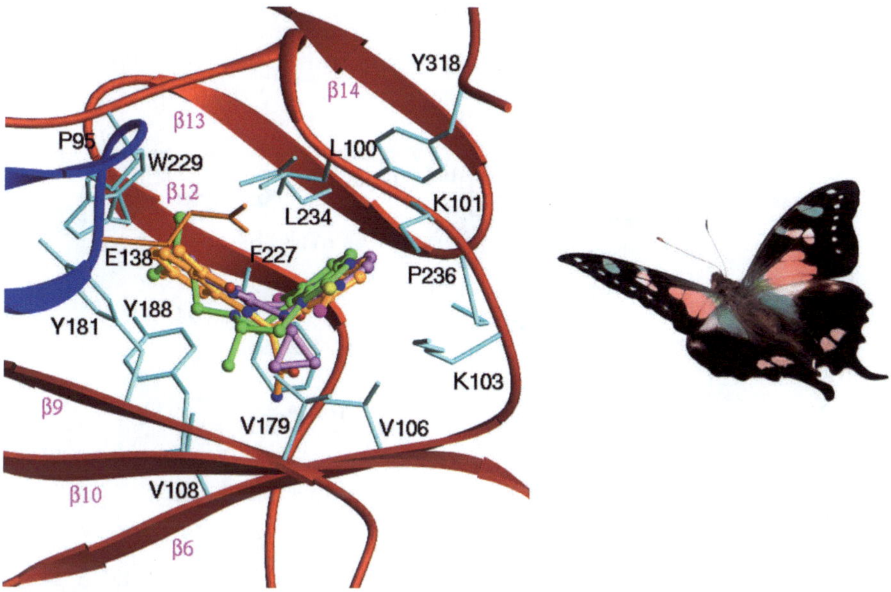


**Fig. 6.3** First-generation NNRTIs bind in a "butterfly-like" configuration. The figure shows the strikingly similar binding modes of three NNRTIs: TIBO (*green*), α-APA (*yellow*), and nevirapine (*purple*). The orientation of the pocket is similar to the orientation shown in Fig. 6.2; however, in Fig. 6.3, the loop of p51 that makes a small contribution to the formation of the NNRTI-binding pocket is shown in *blue*, and the rest of the pocket is formed by p66 (shown in *red*) (Figure from Ding et al. 1995)

the therapeutic effect of NNRTIs rests on their ability to block the polymerase activity of RT, some NNRTIs also have a more subtle, but measureable, effect on the RNase H activity of RT. This is because the binding of an NNRTI can affect the trajectory of the nucleic acid as it passes through RT (Das et al. 2012). Because the NNRTI-binding site is closer to the polymerase active site, the distortion is greater there, and the NNRTI binding affects the polymerase activity of RT to a greater extent than it affects the RNase H activity; however, less is not none, and there is good evidence that NNRTI binding can affect RNase H cleavage (Gopalakrishnan and Benkovic 1994; Palaniappan et al. 1995). Additionally, single-molecule experiments showed that binding of an NNRTI can affect the orientation of RT on the nucleic acid substrate, decreasing the amount of the polymerase-competent RT/nucleic acid complex (Abbondanzieri et al. 2008).

NNRTIs can also affect the stability and activity of some RT mutants. HIV-1 RT is a heterodimer, composed to two subunits, p66 and p51. Although there are retroviral RTs that appear to be active as monomers (e.g., MLV RT), HIV-1 RT must be a dimer to be active (Restle et al. 1990). For this reason, mutations that affect dimerization also affect the polymerase activity of RT. There are data to show that some NNRTIs can enhance the dimerization of some RT mutants and thus can, if present at just the right concentration, actually enhance the activity of some RT mutants

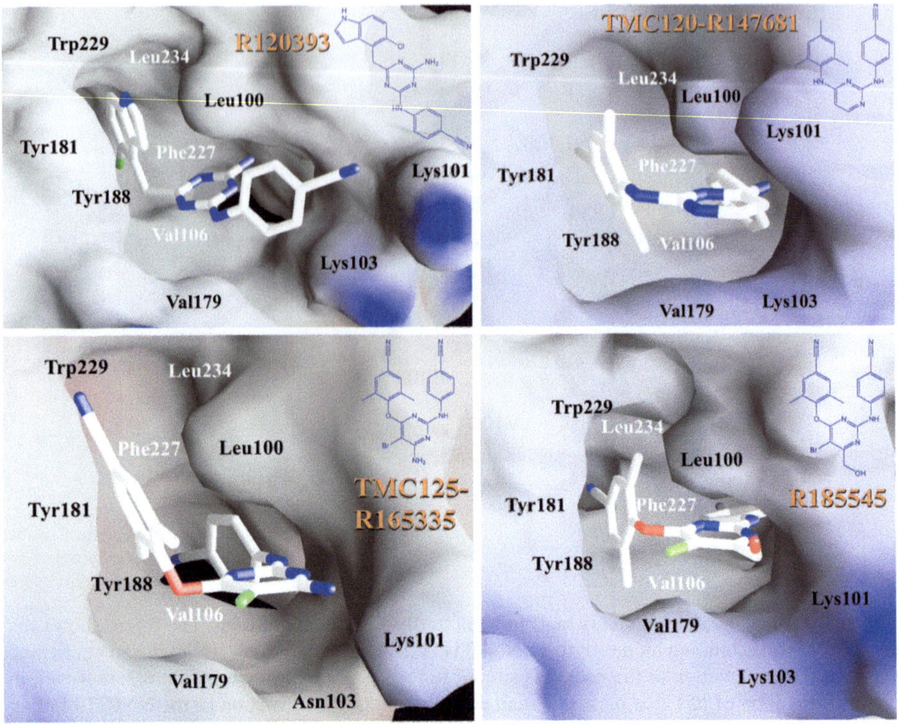


**Fig. 6.4** Flexible NNRTIs that share a common chemical scaffold can bind to HIV-1 RT in different conformations. The figure shows the binding of four related compounds in the series that includes etravirine (TMC-125) and dapivirine (TMC-120) bound to HIV-1 RT. The binding pocket is shown as a electrostatic potential surface (*blue* – positive charge, *red* – negative charge, and *gray* – hydrophobic parts of the surface), and the NNRTIs are shown as *stick diagrams* to illustrate that the orientation of the components of the four compounds differs from the butterfly configuration seen with first-generation inhibitors (see Fig. 6.3) and that these similar compounds can also bind in several different ways (From Das et al. 2004)

(Tachedjian and Goff 2003; Tachedjian et al. 2005, 2001). In the context of the viral replication, this process may be complex because, in an infected cell, RT is synthesized as part of the Gag-Pol polyprotein and the NNRTI may bind to the RT component of Gag-Pol and/or some processing intermediate, as well as the mature, heterodimeric RT. In a simple model assay system, presence of some NNRTIs increased the ability of some RT mutants to dimerize, presumably by helping the p66 subunit (and/or some of its precursors) to fold and associate with p51 (and/or its precursors). Intuitively, this makes sense; the NNRTIs can be thought of as stabilizing the hydrophobic core in ways that could assist the folding of the two subunits. However, a large NNRTI, TSAO, has a destabilizing effect on the HIV-1 RT heterodimer, presumably because of the magnitude of the distortion and expansion of the pocket region (Das et al. 2011).

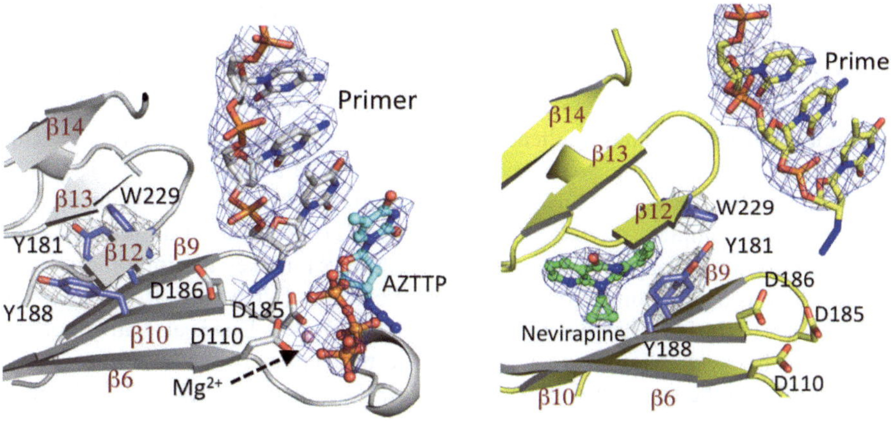


**Fig. 6.5** NNRTI binding affects the position of the end of the primer stand, blocking DNA synthesis. The figure shows the relationship of the side chains of the active site amino acids (110, 185, and 186) which chelate the essential $Mg^{2+}$ ions involved in catalysis to the end of the primer strand. In the absence of a bound NNRTI (*left panel*), the incoming triphosphate (in this case an AZTTP) is positioned near the catalytic triad and the end of the primer strand (in this case terminated with an AZTMP, which prevents the incorporation of the incoming AZTTP). The portion of the RT where an incoming NNRTI would bind is shown as a ribbon diagram and the β-strands are marked. The positions of the side chains of Y181, Y188, and W229 are also shown. The electron density for these side chains, for the incoming AZTTP, for the nucleic acid, and for nevirapine is shown as blue mesh. The right panel is similar, except that there is no incoming AZTTP at the active site, the 3′-end of the primer has moved away from the polymerase active site, and the NNRTI-binding pocket residues are repositioned by the binding of nevirapine to the RT/dsDNA complex (From Das et al. 2012)

## 6.2 Inhibition of Viral Replication

Thus far, we have only considered the ability of an NNRTI to affect the behavior and activity of a single molecule of RT. This is in some sense appropriate, because a single molecule of RT is, ultimately, the level at which an NNRTI acts. However, the true goal of NNRTI therapy is to block the replication of the HIV-1 virus, and NNRTIs are used therapeutically because they are able to block the replication of the virus in patients. Although it would seem, at first glance, that blocking the polymerase activity of RT and blocking viral replication are the same thing, this is an oversimplification. The complication is that there are approximately 50 RT heterodimers in an HIV-1 virion of which approximately half must be enzymatically active for efficient viral replication (Julias et al. 2001). This has several consequences. First, because relative to the amount of RT that is needed to complete the synthesis of the DNA copy of the viral genome, there is a modest excess of RT in the virion, an NNRTI has little impact on viral replication until the "excess" RTs are inhibited. Once the number of active RTs is reduced below the threshold, the inhibition curve is relatively steep because, despite the fact that the interaction of RT and an NNRTI is one-to-one, the inhibition of reverse transcription is multivalent. This is the result of the fact that blocking reverse transcription involves inhibiting a sufficient fraction of the RTs that are participating in the conversion of the viral RNA into DNA to stop the process of reverse transcription before it is completed. This affects the shape of the inhibition curve and has implications for finding a clinically relevant relationship between the $IC_{50}$ and the dose of the drug that should be used in patients to provide an optimal reduction in viral replication (Jilek et al. 2012). Although the underlying data are more complex, the basic idea is simple. The fact that, in an infected cell, reverse transcription is multivalent with respect to RT makes the inhibition curve steep, which is a good thing from the therapeutic point of view; the effective dose is much closer to the $IC_{50}$ than it is for a monovalent inhibitor. This is also an important distinction between NRTIs and NNRTIs. The true target of an NRTI is not the enzyme RT, but the viral DNA into which the NRTI is incorporated by RT, which is, when considered in this way, a monovalent, not a multivalent target.

The fact that reverse transcription involves multiple RTs also has implications for the effects of mutations that reduce the polymerase activity of RT. As the previous paragraph implies, the susceptibility of viral replication to inhibition by NNRTIs depends on the number of RTs present in a virion that can contribute to the synthesis of the viral DNA. Virions that have fewer enzymatically active RTs that can bind an NNRTI are more susceptible to inhibition by NNRTIs (Ambrose et al. 2006). Alternatively, mutations in RT that reduce its polymerase activity can increase the number of RTs that are needed to complete the synthesis of viral DNA. This, in turn, decreases the number of RTs that must be inhibited to block the synthesis of the viral DNA, which makes the virus more susceptible to inhibition by NNRTIs.

### *6.2.1 The Problem of Resistance*

Resistance is a problem for any anti-HIV drug; NNRTIs are not an exception. As has already been mentioned, the NNRTI-binding site is not evolutionarily conserved, and although some of the best evidence is indirect, some of the key amino acids that participate in the binding of NNRTIs can be mutated with only a minimal cost to the virus's ability to replicate (Armstrong et al. 2011; Deeks 2001; Joly et al. 2004; Wang et al. 2010). Broadly speaking, NNRTI-resistance mutations can act in three different ways. (1) Loss of contact between the amino acid side chain and the drug (e.g., mutations that replace either Y181 or Y188 with smaller, nonaromatic amino acids) which was commonly selected by the first-generation NNRTIs (Das et al. 1996; Hsiou et al. 1998; Ren et al. 1999). Many NNRTIs have aromatic moieties that interact with these tyrosines via π-stacking interactions; the loss of this interaction reduces the binding of the drugs. (2) Interference with the ability of the drug to enter the drug-binding pocket. There is one clear example, K103N (Hsiou et al. 2001). This mutation creates a new hydrogen bond with Y188. Drugs enter the pocket in this region, and the presence of a hydrogen bond between N103 and Y188 makes entry more difficult. Because most NRTIs use a common entry pathway, the

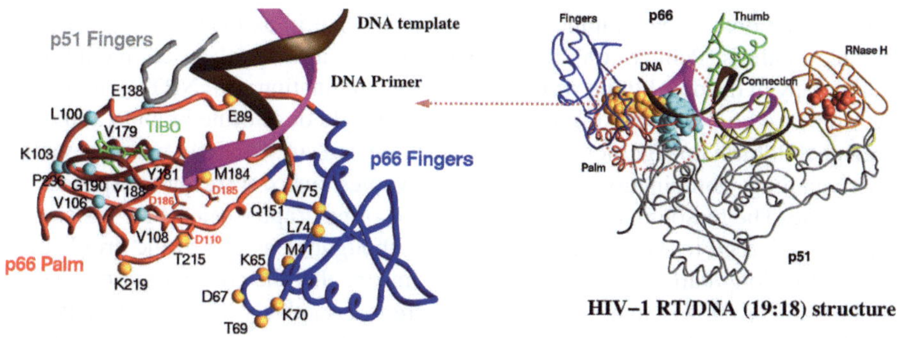


**Fig. 6.6** Locations of the sites for common NNRTI- and NRTI-resistance mutations. On the right is a small ribbon diagram of the structure of HIV-1 RT which shows the binding sites for NRTI-TPs in *gold* and NNRTIs in *light blue*. The p51 subunit is *gray*, and the subdomains of the p66 subunit are color coded: fingers, *blue*; palm, *red*; thumb, *green*; and connection, *yellow*. The DNA is shown as two ribbons that indicate the trajectory of the sugar-phosphate backbone. The template is *brown*, and the primer is *purple*. The RNase H domain is shown in *orange*. The *left panel* shows a close-up of the portion of RT that carries the bulk of the NNRTI- and NRTI-resistance mutations. The color coding for the DNA and the subdomains is the same as shown in the *right panel*. The sites of the common mutations that give rise to NRTI resistance are shown in *light blue*; the sites of NRTI-resistance mutations are shown in *gold* (From Arnold et al. 1996)

K103N mutation confers a similar level of cross resistance to a number of different NNRTIs. Although the data are not as clear, mutations at other positions, including positions 101 and 138, may also affect the entry of NNRTIs into the binding pocket. (3) Mutations that cause the side chain of the amino acid to intrude into the drug-binding pocket, causing steric hindrance. The mutations at G190 are examples of this mechanism; any substitution at this position will substitute, for the glycine normally present at this position, an amino acid with a larger side chain. Mutations at position 100 may also cause steric hindrance. A catalog of the mutations that give rise to NNRTI resistance essentially defines the shape of the pocket (Fig. 6.6). With one interesting exception, these mutations cause resistance because they alter an amino acid in the p66 subunit. The reason for this is simple: the drug-binding pocket is composed primarily of amino acids from the p66 subunit. There is, however, one small portion of the drug-binding pocket that is contributed by the p51 subunit, and there is one mutation, E138K, whose action depends on the change in p51, and not the change in p66 (Boyer et al. 1994).

There are other considerations that affect how resistance arises and how a given drug can be used successfully in patients. The first NNRTI to be widely used, nevirapine, is now off patent, is easy to synthesize, and has a long half-life in patients. Unfortunately, many of the common NNRTI-resistance mutations have a profound effect on the susceptibility of HIV-1 to nevirapine and it is no longer widely used in the developed world. However, it is still used extensively in the developing world, both therapeutically and for the prevention of mother-to-child transmission at birth. Although a single dose of nevirapine, given to the mother shortly before birth, significantly reduces the fraction of newborns that are infected, the administration of just this one dose of nevirapine selects for resistance in as many as 80 % of the infected mothers (Palmer et al. 2006). Nevirapine is also widely used in combination therapy in the developing world (usually in combination with NRTIs); however, the women who developed resistance from the single dose of nevirapine administered at childbirth do significantly worse when treated subsequently with a nevirapine-containing combination therapy than women who were not exposed to nevirapine. The reason that giving a single dose of nevirapine selects for resistance that can be traced to what would normally be thought of as one of the drug's best features: its long half-life. The long half-life means that the concentration of nevirapine trails off slowly in the patient, and there is a relatively long period (days) in which the virus in the patient is exposed to a suboptimal dose of the drug. Not surprisingly, this selects resistant viruses. This selection can be significantly reduced by administering a second drug to the mothers for a short time after they have received the single dose of nevirapine (Farr et al. 2010; Micek et al. 2012; Palmer et al. 2012). The second drug blocks the replication of NNRTI-resistant viruses during the time when there is a suboptimal dose of nevirapine in the mother. The addition of a second drug permits the use of single-dose nevirapine therapy to prevent mother-to-child transmission by suppressing the development of nevirapine resistance, which allows the mother to have the option of using a nevirapine-containing combination therapy later.

Obviously, another alternative would be to use a better drug, in particular a better NNRTI. Efavirenz has a better resistance profile than nevirapine, although it is still susceptible to a number of common NNRTI-resistance mutations. However, it has been reported that efavirenz might be a teratogen (Watts 2007), and despite some questions about the importance of the problem (Ford et al. 2011), efavirenz is not recommended for women of child-bearing age. Efavirenz also affects the brain, and it can cause sleeplessness and particularly vivid dreams. Crushing and smoking the drug enhances its effects on the brain, and it is used as a recreational drug in parts of Africa. This means that some of the efavirenz intended for therapeutic use has been diverted, either because some of the drug was misappropriated before it got into the hands of the patients or because some of the patients are poor enough, and desperate enough, to sell the drug, rather than take it themselves. The third approved NNRTI, delavirdine, is not as potent as efavirenz and must be taken three times a day. For these reasons, it is not widely used today.

## 6.3 Development of Improved NNRTIs

The fact that the first three NNRTIs to be approved have obvious drawbacks points to the need for better NNRTIs. Two NNRTIs, etravirine and rilpivirine, which have better properties, have recently been approved (in 2008 and 2011, respectively). Both drugs came from the same collaborative program organized by Janssen Pharmaceuticals (Janssen et al. 2005), and both were specifically designed to be effective against viruses that carry the common NNRTI-resistance mutations

(Das et al. 2004; Janssen et al. 2005). Based on the preclinical analysis, rilpivirine should be the better drug; however, there has been relatively little experience with either of the drugs in patients, so it is not yet clear how well this promising preclinical data will translate. The available clinical data will be discussed later.

The development of etravirine and rilpivirine is worth reviewing because it provides a guide to the larger problem of how to develop drugs that are effective against known resistance mutations. The first step was to understand how the known resistance mutations worked. This involved careful structural analyses of complexes of NNRTIs with WT RT and RTs carrying the important drug-resistant mutations. These structures were used to develop models and, from the models, new compounds were designed and tested for their effectiveness against both WT and drug-resistant viruses. Structures of RT with the most promising compounds bound were generated and a new cycle of modeling, synthesis, testing against WT and resistant viruses, and structural analysis was carried out. This eventually led to the development of etravirine and rilpivirine. Several important ideas emerged from this effort (Das et al. 2008, 2004, 2005; Janssen et al. 2005). Because the shape of the drug-binding pocket is defined by the interaction between the drug and RT, using modeling to reliably predict how a new compound will be bound is quite difficult. The best modeling results were obtained when a structure or structures of a closely related compound bound to RT were available. There were some design principles that were evident from the beginning. Because NNRTI-resistance mutations at Y181 and Y188 are common, a successful compound should have modest or minimal interactions with these aromatic side chains or should be designed so that the compound could be able to make compensatory interactions if these side chains are replaced by mutation(s). The K103N mutation has a relatively similar cost to the binding of most NNRTIs. The underlying explanation could involve (1) the unfavorable cost of breaking the hydrogen bond between Y188 and N103 side chains or (2) an increase in the rate at which an NNRTI would exit the pocket (off rate) or (3) both. However, the binding of NNRTIs, like rilpivirine, that interact with an N103 side chain is minimally to modestly affected by the mutation. All potent NNRTIs make a hydrogen bond with the main-chain carbonyl of K101. This interaction appears to be an important pharmacophoric determinant for the design of new NNRTIs. Another site to target in NNRTI design is W229. Any mutation at W229 has a considerable cost for RT (Pelemans et al. 2000); for this reason, a successful compound should interact extensively with this residue. Similarly, because the position of the amino acid backbone is less variable and the chemical groups are unaltered compared to the side chains, interactions with the backbone are better, from the point of view of evading drug resistance, than are interactions with the more variable amino acid side chains.

There were also some key ideas that were not so obvious at the outset, but which emerged as the project developed. Although the drug-binding pocket is flexible, there are limits to its size. There are mutations, like the mutations at G190, which change the size and shape of the pocket. These mutations can interfere with the binding of large, rigid NNRTIs. The most successful compounds are somewhat smaller than the maximum size of the pocket and have some flexibility. This allows these compounds to adapt both the position in which they bind and the orientation

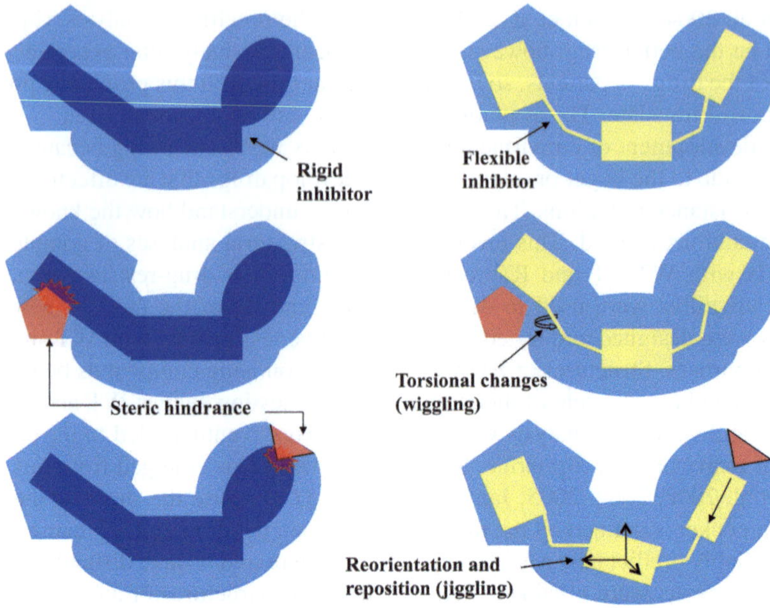


**Fig. 6.7** Strategic flexibility allows NNRTIs to cope with common resistance mutations. On the *left* is shown what happens when a large rigid NNRTI (shown in *dark blue*) is confronted with either of two mutations (shown in *red*) that make the binding pocket (shown in *light blue*) smaller. In both cases this leads to steric hindrance and loss of binding. On the right a somewhat smaller and more flexible inhibitor (shown in *yellow*) can either alter its configuration (wiggling, *middle panel*) or reposition itself (jiggling, *bottom panel*) in ways that allow it to be able to bind to either of the two mutant RTs (From Das et al. 2004)

of their moieties so that they can adapt their binding to match the structures of the different binding pockets present in WT RT and drug-resistant mutants (Fig. 6.7). These adaptive changes in the position and conformation of the drugs have been called "wiggling" and "jiggling." The ability of the best compounds, including rilpivirine, to wiggle and jiggle has been documented by structural analysis of complexes of the drug with WT and mutant RTs (Das et al. 2008). In effect, rilpivirine has the ability to present itself differently to WT and mutant RTs; this allows it to bind effectively to WT RT and the drug-resistant mutants, making rilpivirine much more effective against the common drug-resistant viruses than any of the first-generation NNRTIs. What happens when these compounds are used in patients?

Although it is too early to have really definitive clinical data, the available data look quite promising. When compared to efavirenz in a combination therapy trial, etravirine was as effective as efavirenz in fully suppressing the virus (Gazzard et al. 2011). At 48 weeks, a slightly higher percentage of the patients in the etravirine arm had viral loads below 50 copies/ml, although there were too few patients in the trial for this difference to be significant. Perhaps the most gratifying result of the study was that, in contrast to the patients who failed therapy in the efavirenz arm of the trial, several of whom had NNRTI-resistant virus, none of the patients in the etravirine arm who failed had drug-resistant virus. Viral failures that do not involve the selection of resistance are usually thought to be the result on noncompliance, rather than a problem with the drug. The patients in the etravirine arm also reported fewer side effects. As expected based on the preclinical data, etravirine can be used successfully in patients infected with virus that carries some of the common NNRTI-resistance mutations. Because rilpivirine was approved more recently, less clinical data are available (Cohen et al. 2011; Molina et al. 2011); however, here too, the preliminary data appear to be promising. The limited data that is available suggests that a rilpivirine-based combination had a marginally higher treatment failure than an efavirenz-based therapy; however, there were fewer adverse side effects in the rilpivirine arm. One puzzling aspect of the data from the rilpivirine trials is that, in some of the patients that failed therapy, mutations were seen that rilpivirine is able to overcome in in vitro assays. There is, at this time, no obvious explanation for these results.

### 6.3.1 The Future

Attempting to imagine what is going to happen in the future is always difficult, and what will happen in terms of how important NNRTIs will be, and how they will be used, will depend not only what will happen in terms of NNRTI research but also other factors that will affect the future course of HIV therapy and on progress towards developing better ways to reduce, or prevent, HIV transmission. Although an effective preventive vaccine is a laudable goal, if anything, the possibility of developing an effective vaccine appears to be more remote than it did in the early days of the HIV epidemic. In contrast, there are compelling reasons to believe that drugs, if they are used appropriately, can reduce and/or prevent the spread of the virus. There are two obvious ways drug therapy can be used to reduce the spread of HIV. First, if a patient is successfully treated such that his/her virus load remains below 50 copies/ml in the blood, there is essentially little chance that the patient will transmit the virus. For this reason the first step is to get patients on effective therapies. This approach has greatly slowed the spread of the virus in the developed world and, as more effective drugs become more widely available in the developing world, has begun to blunt the epidemic there as well. It is likely that NNRTIs will continue to play an important role in HAART, and the development of the second-generation NNRTIs, and the willingness of the pharmaceutical industry to allow the new NNRTIs to be used without fee in the developing world, strongly suggests that these drugs will continue to have a significant role in future drug therapy.

Second, it is also possible that NNRTIs could be used to prophylactically treat uninfected people who are obviously at risk for infection (partners of infected individuals, sex workers, etc.). This is the subject of Chap. 15 and will be touched on here only briefly. Because this approach is still in a developmental stage, it is unclear how best to proceed, not only in terms of what drug classes to choose but also how to give the drug (orally, vaginally, and/or rectally). A simple consideration of the viral life cycle suggests that it would be better to block the virus before the viral

DNA integrates; this implies that entry inhibitors, integrase inhibitors, and/or RT inhibitors would be better than protease inhibitors. A prophylactic approach would require that the drug(s), which would be given to uninfected people, would have minimal toxicity, and NNRTIs are good candidates from this point of view. Because the majority (approximately 80 %) of sexually transmitted HIV-1 infections involve infection with a single virus (Keele et al. 2008), transmission is the point at which the virus is the most vulnerable. This means that a drug that has relatively modest efficacy (reducing infection by, e.g., tenfold), which is far below the therapeutic threshold, would still be very useful prophylactically, and, as long as the virus that was likely to be encountered was not resistant to the drug that was used prophylactically, a single drug might well be sufficient. However, given the caveat about the possibility of encountering a resistant virus, it would be better if different drugs, with different resistance profiles, would be used therapeutically and prophylactically. All of this points toward the need to develop new drugs. Impressive progress has been made; however, the twin problems of toxicity and viral resistance are not going to go away. We need to work on anti-HIV drug development now, so that the drugs we will need in the future will be there when we need them. Based on the data we have now, and reasonable guesses about what the future might hold, some of these new anti-HIV drugs should be NNRTIs.

**Acknowledgments** This work was supported (in part) by the Intramural Research Program of the National Institutes of Health, National Cancer Institute, Center for Cancer Research and an NIH R37 Merit Award AI 27690 to Eddy Arnold.

## References

Abbondanzieri EA, Bokinsky G, Rausch JW, Zhang JX, Le Grice SF, Zhuang X (2008) Dynamic binding orientations direct activity of HIV reverse transcriptase. Nature 453(7192):184–189

Ambrose Z, Julias JG, Boyer PL, Kewalramani VN, Hughes SH (2006) The level of reverse transcriptase (RT) in human immunodeficiency virus type 1 particles affects susceptibility to nonnucleoside RT inhibitors but not to lamivudine. J Virol 80(5):2578–2581

Armstrong KL, Lee TH, Essex M (2011) Replicative fitness costs of nonnucleoside reverse transcriptase inhibitor drug resistance mutations on HIV subtype C. Antimicrob Agents Chemother 55(5):2146–2153

Arnold E, Das K, Ding J, Yadav PNS, Hsiou Y, Boyer PL, Hughes SH (1996) Targeting HIV reverse transcriptase for anti-AIDS drug design: Structural and biological considerations for chemotherapeutic strategies. Drug Design and Discovery 13(3–4):29–47

Barber AM, Hizi A, Maizel JV Jr, Hughes SH (1990) HIV-1 reverse transcriptase: structure predictions for the polymerase domain. AIDS Res Hum Retroviruses 6(9):1061–1072

Boyer PL, Ding J, Arnold E, Hughes SH (1994) Subunit specificity of mutations that confer resistance to nonnucleoside inhibitors in human immunodeficiency virus type 1 reverse transcriptase. Antimicrob Agents Chemother 38(9):1909–1914

Cohen CJ, Andrade-Villanueva J, Clotet B, Fourie J, Johnson MA, Ruxrungtham K et al (2011) Rilpivirine versus efavirenz with two background nucleoside or nucleotide reverse transcriptase inhibitors in treatment-naive adults infected with HIV-1 (THRIVE): a phase 3, randomised, non-inferiority trial. Lancet 378(9787):229–237

Das K, Ding J, Hsiou Y, Clark AD Jr, Moereels H, Koymans L et al (1996) Crystal structures of 8-Cl and 9-Cl TIBO complexed with wild-type HIV-1 RT and 8-Cl TIBO complexed with the Tyr181Cys HIV-1 RT drug-resistant mutant. J Mol Biol 264(5):1085–1100

Das K, Clark AD Jr, Lewi PJ, Heeres J, De Jonge MR, Koymans LM et al (2004) Roles of conformational and positional adaptability in structure-based design of TMC125-R165335 (etravirine) and related non-nucleoside reverse transcriptase inhibitors that are highly potent and effective against wild-type and drug-resistant HIV-1 variants. J Med Chem 47(10): 2550–2560

Das K, Lewi PJ, Hughes SH, Arnold E (2005) Crystallography and the design of anti-AIDS drugs: conformational flexibility and positional adaptability are important in the design of non-nucleoside HIV-1 reverse transcriptase inhibitors. Prog Biophys Mol Biol 88(2):209–231

Das K, Bauman JD, Clark AD Jr, Frenkel YV, Lewi PJ, Shatkin AJ et al (2008) High-resolution structures of HIV-1 reverse transcriptase/TMC278 complexes: strategic flexibility explains potency against resistance mutations. Proc Natl Acad Sci USA 105(5):1466–1471

Das K, Bauman JD, Rim AS, Dharia C, Clark AD Jr, Camarasa MJ et al (2011) Crystal structure of tert-butyldimethylsilyl-spiroaminooxathioledioxide-thymine (TSAO-T) in complex with HIV-1 reverse transcriptase (RT) redefines the elastic limits of the non-nucleoside inhibitor-binding pocket. J Med Chem 54(8):2727–2737

Das K, Martinez SE, Bauman JD, Arnold E (2012) HIV-1 reverse transcriptase complex with DNA and nevirapine reveals non-nucleoside inhibition mechanism. Nat Struct Mol Biol 19(2):253–259

Deeks SG (2001) International perspectives on antiretroviral resistance. Nonnucleoside reverse transcriptase inhibitor resistance. J Acquir Immune Defic Syndr 26(Suppl 1):S25–S33

Ding J, Das K, Moereels H, Koymans L, Andries K, Janssen PA et al (1995) Structure of HIV-1 RT/TIBO R 86183 complex reveals similarity in the binding of diverse nonnucleoside inhibitors. Nat Struct Biol 2(5):407–415

Farr SL, Nelson JA, Ng'ombe TJ, Kourtis AP, Chasela C, Johnson JA et al (2010) Addition of 7 days of zidovudine plus lamivudine to peripartum single-dose nevirapine effectively reduces nevirapine resistance postpartum in HIV-infected mothers in Malawi. J Acquir Immune Defic Syndr 54(5):515–523

Ford N, Calmy A, Mofenson L (2011) Safety of efavirenz in the first trimester of pregnancy: an updated systematic review and meta-analysis. AIDS 25(18):2301–2304

Gazzard B, Duvivier C, Zagler C, Castagna A, Hill A, van Delft Y et al (2011) Phase 2 double-blind, randomized trial of etravirine versus efavirenz in treatment-naive patients: 48-week results. AIDS 25(18):2249–2258

Gopalakrishnan V, Benkovic S (1994) Effect of a thiobenzimidazolone derivative on DNA strand transfer catalyzed by HIV-1 reverse transcriptase. J Biol Chem 269(6):4110–4115

Hopkins AL, Ren J, Esnouf RM, Willcox BE, Jones EY, Ross C et al (1996) Complexes of HIV-1 reverse transcriptase with inhibitors of the HEPT series reveal conformational changes relevant to the design of potent non-nucleoside inhibitors. J Med Chem 39(8):1589–1600

Hsiou Y, Ding J, Das K, Clark AD Jr, Hughes SH, Arnold E (1996) Structure of unliganded HIV-1 reverse transcriptase at 2.7 A resolution: implications of conformational changes for polymerization and inhibition mechanisms. Structure 4(7):853–860

Hsiou Y, Das K, Ding J, Clark AD Jr, Kleim JP, Rosner M et al (1998) Structures of Tyr188Leu mutant and wild-type HIV-1 reverse transcriptase complexed with the non-nucleoside inhibitor HBY 097: inhibitor flexibility is a useful design feature for reducing drug resistance. J Mol Biol 284(2):313–323

Hsiou Y, Ding J, Das K, Clark AD Jr, Boyer PL, Lewi P et al (2001) The Lys103Asn mutation of HIV-1 RT: a novel mechanism of drug resistance. J Mol Biol 309(2):437–445

Huang H, Chopra R, Verdine GL, Harrison SC (1998) Structure of a covalently trapped catalytic complex of HIV-1 reverse transcriptase: implications for drug resistance. Science 282(5394):1669–1675

Jacobo-Molina A, Ding J, Nanni RG, Clark AD Jr, Lu X, Tantillo C et al (1993) Crystal structure of human immunodeficiency virus type 1 reverse transcriptase complexed with double-stranded DNA at 3.0 A resolution shows bent DNA. Proc Natl Acad Sci USA 90(13):6320–6324

Janssen PA, Lewi PJ, Arnold E, Daeyaert F, de Jonge M, Heeres J et al (2005) In search of a novel anti-HIV drug: multidisciplinary coordination in the discovery of 4-[[4-[[4-[(1E)-2-cyanoethenyl]-2,6-dimethylphenyl]amino]-2- pyrimidinyl]amino]benzonitrile (R278474, rilpivirine). J Med Chem 48(6):1901–1909

Jilek BL, Zarr M, Sampah ME, Rabi SA, Bullen CK, Lai J et al (2012) A quantitative basis for antiretroviral therapy for HIV-1 infection. Nat Med 18(3):446–451

Joly V, Descamps D, Peytavin G, Touati F, Mentre F, Duval X et al (2004) Evolution of human immunodeficiency virus type 1 (HIV-1) resistance mutations in nonnucleoside reverse transcriptase inhibitors (NNRTIs) in HIV-1-infected patients switched to antiretroviral therapy without NNRTIs. Antimicrob Agents Chemother 48(1):172–175

Julias JG, Ferris AL, Boyer PL, Hughes SH (2001) Replication of phenotypically mixed human immunodeficiency virus type 1 virions containing catalytically active and catalytically inactive reverse transcriptase. J Virol 75(14):6537–6546

Keele BF, Giorgi EE, Salazar-Gonzalez JF, Decker JM, Pham KT, Salazar MG et al (2008) Identification and characterization of transmitted and early founder virus envelopes in primary HIV-1 infection. Proc Natl Acad Sci USA 105(21):7552–7557

Kohlstaedt LA, Wang J, Friedman JM, Rice PA, Steitz TA (1992) Crystal structure at 3.5 A resolution of HIV-1 reverse transcriptase complexed with an inhibitor. Science 256(5065): 1783–1790

Lansdon EB, Samuel D, Lagpacan L, Brendza KM, White KL, Hung M et al (2010) Visualizing the molecular interactions of a nucleotide analog, GS-9148, with HIV-1 reverse transcriptase-DNA complex. J Mol Biol 397(4):967–978

Micek MA, Blanco AJ, Carlsson J, Beck IA, Dross S, Matunha L et al (2012) Effects of short-course zidovudine on the selection of nevirapine-resistant HIV-1 in women taking single-dose nevirapine. J Infect Dis 205(12):1811–1815

Molina JM, Cahn P, Grinsztejn B, Lazzarin A, Mills A, Saag M et al (2011) Rilpivirine versus efavirenz with tenofovir and emtricitabine in treatment-naive adults infected with HIV-1 (ECHO): a phase 3 randomised double-blind active-controlled trial. Lancet 378(9787): 238–246

Palaniappan C, Fay PJ, Bambara RA (1995) Nevirapine alters the cleavage specificity of ribonuclease H of human immunodeficiency virus 1 reverse transcriptase. J Biol Chem 270(9): 4861–4869

Palmer S, Boltz V, Martinson N, Maldarelli F, Gray G, McIntyre J et al (2006) Persistence of nevirapine-resistant HIV-1 in women after single-dose nevirapine therapy for prevention of maternal-to-fetal HIV-1 transmission. Proc Natl Acad Sci USA 103(18):7094–7099

Palmer S, Boltz VF, Chow JY, Martinson NA, McIntyre JA, Gray GE et al (2012) Short-course Combivir after single-dose nevirapine reduces but does not eliminate the emergence of nevirapine resistance in women. Antivir Ther 17(2):327–336

Pelemans H, Esnouf R, De Clercq E, Balzarini J (2000) Mutational analysis of trp-229 of human immunodeficiency virus type 1 reverse transcriptase (RT) identifies this amino acid residue as a prime target for the rational design of new non-nucleoside RT inhibitors. Mol Pharmacol 57(5):954–960

Ren J, Esnouf R, Hopkins A, Ross C, Jones Y, Stammers D et al (1995) The structure of HIV-1 reverse transcriptase complexed with 9-chloro-TIBO: lessons for inhibitor design. Structure 3(9):915–926

Ren J, Esnouf RM, Hopkins AL, Stuart DI, Stammers DK (1999) Crystallographic analysis of the binding modes of thiazoloisoindolinone non-nucleoside inhibitors to HIV-1 reverse transcriptase and comparison with modeling studies. J Med Chem 42(19):3845–3851

Ren J, Bird LE, Chamberlain PP, Stewart-Jones GB, Stuart DI, Stammers DK (2002) Structure of HIV-2 reverse transcriptase at 2.35-A resolution and the mechanism of resistance to non-nucleoside inhibitors. Proc Natl Acad Sci USA 99(22):14410–14415

Restle T, Muller B, Goody RS (1990) Dimerization of human immunodeficiency virus type 1 reverse transcriptase. A target for chemotherapeutic intervention. J Biol Chem 265(16): 8986–8988

Rittinger K, Divita G, Goody RS (1995) Human immunodeficiency virus reverse transcriptase substrate-induced conformational changes and the mechanism of inhibition by nonnucleoside inhibitors. Proc Natl Acad Sci USA 92(17):8046–8049

Sarafianos SG, Das K, Tantillo C, Clark AD Jr, Ding J, Whitcomb JM et al (2001) Crystal structure of HIV-1 reverse transcriptase in complex with a polypurine tract RNA:DNA. EMBO J 20(6):1449–1461

Sarafianos SG, Clark AD Jr, Das K, Tuske S, Birktoft JJ, Ilankumaran P et al (2002) Structures of HIV-1 reverse transcriptase with pre- and post-translocation AZTMP-terminated DNA. EMBO J 21(23):6614–6624

Spence RA, Kati WM, Anderson KS, Johnson KA (1995) Mechanism of inhibition of HIV-1 reverse transcriptase by nonnucleoside inhibitors. Science 267(5200):988–993

Tachedjian G, Goff SP (2003) The effect of NNRTIs on HIV reverse transcriptase dimerization. Curr Opin Investig Drugs 4(8):966–973

Tachedjian G, Orlova M, Sarafianos SG, Arnold E, Goff SP (2001) Nonnucleoside reverse transcriptase inhibitors are chemical enhancers of dimerization of the HIV type 1 reverse transcriptase. Proc Natl Acad Sci USA 98(13):7188–7193

Tachedjian G, Moore KL, Goff SP, Sluis-Cremer N (2005) Efavirenz enhances the proteolytic processing of an HIV-1 pol polyprotein precursor and reverse transcriptase homodimer formation. FEBS Lett 579(2):379–384

Tu X, Das K, Han Q, Bauman JD, Clark AD Jr, Hou X et al (2010) Structural basis of HIV-1 resistance to AZT by excision. Nat Struct Mol Biol 17(10):1202–1209

Tuske S, Sarafianos SG, Clark AD Jr, Ding J, Naeger LK, White KL et al (2004) Structures of HIV-1 RT-DNA complexes before and after incorporation of the anti-AIDS drug tenofovir. Nat Struct Mol Biol 11(5):469–474

Wang J, Bambara RA, Demeter LM, Dykes C (2010) Reduced fitness in cell culture of HIV-1 with nonnucleoside reverse transcriptase inhibitor-resistant mutations correlates with relative levels of reverse transcriptase content and RNase H activity in virions. J Virol 84(18):9377–9389

Watts DH (2007) Teratogenicity risk of antiretroviral therapy in pregnancy. Curr HIV/AIDS Rep 4(3):135–140

# Part III
# Alternative Strategies to Interfere with the Function of HIV RT

# Chapter 7
# Ribonuclease H Inhibitors: Structural and Molecular Biology

**Jason W. Rausch**

## Abbreviations

| | |
|---|---|
| ddNTP | Dideoxynucleoside triphosphate |
| dNTP | Deoxynucleoside triphosphate |
| ds | Double stranded |
| HIV-1 | Human immunodeficiency virus type 1 |
| HIV-2 | Human immunodeficiency virus type 2 |
| IN | Integrase |
| Mo-MLV | Moloney murine leukemia virus |
| NNRTI | Non-nucleoside RT inhibitor |
| NRTI | Nucleoside RT inhibitor |
| nt | Nucleotide |
| PPT | Polypurine tract |
| PR | Protease |
| RNase H | Ribonuclease H |
| RT | Reverse transcriptase |

*Conflict of interest statement*: None declared

J.W. Rausch (✉)
Reverse Transcriptase Biochemistry Section, HIV Drug Resistance Program,
Frederick National Laboratory for Cancer Research, Building 535, Room 325,
Frederick, MD 21702, USA
e-mail: rauschj@mail.nih.gov

S.F.J. LeGrice and M. Götte (eds.), *Human Immunodeficiency Virus Reverse Transcriptase: A Bench-to-Bedside Success*, DOI 10.1007/978-1-4614-7291-9_7,

## 7.1 Introduction

PR, IN, and RT are the only enzymes encoded by the HIV genome, and all are targeted by antiviral therapeutic agents (De Clercq 2009). HIV PR cleaves viral polyprotein precursors to produce mature virus and is targeted by PR inhibitors saquinavir, ritonavir, and several others. IN is required for assimilation of viral DNA into the genome of the infected cell and is targeted by raltegravir. A second IN inhibitor, elvitegravir, is currently in the latter stages of clinical trials. RT possesses two enzymatic activities: DNA polymerase and RNase H. The former is targeted by both NRTIs (e.g., didanosine, or 2′,3′-dideoxyinosine) and NNRTIs (e.g., nevirapine, efavirenz) (Ruane and DeJesus 2004). Of the 25 anti-HIV compounds in clinical use today, 12 target the multifunctional RT enzyme. However, none of these inhibitors target the RNase H activity of RT, despite it being absolutely essential for synthesis of preintegrative viral DNA (Schatz et al. 1989). In this chapter, various aspects of HIV-1 RT-associated RNase H and its inhibition will be discussed, including (i) the structure of HIV-1 RT, with emphasis on the RNase H domain; (ii) the two-metal-ion-dependent mechanism of RNase H cleavage; and (iii) efforts and progress toward developing potent and specific inhibitors of the lone untargeted HIV enzymatic function.

## 7.2 Structure of HIV-1 RT and the RNase H Domain

HIV-1 RT is a heterodimer of 66- and 51-kD subunits (p66 and p51, respectively) cleaved from the *gag-pol* precursor by HIV-1 PR during virus maturation (Di Marzo Veronese et al. 1986). Many X-ray crystal structures of HIV-1 RT have been resolved, including apoenzyme and enzyme liganded with double-stranded nucleic acid, dNTP, and/or any of several RT inhibitors (Huang et al. 1998; Jacobo-Molina et al. 1993; Kohlstaedt et al. 1992; Ren et al. 2001; Rodgers et al. 1995; Sarafianos et al. 2001). Although amino acids 1–440 are common to p66 and p51, the two subunits assume quite different tertiary structures in the context of heterodimeric RT (Fig. 7.1). The larger subunit houses the DNA polymerase and RNase H active centers of the enzyme toward its amino- and carboxy-termini, respectively (Rothwell and Waksman 2005). The polymerase domain resembles a grasping right hand, comprised of fingers (residues 1–85, 118–155), palm (86–117, 155–237), and thumb (238–318) subdomains, and is linked to the RNase H domain via a connection subdomain (residues 319–426) (Fig. 7.1a). In contrast, p51 is relatively compact, serving primarily to support the more extended structure of its catalytic counterpart (Fig. 7.1b). Together, the two subunits form a template-primer binding cleft capable of accommodating dsDNA, dsRNA, or an RNA/DNA hybrid, all of which are required for synthesis of the preintegrative DNA intermediate.

The RNase H domain (p66 residues 427–560) is comprised of a central five-stranded mixed β-sheet surrounded by four α-helices and eight connecting loops (Davies et al. 1991). It is stabilized by interactions with the p66 connection

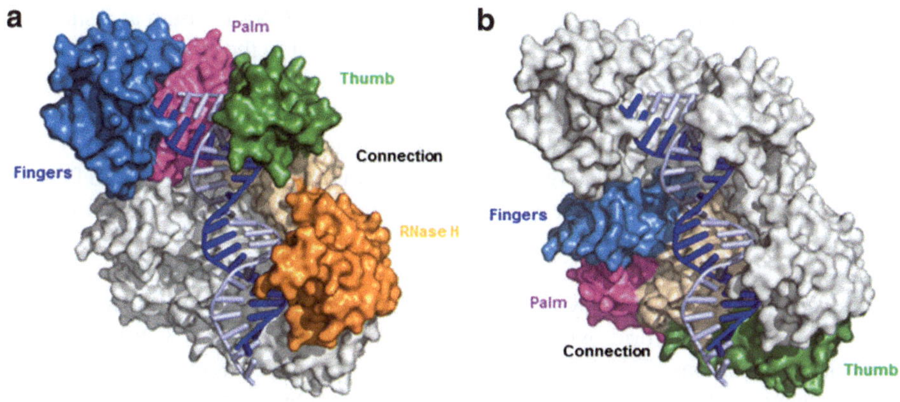


**Fig. 7.1** Structure of HIV-1 RT in complex with an RNA/DNA hybrid (Sarafianos et al. 2001). RNA and DNA strand ladders are depicted in *navy* and *blue gray*, respectively. RT subdomains in (**a**) p66 and (**b**) p51 are highlighted

subdomain, as well as with the p51 thumb and connection subdomains (Kohlstaedt et al. 1992), rendering the HIV-1 RNase H domain less flexible than its counterparts in Mo-MLVRT (Das and Georgiadis 2004). The RNase H active site is located on a highly solvent-exposed surface facing the same direction as the interior of the template-primer binding cleft. It is comprised of four highly conserved acidic residues (D443, E478, D498, and D549) that bind the Mg ions necessary for catalysis, as well as a flexible "His loop" near the interface with the p51 thumb. This DEDD motif is common to members of the polynucleotidyltransferase family of enzymes, which includes relatives such as *E. coli* and *B. halodurans* RNases H, HIV IN, *E. coli* RuvC, and the PIWI domain of argonaute, the catalytic component of the RISC complex (Ma et al. 2005; Song et al. 2004).

In X-ray crystal structures in which RT is in complex with duplex DNA (Huang et al. 1998; Jacobo-Molina et al. 1993), RT-nucleic acid contacts occur at multiple sites throughout the polymerase domain, especially around the polymerase active center, and involving the so-called primer and template grip residues. As with most nonspecific nucleic acid-binding proteins, contacts outside of the polymerase active center occur primarily along the sugar-phosphate backbones of both strands of nucleic acid. Duplex DNA bound in the polymerase domain is constrained in an A-form geometry; however, it relaxes into a B-form conformation as it extends past the connection subdomain and RNase H domain. The transition from A- to B-form is marked by a ~40° bend in the substrate and occurs 6–8 bp from the primer 3′ terminus. The polymerase and RNase H domains are separated by approximately 18 bp of dsDNA – a structural observation confirmed by biochemical experimentation (Gotte et al. 1998).

Crystal structures of RT in complex with dsDNA and an incoming dNTP (Huang et al. 1998) or an RT inhibitor (Das et al. 2012; Kohlstaedt et al. 1992; Ren et al. 2001) have greatly clarified how these ligands/inhibitors are bound by the enzyme. Binding sites for dNTPs and NRTIs have been shown to involve primarily fingers and palm subdomain residues. In contrast, the allosteric NNRTIs bind at a site hidden at the base of the p66 thumb. This site is not apparent in the absence of inhibitor but is revealed only when the thumb is displaced by NNRTI binding (Das et al. 2012; Kohlstaedt et al. 1992). Interestingly, the NNRTI binding site does not exist at all in HIV-2 RT, rendering this class of inhibitors ineffective against HIV-2 (Shih et al. 1991). While primarily targeting the polymerase activity of RT, NNRTIs have also been shown to affect RNase H cleavage via an allosteric mechanism, an inhibitory activity that may contribute to antiviral function of this class of inhibitors (Hang et al. 2007).

To date, only a single X-ray crystal structure has been solved for RT in complex with an RNA/DNA hybrid, which in this case contains the HIV-1 3′ PPT together with flanking regions of the viral genomic sequence (Sarafianos et al. 2001). While this segment of RNA is alternately recognized as both template and primer during the course of viral DNA synthesis, it is oriented in the position of template in the co-crystal structure. Several contacts between RT and RNA 2′ hydroxyl groups are revealed in this structure, as are contacts between the DNA primer strand and RT residues in the vicinity of the connection and RNase H domains. These contacts are made by residues collectively referred to as the RNase H primer grip, including G359, A360, H361, T473, N474, Q475, K476, Y501, and I505 from the p66 subunit as well as K395 and E396 of p51. Although the apparent role of the RNase H primer grip is to guide the hybrid toward the RNase H active center, the template is not properly aligned for hydrolysis in this RT co-crystal structure. Specifically, as is the case with DNA template strands in RT-dsDNA co-crystal structures, the RNA strand in the RT-RNA/DNA structure does not pass into the RNase H catalytic center, and the purported scissile phosphate resides ~4 Å from the optimal location for cleavage. This may be because the hybrid duplex assumes a DNA-like B-form conformation in the vicinity of the connection subdomain and RNase H domain, contains localized disruptions in rA:dT base pairing, and is unresolved (and base pairing is likely disrupted altogether) as it emerges past the RNase H domain.

Structures of the isolated HIV-1 RNase H domain, as well as *E. coli* and *B. halodurans* RNases H, have also been resolved (Davies et al. 1991; Katayanagi et al. 1990; Nowotny et al. 2005). Unlike its *E. coli* counterpart, the HIV-1 RNase H domain lacks a basic loop-helix motif (α-helix C) that in the bacterial and some retroviral enzymes contributes to substrate binding. The HIV element relies instead upon extensive contacts in the polymerase domain in addition to those within the RNase H primer grip. Consequently, the isolated HIV-1 domain is highly deficient in substrate binding and is only weakly active or inactive as an RNase H (Cirino et al. 1993; Hansen et al. 1988). The addition of flanking connection subdomain residues or a polyhistidine tag, insertion of an *E. coli* RNase H α-helix C, and/or use of $Mn^{+2}$ instead of $Mg^{+2}$ in RNase H assays restore activity to the isolated HIV-1 domain, allowing it to be used for RNase H inhibitor screening (Hansen et al. 1988; Keck and Marqusee 1995; Stahl et al. 1994).

Finally, active site mutants of *B. halodurans* RNase H were crystallized in complex with an RNA/DNA hybrid, greatly elucidating the mechanisms of substrate binding and RNase H cleavage (Nowotny et al. 2005). In contrast to substrate within the HIV-1 RT-RNA/DNA co-crystal structure, the *B. halodurans* RNase H substrate largely assumes an A-like or mixed conformation. Moreover, a four-residue motif designated the phosphate-binding pocket (PBP) was shown in the bacterial enzyme to bind and distort the DNA strand two base pairs from the scissile phosphate, thereby restricting substrates to RNA/DNA hybrids alone. Homology between the *B. halodurans* and HIV-1 PBPs is incomplete, however, with only T473 and Q475 likely to serve a similar role as the HIV-1 enzyme. This may explain why the cleavage specificity of the HIV-1 RNase H is somewhat relaxed, with RT capable of cleaving one strand in an RNA duplex in certain specialized substrates (Gotte et al. 1995).

## 7.3 Mechanism of RNase H Cleavage

During the course of viral DNA synthesis, HIV-1 RT encounters numerous hybrid nucleic acid duplexes in many different structural contexts (Telesnitsky and Goff 1997). RNA hydrolysis occurs both throughout the course of minus-strand DNA synthesis and during more specialized events such as minus-strand and plus-strand transfer, PPT selection, and minus- and plus-strand primer removal. Detailed reviews encompassing these events, structural variations among RNA/DNA hybrids, and the manner in which RT accommodates these substrates are available elsewhere (Champoux and Schultz 2009; Rausch and Le Grice 2004; Schultz and Champoux 2008). Here, however, a brief summary of the different modes of RNase H cleavage is provided.

"Polymerase-dependent" cleavage refers to hydrolytic events catalyzed during the course of DNA synthesis by the polymerizing enzyme (Furfine and Reardon 1991; Peliska and Benkovic 1992). These generally occur ~18 bp from the primer terminus either prior to initiation, during pausing, or after termination of synthesis while the enzyme remains bound to the nascent hybrid. Hydrolysis products generated by this mode of cleavage are both consistent with and identified by the spacing between polymerase and RNase H active sites in HIV-1 RT (~18 bp). This spacing suggests that polymerase-dependent cleavage occurs while the polymerase domain is engaged at or near the 3′ terminus of the DNA primer. Moreover, the tendency for most hydrolytic events to occur when RT is not actively synthesizing DNA suggests that nucleotide incorporation/enzyme translocation and RNase H cleavage do not occur concurrently. The notion that HIV-1 RT cannot simultaneously catalyze both DNA synthesis and RNase H cleavage is consistent with a structural model of RT in complex with an RNA/DNA hybrid spanning the two active sites (Nowotny et al. 2007).

In arrested RT-RNA/DNA complexes, "polymerase-dependent" cleavage events can be subdivided into those occurring "pre-" and "post-translocation," where translocation refers to the incremental advance of the enzyme that occurs upon incorporation of a dNTP into the primer strand (Gotte et al. 2010). More specifically, incorporation of (i) a ddNTP at the 3′ terminus of the primer strand followed by (ii) inclusion of the next dNTP to be incorporated in the reaction mixture locks the RT-RNA/DNA complex into the "post-translocation" state. Conversely, inclusion of the pyrophosphate analog phosphonoformic acid (PFA, foscarnet) in the reaction mixture stabilizes the "pre-translocation" state of the enzyme-substrate complex. In both "pre-" and "post-translocation" complexes, RNase H-mediated hydrolysis occurs ~18 bp from the primer terminus; however, the precise cleavage site in the former complex is one nucleotide further removed from the primer terminus than in the latter. This observation becomes particularly useful when examining the mechanisms of RNase H inhibition and/or the functional relationship(s) between RNase H and DNA polymerase inhibitors.

Likewise, in the absence of dNTPs, ddNTPs, or other modulators of polymerase domain function, cleavage of the RNA strand occurs primarily ~18 bp from the recessed 3′ terminus, again suggesting engagement of the primer terminus by the polymerase domain of the enzyme (Champoux and Schultz 2009; Furfine and Reardon 1991). This "primary" cleavage is followed by one or more "secondary" hydrolytic events 7–12 bp from the primer terminus (Palaniappan et al. 1996). The latter are also classified as "polymerase-independent" cleavage events, since the spacing between the primer terminus and site(s) of secondary cleavage is not consistent with their simultaneous occupancy within the two active sites of the enzyme (Furfine and Reardon 1991; Peliska and Benkovic 1992).

"Primary" cleavage events are also referred to as "3′-end-directed" cleavages since they occur when the polymerase domain is oriented over the recessed 3′ terminus of the primer strand. Conversely, when the substrate contains a 3′ DNA overhang, cleavage is observed 13–19 nt from the recessed RNA 5′ terminus (DeStefano et al. 1993; Palaniappan et al. 1996). These hydrolytic events are dubbed "5′-end-directed" cleavages since the polymerase domain is located over the DNA strand opposite the recessed RNA 5′ terminus. Moreover, on longer substrates, "internal" cleavage occurs at sites well removed from the 3′ and 5′ termini of either strand (Schultz et al. 2004). While less efficient than either end-directed mode of cleavage, "internal" cleavage is likely at least partly responsible for clearance of genomic RNA following minus-strand DNA synthesis. For this mode of substrate binding and hydrolysis in particular, RNA sequence appears to play an important role in cleavage site selection. While not absolute, a set of specificity rules for predicting the sites at which RNase H-mediated hydrolysis is likely to occur has been established, based on hydrolysis patterns obtained from cleavage of a wide variety of RNA/DNA hybrids (Champoux and Schultz 2009; Schultz and Champoux 2008; Schultz et al. 2004).

Nucleic acid sequence and structure, as well as numerous protein-nucleic acid contacts throughout RT and the template-primer, serve as determinants affecting both substrate binding and RNase H cleavage specificity and efficiency. Nevertheless, the chemistry of catalysis is likely common to all modes of RNase H-mediated hydrolysis. The cleavage mechanism was derived from biochemical experiments (Klumpp et al. 2003) and validated in a series of *B. halodurans* RNase H-RNA/DNA co-crystal structures in which enzymes containing mutations in conserved active site residues were used to capture various catalytic intermediates (Nowotny et al. 2005). Moreover, a molecular model integrating the catalytic mechanisms of *B. halodurans* RNase H and human RNase H1 into an HIV RT-RNA/DNA complex has been generated (Nowotny et al. 2007), although the structural coordinates for this model are

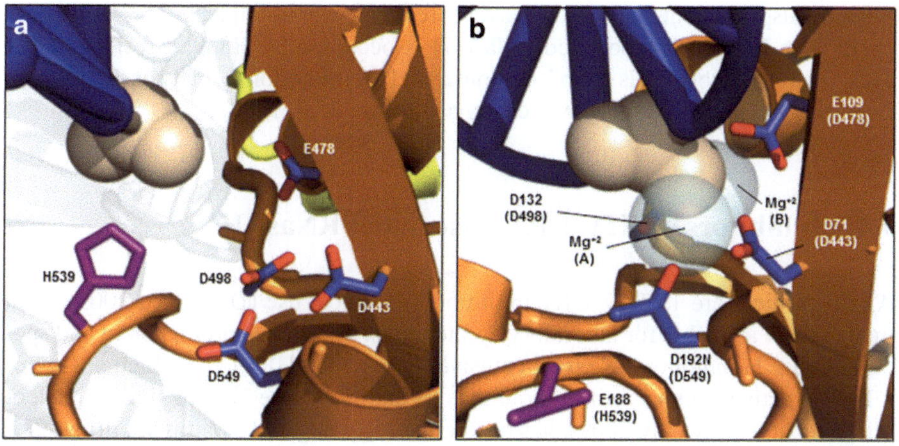


**Fig. 7.2** RNase H active sites. (**a**) Image of RT-associated RNase H active site generated using RT-RNA/DNA co-crystal structure coordinates (Sarafianos et al. 2001). Catalytic residues are highlighted. The scissile phosphate (*wheat spheres*) of the RNA strand (*navy ladder*) is not properly positioned for cleavage in this complex nor are the two metal-ion cofactors present. (**b**) Active site of *Bacillus halodurans* RNase H (Nowotny et al. 2005). Catalytic amino acids are labeled together with the homologous HIV residues (parentheses), as are the A- and B-site Mg ions. The crystallized enzyme was engineered to contain a D192N mutation in order to ensure that the RNA strand would not be hydrolyzed

not publicly available. Here, detailed structures of the HIV-1 RT-associated and *B. halodurans* RNase H active sites are provided in Fig. 7.2a, b, respectively.

Extrapolating the *B. halodurans* RNase H mechanism to HIV-1 RT-associated RNase H, two $Mg^{+2}$ ions (A and B) are coordinated in part by conserved acidic residues at adjacent sites in the RNase H active center: $Mg^{+2}$ A by D443 and D549, and $Mg^{+2}$ B by D443, E478, and D498. Other coordination partners for these ions at various stages of the reaction include water and both bridging and non-bridging oxygen atoms of the scissile phosphodiester bond. Involvement of the substrate in metal-ion coordination suggests that $Mg^{+2}$ binding at one or both sites may be substrate dependent, a notion supported by an isothermal calorimetric study of HIV-1 RT demonstrating that only one Mg ion is bound at the RNase H active center in the apoenzyme (Cowan et al. 2000).

In the initial stage of the cleavage reaction, the two catalytic metal ions are 4 Å apart. Mg ion A is coordinated in perfect octahedral geometry and activates a water molecule for nucleophilic attack on the scissile phosphate. In contrast, the coordination geometry of Mg ion B is slightly distorted, perhaps serving to destabilize the enzyme-substrate complex and drive it toward catalysis. Nucleophilic attack on the scissile phosphate ensues, bringing the Mg ions closer together (3.5 Å) and creating a pentavalent phosphate intermediate. This intermediate is quickly converted to the 5′-phosphate and 3′-hydroxyl cleavage products, completing the Sn2 substitution reaction and once again separating the Mg ions (4.8 Å). For *B. halodurans* RNase H, it was speculated that repositioning of residue D188 facilitated metal-ion and/or product displacement once the cleavage reaction was complete. The equivalent position in HIV-1 RT-associated RNase H is occupied by H539, a residue at the apex of the flexible "His loop" demonstrated by mutational analysis to play an important but nonessential role in catalysis (Schatz et al. 1989).

## 7.4 Inhibitors of HIV-1 RT-Associated RNase H

Although there are 12 HIV-1 RT inhibitors in clinical use (De Clercq 2009), with many others available for laboratory applications, development of a potent and specific RNase H inhibitor has proven elusive. In contrast, HIV-1 IN, which possesses a DEDD active center motif homologous to that of RNase H with similar nucleolytic function (Nowotny 2009), has been effectively targeted by antiviral drugs. The difficulty in targeting RNase H, an activity shown to be as essential to virus replication as that of any other viral enzyme (Schatz et al. 1989), may lie in the relatively open topology of its active center compared to its viral enzyme counterparts (Fig. 7.3). While potential inhibitor binding pockets exist in and around the RNase H domain, there are few near the active center, suggesting that effective small molecule inhibition of RNase H activity will require binding directly either to the catalytic metal ions present at the enzyme active site or to a remote site for allosteric inhibition of RNase H.

### *7.4.1 Active Site Inhibitors*

Direct targeting of the RNase H active site typically requires that compounds contain a three-oxygen pharmacophore that, in conjunction with the four acidic active site residues, effectively and simultaneously coordinates the two catalytic Mg ions (Klumpp and Mirzadegan 2006). Binding the active site metal ions in this manner would be expected to prevent accommodation of substrate within the RNase H domain, inhibit catalytic chemistry, and/or prevent hydrolysis. Optimal ionic interaction between metal ions and inhibitor requires that compounds have at least one negative charge, preferably on the bridging oxygen that interacts with both ions simultaneously. Unfortunately, compounds with such a charge would also be expected to have limited cellular permeability – a deficiency commonly observed in this class of inhibitors. Moreover, in contrast to other enzymes that utilize tandem metal ions, the relatively open surface of the RNase H domain limits secondary interactions between enzyme and inhibitor that might be expected to increase binding affinity and inhibitor specificity (Fig. 7.3). Despite these challenges, a number of submicromolar inhibitors that directly target the RNase H active site have been developed (Fig. 7.4).

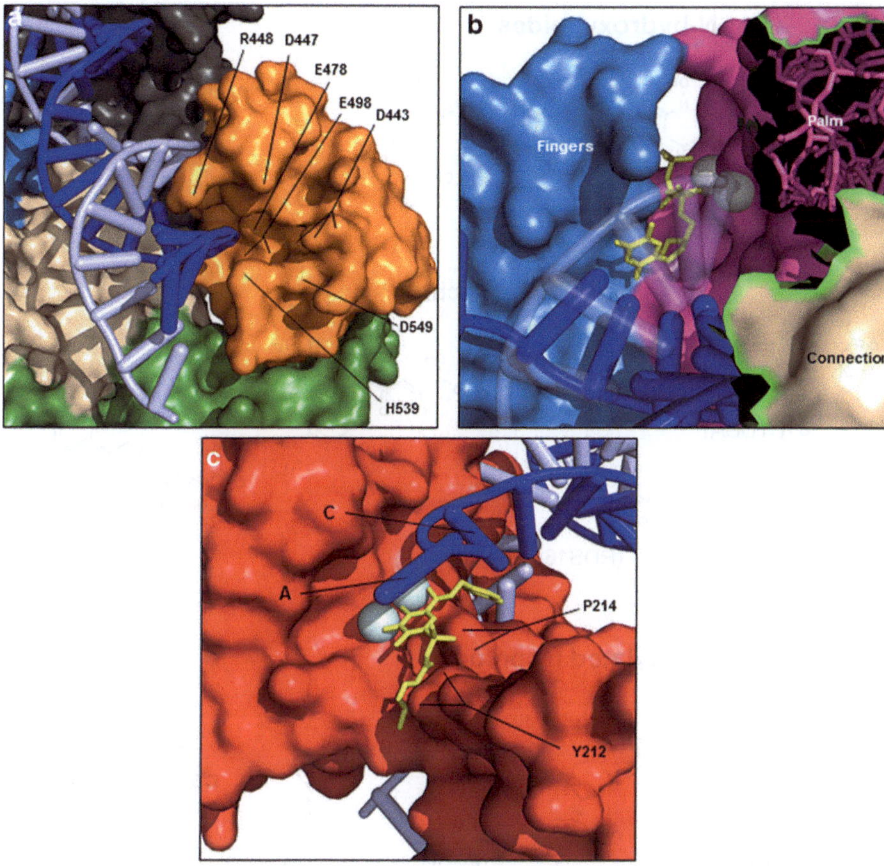


**Fig. 7.3** Comparison of RNase H, DNA polymerase, and integrase active sites. (**a**) Surface representation of the RNase H active site from the RT-RNA/DNA co-crystal structure (Sarafianos et al. 2001). Locations of catalytic and other active site residues are highlighted. Although active site metal ions are not present in the co-crystal, the relatively open structure of the RT-associated RNase H active site is revealed. (**b**) Polymerase active site of RT in complex with dsDNA and an incoming dNTP (Huang et al. 1998). Select p66 subdomains are highlighted, as are the DNA template (*navy ladder*), DNA primer (*blue-gray, semitransparent ladder*), and catalytic Mg ions (*spheres*). The incoming dNTP (*yellow sticks*) coordinates the two active site metal ions, is nestled tightly among fingers and palm subdomain residues, and is engaged in a stacking interaction with the 3′-terminal primer nucleotide. Note that NRTIs would be expected to be accommodated in the same, or a similar, manner prior to incorporation into nascent DNA. To facilitate visualization of the polymerase active site, the p66 thumb subdomain has been removed. (**c**) Surface representation of the prototype spuma viral integrase active center in a complex containing divalent metal ions (*blue-white spheres*), dsDNA (*navy, blue-gray ladders*), and the metal-binding inhibitor raltegravir (Hare et al 2010). The inhibitor is shown to coordinate both active site metal ions, form close contacts with integrase residues P214 and Y212, and form π-stacking interactions with the invariant C-A nucleotide tandem at the 3′ terminus of the short DNA strand

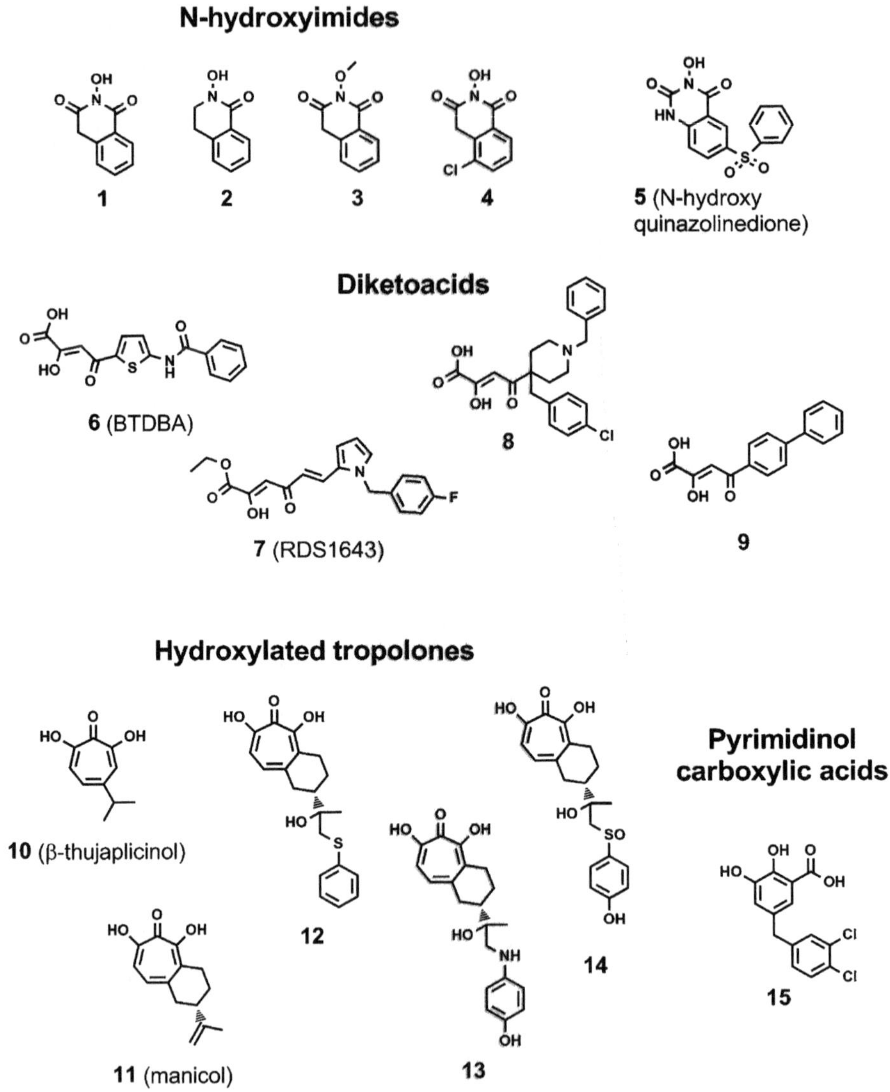


**Fig. 7.4** Chemical structures of RNase H active site inhibitors (Note that **3**, **4**, **8**, and **9** were synthesized to evaluate structure-activity relationships within their respective chemical classes and are non-inhibitory)

#### 7.4.1.1 N-Hydroxyimides

Originally identified as inhibitors of influenza virus endonuclease (Parkes et al. 2003), N-hydroxyimides were the first compounds to be specifically selected and/or designed to target the HIV-1 RNase H active center (Klumpp et al. 2003). As is now the standard for this class of inhibitor, N-hydroxyimides contain three oxygen ligands optimally oriented to simultaneously interact with two metal ions spaced ~4 Å apart. In the original experiments, compounds were designed to have minimal molecular weight, complexity, and potential for interaction with protein so that their inhibitory potency would be based solely on their interaction with metal ions. Moreover, it was intended that binding to enzymes requiring fewer than two metal ions should be minimal, and catalytic activity should not be affected. *E. coli* RNase H was to serve as a control in this regard, as this enzyme is thought to utilize a one-metal-ion-dependent mechanism for catalysis (Keck et al. 1998; Tsunaka et al. 2003, 2005).

As the dual metal-ion-binding model predicted, the prototype compound (**1**,2-hydroxy-4H-isoquinoline-1,3-dione) potently inhibited HIV RNase H activity (IC50 = 0.6–1 μM) yet was completely ineffective against *E. coli* RNase H (IC50 > 50 μM) (Hang et al. 2004; Klumpp et al. 2003). This specificity was ascribed to a difference in the number of metal ions present at the active sites of the two enzymes, although it is possible that differences in substrate affinity and turnover between HIV-1 and *E. coli* RNases H were influential as well. In chemical substitution experiments, each of the three N-hydroxyimide oxygen was found to be essential for inhibition, and derivatization of the hydroxyl group was not tolerated (e.g., compounds **2, 3**) (Klumpp and Mirzadegan 2006). However, select substitutions on the phenyl ring were found to increase the potency of N-hydroxyimide inhibitors (compound **4**), perhaps via interaction with amino acid residues at the periphery of the RNase H active site.

NMR and fluorescence quenching analysis using the isolated HIV-1 RNase H domain established that the binding of N-hydroxyimides to the domain was metal-ion dependent (Hang et al. 2004). Moreover, the inhibitory potencies of these compounds against RT-associated and isolated RNases H were similar, validating the notion that the N-hydroxyimides bind at the RNase H active site. In buffered aqueous solution, metal ions were found to shift the structural equilibrium of N-hydroxyimides toward the tautomeric enol form, although the affinity of these compounds for free $Mg^{+2}$ was low. It is likely that the enol form is also assumed when inhibitors bind metal ions at the RNase H active site. The nature of inhibitor binding was confirmed in high-resolution crystal structures of the isolated RNase H domain together with $Mn^{+2}$ ions and N-hydroxyimide compounds (Klumpp and Mirzadegan 2006). In these studies, co-crystallization was found to be metal-ion dependent, and the resulting structure showed the two metal ions in coordination with the hydroxyimide three-oxygen pharmacophore. In addition, the flexible "His loop," which is unresolved in some unliganded RNase H structures, appeared to be stabilized by direct interaction with the inhibitor.

Although coordinates for the HIV-1 RNase H-N-hydroxyimide co-crystal structure are not publicly available, a related active site inhibitor N-hydroxy quinazolinedione (compound **5**) has been co-crystallized with $Mn^{+2}$ and an isolated HIV-1 RNase H domain containing the basic turn helix motif of *E. coli* RNase H (Fig. 7.5a) (Lansdon et al. 2011). Arrangement of the three-oxygen pharmacophore in N-hydroxy quinazolinedione exactly matches that of hydroxyimides; however, a nitrogen-for-carbon substitution in the former compound creates a bicyclic aromatic

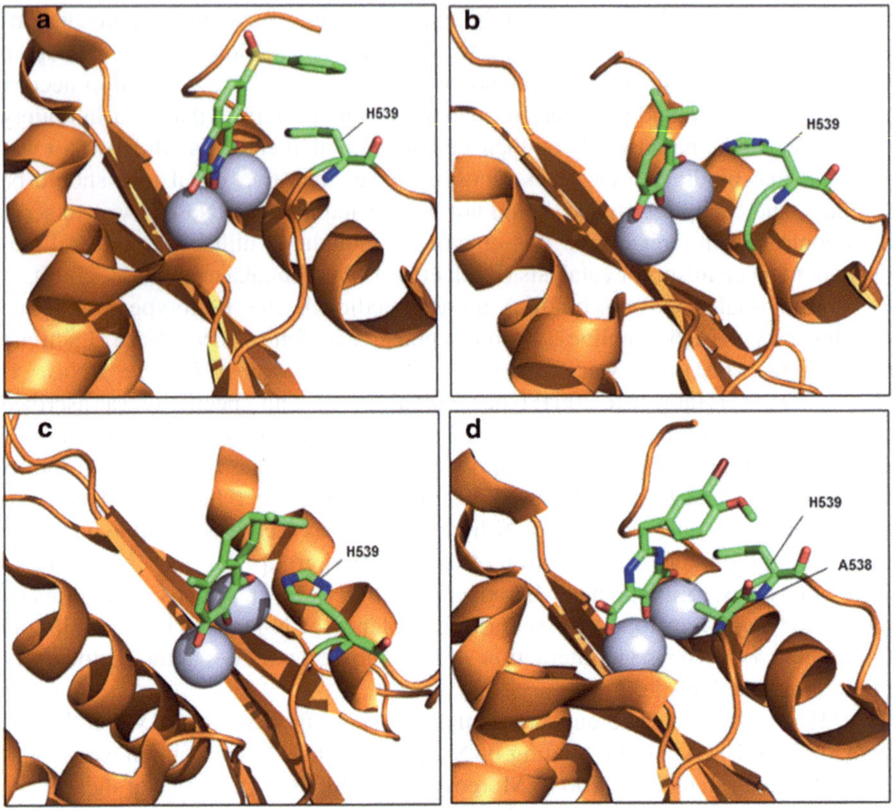


**Fig. 7.5** Metal-binding inhibitors bound at the HIV-1 RNase H active site. $Mn^{+2}$ ions are depicted as *blue-white spheres*, while inhibitors and "His loop" residues H539 and A538 are colored according to their atomic constituents. (**a**) **5** (N-hydroxy quinazolinedione) both coordinates active site metal ions and contributes to an edge-on π-stacking interaction with H539 (Lansdon et al. 2011). The hydroxylated tropolones (**b**) **10** (β-thujaplicinol) (Himmel et al. 2009) and (**c**) **11** (manicol) (Chung et al. 2011) likewise bind metal ions and participate in hydrogen bonding/ionic and edge-to-face π-stacking interactions with H539, respectively. An ionic interaction between the imidazole side chain H539 and a manicol tropylium ion has also been proposed. (**d**) The distal ring in **15**, a pyrimidinol carboxylic acid, may also interact with H539 by means of an edge-to-face π-stacking interaction (Kirschberg et al. 2009)

scaffold that is more stable in aqueous solution (Kirschberg et al. 2009). While coordination of $Mn^{+2}$ ions by **5** closely resembles that observed with hydroxyimides and other active site inhibitors, the distal phenyl ring assumes an additional edge-on π-stacking interaction with the imidazole side chain of H539. This interaction likely serves to increase the binding affinity of this inhibitor, as well as stabilize the flexible "His loop" in a position close to the active site metal ions.

#### 7.4.1.2 Diketoacids

An alternative form of the three-oxygen pharmacophore targeting two metal-ion active sites can be found in diketoacids. These compounds have been described as inhibitors of polynucleotide transferases such as influenza endonuclease, HCV polymerase, and HIV-1 IN as well as HIV-1 RNase H (Hazuda et al. 2000; Shaw-Reid et al. 2003; Summa et al. 2004; Tomassini et al. 1994). The first diketoacid reported to inhibit HIV-1 RNase H (IC50 = 3.2 μM) was 4-[5-(benzoylamino)thien-2-yl]-2,4-dioxobutanoic acid (BTDBA, compound **6**) (Shaw-Reid et al. 2003). Although this and similar diketoacids have also been shown to inhibit HIV-1 IN, BTDBA is clearly a selective inhibitor, with reported IC50 values against *E. coli* RNase H and the polymerase activity of HIV-1 RT both exceeding 50 μM. As with N-hydroxyimides, there is substantial evidence that BTDBA inhibits RNase H by binding to metal ions at the RNase H active site. IC50s of BTDBA against (i) an isolated, recombinant RNase H domain containing the *E. coli* RNase H basic loop-helix motif and (ii) RT containing a D185N polymerase active site mutation were 4.7 and 8.8 μM, respectively, indicating that RNase H inhibition by BTDBA is polymerase domain independent (Shaw-Reid et al. 2003). Moreover, isothermal calorimetric studies demonstrated that inhibitor binding is metal-ion dependent and does not require the presence of substrate. In in vitro strand transfer activity assays, BTDBA acted synergistically with both NRTIs and NNRTIs, confirming that strand transfer requires both activities and that the inhibitory mechanisms of these compounds are independent and mutually compatible (Shaw-Reid et al. 2005). Despite its in vitro potency, however, BTDBA did not inhibit virus replication in cell culture.

RDS1643 (compound **7**), a diketoacid ethyl ester, was also demonstrated to be a selective inhibitor of HIV-1 RNase H, with IC50 values of 13, >100, and >100 μM against RT-associated RNase H, *E. coli* RNase H, and RT polymerase activities, respectively (Tramontano et al. 2005). However, this inhibitor is notably less potent than BTDBA and like compounds and only weakly inhibited avian myeloblastosis virus (AMV) RT-associated RNase H and HIV-1 IN activities (IC50 = 92 μM and 98 μM, respectively). This is likely due to the masking of the carboxylic acid group with an ethyl ester, a derivatization shown to reduce or destroy the inhibitory activity of diketoacid inhibitors of influenza endonuclease (Tomassini et al. 1994). However, perhaps owing to this ester modification, RDS1643 was found to have antiviral activity in MT-4 cell culture (EC50 = 14 μM). Viruses containing various NNRTI resistance mutations in the RT gene were similarly susceptible, indicating that the mechanisms of inhibition differ between these two types of inhibitors.

Kinetic analysis demonstrated that RDS1643 inhibits RNase H activity via a classical noncompetitive mechanism (Tramontano et al. 2005). Moreover, biochemical studies showed that inhibitor binding to the HIV-1 RNase H domain is reversible and does not require substrate but does require divalent metal ions. These results suggest that, like BTDBA, RDS1643 coordinates the tandem metal ions at the RNase H active center. However, the superior antiviral potency of RDS1643 compared to true diketoacids may be explained by greater cellular permeability due to charge neutralization of the carboxylic acid moiety. Potential intracellular removal of the ester group by cellular esterases might then restore the pharmacophore and convert RDS1643 into a more active compound inside the cell.

Other diketoacid compounds demonstrated to inhibit influenza endonuclease and HIV-1 IN in vitro and/or in vivo were shown to be inactive against HIV-1 RNase H (Parkes et al. 2003; Tomassini et al. 1994) (compounds **8**, **9**). To understand the structural basis for this specificity, these compounds, along with BTDBA and RDS1643, were superimposed upon the HIV-1 RNase H-$Mn^{+2}$-hydroxyimide co-crystal structure by molecular modeling (Klumpp and Mirzadegan 2006). Interestingly, for those compounds that were inactive against HIV-1 RNase H, portions of the inhibitor outside of the three-oxygen pharmacophore were predicted to be in steric clash with P537 and/or A538 of the flexible "His loop." This suggests that the "His loop" motif may play an important role in determining inhibitor specificity and potency while providing a potential secondary target for design of next-generation RNase H active site inhibitors.

#### 7.4.1.3 Hydroxylated Tropolones

Using a fluorescence-based assay system (Parniak et al. 2003; Shaw-Reid et al. 2005) to screen a large natural product library, β-thujaplicinol was identified as a potent and selective inhibitor of HIV-1 RNase H (IC50=0.21 μM) (Budihas et al. 2005). Derived from the plant *Thuja plicata*, β-thujaplicinol is an α-hydroxytropolone and, as such, possesses a three-oxygen pharmacophore similar to those of hydroxyimides and diketoacids. β-Thujaplicinol is also active against HIV-2 RNase H, which precludes binding of the inhibitor to the NNRTI binding site, but is only weakly active or inactive against *E. coli* and human RNases H, as well as the polymerase activity of HIV-1 RT. Despite its in vitro potency and selectivity, β-thujaplicinol lacks antiviral activity in cell culture.

The mechanism of β-thujaplicinol-mediated inhibition of HIV-1 RNase H activity has been investigated in detail. One study examines the effects of the inhibitor on stabilized RT-RNA/DNA complexes, as well as in experiments in which the order of addition of enzymatic reaction components is varied (Beilhartz et al. 2009). In HIV-1 RT-RNA/DNA complexes stabilized in either a "pre-" or "post-translocation" complex, β-thujaplicinol did not inhibit RNase H cleavage, suggesting that it can neither bind RT nor inhibit RT-associated RNase H when the RNase H active center is occupied by substrate. In order-of-addition experiments, no inhibition of RNase H activity was observed unless RT, $Mg^{+2}$, and inhibitor were pre-incubated and the reaction initiated by addition of substrate. Moreover, cleavage reactions, although slowed, eventually ran to completion, regardless of the order in which the reaction components were added. Taken together, these data strongly suggest that β-thujaplicinol (i) inhibits HIV-1 RNase H by binding the Mg ions at the RNase H active site, (ii) competes with substrate for active site occupancy, and (iii) will eventually be displaced by substrate over the course of the reaction.

This apparently competitive mode of inhibition differs from previous reports of β-thujaplicinol and other active site inhibitors behaving noncompetitively (i.e., are able to bind enzyme in the absence or presence of substrate) (Budihas et al. 2005; Tramontano et al. 2005). To reconcile this apparent discrepancy, it has been suggested that β-thujaplicinol selectively inhibits secondary RNase H cleavage and/or binds the enzyme-product complex once the primary cleavage is complete. Moreover, independent biochemical data suggest that an RNA/DNA hybrid can associate with a preformed RT-β-thujaplicinol complex, even if the reverse is not possible (Beilhartz et al. 2009; Budihas et al. 2005; Himmel et al. 2009). These findings are consistent with both classical noncompetitive kinetics and the observation that β-thujaplicinol cannot bind the enzyme-substrate complex.

Crystal structures of either HIV-1 RT or the isolated RNase domain in complex with β-thujaplicinol confirm that the inhibitor binds at the RNase H active site via metal-ion coordination (Fig. 7.5b) (Himmel et al. 2009). As was the case with the N-hydroxyimide and compound **5**, β-thujaplicinol forms secondary contacts with the flexible "His loop," thereby stabilizing it in a position near the RNase H active site. The side chain of R557 is also in close proximity to bound β-thujaplicinol, albeit on the side opposite that of the "His loop." Superimposition of RT-β-thujaplicinol and human RNase H1-RNA/DNA co-crystal structures (Himmel et al. 2009; Nowotny et al. 2007) predicts that β-thujaplicinol binding would be sterically incompatible with both the scissile phosphate in the RNA strand and the water molecule positioned for nucleophilic attack. Consequently, it appears that a molecular complex containing HIV-1 RT, an RNA/DNA hybrid, and β-thujaplicinol could only exist if the trajectory of the hybrid was altered so that it was accommodated normally, or nearly so, in the template-primer binding cleft, but was displaced relative to the RNase H active site.

Another hydroxytropolone, manicol (compound **11**), is structurally similar to β-thujaplicinol except for a bulky aliphatic substituent on the tropolone ring. Manicol exhibits a high degree of inhibitory potency against HIV-1 RT-associated RNase H in vitro (IC50=0.6 μM) but, like β-thujaplicinol, has no antiviral activity in cell culture (Chung et al. 2011). In an HIV RT co-crystal structure, TMC278 (rilpivirine), an NNRTI, is bound in the NNRTI binding site, while manicol coordinates $Mn^{+2}$ in the RNase H active site (Chung et al. 2011) (Fig. 7.5c). Although the aliphatic substituent of manicol generally projects away from the RNase H active site, multiple carbon atoms are within 4 Å of the H539 imidazole side chain in the stabilized "His loop." Accordingly, the 2-isopropanyl group at the apex of the aliphatic substituent was selected for derivatization for the purpose of increasing the number of secondary enzyme-inhibitor contacts and improving the binding/efficacy of the inhibitor.

While differences in inhibitor potency among manicol and manicol derivatives were modest, several derivatives displayed antiviral activity in cell culture – the most notable being compound **12** (Chung et al. 2011). The mechanistic basis for this improved antiviral function is unclear, although increased hydrophobicity resulting in improved cellular entry may have been instrumental. Compounds **13** and **14** also inhibited the DNA polymerase activity of HIV-1 RTs containing either an E478Q or a D549A mutation. Since both of these mutations preclude binding of one or both metals at the RNase H active site and binding of hydroxytropolones to the active site is metal-ion dependent, it is possible that inhibition of DNA synthesis by these manicol derivatives is the result of binding to an unknown second site within RT.

#### 7.4.1.4 Pyrimidinol Carboxylic Acids

A fourth scaffold for the metal-binding three-oxygen pharmacophore can be found in pyrimidinol carboxylic acid derivatives (Kirschberg et al. 2009). Several of these compounds are active against HIV-1 RT-associated RNase H, with IC50s in the low to submicromolar range, and are only weakly active or inactive against human RNase H1. One derivative (compound **15**) was successfully crystallized together with $Mn^{+2}$ and the isolated RNase H domain of HIV-1 RT containing the basic loop-helix motif of *E. coli* RNase H (Fig. 7.5d). In this structure, resolved to 1.7 Å, the carboxyl group of the inhibitor is shown to coordinate metal ion B, while metal ion A is coordinated by two phenolic oxygen atoms. Moreover, C2 in the phenyl substituent interacts directly with imidazole side chain of H539, possibly through an edge-to-face π-π interaction. None of the pyrimidinol carboxylic acid derivatives possess antiviral activity in cell culture.

### *7.4.2 Allosteric Inhibitors*

Allosteric inhibition of RNase H activity is an attractive alternative to active site inhibitors. Allosteric inhibitors need not contain the negatively charged three-oxygen pharmacophore necessary for metal-ion coordination and therefore have potential for greater cell permeability. Moreover, binding of such inhibitors is not limited to the face of the active site, or even the RNase H domain, so sites more amenable to high-affinity interaction with small molecules can be utilized.

As previously noted, NNRTIs are excellent examples of allosteric inhibitors. Originally identified as potent inhibitors of the HIV-1 RT polymerase activity, NNRTIs bind near the base of the p66 thumb subdomain 10–15 Å from the DNA polymerase active site (K. Das et al. 2012; Kohlstaedt et al. 1992). This binding site is created by an induced fit mechanism in the presence of inhibitor and is not apparent otherwise. NNRTIs have also been shown to modulate RT-associated RNase H activity, indicating that inhibitor binding may induce global structural changes affecting positioning of the RNase H domain, substrate binding, or both. Two other classes of allosteric RNase H inhibitors have been identified: N-acyl hydrazones and vinylogous ureas, the latter of which affect RNase H activity almost exclusively (Fig. 7.6).

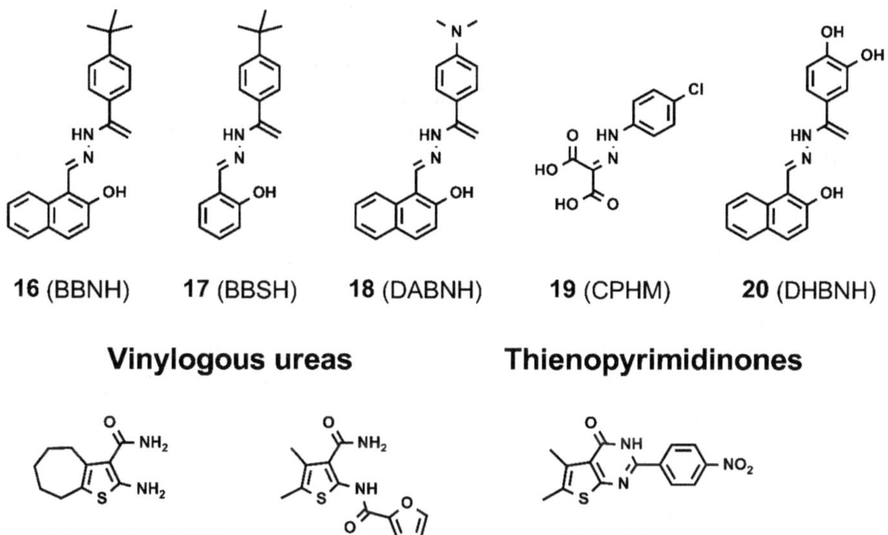

N-acyl hydrazones

**16** (BBNH) **17** (BBSH) **18** (DABNH) **19** (CPHM) **20** (DHBNH)

Vinylogous ureas

Thienopyrimidinones

**21** (NSC727447) **22** (NSC747448) **23** (NSC747665)

**Fig. 7.6** Chemical structures of allosteric RNase H inhibitors

### 7.4.2.1 N-Acyl Hydrazones

N-(4-tert-butylbenzoyl)-2-hydroxy-1-naphthaldehyde hydrazine (BBNH, compound **16**) is an N-acyl hydrazone shown to inhibit HIV-1 RT polymerase and RNase H activities with equal efficiency (IC50s range from 0.8 to 15 μM, depending on the assay) (Borkow et al. 1997). The compound also inhibits *E. coli* and Moloney murine leukemia virus (Mo-MLV) RT-associated RNases H (IC50s = 2.7 and 0.8 μM), but not the polymerase and RNase H activities of HIV-2 RT (IC50s > 50 μM). BBNH displays moderate antiviral activity in MT4 cell culture (IC50 = 5 μM) but is cytostatic or cytotoxic at concentrations above 10 or 25 μM, respectively.

Inhibition of HIV-1 RNase H by BBNH requires pre-incubation of RT and inhibitor, is reversible, and is linearly competitive with nucleic acid substrate (Borkow et al. 1997). In an experiment designed to explore potential interaction(s) between BBNH and NNRTIs, BBNH was found to inhibit inactivation of HIV-1 RT polymerase function by a photoactivatable analog of nevirapine. With respect to RNase H activity, however, the potency of BBNH was not diminished in the presence of NNRTIs UC38 and TIBO nor was it affected after photo-cross-linking a nevirapine derivative to the catalytic RT. Moreover, BBNH inhibited the polymerase and RNase H activities of RTs containing any of a number of NNRTI resistance mutations (K103N, Y181I, Y188H, or Y188L) to approximately the same degree as the wild-type enzyme – with one exception. Although BBNH inhibited the polymerase activities of wild-type and Y181C RT with equal potency, the inhibitor was completely ineffective against the RNase H activity of the mutant enzyme (IC50 > 50 μM).

Taken together, these data suggested that the BBNH binds RT at a site different from, but perhaps overlapping, the NNRTI binding site. Certainly, NNRTI binding and/or activity is affected by BBNH, although it is not clear that the reverse is true. It was further proposed that BBNH-mediated inhibition of RNase H activity is caused by binding of the inhibitor to a second site closer to or within the RNase H domain (Borkow et al. 1997). This notion is supported by the apparent independence of BBNH and NNRTI function with respect to inhibition of RNase H activity, as well as the observation that BBNH inhibits *E. coli* RNase H, an enzyme with little homology to the polymerase domain or NNRTI binding site within HIV-1 RT.

A molecular model of the putative second binding site was generated in which the hydrazone moiety of BBNH was proposed to interact with the metal ions at the RNase H active site, while the distal aromatic ring of the inhibitor participated in a stacking interaction with the side chain of RNase H primer grip residue Y501 (Arion et al. 2002). To test this model, HIV-1 RTs containing a series of Y501 substitution mutations were expressed, purified, and assayed for enzyme activity in the presence and absence of inhibitor. In large part, the study supported involvement of Y501 in inhibitor binding. Although the RNase H activity of Y501F RT remained sensitive to BBNH, RT containing a Y501R mutation, which would likely be less amenable to π stacking, was completely resistant.

The activity profiles of BBNH analogs were in some cases shown to be dramatically different from those of the parent compound (Arion et al. 2002; Klumpp and Mirzadegan 2006; Sluis-Cremer et al. 2002). For example, BBSH (compound **17**), which contains a phenol group in place of the naphthol moiety, inhibited HIV-1 RT polymerase activity (IC50 = 2.5 μM) but not RNase H (IC50 > 50 μM). Conversely, IC50s for DABNH (compound **18**) against RT-associated polymerase and RNase H activities were >40 and 4 μM, respectively. The effects of these compounds on the stability of the HIV-1 RT heterodimer were also variable. Specifically, BBNH and BBSH reduced the stability of the HIV-1 RT heterodimer, while DABNH appeared to have no effect (Sluis-Cremer et al. 2002). Clearly, the inhibitory profiles of N-acyl hydrazones can be greatly affected by relatively small structural changes.

The chlorophenylhydrazone of mesoxalic acid (CPHM, compound **19**) was identified as an inhibitor of strand transfer catalyzed by HIV-1 RT (Gabbara et al. 1999). This compound inhibits strand transfer, polymerase, and RNase H activities with approximately equal potency (IC50s = 4.5, 2.3, and 3.0 μM, respectively) but possesses no antiviral activity. Interestingly, CPHM did not inhibit the RNase H activity of RT containing a polymerase active site mutation (D185N), while in the reverse experiment, inhibition of RT polymerase activity was unaffected by mutation of the RNase H active site (D443N) (Shaw-Reid et al. 2003). These data suggest the CPHM inhibits both RT activities by binding in the polymerase domain, though perhaps at a site(s) different from that recognized by BBNH. The latter supposition is supported by the markedly different chemical composition of CPHM compared to BBNH, BBSH, and DABNH, which are all relatively similar by comparison (Fig. 7.6).

Another hydrazone (E)-3,4-dihydroxy-N′-((2-methoxynaphthalen-1-yl)methylene)benzohydrazide (DHBNH, compound **20**) potently inhibits HIV-1 RT-associated

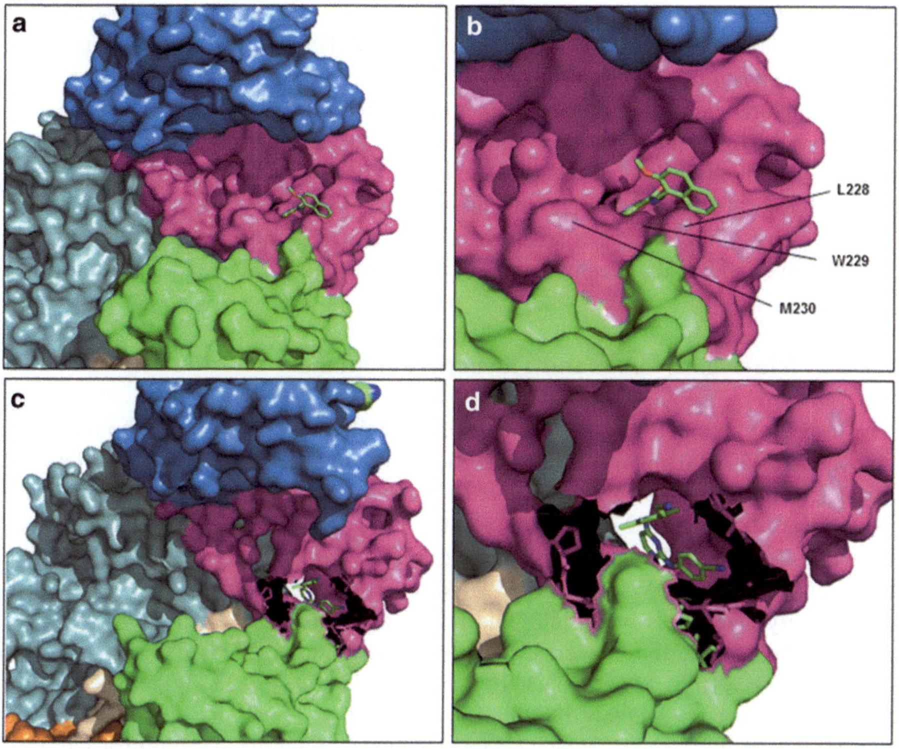


**Fig. 7.7** Allosteric inhibitors bound to HIV-1 RT. (**a**) **20** (DHBNH) binds to the inner surface of the p66 palm subdomain near the base of the p66 thumb (Himmel et al. 2006). The p51 subunit (*gray*) as well as the fingers (*blue*), palm (*magenta*), and thumb subdomains of p66 are highlighted. (**b**) A closer view of the **20** (DHBNH) binding site. Primer grip residues comprising or adjacent to the binding site are highlighted. (**c**) The NNRTI rilpivirine is enveloped within the p66 palm subdomain at the base of the p66 thumb subdomain (Das et al. 2008). In contrast to DHBNH, bound rilpivirine is not directly exposed to the nucleic acid-binding cleft. Instead, this inhibitor accesses its binding site from the opposite side of the enzyme. (**d**) A closer view of rilpivirine binding. Note that in both (**c**) and (**d**), it was necessary to remove p66 residues 227–234 in order to reveal the NNRTI binding site

RNase H (IC50 = 0.5 μM)and is inactive against RT polymerase activity (Himmel et al. 2006). DHBNH also possesses antiviral activity (IC50 4.5 μM), with an improved selectivity index relative to BBNH (CC50 > 100 μM). A co-crystal structure with HIV-1 RT shows DHBNH bound among residues D186, Y188, and W229 of the p66 subunit, approximately 15 Å from the NNRTI binding site and ~50 Å from the RNase H active site (Fig. 7.7). No association between DHBNH and either the RNase H domain or metal ions was observed. Moreover, DHBNH only weakly inhibits RNase H activity catalyzed by the isolated HIV-1 RNase H domain (Himmel et al. 2006), suggesting that its inhibitory mechanism is truly allosteric.

It is difficult to reconcile the single DHBNH binding site revealed in the co-crystal structure with either exclusive inhibition of HIV-1 RNase H activity or the two-binding-site model proposed for BBNH. One explanation is that crystal packing prevented DHBNH binding to the second site or that binding to the second site is somehow substrate dependent. Alternatively, DHBNH and other N-acyl hydrazones may bind only to the site identified in the crystal structure, affecting subtle changes in RT structure and/or flexibility that alter the way the substrate is accommodated by the enzyme (Himmel et al. 2006). These changes might then manifest themselves in either reduced polymerase or RNase H activity or both, depending on the structure of the inhibitor, its effect on RT, and the nature of the perturbation in substrate binding. Regardless of which hypothesis is proven correct, it is clear that additional study is required to elucidate the mechanism(s) of action of this complex class of inhibitors.

#### 7.4.2.2 Vinylogous Ureas and Thienopyrimidinones

The lead vinylogous urea compounds were identified in a high-throughput robotic screen of ~230,000 natural and synthetic compounds from a series of National Cancer Institute libraries. 2-Amino-5,6,7,8-tetrahydro-4H-cyclohepta[b]thiophene-3-carboxamide (NSC727447, compound **21**) and N-[3-(aminocarbonyl)-4,5-dimethyl-2-thienyl]-2-furancarboxamide (NSC747448, compound **22**) were shown to potently inhibit HIV-1 and HIV-2 RNases H, with IC50s of 2.0 and 3.2 μM, respectively, but only weakly inhibited HIV-1 replication in cell culture (Wendeler et al. 2008). Both inhibitors were also active against human RNase H1 to a moderate degree (IC50s = 10.6 μM and 29 μM for NSC727447 and NSC727448, respectively), although neither appreciably inhibited *E. coli* RNase H activity (IC50s >100 and 73 μM). Yonetani-Theorell analysis demonstrated that while the inhibitory activities of NSC727447 and β-thujaplicinol are mutually exclusive, only the vinylogous ureas bind RT in the presence of substrate. These data suggest that vinylogous ureas and RNase H active site inhibitors differ in their inhibitory mechanisms.

Structure-activity relationship analysis of vinylogous ureas and their cyclized counterparts, thienopyrimidinones, demonstrated that in addition to the 2-amino-thiophene-3-carboxamide core shared by NSC727447 and NSC727448, peripheral functional groups were important for inhibition of RNase H activity as well (Chung et al. 2010). For example, removing or changing the size of the heptane ring in NSC727447 was extremely detrimental to inhibitory function, as was substituting the furan group of NSC727448 with a 3-methyl-4-ethyl-chlorobenzene moiety. However, one thienopyrimidinone, NSC727665 (compound **23**), was found to be more active than either of the lead compounds, with an IC50 of 0.85 μM against HIV-1 RT-associated RNase H. In this compound, the nitro group was found to be essential, as derivatives in which the group was either removed or replaced with a methoxy group were completely devoid of inhibitory function.

Mass spectrometric and mutational studies suggest that the vinylogous ureas bind near the p51-RNase H domain interface at the base of the p51 thumb (Wendeler et al. 2008). Molecular docking of NSC727447 and NSC727665 into this site suggests that the binding pocket is formed by p51 residues K275, V276, C280, and R284, among others, as well as p66 residues G541 and H539 (Chung et al. 2010). The potential involvement of H539 suggests a mechanism of inhibition that may involve interference with the function of the flexible, catalytically important "His loop." Alternatively, because the putative binding site is located at the p66-p51 interface, it is possible that binding of vinylogous ureas and/or thienopyrimidinones affects dimer stability. Scanning mutagenesis of residues K275 through T286 of the p51 thumb subdomain indicates that C280 and T286 are important for binding of NSC727665, as mutation of these residues produced enzymes that were resistant to RNase H inhibition (Chung et al. 2012). Interestingly, these mutations did not affect inhibitory susceptibility to NSC727447, suggesting that this inhibitor may bind to an alternative site or form different contacts with the enzyme.

### *7.4.3 RNase H Inhibitors with Unknown Binding Sites*

For a subset of RNase H inhibitors, binding sites and mechanisms of action remain unknown. These include quinones and naphthoquinones, nucleotides and dinucleotides, mappicine analogs, thiocarbamates, and triazoles (Fig. 7.8).

#### 7.4.3.1 Quinones and Naphthoquinones

Illimaquinone (compound **24**), a marine sponge metabolite, is one of the first RNase H inhibitors to be identified. This compound weakly inhibits HIV-1-, HIV-2-, and *E. coli*-associated RNases H, with IC50 values in the 15–50 μM range, and is virtually inactive against the polymerase activities of HIV-1 and HIV-2 RT (Loya et al. 1990; Min et al. 2002). In contrast, the related compounds 6′-hydroxy avarol (compound **25**) and 6′-hydroxy-4-methoxy avarone (compound **26**) inhibit both activities of HIV-1 RT with similar potency (Loya and Hizi 1990).

Limited structure-activity relationship studies leave the mechanism(s) of action of this class of inhibitors unclear. Among naphthoquinone derivatives, the unsubstituted compound (compound **27**) was the most potent against HIV-1 RT-associated RNase H activity, while hydroxylated derivatives (e.g., compound **28**) demonstrated both increased potency against polymerase activity and reduced effectiveness against RNase H (Min et al. 2002). Illimaquinone was shown to be antagonistic with N-ethylmaleimide (NEM), a reagent shown to inhibit RT-associated DNA polymerase activity by covalently modifying residue C280 (Loya and Hizi 1993). This suggests an inhibitory mechanism involving binding to or modification of cysteine side chains, a notion supported by the observation that RT containing a C280S mutation is resistant to illimaquinone-mediated inhibition.

#### 7.4.3.2 Nucleotides and Dinucleotides

The nucleotide analog AZT-MP (compound 29) was demonstrated to weakly inhibit HIV-1 RT-associated RNase H activity, while the di- and triphosphate derivatives of

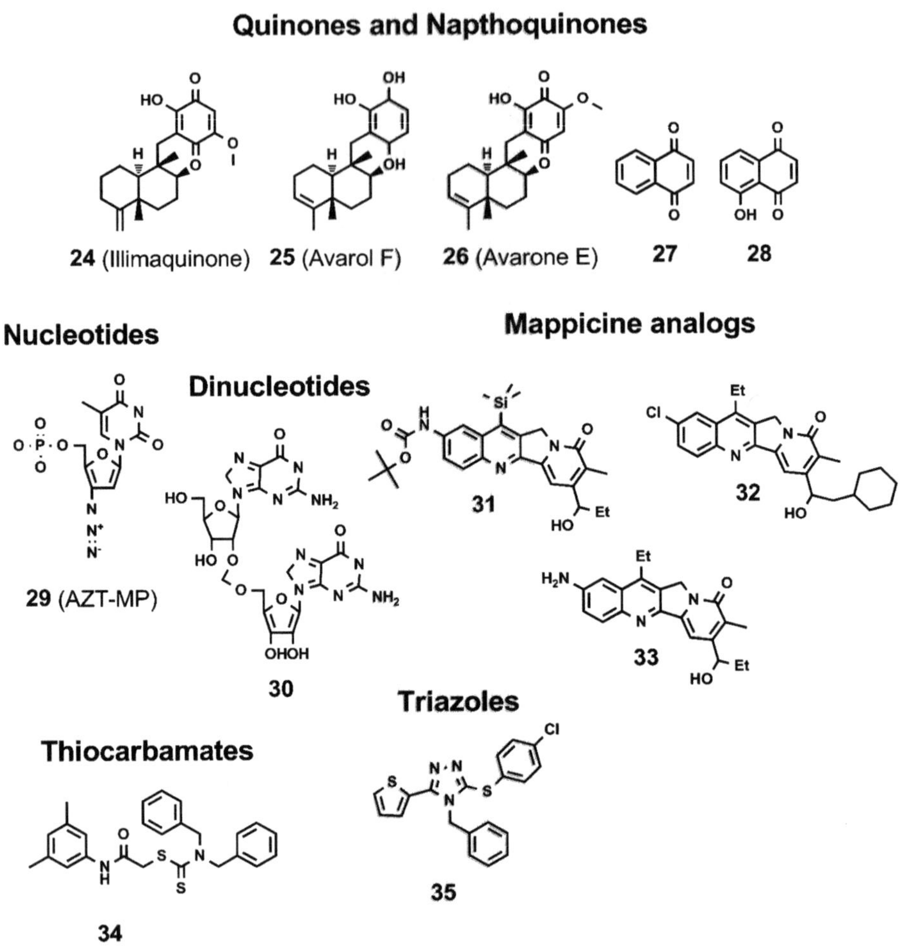


**Fig. 7.8** Chemical structures of RNase H inhibitors having unknown binding sites and mechanisms of action

this compound (AZT-DP and AZT-TP) are completely ineffectual as RNase H inhibitors(Tan et al. 1991). Interestingly, the potency and mechanism of action of AZT-MP was shown to be dependent upon both substrate and the metal ion used to catalyze the reaction (Zhan et al. 1994). Specifically, cleavage of poly-rG:dC was inhibited more effectively in the presence of $Mn^{+2}$ ions (IC50 = 25 μM) than when $Mg^{+2}$ ions were utilized as the RT cofactor (IC50 = 250 μM). Moreover, kinetic analysis revealed that the mechanisms for inhibiting RNase H in reactions utilizing poly-rG:dC or poly-rA:dT as substrate were uncompetitive/mixed or competitive, respectively. The AZT-MP binding site for inhibition of RT-associated RNase H activity remains unresolved.

Among unmodified nucleotide dimers, diguanosine (rGrG) was found to be the most potent against HIV-1 RT-associated RNase H activity (IC50 = 15 μM) (Allen et al. 1996). However, this dinucleotide did not inhibit RT-associated polymerase activity nor did it inhibit either *E. coli* RNase H1 or RNase H activity in HeLa cell nuclear extracts. Although not a particularly robust inhibitor, the potency of diguanosine against RT-associated RNase H could be improved by substitution of the 5′-3′ diester linkage with a 5′-2′-formacetal (IC50 = 5 μM) (compound **30**).

#### 7.4.3.3 Mappicine Analogs, Thiocarbamates, and Triazoles

Select mappicine analogs (e.g., compounds **31–33**), thiocarbamates (e.g., compound **34**), and triazoles (e.g., compound **35**) have been described as selective HIV-1 RNase H inhibitors, some of which also possess antiviral activity in cell culture (Curran et al. 2004; Di Grandi et al. 2010). In the case of mappicine analogs, for example, antiviral IC50 values ranging from 2.1 to 10 μM have been reported. In addition, virus containing NNRTI resistance mutations (i.e., K103N/Y181C or Y106A/Y181C) remained susceptible to mappicine analogs, suggesting that these RNase H inhibitors do not bind at the NNRTI binding site.

Since none of these compounds contain substituents that resemble the three-oxygen pharmacophores common among RNase H active site inhibitors, it is likely that they act allosterically. However, the actual binding sites and mechanisms of action of these inhibitors remain unknown.

## 7.5 RNase H and DNA Polymerase Inhibitor Interaction

Highly active antiretroviral therapy, or HAART, refers to multidrug regimens that target multiple HIV proteins/functions simultaneously (Asahchop et al. 2012). Because HAART is currently the best clinical options for controlling HIV infection, it is important to determine whether inhibitors developed to specifically target HIV RT-associated RNase H are also compatible with the NRTI and NNRTI components of these regimens.

While there is no evidence to suggest that NRTIs affect the function of RNase H inhibitors or vice versa, it has been speculated that inhibition of RNase H activity by any means may decrease the sensitivity of RT to the thymidine analog AZT (azidothymidine, zidovudine) (Delviks-Frankenberry et al. 2007; Ehteshami et al. 2008; Nikolenko et al. 2007). Sequencing of the RT gene in AZT-resistant clinical isolates revealed a common set of mutations proximal to the polymerase active site. These mutations, collectively referred to as thymidine analog mutations, or TAMs, were determined to be sufficient to confer AZT resistance to RT by increasing the rate at which the enzyme catalyzed AZT excision. Other mutations (e.g., N348I, A360V, and D549N) within the connection subdomain and RNase H domain of the enzyme were subsequently found to enhance AZT resistance in these enzymes, although the mechanism by which this occurred was unclear. It was speculated that the mutations in the connection subdomain and RNase H domain reduced RNase H activity in the AZT-resistant enzymes and that doing so prolonged the residency time of RT on the RNA/DNA substrate when minus-strand synthesis was prematurely terminated by incorporation of AZT. This, in turn, would allow more time for RT to excise the chain terminator and resume DNA synthesis (Delviks-Frankenberry et al. 2010).

Kinetic analysis of recombinant enzymes containing the connection and RNase H domain mutations partially confirmed this hypothesis. Several of these mutations did, in fact, reduce RNase H activity, and this reduction generally correlated with increased AZT resistance independent of TAMs (Delviks-Frankenberry et al. 2008). Some connection subdomain mutations, however, appeared to directly affect the rate of AZT excision, possibly by affecting how the RNA/DNA substrate is positioned within the template-primer binding cleft. The relative importance of these RNase H-dependent and RNase H-independent mechanisms of enhancing AZT resistance remains to be determined, as does the potential for antagonism between RNase H inhibitors and AZT. In a recent study, it was determined that four of five RNase H inhibitors tested had little effect on the AZT susceptibility of HIV in cell culture, while the fifth decreased susceptibility approximately fivefold (Davis et al. 2011). None of the RNase H inhibitors had any effect on HIV susceptibility to another NRTI, the cytidine analog 3TC (lamivudine).

It is worth noting that resistance to NRTIs other than AZT is achieved not by increasing the rate at which chain terminators are excised from nascent DNA but rather by active site discrimination, which reduces the frequency with which nucleoside analogs are incorporated (Sluis-Cremer et al. 2000). Consequently, impairing RNase H function by inhibition or mutation should have no effect on resistance to these inhibitors, given that there would be no obvious advantage to increasing residency time on a chain-terminated substrate. This notion is confirmed in experiments with d4T (stavudine)-resistant RT, where introduction of N348I and A360V connection subdomain mutations had no effect on inhibitor susceptibility (Ehteshami et al. 2008).

Studies have shown that interactions between NNRTIs and active site-binding RNase H inhibitors vary with the compounds being evaluated, the activity being measured, and the assay being utilized. In addition to the aforementioned mixed interactions between NNRTIs and the acyl hydrazone BBNH (Borkow et al. 1997), efavirenz and diketoacid analogs were found to behave antagonistically with respect to RNase H inhibition, additively in an RNA-dependent DNA synthesis assay, and synergistically in an assay system requiring the coordinated completion of multiple stages of reverse transcription (Shaw-Reid et al. 2005). In contrast, another NNRTI, calanolide A, and β-thujaplicinol were shown to behave synergistically with respect to RNase H inhibition (Budihas et al. 2005). These mixed results are perhaps not surprising, given the allosteric mechanism of NNRTIs and their capacity to either agonize or antagonize RNase H function, depending on the compound. Moreover, the inhibitory potencies of NNRTIs have been shown to be partially dependent upon substrate sequence and structure (Hang et al. 2007), rendering potential interactions between NNRTIs and RNase H inhibitors highly context dependent.

## 7.6 Conclusions and Perspectives

Success in developing potent and specific RNase H inhibitors has been mixed. Active site inhibitors have been developed with submicromolar IC50s against HIV-1 RT-associated RNase H, and most are only weakly active against RT polymerase and *E. coli* RNase H activities. Several X-ray co-crystal structures have elucidated how and where these inhibitors bind in the RNase H active center, as well as how secondary contacts with the "His loop" might be exploited in future rounds of derivatization. However, most of the current RNase H active site inhibitors have low selectivity indices, most likely due to poor cellular entry and/or cytotoxicity, and possess minimal to moderate antiviral activity.

Because they do not require the three-oxygen pharmacophore necessary for coordination of tandem metal ions and can theoretically bind any site on the surface of HIV-1 RT, allosteric RNase H inhibitors perhaps offer greater promise as therapeutic agents. This potential is further accentuated by the observation that some compounds (NNRTIs, BBNH) inhibit the polymerase and RNase H activities of RT simultaneously. Both N-acyl hydrazones and thienopyrimidinones have been shown to possess strong to moderate antiviral activities in cell culture; however, no member of either class of compounds has as yet been proven sufficiently effective to be entered into clinical trials.

Inclusion of both an IN active site inhibitor and NNRTIs in current HAART regimens offers hope that comparable active site and allosteric RNase H inhibitors may be developed. Engineering RNase H active site inhibitors to more actively engage the flexible "His loop," or even select bases within the RNA substrate, could increase both the potency and selectivity of these compounds. Pharmacokinetics might also be improved by synthesizing charge-neutral, inert derivatives of active site inhibitors that might be activated by cellular enzymes after passing through the plasma membrane. With respect to allosteric RNase H inhibitors, identifying the binding sites and determining the mechanisms of action of thienopyrimidinones, mappicine analogs, and other inhibitory compounds is an important first step toward improving their efficacy. Moreover, the development of fluorescent assay systems for measuring RT-associated RNase H activity will undoubtedly accelerate identification of new RNase H inhibitors.

## References

Allen SJW, Krawczyk SH, McGee LR, Bischofberger N, Mulato AS, Cherrington JM (1996) Inhibition of HIV-1 RNase H activity by nucleotide dimers and monomers. Antivir Chem Chemother 7(1):37–45

Arion D, Sluis-Cremer N, Min KL, Abram ME, Fletcher RS, Parniak MA (2002) Mutational analysis of Tyr-501 of HIV-1 reverse transcriptase. Effects on ribonuclease H activity and inhibition of this activity by N-acylhydrazones. J Biol Chem 277(2):1370–1374

Asahchop EL, Wainberg MA, Sloan RD, Tremblay CL (2012) Antiviral drug resistance and the need for development of new HIV-1 reverse transcriptase inhibitors. Antimicrob Agents Chemother. doi:10.1128/AAC.00591-12

Beilhartz GL, Wendeler M, Baichoo N, Rausch J, Le Grice S, Gotte M (2009) HIV-1 reverse transcriptase can simultaneously engage its DNA/RNA substrate at both DNA polymerase and RNase H active sites: implications for RNase H inhibition. J Mol Biol 388(3):462–474

Borkow G, Fletcher RS, Barnard J, Arion D, Motakis D, Dmitrienko GI, Parniak MA (1997) Inhibition of the ribonuclease H and DNA polymerase activities of HIV-1 reverse transcriptase by N-(4-tert-butylbenzoyl)-2-hydroxy-1-naphthaldehyde hydrazone. Biochemistry 36(11): 3179–3185

Budihas SR, Gorshkova I, Gaidamakov S, Wamiru A, Bona MK, Parniak MA et al (2005) Selective inhibition of HIV-1 reverse transcriptase-associated ribonuclease H activity by hydroxylated tropolones. Nucleic Acids Res 33(4):1249–1256

Champoux JJ, Schultz SJ (2009) Ribonuclease H: properties, substrate specificity and roles in retroviral reverse transcription. FEBS J 276(6):1506–1516. doi:10.1111/j.1742-4658.2009.06909.x

Chung S, Wendeler M, Rausch JW, Beilhartz G, Gotte M, O'Keefe BR et al (2010) Structure-activity analysis of vinylogous urea inhibitors of human immunodeficiency virus-encoded ribonuclease H. Antimicrob Agents Chemother 54(9):3913–3921. doi:10.1128/AAC.00434-10

Chung S, Himmel DM, Jiang JK, Wojtak K, Bauman JD, Rausch JW et al (2011) Synthesis, activity, and structural analysis of novel alpha-hydroxytropolone inhibitors of human immunodeficiency virus reverse transcriptase-associated ribonuclease H. J Med Chem 54(13):4462–4473. doi:10.1021/jm2000757

Chung S, Miller JT, Johnson BC, Hughes SH, Le Grice SF (2012) Mutagenesis of human immunodeficiency virus reverse transcriptase p51 subunit defines residues contributing to vinylogous urea inhibition of ribonuclease H activity. J Biol Chem 287(6):4066–4075. doi:10.1074/jbc.M111.314781

Cirino NM, Kalayjian RC, Jentoft JE, Le Grice SF (1993) Fluorimetric analysis of recombinant p15 HIV-1 ribonuclease H. J Biol Chem 268(20):14743–14749

Cowan JA, Ohyama T, Howard K, Rausch JW, Cowan SM, Le Grice SF (2000) Metal-ion stoichiometry of the HIV-1 RT ribonuclease H domain: evidence for two mutually exclusive sites leads to new mechanistic insights on metal-mediated hydrolysis in nucleic acid biochemistry. J Biol Inorg Chem JBIC A Publ Soc Biol Inorg Chem 5(1):67–74

Curran DP, Parniak MA, Gabarda A (2004) USA Patent no. US 2004/0058948 A1. United States Patent Application Publication: U. S. P. Office

Das D, Georgiadis MM (2004) The crystal structure of the monomeric reverse transcriptase from Moloney murine leukemia virus. Structure 12(5):819–829. doi:10.1016/j.str.2004.02.032

Das K, Bauman JD, Clark AD Jr, Frenkel YV, Lewi PJ, Shatkin AJ et al (2008) High-resolution structures of HIV-1 reverse transcriptase/TMC278 complexes: strategic flexibility explains potency against resistance mutations. Proc Natl Acad Sci USA 105(5):1466–1471. doi:10.1073/pnas.0711209105

Das K, Martinez SE, Bauman JD, Arnold E (2012) HIV-1 reverse transcriptase complex with DNA and nevirapine reveals non-nucleoside inhibition mechanism. Nat Struct Mol Biol 19(2):253–259. doi:10.1038/nsmb.2223

Davies JF 2nd, Hostomska Z, Hostomsky Z, Jordan SR, Matthews DA (1991) Crystal structure of the ribonuclease H domain of HIV-1 reverse transcriptase. Science 252(5002):88–95

Davis CA, Parniak MA, Hughes SH (2011) The effects of RNase H inhibitors and nevirapine on the susceptibility of HIV-1 to AZT and 3TC. Virology 419(2):64–71. doi:10.1016/j.virol.2011.08.010

De Clercq E (2009) Anti-HIV drugs: 25 compounds approved within 25 years after the discovery of HIV. Int J Antimicrob Agents 33(4):307–320. doi:10.1016/j.ijantimicag.2008.10.010

Delviks-Frankenberry KA, Nikolenko GN, Barr R, Pathak VK (2007) Mutations in human immunodeficiency virus type 1 RNase H primer grip enhance 3′-azido-3′-deoxythymidine resistance. J Virol 81(13):6837–6845

Delviks-Frankenberry KA, Nikolenko GN, Boyer PL, Hughes SH, Coffin JM, Jere A, Pathak VK (2008) HIV-1 reverse transcriptase connection subdomain mutations reduce template RNA degradation and enhance AZT excision. Proc Natl Acad Sci USA 105(31):10943–10948. doi:10.1073/pnas.0804660105

Delviks-Frankenberry KA, Nikolenko GN, Pathak VK (2010) The "connection" between HIV drug resistance and RNase H. Viruses 2(7):1476–1503. doi:10.3390/v2071476

DeStefano JJ, Bambara RA, Fay PJ (1993) Parameters that influence the binding of human immunodeficiency virus reverse transcriptase to nucleic acid structures. Biochemistry 32(27): 6908–6915

Di Grandi M, Olson M, Prashad AS, Bebernitz G, Luckay A, Mullen S et al (2010) Small molecule inhibitors of HIV RT Ribonuclease H. Bioorg Med Chem Lett 20(1):398–402. doi:10.1016/j.bmcl.2009.10.043

Di Marzo Veronese F, Copeland TD, DeVico AL, Rahman R, Oroszlan S, Gallo RC, Sarngadharan MG (1986) Characterization of highly immunogenic p66/p51 as the reverse transcriptase of HTLV-III/LAV. Science (Washington, DC 1883-) 231(4743):1289–1291

Ehteshami M, Beilhartz GL, Scarth BJ, Tchesnokov EP, McCormick S, Wynhoven B et al (2008) Connection domain mutations N348I and A360V in HIV-1 reverse transcriptase enhance resistance to 3′-azido-3′-deoxythymidine through both RNase H-dependent and -independent mechanisms. J Biol Chem 283(32):22222–22232

Furfine ES, Reardon JE (1991) Reverse transcriptase. RNase H from the human immunodeficiency virus. Relationship of the DNA polymerase and RNA hydrolysis activities. J Biol Chem 266(1):406–412

Gabbara S, Davis WR, Hupe L, Hupe D, Peliska JA (1999) Inhibitors of DNA strand transfer reactions catalyzed by HIV-1 reverse transcriptase. Biochemistry 38(40):13070–13076

Gotte M, Fackler S, Hermann T, Perola E, Cellai L, Gross HJ et al (1995) HIV-1 reverse transcriptase-associated RNase H cleaves RNA/RNA in arrested complexes: implications for the mechanism by which RNase H discriminates between RNA/RNA and RNA/DNA. EMBO J 14(4):833–841

Gotte M, Maier G, Gross HJ, Heumann H (1998) Localization of the active site of HIV-1 reverse transcriptase-associated RNase H domain on a DNA template using site-specific generated hydroxyl radicals. J Biol Chem 273(17):10139–10146

Gotte M, Rausch JW, Marchand B, Sarafianos S, Le Grice SF (2010) Reverse transcriptase in motion: conformational dynamics of enzyme-substrate interactions. Biochim Biophys Acta 1804(5):1202–1212. doi:10.1016/j.bbapap.2009.07.020

Hang JQ, Rajendran S, Yang Y, Li Y, In PW, Overton H et al (2004) Activity of the isolated HIV RNase H domain and specific inhibition by N-hydroxyimides. Biochem Biophys Res Commun 317(2):321–329

Hang JQ, Li Y, Yang Y, Cammack N, Mirzadegan T, Klumpp K (2007) Substrate-dependent inhibition or stimulation of HIV RNase H activity by non-nucleoside reverse transcriptase inhibitors (NNRTIs). Biochem Biophys Res Commun 352(2):341–350

Hansen J, Schulze T, Mellert W, Moelling K (1988) Identification and characterization of HIV-specific RNase H by monoclonal antibody. EMBO J 7(1):239–243

Hare S, Vos AM, Clayton RF, Thuring JW, Cummings MD, Cherepanov P (2010) Molecular mechanisms of retroviral integrase inhibition and the evolution of viral resistance. [Research support, Non-U.S. Gov't]. Proc Natl Acad Sci USA 107(46):20057–20062. doi:10.1073/pnas.1010246107

Hazuda DJ, Felock P, Witmer M, Wolfe A, Stillmock K, Grobler JA et al (2000) Inhibitors of strand transfer that prevent integration and inhibit HIV-1 replication in cells. Science 287(5453):646–650

Himmel DM, Sarafianos SG, Dharmasena S, Hossain MM, McCoy-Simandle K, Ilina T et al (2006) HIV-1 reverse transcriptase structure with RNase H inhibitor dihydroxy benzoyl naphthyl hydrazone bound at a novel site. ACS Chem Biol 1(11):702–712

Himmel DM, Maegley KA, Pauly TA, Bauman JD, Das K, Dharia C et al (2009) Structure of HIV-1 reverse transcriptase with the inhibitor beta-Thujaplicinol bound at the RNase H active site. Structure 17(12):1625–1635

Huang H, Chopra R, Verdine GL, Harrison SC (1998) Structure of a covalently trapped catalytic complex of HIV-1 reverse transcriptase: implications for drug resistance. Science 282(5394): 1669–1675

Jacobo-Molina A, Ding J, Nanni RG, Clark AD Jr, Lu X, Tantillo C et al (1993) Crystal structure of human immunodeficiency virus type 1 reverse transcriptase complexed with double-stranded DNA at 3.0 A resolution shows bent DNA. Proc Natl Acad Sci USA 90(13):6320–6324

Katayanagi K, Miyagawa M, Matsushima M, Ishikawa M, Kanaya S, Ikehara M et al (1990) Three-dimensional structure of ribonuclease H from *E. coli*. Nature 347(6290):306–309

Keck JL, Marqusee S (1995) Substitution of a highly basic helix/loop sequence into the RNase H domain of human immunodeficiency virus reverse transcriptase restores its Mn(2+)-dependent RNase H activity. Proc Natl Acad Sci USA 92(7):2740–2744

Keck JL, Goedken ER, Marqusee S (1998) Activation/attenuation model for RNase H. A one-metal mechanism with second-metal inhibition. J Biol Chem 273(51):34128–34133

Kirschberg TA, Balakrishnan M, Squires NH, Barnes T, Brendza KM, Chen X et al (2009) RNase H active site inhibitors of human immunodeficiency virus type 1 reverse transcriptase: design, biochemical activity, and structural information. J Med Chem 52(19):5781–5784. doi:10.1021/jm900597q

Klumpp K, Mirzadegan T (2006) Recent progress in the design of small molecule inhibitors of HIV RNase H. Curr Pharm Des 12(15):1909–1922

Klumpp K, Hang JQ, Rajendran S, Yang Y, Derosier A, Wong Kai In P et al (2003) Two-metal ion mechanism of RNA cleavage by HIV RNase H and mechanism-based design of selective HIV RNase H inhibitors. Nucleic Acids Res 31(23):6852–6859

Kohlstaedt LA, Wang J, Friedman JM, Rice PA, Steitz TA (1992) Crystal structure at 3.5 A resolution of HIV-1 reverse transcriptase complexed with an inhibitor. Science 256(5065): 1783–1790

Lansdon EB, Liu Q, Leavitt SA, Balakrishnan M, Perry JK, Lancaster-Moyer C et al (2011) Structural and binding analysis of pyrimidinol carboxylic acid and N-hydroxy quinazolinedione HIV-1 RNase H inhibitors. Antimicrob Agents Chemother 55(6):2905–2915. doi:10.1128/AAC.01594-10

Loya S, Hizi A (1990) The inhibition of human immunodeficiency virus type 1 reverse transcriptase by avarol and avarone derivatives. FEBS Lett 269(1):131–134

Loya S, Hizi A (1993) The interaction of illimaquinone, a selective inhibitor of the RNase H activity, with the reverse transcriptases of human immunodeficiency and murine leukemia retroviruses. J Biol Chem 268(13):9323–9328

Loya S, Tal R, Kashman Y, Hizi A (1990) Illimaquinone, a selective inhibitor of the RNase H activity of human immunodeficiency virus type 1 reverse transcriptase. Antimicrob Agents Chemother 34(10):2009–2012

Ma JB, Yuan YR, Meister G, Pei Y, Tuschl T, Patel DJ (2005) Structural basis for 5′-end-specific recognition of guide RNA by the *A. fulgidus* Piwi protein. Nature 434(7033):666–670

Min BS, Miyashiro H, Hattori M (2002) Inhibitory effects of quinones on RNase H activity associated with HIV-1 reverse transcriptase. Phytother Res PTR 16(Suppl 1):S57–S62

Nikolenko GN, Delviks-Frankenberry KA, Palmer S, Maldarelli F, Fivash MJ Jr, Coffin JM, Pathak VK (2007) Mutations in the connection domain of HIV-1 reverse transcriptase increase 3′-azido-3′-deoxythymidine resistance. Proc Natl Acad Sci USA 104(1):317–322

Nowotny M (2009) Retroviral integrase superfamily: the structural perspective. EMBO Rep 10(2):144–151. doi:10.1038/embor.2008.256

Nowotny M, Gaidamakov SA, Crouch RJ, Yang W (2005) Crystal structures of RNase H bound to an RNA/DNA hybrid: substrate specificity and metal-dependent catalysis. Cell 121(7):1005–1016

Nowotny M, Gaidamakov SA, Ghirlando R, Cerritelli SM, Crouch RJ, Yang W (2007) Structure of human RNase H1 complexed with an RNA/DNA hybrid: insight into HIV reverse transcription. Mol Cell 28(2):264–276

Palaniappan C, Fuentes GM, Rodriguez-Rodriguez L, Fay PJ, Bambara RA (1996) Helix structure and ends of RNA/DNA hybrids direct the cleavage specificity of HIV-1 reverse transcriptase RNase H. J Biol Chem 271(4):2063–2070

Parkes KE, Ermert P, Fassler J, Ives J, Martin JA, Merrett JH et al (2003) Use of a pharmacophore model to discover a new class of influenza endonuclease inhibitors. J Med Chem 46(7): 1153–1164

Parniak MA, Min KL, Budihas SR, Le Grice SF, Beutler JA (2003) A fluorescence-based high-throughput screening assay for inhibitors of human immunodeficiency virus-1 reverse transcriptase-associated ribonuclease H activity. Anal Biochem 322(1):33–39. doi:10.1016/j.ab.2003.06.001

Peliska JA, Benkovic SJ (1992) Mechanism of DNA strand transfer reactions catalyzed by HIV-1 reverse transcriptase. [Research Support, U.S. Gov't, P.H.S.]. Science 258(5085):1112–1118

Rausch JW, Le Grice SF (2004) 'Binding, bending and bonding': polypurine tract-primed initiation of plus-strand DNA synthesis in human immunodeficiency virus. Int J Biochem Cell Biol 36(9):1752–1766. doi:10.1016/j.biocel.2004.02.016

Ren J, Nichols C, Bird L, Chamberlain P, Weaver K, Short S et al (2001) Structural mechanisms of drug resistance for mutations at codons 181 and 188 in HIV-1 reverse transcriptase and the improved resilience of second generation non-nucleoside inhibitors. J Mol Biol 312(4):795–805. doi:10.1006/jmbi.2001.4988

Rodgers DW, Gamblin SJ, Harris BA, Ray S, Culp JS, Hellmig B et al (1995) The structure of unliganded reverse transcriptase from the human immunodeficiency virus type 1. Proc Natl Acad Sci USA 92(4):1222–1226

Rothwell PJ, Waksman G (2005) Structure and mechanism of DNA polymerases. Adv Protein Chem 71:401–440. doi:10.1016/S0065-3233(04)71011-6

Ruane PJ, DeJesus E (2004) New nucleoside/nucleotide backbone options: a review of recent studies. J Acquir Immune Defic Syndr 37(Suppl 1):S21–S29

Sarafianos SG, Das K, Tantillo C, Clark AD Jr, Ding J, Whitcomb JM et al (2001) Crystal structure of HIV-1 reverse transcriptase in complex with a polypurine tract RNA:DNA. EMBO J 20(6):1449–1461

Schatz O, Cromme FV, Gruninger-Leitch F, Le Grice SF (1989) Point mutations in conserved amino acid residues within the C-terminal domain of HIV-1 reverse transcriptase specifically repress RNase H function. FEBS Lett 257(2):311–314

Schultz SJ, Champoux JJ (2008) RNase H activity: structure, specificity, and function in reverse transcription. Virus Res 134(1–2):86–103. doi:10.1016/j.virusres.2007.12.007

Schultz SJ, Zhang M, Champoux JJ (2004) Recognition of internal cleavage sites by retroviral RNases H. J Mol Biol 344(3):635–652. doi:10.1016/j.jmb.2004.09.081

Shaw-Reid CA, Munshi V, Graham P, Wolfe A, Witmer M, Danzeisen R et al (2003) Inhibition of HIV-1 ribonuclease H by a novel diketo acid, 4-[5-(benzoylamino)thien-2-yl]-2,4-dioxobutanoic acid. J Biol Chem 278(5):2777–2780. doi:10.1074/jbc.C200621200

Shaw-Reid CA, Feuston B, Munshi V, Getty K, Krueger J, Hazuda DJ et al (2005) Dissecting the effects of DNA polymerase and ribonuclease H inhibitor combinations on HIV-1 reverse-transcriptase activities. Biochemistry 44(5):1595–1606. doi:10.1021/bi0486740

Shih CK, Rose JM, Hansen GL, Wu JC, Bacolla A, Griffin JA (1991) Chimeric human immunodeficiency virus type 1/type 2 reverse transcriptases display reversed sensitivity to nonnucleoside analog inhibitors. Proc Natl Acad Sci USA 88(21):9878–9882

Sluis-Cremer N, Arion D, Parniak MA (2000) Molecular mechanisms of HIV-1 resistance to nucleoside reverse transcriptase inhibitors (NRTIs). Cell Mol Life Sci CMLS 57(10):1408–1422

Sluis-Cremer N, Arion D, Parniak MA (2002) Destabilization of the HIV-1 reverse transcriptase dimer upon interaction with N-acyl hydrazone inhibitors. Mol Pharmacol 62(2):398–405

Song JJ, Smith SK, Hannon GJ, Joshua-Tor L (2004) Crystal structure of argonaute and its implications for RISC slicer activity. Science 305(5689):1434–1437

Stahl SJ, Kaufman JD, Vikic-Topic S, Crouch RJ, Wingfield PT (1994) Construction of an enzymatically active ribonuclease H domain of human immunodeficiency virus type 1 reverse transcriptase. Protein Eng 7(9):1103–1108

Summa V, Petrocchi A, Pace P, Matassa VG, De Francesco R, Altamura S et al (2004) Discovery of alpha, gamma-diketo acids as potent selective and reversible inhibitors of hepatitis C virus NS5b RNA-dependent RNA polymerase. J Med Chem 47(1):14–17. doi:10.1021/jm0342109

Tan CK, Civil R, Mian AM, So AG, Downey KM (1991) Inhibition of the RNase H activity of HIV reverse transcriptase by azidothymidylate. Biochemistry 30(20):4831–4835

Telesnitsky A, Goff SP (1997) Reverse transcriptase and the generation of retroviral DNA. In: Coffin JM, Hughes SH, Varmus HE (eds) Retroviruses. Cold Spring Harbor Laboratory Press, Plainview, pp. 121–160

Tomassini J, Selnick H, Davies ME, Armstrong ME, Baldwin J, Bourgeois M et al (1994) Inhibition of cap (m7GpppXm)-dependent endonuclease of influenza virus by 4-substituted 2,4-dioxobutanoic acid compounds. Antimicrob Agents Chemother 38(12):2827–2837

Tramontano E, Esposito F, Badas R, Di Santo R, Costi R, La Colla P (2005) 6-[1-(4-Fluorophenyl) methyl-1H-pyrrol-2-yl)]-2,4-dioxo-5-hexenoic acid ethyl ester a novel diketo acid derivative which selectively inhibits the HIV-1 viral replication in cell culture and the ribonuclease H activity in vitro. Antiviral Res 65(2):117–124

Tsunaka Y, Haruki M, Morikawa M, Oobatake M, Kanaya S (2003) Dispensability of glutamic acid 48 and aspartic acid 134 for Mn2+–dependent activity of *E. coli* ribonuclease HI. Biochemistry 42(11):3366–3374. doi:10.1021/bi0205606

Tsunaka Y, Takano K, Matsumura H, Yamagata Y, Kanaya S (2005) Identification of single Mn(2+) binding sites required for activation of the mutant proteins of *E. coli* RNase HI at Glu48 and/or Asp134 by X-ray crystallography. J Mol Biol 345(5):1171–1183

Wendeler M, Lee HF, Bermingham A, Miller JT, Chertov O, Bona MK et al (2008) Vinylogous ureas as a novel class of inhibitors of reverse transcriptase-associated ribonuclease H activity. ACS Chem Biol 3(10):635–644

Zhan X, Tan CK, Scott WA, Mian AM, Downey KM, So AG (1994) Catalytically distinct conformations of the ribonuclease H of HIV-1 reverse transcriptase by substrate cleavage patterns and inhibition by azidothymidylate and N-ethylmaleimide. Biochemistry 33(6):1366–1372

# Chapter 8
# Targeting Small Molecules and Peptides to the p66-p51 Reverse Transcriptase Interface

**Daouda Abba Moussa, Audrey Agopian, and Gilles Divita**

## 8.1 Introduction

Thirty years after the discovery of human immunodeficiency virus (HIV) (Barre-Sinoussi et al. 1983; Sepkowitz 2001), the causative agent of acquired immunodeficiency syndrome (HIV/AIDS), and the successful development and approval of antiretroviral drugs, no curative treatment is available (Simon et al. 2006; Saliba and Yeni 2006; Sarafianos et al. 2004). Therefore, there remains an urgent need for new and less toxic drugs that are either active against the emerging drug-resistant viruses or directed to novel targets in the replication cycle, which may complement multidrug combinations. A better understanding of individual steps of the viral replication cycle and of the dynamics during infection has provided major breakthroughs for the development of a wide spectrum of antiviral strategies (Sarafianos et al. 2004; Camarasa et al. 2006). Recently, in order to offer new perspectives for the design of inhibitors, extensive efforts have been made in the synthesis of molecules that target the interface of multi-subunit proteins required for virus entry, replication, and maturation (Nielsen et al. 2005; Camarasa et al. 2006; Divita et al. 1994; Mori et al. 2011; Esposito et al. 2012; Warrilow et al. 2009).

Reverse transcriptase (RT) is a key component of the reverse transcription complex (RTC), catalyzing both RNA-and DNA-dependent DNA polymerase and RNAse H activities required to convert single-stranded viral RNA into double-stranded DNA (Telesnitsky and Goff 1997; Le Grice 1993; Katz and Skalka 1994). Since RT plays a central role in replication, it has been one of the main targets for therapeutic treatment of AIDS (Richman 1996; Parniak and Sluis-Cremer 2000; Esnouf et al. 1995). Currently used HIV RT inhibitors are directed against the polymerase activity of this enzyme and divided in two main classes according to their molecular mechanism: nucleoside/nucleotide RT inhibitors (NRTIs/NtRTIs) that

D.A. Moussa • A. Agopian • G. Divita (✉)
Chemical Biology and Nanotechnology for Therapeutics, Centre de Recherches de Biochimie Macromoléculaire, UMR 5237 CNRS, 1919 Route de Mende, 34293 Montpellier, France
e-mail: gilles.divita@crbm.cnrs.fr

S.F.J. LeGrice and M. Götte (eds.), *Human Immunodeficiency Virus Reverse Transcriptase: A Bench-to-Bedside Success*, DOI 10.1007/978-1-4614-7291-9_8,

act as chain terminators, and non-nucleoside RT inhibitors (NNRTIs) that act as conformational and noncompetitive inhibitors (De Clercq 2002). However, one of the major limitations of RT inhibitors currently administered in the clinic is the rapid emergence of drug-resistant strains (De Clercq 2004). Together, the poor fidelity of DNA polymerase and the high level of errors made by HIV-1 RT during virus replication are responsible for the high genetic variation of HIV and therefore the rapid selection for mutant variants (De Clercq 2004; Coffin 1995).

The biologically active form of RT is an asymmetric heterodimer containing p66 and p51 subunits, each sharing four common subdomains designated palm, fingers, thumb, and connection, while the p66 subunit contains an additional C-terminal RNase H domain (di Marzo Veronese et al. 1986; Fauci 1988; Restle et al. 1990). The p51 subunit is derived from p66 by cleavage between residues Phe440 and Tyr441 by the virus-coded protease. As such both p66 and p51 harbor a DNA polymerase domain, but these individual subunits are catalytically inert as monomers, i.e., a dimeric organization is a prerequisite for activation of both polymerase and RNAse H activities (Katz and Skalka 1994). Since enzymatic activities of HIV-1 RT are entirely dependent on its oligomeric structures, this suggests that inhibiting subunit assembly, or modulation of their quaternary structure, might provide alternative antiviral targets. Indeed, several groups (Divita et al. 1994; Sluis-Cremer and Tachedjian 2002; Wapling et al. 2007; Srivastava et al. 2006) have postulated that the heterodimeric organization of RT constitutes target for the design of new inhibitors and have demonstrated that preventing or controlling RT dimerization has a major impact on virus replication (Divita et al. 1994; Morris et al. 1999a; Depollier et al. 2005). Molecules have been described that modulate RT dimer stability or mimic the protein-protein interface. Several small molecules (Grohmann et al. 2008) or non-nucleoside RT inhibitors (NNRTIs) have been shown to stabilize the dimer (Tachedjian et al. 2001), whereas others impair the stability of RT (Sluis-Cremer et al. 2002; Sluis-Cremer et al. 2000). Short peptides have been described to prevent dimerization and inhibit HIV-1 RT by destabilizing or disrupting subunit interaction (Divita et al. 1994; Morris et al. 1999a; Agopian et al. 2009).

The present chapter will focus on strategies that have been developed to target HIV-1 RT activation and dimerization. Recent advances in the design and mechanism of small molecules and peptides that specifically target the dimer interface, resulting in enzyme inactivation, will be highlighted.

## 8.2 Structural Organization of Heterodimeric HIV-1 RT

### *8.2.1 Reverse Transcriptase: Asymmetric Polymerase Structure*

HIV-1 RT structure and mechanism have been investigated in detail by combining crystallography, steady-state/transient kinetics, and single-molecule assay (Rittinger et al. 1995; Spence et al. 1995; Kati et al. 1992; Gotte et al. 1999; Sarafianos et al. 2009, 2002). RT presents a highly dynamic asymmetric structural organization essential for discriminating between different nucleic acid substrates and to

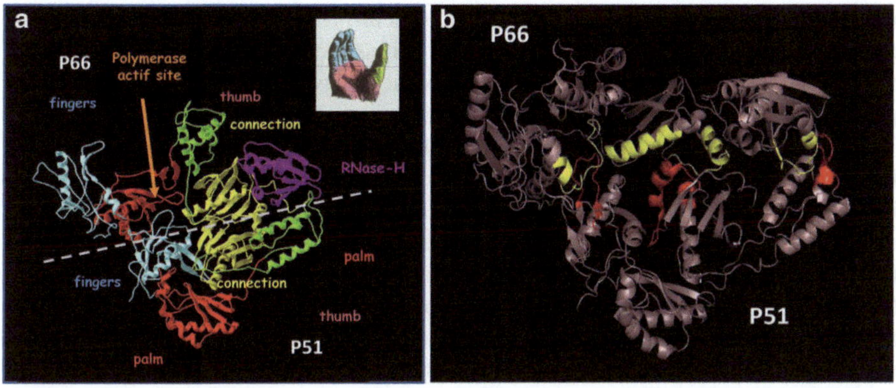


**Fig. 8.1** *Structure of HIV-1 reverse transcriptase. Panel a*: Structure of the open conformation of HIV-1 RT p66 and p51 subunits and location on the structure and on the sequence of p66 of the different subdomains: fingers (*cyan*), palm (*red*), thumb (*green*), connection (*yellow*), and RNase H (*purple*). *Panel b*: Location of the interface "hot spots" on the structure of p66 (*yellow*) and p51 (*red*) of the closed conformation of HIV-1 RT

correctly position them in DNA polymerase and RNase H catalytic sites (Divita et al. 1993a; Liu et al. 2010; Abbondanzieri et al. 2008). The structures of HIV-1 and HIV-2 RTs have been solved in different conformational states (Kohlstaedt et al. 1992; Jacobo-Molina et al. 1993; Huang et al. 1998; Rodgers et al. 1995; Esnouf et al. 1995; Ren et al. 2002) in the absence and presence of substrates or inhibitors, revealing important features of enzyme structure as it pertains to function (Fig. 8.1A). Although the folding of individual p66 and p51 subdomains is similar, their spatial organization differs markedly, especially in the arrangement of their DNA polymerase subdomains. The p51 subunit exhibits a closed compact structural organization with its catalytic site buried, rendering it catalytically inactive. In contrast, the p66 subunit, which contains the active sites for both DNA polymerase and RNase H function, folds into an open, extended structure containing a large active site cleft within the palm subdomain exposing the nucleic acid binding site. Highly conserved motifs in the fingers and palm domains of p66, together with the thumb, act as a clamp that favors correct binding of the primer/template. The primer grip motif is responsible for the appropriate placement of the primer terminus at the DNA polymerase active site and is involved in the translocation of the primer/template following nucleotide incorporation (Sarafianos et al. 2002; Jacques et al. 1994). Although the p51 subunit plays predominantly a structural role in the RT heterodimer, it may promote loading of the p66 subunit on the primer/template and maintain the appropriate enzyme conformation during initiation of reverse transcription.

### 8.2.2 Reverse Transcriptase Dimeric Interface Structure

Heterodimeric RT exhibits a large subunit interface; about 4,800 $A^2$ of protein surface is buried in the RT dimer complex with more than 80 % corresponding to nonpolar atoms (Wang et al. 1994). From a protein-protein interaction point of view, the binding energy is not equally distributed across the dimer interface but localized at a small number of "hot spot" residues, which play a crucial role in stabilizing protein-protein interactions (Fig. 8.1B) (Srivastava et al. 2006; Wang et al. 1994; Tachedjian et al. 2000, 2003). In the p66 subunit, residues forming the dimer interface are derived primarily from the palm, connection, and RNase H domain, while in the p51 they are located in the fingers, thumb, and connection domains. The residues contributing to the total buried surface area upon dimerization are located in three hydrophobic clusters making the largest contribution to total binding strength (Fig. 8.1B). A single region in the palm domain of p66 (D86–L92) interacts with two regions of the fingers of p51 (P52–P55 and S135–P140). The RNase H residues (537–546) in the p66 subunit interact with the p51 thumb residues (280–290), and the p66 connection residues (W406-W410) interact with residues in the p51 connection domain residues (P392-W401) (Fig. 8.1B) (Wang et al. 1994; Divita et al. 1995a). The connection subdomain of HIV-1 RT encompassed a tryptophan repeat motif (residues 398–414) located near the RT dimer interface which is highly conserved in HIV-1, HIV-2, as well as in simian immunodeficiency virus RTs. This tryptophan cluster contains the six residues $W^{398}$, $W^{401}$, $W^{402}$, $W^{406}$, $W^{410}$, and $W^{414}$, which represent hot spots for RT subunit interactions and heterodimerization (Tachedjian et al. 2000, 2003; Divita et al. 1995a, b). Contacts between the two connection subdomains are intrinsically asymmetric and form the only interactions between equivalent subdomains from each subunit.

## 8.3 RT Dimerization Mechanism

### *8.3.1 Dimerization Molecular Mechanism*

Although both subunits carry the catalytic polymerase domain, they are catalytically inert at the monomer state (di Marzo Veronese et al. 1986; Restle et al. 1990; Müller et al. 1989). The two subunits form a highly thermodynamically stable heterodimer with a Gibbs free energy of dimer stabilization of approximately 10–12 kcal/mol and an association constant of about $10^9$ M, which is further stabilized by the binding of primer/template (Divita et al. 1995b). The p66/p51 is the biologically relevant form of RT found, and although homodimers have been constructed in vitro, they are less stable with association constants 1 or 2 log lower (p66/p51 (nM) > p66/p51 (μM) > p51/p51 (mM)) (Restle et al. 1990; Divita et al. 1995b; Müller et al. 1989). Formation of heterodimeric RT has been characterized in detail via structural, kinetic, and functional approaches and was reported to involve a two-step process (Fig. 8.2) (Divita et al. 1995a, b). The first "dimerization" step involves rapid concentration-dependent association of the two subunits mediated primarily by contact between the two connections subdomains and involving the "Trp-cluster", yielding an inactive heterodimeric intermediate. The rate of the first dimerization step increases in the presence of primer/template and $Mg^{++}$ or $Mn^{++}$ (Divita et al. 1993b).

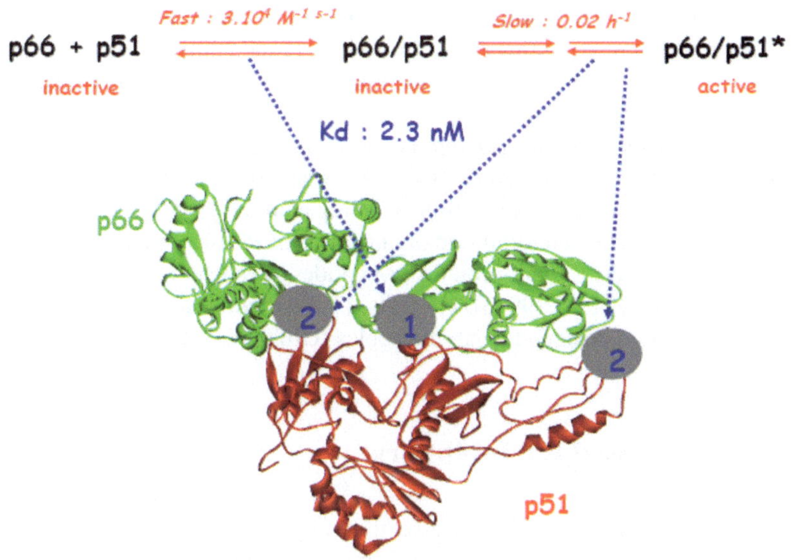


**Fig. 8.2** *Mechanism of RT dimerization.* The dimerization mechanism is a two-step process and the different structural domains involved in each are reported on the structure of RT. (1) The first step is a rapid association of the two subunits mediated by interaction between the two connection subdomains yielding an inactive heterodimeric intermediate enzyme. (2) The second step is a slow isomerization of the intermediate to the mature heterodimeric form of RT, involving the interaction of the thumb domain of p51 with the p66 RNase H domain and placement of the p51 fingers within the p66 palm

The second "maturation" step involves slow isomerization of the intermediate to the mature heterodimeric form of RT, involving an interaction of the p51 thumb subdomain with the RNAse H of p66 and the placement of the p51 fingers within the p66 palm subdomain, stabilizing the polymerase catalytic site. During virus maturation, the RT dimerization scenario is probably more complicated and involves direct multimerization of Gag-Pol precursor polyproteins (Liao and Wang 2004; Zybarth and Carter 1995; Cherry et al. 1998). It was postulated that due to its size and propensity for dimerization, RT is the driving force behind Gag-Pol dimerization and PR activation. Gag-Pol dimerization mechanism initiates by RNAse H domain-mediated p66/p66 association and is followed by PR cleavage (Figueiredo et al. 2006; Sluis-Cremer et al. 2004a; Hornak and Simmerling 2007).

### *8.3.2 Targeting the Protein-Protein Interface: Inhibition of RT Dimerization*

Information-poor small molecules suffer from lack of specificity, side effects, and toxicity, and the therapeutic paradigm that "small is beautiful" is insufficient to overcome the complexity of diseases like AIDS or cancer. Future generations of drug should be based on rational design of information-rich macromolecules that exhibit protein modulatory functions. In protein-protein interactions, the binding energy is not evenly distributed across the dimer interface but involves specific "hot spots" that stabilize protein complexes. With that aim, in 1990, Goody and coworkers suggested that, as the dimeric structure is a prerequisite for activation of both DNA polymerase and RNAse H activities of RT, this was a potential target for the design of a new class of inhibitors (Fauci 1988; Wapling et al. 2007; Srivastava et al. 2006; Müller et al. 1989). The heterodimeric organization of RT has since been a target for new inhibitors, including small proteins or peptides (Restle et al. 1992; Divita et al. 1994, 1995c; Morris et al. 1999a), NNRTIs, (Tachedjian et al. 2001, 2005) and in silico structure-based designed small molecules (Grohmann et al. 2008).

## 8.4 Peptide-Based RT Inhibitors

Divita and coworkers first demonstrated that, as already successfully applied for HIV-1 protease, modulating protein-protein interaction represented a new strategy to inhibit HIV-1 RT by direct binding of a peptide to the subunit interface (Divita et al. 1994; Restle et al. 1990; Menendez-Arias 2002). The three-dimensional structure of RT has been used for rational design of two families of peptides able to interfere with either dimerization (Divita et al. 1994; Restle et al. 1990; Morris et al. 1999a) or maturation (Agopian et al. 2009; Abba-Moussa and Divita, submitted). The use of small peptides targeting hot spots required for RT dimerization constitutes a new strategy to inhibit HIV-1 RT (Divita et al. 1994; Restle et al. 1990) and block viral replication (Morris et al. 1999a; Depollier et al. 2005; Agopian et al. 2009).

### *8.4.1 Peptide-Based RT Dimerization Inhibitors*

The heterodimeric architecture of HIV-1 RT can be altered by small peptides derived from the Trp-cluster of the connection subdomains which mimic interfacial domains and therefore constitute potent antiviral molecules (Divita et al. 1994; Restle et al. 1990; Menendez-Arias 2002). As proof of concept, small peptides (15–19 amino acids) derived from highly conserved sequences within the RT connection subdomain were shown to block RT dimerization and to inhibit its enzymatic activities (Divita et al. 1994; Restle et al. 1990). Peptides corresponding to the Trp repeat motif of the connection subdomain were foreseen as potential inhibitors of dimerization of the enzyme. A peptide containing the four first tryptophan residues $W^{398}$, $W^{401}$, $W^{402}$, and $W^{406}$ of the tryptophan repeat motif (sequence 389–407: FKLPIQKETWETWWTEYWE) was reported to bind both p66 and p51 subunits with a micromolar dissociation constant and reduce RT dimerization in a concentration-dependent manner. That first-generation peptide was used as lead compound for further development of peptide-based inhibition, and a minimum "inhibitory" motif, Pep-7, (KETWETWWTE) was selected. This decapeptide tightly binds RT with a dissociation constant of 100 nM and abolished dimerization in vitro. When associated to nanoparticle carrier (Morris et al. 1997), nanomolar concentrations of Pep-7 ($IC_{50}$: 1.6 nM) were reported to block and to abolish production of viral particles, without any adverse toxic side effects and a selectivity index tenfold higher than for AZT, the reference RT inhibitor used in clinic (Morris et al. 1999a). More recently a cyclic version of Pep-7 has been designed and showed to be more stable and efficient in vivo in a murine model. A major issue in the development of new anti-HIV drugs is their ability to block replication of resistant strains. Pep-7 was reported to be efficient on all HIV-1 and HIV-2 subgroups and to inhibit viral replication of NRTI- and NNRTI-resistant strains. The broad-spectrum activity of Pep-7 is associated to the fact that the Trp-cluster of the connection domain of RT is highly conserved.

From a mechanistic point of view, Pep-7 altered the first step of RT heterodimer formation, i.e., monomer-monomer association. However, this peptide does not induce dissociation of the heterodimer, which was not surprising, due to its very low equilibrium dissociation constant (μM) compared with that of heterodimeric RT (<nM) (Restle et al. 1990; Morris et al. 1999a; Depollier et al. 2005). Pep-7 preferentially interacts with p51 subunit within the heterodimer, and its destabilizing effect is associated with the disruption of close contacts between the fingers and the connection subdomains of the p51 subunit (Depollier et al. 2005). The Pep-7 binding site of Pep-7 was located in a cleft between the fingers, and the connection subdomains of p51 involving the tryptophan cluster of the connection domain of p51 extended to two other highly conserved aromatic residues, $Trp^{24}$ and $Phe^{61}$ (Fig. 8.3A, B). $Trp^{24}$ and $Phe^{61}$ form close contacts with residues of the connection subdomain including $W^{402}$ and $E^{399}$ for $W^{24}$ and $W^{402}$, $T^{403}$, $I^{411}$ for $F^{61}$. These two residues have been shown to be essential for structural organization in the dimeric form of RT and for its biological activities (Depollier et al. 2005; Fisher et al. 2003; Agopian et al. 2007). They are essential for the proper binding of the primer/template and control conformational changes required for RT ordered mechanism. The emergence of resistance to new drug is a major issue to consider when designing new RT inhibitors. So far no resistance to Pep-7 or derived peptides has been selected which is not surprising, taking into account the major role of the Trp-cluster of the connection domain on RT structural integrity and of the residues $Trp^{24}$ and $Phe^{61}$ on RT processing and fidelity (Agopian et al. 2007; Fisher and Prasad 2002). The Pep-7 concentration required to abolish viral replication is about 100-fold lower than the inhibition constant for RT polymerase, suggesting that in vivo and in vitro mechanism of Pep-7 is probably different. Accordingly, Pep-7 was reported to reduce Gag-Pol polyprotein maturation, suggesting that p66/p66 homodimer would be the major target.

### *8.4.2 Peptide Inhibitors of RT Maturation*

Conformational flexibility of HIV-1 RT during maturation constitutes also an attractive target. A major advantage in targeting RT maturation is that it does not

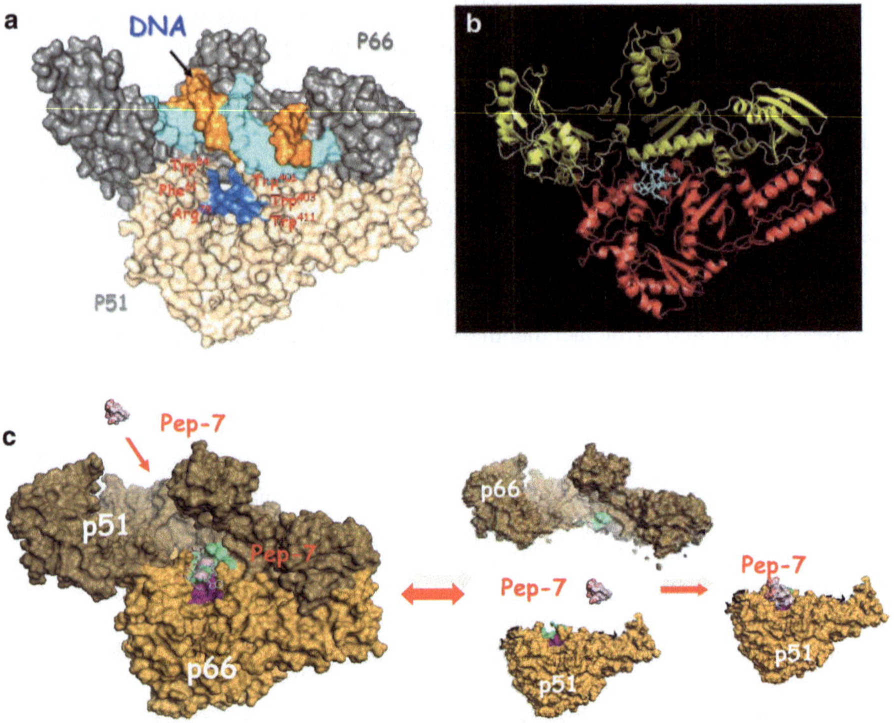


**Fig. 8.3** *Peptide inhibitors of RT dimerization. Panel a*: Location of the binding pocket of Pep-7 (*blue*) and major residues involved in Pep-7 binding (*in red*) on the structure of HIV-1 RT in complex with a primer/template. *Panel b*: Docking of Pep-7 in the structure of HIV-1 RT. Pep-7 (*in cyan*) binds in a newly identified pocket on p51 subunit (*red*) and major contacts involved the Trp-cluster of the connection subdomain. *Panel c*: Scheme of the molecular mechanism of Pep-7. Pep-7 binds the p51 subunit within the heterodimer and disrupts the subunit interface

necessarily require complete dissociation of the p66 and p51 subunits of the heterodimer to interrupt enzymatic function. RT maturation involves contacts between the thumb of p51 and the RNase H of p66 as well as between the fingers of p51 and the p66 palm (Divita et al. 1995a, b; Morris et al. 1999b). The p66 thumb subdomains play an important role in catalysis and integrity of the dimeric RT form. While the p66 thumb is involved in primer/template binding and DNA polymerase activity (Kohlstaedt et al. 1992; Jacobo-Molina et al. 1993; Huang et al. 1998; Jacques et al. 1994), the p51 counterpart is required for conformational changes associated with heterodimer maturation (Morris et al. 1999b) and stabilizing the RNase H domain.

The RT maturation peptide inhibitor $P_{AW}$ (seq: GTKWLTEWIPLTAEAE) was selected from an extremely well-conserved structural motif located between amino acids 284 and 300 of the thumb subdomain, corresponding to the end of helix αI, the loop connecting helices αI and αJ and a part of helix αJ (Agopian et al. 2009; Morris et al. 1999b). $P_{AW}$ is a conformational inhibitor of HIV-1 RT, interfering with the conformational change associated with full enzyme activation (Agopian et al. 2009; Morris et al. 1999b), and is also a potent inhibitor of DNA polymerase function in vitro (Ki: 120 nM). When linked to the carrier peptide Pep-1 (Morris et al. 2001), this peptide also abolishes viral replication with an $IC_{50}$ in the nanomolar range and without any toxic side effects. More recently, $P_{27}$ (GTKWLTEWIPLTA) a truncated version of $P_{AW}$, similarly inhibited DNA-and RNA-dependent DNA polymerase activity (Ki: 50 nM). In comparison to $P_{AW}$, $P_{27}$ exhibits a 15-fold higher antiviral activity, with an $IC_{50}$ of 0.1 nM and does not induce a toxic response with a selectivity index (>5,000) tenfold higher that AZT. Moreover, $P_{27}$ is active against all HIV-1 isolates and multiresistant strains as subnanomolar concentrations of $P_{27}$ were reported to inhibit viral replication of NRTI-and NNRTI-resistant strains (Abba-Moussa et al., submitted).

From a mechanistic standpoint, $P_{AW}$ and $P_{27}$ selectively bind dimeric RT and stabilize the enzyme in an inactive/non-processive conformation. $P_{AW}$ and $P_{27}$ prevent proper binding of both primer/template and dNTP, reducing the affinity of RT for primer/template and dNTP, four- and 62-fold, respectively (Agopian et al. 2009; Abba-Moussa et al., submitted). The dramatic reduction of dNTP affinity for RT by two orders of magnitude and the tenfold increase in the rate of misincorporation suggest that $P_{27}$ altered the conformation of the dNTP binding site. $P_{27}$ and $P_{AW}$ interact preferably with the open conformation of RT and the RT–primer/template complex. The $P_{AW}$ binding site is located between the p66 fingers and thumb and involves residues $W^{24}$ and $F^{61}$ (Agopian et al. 2009). Although both peptides promote RT dimerization, as previously reported for NNRTIs such as efavirenz (EFZ) (Tachedjian et al. 2001; Pauwels 2004; Young et al. 1995), their mechanism and binding site are different, and they are not antagonists of EFZ (Abba-Moussa et al., submitted).

## 8.5 Blocking RT Dimerization or Activation by Small Molecules

Several small nonpeptidic molecules have been described that alter the dimer organization of RT and/or to inhibit dimerization (Fig. 8.4). The binding of small molecules to the dimer interface of HIV-1 RT has been reported to induce conformational changes that affect the overall stability of the heterodimeric complex without causing its dissociation (Warrilow et al. 2009; Parniak and Sluis-Cremer 2000; Sluis-Cremer and Tachedjian 2002; Sluis-Cremer et al. 2002, 2000).

NNRTIs are a class of RT inhibitors that bind in a similar allosteric noncompetitive manner to a hydrophobic pocket of HIV-1 RT located close to the DNA polymerase active site (De Clercq 2004; Kohlstaedt et al. 1992; Jacobo-Molina et al. 1993; Zhan et al. 2011; Sluis-Cremer et al. 2004b). This pocket is exposed upon inhibitor binding and only exists in the HIV-1 enzyme. Upon binding to RT, NNRTIs alters the catalytic activity by maintaining RT in a non-processive conformation (Pauwels 2004; Zhan et al. 2011; Sluis-Cremer et al. 2004b). More than 30 different structures of HIV-1 RT/NNRTI cocrystals have been reported, of which NVP,

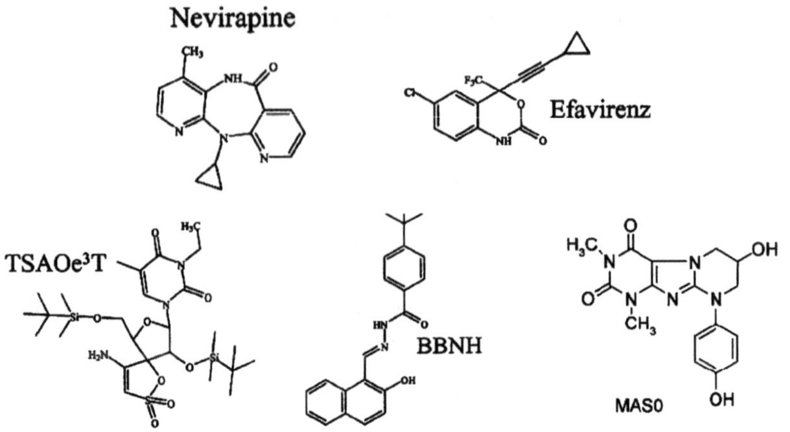


**Fig. 8.4** *Structures of the small molecules that modulate RT dimerization.* EFV and NVP enhance HIV-1 RT dimerization, while TSAOe3T, BBNH, and MAS0 act as destabilizers

delavirdine, and efavirenz are currently used in combination antiretroviral therapy. Combining yeast two-hybrid assays, in vitro binding assays, urea denaturation, and size exclusion chromatography demonstrated that several NNRTIs affect the dimerization of RT subunits. NNRTIs have been reported to modulate the overall stability of the heterodimeric RT depending on their binding site on RT. While EFV and nevirapine (NVP) enhance HIV-1 RT dimerization, the NNRTIs TSAOe3T, BBNH, and BBSH act as destabilizers of dimeric RT (Sluis-Cremer and Tachedjian 2002; Srivastava et al. 2006; Tachedjian et al. 2001; Tachedjian and Goff 2003; Venezia et al. 2006; Sluis-Cremer and Tachedjian 2008).

### *8.5.1 NNRTIs That Promote RT Dimerization*

Several NNRTIs including EFV, NVP, TMC120, TMC125, UC781, 8-Cl-TIBO, HBY 097, and α-APA have been reported to dramatically increase the association of the p66 and p51 subunits, therefore promoting RT dimerization both in vitro and in vivo (Tachedjian et al. 2001; Tachedjian and Goff 2003; Figueiredo et al. 2006; Sluis-Cremer et al. 2004a). Such NNRTI-mediated enhancement of dimerization contributes in part to their inhibitory activity, although the mechanism by which these small molecules enhance dimerization remains unclear as well as their mode of binding. EFV has been reported to interact similarly with the p66 and p51 monomers and to enhance dimerization of both homo- and heterodimers (Müller et al. 1989; Venezia et al. 2006). Activation of RT dimerization requires NNRTI to its site in the p66 subunit and is abolished by drug resistance mutations (Tachedjian et al. 2001), suggesting that perturbing the subunit interactions is indirect. Interestingly, the NNRTI delavirdine does not enhance RT dimerization. In contrast to other NNRTI, delavirdine structure is longer and does not fit the NNRTI-binding pocket but protrudes from this site (Sluis-Cremer and Tachedjian 2008). Tachedjian and coworkers demonstrated that residues forming the NNRTI-binding pocket contribute to the stability of p66/p51 HIV-1 RT and that mutating K101A, P225H, Y318F, and Y318W decreased heterodimer stability, whereas K103N, V108I, V108W, Y181C, Y188L, G190A, G190E, G190W, and P225W increased RT stability. G190 of the p66 RT subunit was found to play a major role in heterodimer stability evidenced by a dramatic increase in stability in response to a G190W mutation (Figueiredo et al. 2006, 2008).

Regulation of Gag-Pol polyprotein processing is an essential step in the production of mature viral particles. NNRTIs also promote HIV-1 RT maturation at the level of the Gag-Pol protein. In the context of the virus maturation, NNRTIs have been shown to enhance processing by promoting interactions of separate Gag-Pol polyproteins. This leads to premature activation of the Gag-Pol-embedded HIV-1 PR, resulting in a decrease in full-length viral polyproteins available for assembly and budding (Figueiredo et al. 2006, 2008).

### *8.5.2 NNRTI-Induced Destabilization of the HIV-1 RT Dimer Interface*

NNRTIs such as [2_,5_-bis-*O*-(*tert*butyldimethylsilyl)-beta-D-ribofuranose]-3_-spiro-5__-(4-amino-1,2-oxathiole-2,2-dioxide) derivatives (TSAO) (Camarasa et al. 2006; Balzarini et al. 1992a, b) and N-(4-tert-butylbenzoyl)-2-hydroxy-1-naphthaldehyde hydrazone (BBNH) derivatives (Fig. 8.4) (Sluis-Cremer et al. 2002; Borkow et al. 2002) have been reported to destabilize RT subunit interactions and interfere with heterodimerization-induced conformational changes essential for enzyme function (Sluis-Cremer et al. 2002). The mechanism of action and the structure of TSAO are significantly distinct from conventional NNRTI (Balzarini et al. 1992a, b). This drug makes extensive contact with the $\beta7/\beta8$ loop of the p51 subunit that forms the "floor" of the NNRTI-binding pocket and fits in a groove-like structure that constitutes the primer/template binding site in the p66 subunit. Moreover, the TSAO resistance-conferring mutation E138K occurs in the p51 subunit (Opar 2007). The N3-ethyl derivative of TSAO-T, TSAOe3T, has been shown to destabilize both the p66/p51 and p66/p66 forms of HIV-1 RT (Sluis-Cremer et al. 2000). The binding of TSAO decreases the Gibbs free energy of RT dimer dissociation, resulting in loss of p66/p51 and p66/p66 dimer stability, without inducing subunit dissociation. TSAO seems to induce conformational perturbations that affect the p66/p51 RT interface (Sluis-Cremer et al. 2000).

BBNH was reported as a multitarget inhibitor of HIV-1 RT that binds to both the DNA polymerase and RNase H domains and inhibits both enzymatic activities (Borkow et al. 2002). Binding of BBNH to HIV-1 RT decreases the Gibbs free energy of the RT dimer and therefore alters the stability of the heterodimeric enzyme (Sluis-Cremer et al. 2002). From a mechanistic point of view, it was shown that BBNH binding in the DNA polymerase domain alone was sufficient to affect RT stability and it has been speculated that BBNH binds in a manner analogous to TSAOe3T (Sluis-Cremer et al. 2002).

### *8.5.3 DimRTIs*

A structure-based ligand approach has been applied to identify potent dimerization RT inhibitors (DimRTIs) (Grohmann et al. 2008). Using the p66 tryptophan repeat motif as a target, one efficient compound, MAS0 (7-hydroxy-9-(4-hydroxyphenyl)-1,3- dimethyl-1,6,7,8,9,10a-hexahydropyrimido [2,1-f]purine-2, 4(3H,4aH)-dione), was selected (Fig. 8.4). Analysis of MAS0 mechanism has revealed that this compound inhibits both the DNA polymerase and the RNase H activity of RT in a dose-dependent manner. MAS0 affected p51/p66 subunit association reaction without promoting dissociation and induces a shift in dimer equilibrium from an active dimer to an inactive dimer. MAS0 was showed to increase the apparent rate of subunit dissociation and trap the enzyme in an inactive state (Mori et al. 2011; Grohmann et al. 2008).

## 8.6 Conclusions and Perspectives

AIDS remains one of the major health and socio-epidemiological problems of modern time. Despite the development of HAART, none of the currently used therapies is capable of eradicating HIV and all suffer from the emergence of viral resistances and toxic side effects. Therefore, there is an urgent need of more specific agents and ones that are active against drug-resistant strains. Current research focuses on both the rational design of new inhibitors targeting protein-protein interactions required for virus maturation and the identification of new targets. Along these lines, RT dimerization plays a major role in virus maturation, by regulating Gag-Pol processing, an essential step in the production of mature viral particles (Opar 2007). RT dimerization has been identified as a new interesting target for the design of new drugs and peptide-based strategies (Wapling et al. 2007; Srivastava et al. 2006; Morris et al. 1999a). Various peptide-based inhibitors and NNRTIs have been demonstrated to modulate the overall stability and conformational flexibility of heterodimeric RT and have been validated in vitro and in vivo. This strategy offers several advantages: (1) it does not require complete dissociation of the HIV-1 RT heterodimer to negatively impact the enzymatic function; (2) it targets essential "hot spots" in the dimer interface, limiting the risk of resistance mutation; (3) it is active on all resistance strains; and (4) these inhibitors can be combined with currently used clinical drugs.

**Acknowledgments** This work was supported in part by the Centre National de la Recherche Scientifique (CNRS) and by grant from the Agence Nationale de Recherche sur le SIDA (ANRS) and SIDACTION. This work is part of the program "Targeting Replication and Integration of HIV" (TRIoH) supported by the EC (LSHB-CT-2003-503480). We would like to thank all members of the laboratory and our collaborators for fruitful discussions.

## References

Abba-Moussa D, Divita G Peptide-based antagonist of HIV-1 reverse transcriptase dimerization as potent inhibitors of multidrug-resistant strains propagation. submitted

Abba-Moussa D, Aldrian-Herrada G, Agopian A, Boublick Y, Schinazi R, Clayette P, Divita G Insight into the mechanism of peptide-based inhibitors of HIV-1 reverse transcriptase maturation. Submitted

Abbondanzieri EA, Bokinsky G, Rausch JW, Zhang JX, Le Grice SF, Zhuang X (2008) Dynamic binding orientations direct activity of HIV reverse transcriptase. Nature 453(7192):184–189

Agopian A, Depollier J, Lionne C, Divita G (2007) p66 Trp24 and Phe61 are essential for accurate association of HIV-1 reverse transcriptase with primer/template. J Mol Biol 373(1):127–140, 12

Agopian A, Gros E, Aldrian-Herrada G, Bosquet N, Clayette P, Divita G (2009) A new generation of peptide-based inhibitors targeting HIV-1 reverse transcriptase conformational flexibility. J Biol Chem 284(1):254–264

Balzarini J, Pérez-Pérez MJ, San-Félix A, Camarasa MJ, Bathurst IC, Barr PJ, De Clercq E (1992a) Kinetics of inhibition of human immunodeficiency virus type 1 (HIV-1) reverse transcriptase by the novel HIV-1-specific nucleoside analogue [2',5'-bis-O-(tert-butyldimethylsilyl)-beta-D-ribofuranosyl]-3'-spiro-5 "- (4"-amino-1",2"-oxathiole-2",2"-dioxide)thymine (TSAO-T). J Biol Chem 267(17):11831–11838, Jun 15

Balzarini J, Pérez-Pérez MJ, San-Félix A, Schols D, Perno CF, Vandamme AM, Camarasa MJ, De Clercq E (1992b) 2',5'-Bis-O-(tert-butyldimethylsilyl)-3'-spiro-5"-(4"-amino-1",2"- oxathiole-2",2'-dioxide)pyrimidine (TSAO) nucleoside analogues: highly selective inhibitors of human immunodeficiency virus type 1 that are targeted at the viral reverse transcriptase. Proc Natl Acad Sci USA 89(10):4392–4396, May 15

Barre-Sinoussi F, Chermann JC, Rey F, Nugeyre MT, Chamaret S, Gruest J, Dauguet C, Axler-Blin C, Vezinet-Brun F, Rouzioux C, Rozenbaum W, Montagnier L (1983) Isolation of a T-lymphotropic retrovirus from a patient at risk for acquired immune deficiency syndrome (AIDS). Science 220:868–871

Borkow G, Salomon H, Wainberg MA, Parniak MA (2002) Attenuated infectivity of HIV type 1 from epithelial cells pretreated with a tight-binding nonnucleoside reverse transcriptase inhibitor. AIDS Res Hum Retroviruses 18:711–714

Camarasa MJ, Velazquez S, San-Felix A, Perez-Perez MJ, Gago F (2006) Dimerization inhibitors of HIV-1 reverse transcriptase, protease and integrase: a single mode of inhibition for the three HIV enzymes? Antiviral Res 1:260–267

Cherry E, Liang C, Rong L, Quan Y, Inouye P, Li X, Morin N, Kotler M, Wainberg MA (1998) Characterization of human immunodeficiency virus type-1 (HIV-1) particles that express protease-reverse transcriptase fusion proteins. J Mol Biol 284(1):43–56

Coffin JM (1995) HIV population dynamics in vivo: implications for genetic variation, pathogenesis, and therapy. Science 267(5197):483–489

De Clercq E (2002) Strategies in the design of antiviral drugs. Nat Rev Drug Discov 1:13–25

De Clercq E (2004) Non-nucleoside reverse transcriptase inhibitors (NNRTIs): past, present, and future. Chem Biodivers 1:44–64

Depollier J, Hourdou ML, Aldrian-Herrada G, Rothwell P, Restle T, Divita G (2005) Insight into the mechanism of a peptide inhibitor of HIV reverse transcriptase dimerization. Biochemistry 44:1909–1918

di Marzo Veronese F, Copeland TD, DeVico AL, Rahman R, Oroszlan S, Gallo RC, Sarngadharan MG (1986) Characterization of highly immunogenic p66/p51 as the reverse transcriptase of HTLV-III/LAV. Science 231:1289–1291

Divita G, Müller B, Immendörfer U, Gautel M, Rittinger K, Restle T, Goody RS (1993a) Kinetics of interaction of HIV reverse transcriptase with primer/template. Biochemistry 32(31): 7966–7971, 10

Divita G, Restle T, Goody RS (1993b) Characterization of the dimerization process of HIV-1 reverse transcriptase heterodimer using intrinsic protein fluorescence. FEBS Lett 324(2):153–158, 14

Divita G, Restle T, Goody RS, Chermann JC, Baillon JG (1994) Inhibition of human immunodeficiency virus type 1 reverse transcriptase dimerization using synthetic peptides derived from the connection domain. J Biol Chem 269:13080–13083

Divita G, Rittinger K, Geourjon C, Deleage G, Goody RS (1995a) Dimerization kinetics of HIV-1 and HIV-2 reverse transcriptase: a two step process. J Mol Biol 245:508–521

Divita G, Rittinger K, Restle T, Immendorfer U, Goody RS (1995b) Conformational stability of dimeric HIV-1 and HIV-2 reverse transcriptases. Biochemistry 34:16337–16346

Divita G, Baillon JG, Rittinger K, Chermann JC, Goody RS (1995c) Interface peptides as structure-based human immunodeficiency virus reverse transcriptase inhibitors. J Biol Chem 270:28642–28646

Esnouf R, Ren J, Ross C, Jones Y, Stammers D, Stuart D (1995) Mechanism of inhibition of HIV-1 reverse transcriptase by non-nucleoside inhibitors. Nat Struct Biol 2:303–308

Esposito F, Corona A, Tramontano E (2012) HIV-1 reverse transcriptase still remains a new drug target: structure, function, classical inhibitors, and new inhibitors with innovative mechanisms of actions. Mol Biol Int 2012:586401

Fauci AS (1988) The human immunodeficiency virus: infectivity and mechanisms of pathogenesis. Science 239:617–622

Figueiredo A, Moore KL, Mak J, Sluis-Cremer N, de Bethune MP, Tachedjian G (2006) Potent non-nucleoside reverse transcriptase inhibitors target HIV-1 Gag-Pol. PLoS Pathog 2(11):e119, Nov

Figueiredo A, Zelina S, Sluis-Cremer N, Tachedjian G (2008) Impact of residues in the nonnucleoside reverse transcriptase inhibitor binding pocket on HIV-1 reverse transcriptase heterodimer stability. Curr HIV Res 6(2):130–137, Mar

Fisher TS, Prasad VR (2002) Substitutions of Phe61 located in the vicinity of template 5'-overhang influence polymerase fidelity and nucleoside analog sensitivity of HIV-1 reverse transcriptase. J Biol Chem 277(25):22345–22352

Fisher TS, Darden T, Prasad VR (2003) Substitutions at Phe61 in the beta3-beta4 hairpin of HIV-1 reverse transcriptase reveal a role for the fingers subdomain in strand displacement DNA synthesis. J Mol Biol 325:443–459

Gotte M, Li X, Wainberg MA (1999) HIV-1 reverse transcription: a brief overview focused on structure-function relationships among molecules involved in initiation of the reaction. Arch Biochem Biophys 365:199–210

Grohmann D, Corradi V, Elbasyouny M, Baude A, Horenkamp F, Laufer SD, Manetti F, Botta M, Restle T (2008) Small molecule inhibitors targeting HIV-1 reverse transcriptase dimerization. Chembiochem 9(6):916–922, Apr 14

Hornak V, Simmerling C (2007) Targeting structural flexibility in HIV-1 protease inhibitor binding. Drug Discov Today 12:132–138

Huang H, Chopra R, Verdine GL, Harrison SC (1998) Structure of a covalently trapped catalytic complex of HIV-1 reverse transcriptase: implications for drug resistance. Science 282:1669–1675

Jacobo-Molina A, Ding J, Nanni RG, Clark AD Jr, Lu X, Tantillo C, Williams RL, Kamer G, Ferris AL, Clark P et al (1993) Crystal structure of human immunodeficiency virus type 1 reverse transcriptase complexed with double stranded DNA at 3.0 a resolution shows bent DNA. Proc Natl Acad Sci USA 90:6320–6324

Jacques PS, Wöhrl BM, Ottmann M, Darlix JL, Le Grice SF (1994) Mutating the "primer grip" of p66 HIV-1 reverse transcriptase implicates tryptophan-229 in template-primer utilization. J Biol Chem 269(42):26472–26478

Kati WM, Johnson KA, Jerva LF, Anderson KS (1992) Mechanism and fidelity of HIV reverse transcriptase. J Biol Chem 267:25988–25997

Katz RA, Skalka AM (1994) The retroviral enzymes. Annu Rev Biochem 63:133–173

Kohlstaedt LA, Wang J, Friedman JM, Rice PA, Steitz TA (1992) Crystal structure at 3.5 A resolution of HIV-1 reverse transcriptase complexed with an inhibitor. Science 256:1783–1790

Le Grice SFJ (1993) Human immunodeficiency virus reverse transcriptase. In: Skala A, Goff SP (eds) Reverse transcriptase. Cold Spring Harbor Laboratory Press, Plainview, NY, pp. 163–191

Liao WH, Wang CT (2004) Characterization of human immunodeficiency virus type 1 Pr160 gag-pol mutants with truncations downstream of the protease domain. Virology 2329(1):180–188

Liu S, Harada BT, Miller JT, Le Grice SF, Zhuang X (2010) Initiation complex dynamics direct the transitions between distinct phases of early HIV reverse transcription. Nat Struct Mol Biol 7(12):1453–1460

Menendez-Arias L (2002) Targeting HIV: antiretroviral therapy and development of drug resistance. Trends Pharmacol Sci 23:381–388

Mori M, Manetti F, Botta M (2011) Targeting protein-protein and protein-nucleic acid interactions for anti-HIV therapy. Curr Pharm Des 17(33):3713–3728

Morris MC, Vidal P, Chaloin L, Heitz F, Divita G (1997) A new peptide vector for efficient delivery of oligonucleotides into mammalian cells. Nucleic Acids Res 25(14):2730–2736, Jul 15

Morris MC, Robert-Hebmann V, Chaloin L, Mery J, Heitz F, Devaux C, Goody RS, Divita G (1999a) A new potent HIV-1 reverse transcriptase inhibitor: a synthetic peptide derived from the interface subunit domain. J Biol Chem 274:24941–24946

Morris MC, Berducou C, Mery J, Heitz F, Divita G (1999b) The thumb domain of the P51-subunit is essential for activation of HIV reverse transcriptase. Biochemistry 38(46):15097–15103, Nov 16

Morris MC, Depollier J, Mery J, Heitz F, Divita G (2001) A peptide carrier for the delivery of biologically active proteins into mammalian cells. Nat Biotechnol 19(12):1173–1176, Dec

Müller B, Restle T, Weiss S, Gautel M, Sczakiel G, Goody RS (1989) Co-expression of the subunits of the heterodimer of HIV-1 reverse transcriptase in escherichia coli. J Biol Chem 264:13975–13978

Nielsen MH, Pedersen FS, Kjems J (2005) Molecular strategies to inhibit HIV-1 replication. Retrovirology 2:10

Opar A (2007) New HIV drug classes on the horizon. Nat Rev Drug Discov 6:258–259

Parniak MA, Sluis-Cremer N (2000) Inhibitors of HIV-1 reverse transcriptase. Adv Pharmacol 49:67–196

Pauwels R (2004) New non-nucleoside reverse transcriptase inhibitors (NNRTIs) in development for the treatment of HIV infections. Curr Opin Pharmacol 4(5):437–446

Ren J, Bird LE, Chamberlain PP, Stewart-Jones GB, Stuart DI, Stammers DK (2002) Structure of HIV-2 reverse transcriptase at 2.35 a resolution and the mechanism of resistance to non-nucleoside inhibitors. Proc Natl Acad Sci USA 99:14410–14415

Restle T, Müller B, Goody RS (1990) Dimerization of human immunodeficiency virus type 1 reverse transcriptase. A target for chemotherapeutic intervention. J Biol Chem 265:8986–8988

Restle T, Pawlita M, Sczakiel G, Müller B, Goody RS (1992) Structure-function relationships of HIV-1 reverse transcriptase determined using monoclonal antibodies. J Biol Chem 267(21): 14654–14661, Jul 25

Richman DD (1996) Antiretroviral drug resistance: mechanisms, pathogenesis, clinical significance. Adv Exp Med Biol 394:383–395

Rittinger K, Divita G, Goody RS (1995) Human immunodeficiency virus reverse transcriptase substrate-induced conformational changes and the mechanism of inhibition by nonnucleoside inhibitors. Proc Natl Acad Sci USA 92:8046–8049

Rodgers DW, Gamblin SJ, Harris BA, Ray S, Culp JS, Hellmig B, Woolf DJ, Debouck C, Harrison SC (1995) The structure of unliganded reverse transcriptase from the human immunodeficiency virus type 1. Proc Natl Acad Sci USA 92:1222–1226

Saliba G, Yeni P (2006) Recent and future therapeutic advances in the management of HIV infection. Pathol Biol (Paris) 54(10):545–550

Sarafianos SG, Clark AD Jr, Das K, Tuske S, Birktoft JJ, Ilankumaran P, Ramesha AR, Sayer JM, Jerina DM, Boyer PL, Hughes SH, Arnold E (2002) Structures of HIV-1 reverse transcriptase with pre- and post-translocation AZTMP-terminated DNA. EMBO J 21:6614–6624

Sarafianos SG, Das K, Hughes SH, Arnold E (2004) Taking aim at a moving target: designing drugs to inhibit drug-resistant HIV-1 reverse transcriptases. Curr Opin Struct Biol 14:716–730

Sarafianos SG, Marchand B, Das K, Himmel DM, Parniak MA, Hughes SH, Arnold E (2009) Structure and function of HIV-1 reverse transcriptase: molecular mechanisms of polymerization and inhibition. J Mol Biol 385(3):693–713

Sepkowitz KA (2001) AIDS—the first 20 years. N Engl J Med 344(23):1764–1772

Simon V, Ho DD, Abdool Karim Q (2006) HIV/AIDS epidemiology, pathogenesis, prevention, and treatment. Lancet 368:489–504

Sluis-Cremer N, Tachedjian G (2002) Modulation of the oligomeric structures of HIV-1 retroviral enzymes by synthetic peptides and small molecules. Eur J Biochem 269:5103–5111

Sluis-Cremer N, Tachedjian G (2008) Mechanisms of inhibition of HIV replication by non-nucleoside reverse transcriptase inhibitors. Virus Res 134(1–2):147–156

Sluis-Cremer N, Dmitrienko GI, Balzarini J, Camarasa MJ, Parniak MA (2000) Human immunodeficiency virus type 1 reverse transcriptase dimer destabilization by 1-[Spiro[4"-amino-2",2" -dioxo-1",2" -oxathiole-5",3'-[2', 5'-bis-O-(tert-butyldimethylsilyl)-beta-D-ribofuranosyl]]]-3-ethylthy mine. Biochemistry 39(6):1427–1433

Sluis-Cremer N, Arion D, Parniak MA (2002) Destabilization of the HIV-1 reverse transcriptase dimer upon interaction with N-acyl hydrazone inhibitors. Mol Pharmacol 62(2):398–405

Sluis-Cremer N, Arion D, Abram ME, Parniak MA (2004a) Proteolytic processing of an HIV-1 pol polyprotein precursor: insights into the mechanism of reverse transcriptase p66/p51 heterodimer formation. Int J Biochem Cell Biol 36(9):1836–1847, Sep

Sluis-Cremer N, Temiz NA, Bahar I (2004b) Conformational changes in HIV-1 reverse transcriptase induced by nonnucleoside reverse transcriptase inhibitor binding. Curr HIV Res 2:323–332

Spence RA, Kati WM, Anderson KS, Johnson KA (1995) Mechanism of inhibition of HIV-1 reverse transcriptase by nonnucleoside inhibitors. Science 267(5200):988–993

Srivastava S, Sluis-Cremer N, Tachedjian G (2006) Dimerization of human immunodeficiency virus type 1 reverse – transcriptase as an antiviral target. Curr Pharm Des 12(15):1879–1894

Tachedjian G, Goff SP (2003) The effect of NNRTIs on HIV reverse transcriptase dimerization. Curr Opin Investig Drugs 4(8):966–973

Tachedjian G, Aronson HE, Goff SP (2000) Analysis of mutations and suppressors affecting interactions between the subunits of the HIV type 1 reverse transcriptase. Proc Natl Acad Sci USA 97:6334–6339

Tachedjian G, Orlova M, Sarafianos SG, Arnold E, Goff SP (2001) Nonnucleoside reverse transcriptase inhibitors are chemical enhancers of dimerization of the HIV type 1 reverse transcriptase. Proc Natl Acad Sci USA 98:7188–7193

Tachedjian G, Aronson HE, de los Santos M, Seehra J, McCoy JM, Goff SP (2003) Role of residues in the tryptophan repeat motif for HIV-1 reverse transcriptase dimerization. J Mol Biol 326(2):381–396, Feb 14

Tachedjian G, Moore KL, Goff SP, Sluis-Cremer N (2005) Efavirenz enhances the proteolytic processing of an HIV-1 pol polyprotein precursor and reverse transcriptase homodimer formation. FEBS Lett 579(2):379–384, Jan 17

Telesnitsky A, Goff SP (1997) Reverse transcriptase and the generation of retroviral DNA. In: Coffin JM, Hughes SH, Varmus HE (eds) Retroviruses. Cold Spring Harbor Laboratory Press, Plainview, NY, pp. 121–160

Venezia CF, Howard KJ, Ignatov ME, Holladay LA, Barkley MD (2006) Effects of efavirenz binding on the subunit equilibria of HIV-1 reverse transcriptase. Biochemistry 45(9): 2779–2789, 7

Wang J, Smerdon SJ, Jager J, Kohlstaedt LA, Rice PA, Friedman JM, Steitz TA (1994) Structural basis of asymmetry in the human immunodeficiency virus type 1 reverse transcriptase heterodimer. Proc Natl Acad Sci USA 91(15):7242–7246

Wapling J, Srivastava S, Shehu-Xhilaga M, Tachedjian G (2007) Targeting human immunodeficiency virus type 1 assembly, maturation and budding. Drug Target Insights 2:159–182

Warrilow D, Tachedjian G, Harrich D (2009) Maturation of the HIV reverse transcription complex: putting the jigsaw together. Rev Med Virol 19(6):324–337

Young SD, Britcher SF, Tran LO, Payne LS, Lumma WC, Lyle TA, Huff JR, Anderson PS, Olsen DB, Carroll SS et al (1995) L-743, 726 (DMP-266): a novel, highly potent nonnucleoside inhibitor of the human immunodeficiency virus type 1 reverse transcriptase. Antimicrob Agents Chemother 39(12):2602–2605

Zhan P, Chen X, Li D, Fang Z, De Clercq E, Liu X (2011) HIV-1 NNRTIs: structural diversity, pharmacophore similarity, and implications for drug design. Med Res Rev. doi:10.1002/med.20241, Apr 26

Zybarth G, Carter C (1995) Domains upstream of the protease (PR) in human immunodeficiency virus type 1 Gag-Pol influence PR autoprocessing. J Virol 69(6):3878–3884

# Chapter 9
# Targeting RT Translocation

**Gregory L. Beilhartz, Brian J. Scarth, and Matthias Götte**

## 9.1 Background

Conversion of the RNA genome of human immunodeficiency virus (HIV) into double-stranded DNA requires cyclic incorporation of deoxyribonucleotides by HIV reverse transcriptase (RT). Between two cycles of nucleotide incorporation, RT must move a single position further downstream to accommodate the next nucleotide substrate. This movement is referred to as "translocation." In this chapter, we discuss mechanisms associated with RT translocation and its role in inhibition of DNA synthesis by different classes of RT inhibitors.

## 9.2 Nucleotide Incorporation Cycles

RT translocation is an integral part of the process of nucleotide incorporation (Fig. 9.1). Cycles of nucleotide incorporation begin with either binding of RT to its nucleic acid substrate or a preformed binary complex resulting from the previous cycle. In either case, the enzyme must position the 3′OH group of the primer at the polymerase active site to facilitate nucleophilic attack on the α-phosphate of an incoming deoxynucleotide triphosphate (dNTP) (Liu et al. 2008). Under these conditions, the 3′ end of the primer is located in the priming site (P site), and the adjacent nucleotide binding site (N site) is available for nucleotide binding. Crystallographic studies indicate that dNTP binding and the formation of a ternary complex induces a conformational change in the fingers subdomain of HIV-1 RT

G.L. Beilhartz • B.J. Scarth • M. Götte (✉)
Department of Microbiology and Immunology, McGill University, Montreal, QC, Canada
e-mail: matthias.gotte@mcgill.ca

S.F.J. LeGrice and M. Götte (eds.), *Human Immunodeficiency Virus Reverse Transcriptase: A Bench-to-Bedside Success*, DOI 10.1007/978-1-4614-7291-9_9,

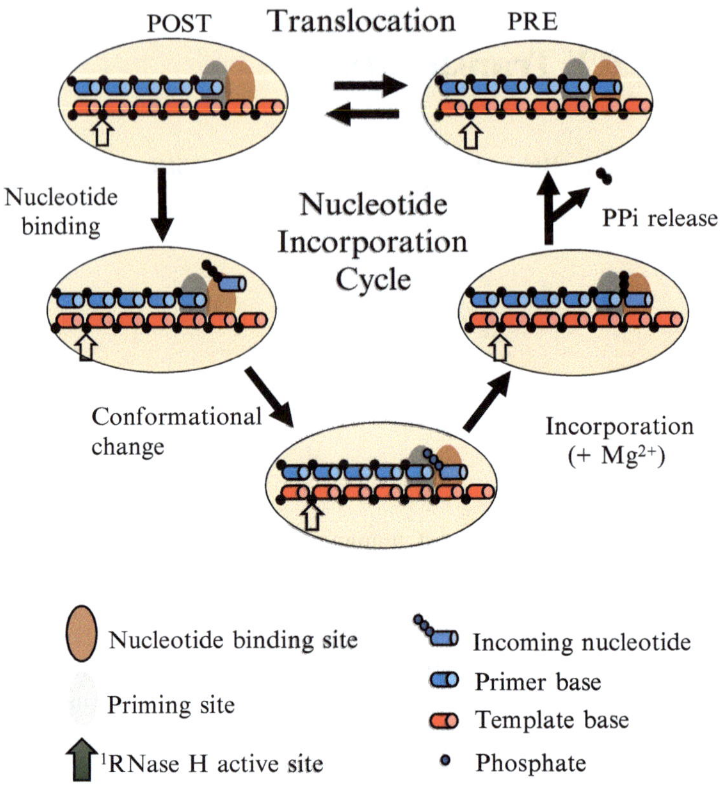


**Fig. 9.1** Nucleotide Incorporation Cycle. Polymerase-competent RT-substrate complexes exist in a thermodynamic equilibrium between two distinct complexes, namely pre- and post-translocation. RT undergoes a one base pair-long Brownian movement on its substrate between these two complexes (*top*). Only post-translocated complexes have a vacant nucleotide binding site that allows for the binding of the next templated nucleotide (*from top left*). Nucleotide binding traps RT in the post-translocated position, and the polymerase domain undergoes a conformation change that facilitates the $Mg^{2+}$-dependent addition of the bound nucleotide onto the 3′ primer terminus. The product of the incorporation reaction, pyrophosphate (PPi), is released following nucleotide incorporation, forming a pre-translocated product complex that will re-establish a new translocational equilibrium. The cycle can then repeat itself, resulting in DNA polymerization

resulting in a closed structure that traps the bound nucleotide (Huang et al. 1998). Phosphodiester bond formation occurs through a mechanism that involves the coordination of two divalent metal ions by the three catalytic residues in the palm subdomain (Steitz 1998, 1999). Immediately following the chemical step, the RT complex exists in its pre-translocational state with the N site occupied by the extended primer terminus. Subsequent binding and incorporation of the next nucleotide requires translocation of RT.

## 9.3 Active and Passive Models of Translocation

Translocation is a rapid process that is kinetically poorly defined (Kati et al. 1992). Pre-steady state kinetics studies, however, point to different populations of binary complexes (Wohrl et al. 1999). Biphasic dissociation kinetics are in agreement with this notion (Ignatov et al. 2005). The different populations of complexes described in these kinetics studies could conceivably represent pre- and post-translocated complexes, respectively.

Based on crystallographic data, it has been proposed that a "translocation track" in HIV-1 RT defines the movement of modular elements against each other. In this model, contacts with the nucleic acid break and reform a single nucleotide further downstream, allowing translocation of the enzyme along its substrate (Ding et al. 1998). It has been suggested that the energy driving translocation is derived from dNTP hydrolysis (Patel et al. 1995). Such a model of translocation, wherein bond formation or release of pyrophosphate (PPi) is energetically required, is often referred to as power-stroke model of translocation. Alternatively, passive models of polymerase translocation predict that thermal energy in the form of random Brownian motion is sufficient to propel translocation in both directions. Directionality is achieved by efficient binding of the next nucleotide, while the affinity of PPi is rather low. The nucleotide acts the pawl of a ratchet. Support for these various models of translocation has been established in different RNA polymerases and will be discussed further.

Both active and passive models have support in the form of structural and biochemical data, respectively, for the single subunit bacteriophage T7 RNA polymerase (T7 RNAP) (Li et al. 1998; Yin and Steitz 2004; Gnatt et al. 2001; Gong et al. 2005; Bar-Nahum et al. 2005). An active model is supported by crystal structures revealing conformational changes between pre- and post-translocated complexes (Gnatt et al. 2001; Gong et al. 2005). The pre-translocated complex is only observed with bound PPi, supporting a mechanism whereby PPi dissociation drives the conformational change to the post-translocated state (Yin and Steitz 2004). In contrast to these findings, exonuclease mapping experiments with T7 RNAP have shown that binding of the next nucleotide causes a single nucleotide shift in the protection pattern of the nucleoprotein complex (Bar-Nahum et al. 2005). These experiments were performed with chain-terminated primers, and therefore, incorporation and release of PPi are not responsible for the observed movement of the enzyme. These results support a passive model for T7 RNAP translocation.

A more complex ratchet model has been suggested based on structural and biochemical data garnered from studies on the multi-subunit *E. coli* RNAP (Bar-Nahum et al. 2005). A bridge helix (F-bridge) located proximal to the polymerase active site is observed in both bent and straight conformations in structures of the polymerase (Gnatt et al. 2001). The authors propose that oscillation between these conformations would drive the forward motion of the translocation process. Additional biochemical studies have shown that binding of templated nucleotides to sites beyond the catalytic $n+1$ position up to $n+3$, and not nucleotide hydrolysis, is responsible for translocation (Gong et al. 2005; Foster et al. 2001).

## 9.4 HIV-1 Reverse Transcriptase Translocation

Crystal structures have captured RT in conformations with a bound nucleotide and structures of RT trapped in the pre- and post-translocated conformations with primer/templates terminated with 3′-azido-3′-deoxythymidine monophosphate (zidovudine, AZT) (Huang et al. 1998; Sarafianos et al. 2002). Comparison of these structures revealed that nucleotide binding leads to a displacement of the YMDD motif within the active site. This movement is compared to the "loading of a springboard," and it is suggested that the "release of the springboard" following catalysis provides the energy required for translocation. However, the unidirectional nature of this active model is difficult to reconcile with the fact that excision of incorporated nucleoside analogue RT inhibitors (NRTIs) occurs in the absence of nucleotide hydrolysis.

In a passive model of translocation, both N and P sites would be accessible to the 3′ end of the primer and could explain the ability of RT to perform the excision reaction. This model of translocation is supported by observations of stalled complexes using fluorescence resonance energy transfer (FRET)-based assays(Rothwell et al. 2003) and chemical footprinting techniques including both site-specific (Marchand and Gotte 2003; Marchand et al. 2007a, b) and DNase I (Meyer et al. 2007). Single-molecule FRET-based experiments revealed two populations that differ by 5 Å, conceivably representing pre- and post-translocation complexes (Rothwell et al. 2003). The presence of dNTP and PPi differentially favors the two conformations, as would be expected for the different states of translocation. Pyrophosphorolysis, however, could not be ruled out under the reaction conditions.

Different methods of site-specific footprinting have allowed direct monitoring of the position of RT on its nucleic acid substrate by specific cleavage at single nucleotide resolution (Marchand and Gotte 2003; Gotte et al. 1998). This level of resolution allows differentiation between pre- and post-translocated complexes. Site-specific footprinting utilizes divalent $Fe^{2+}$ to produce cleavage fragments at positions −18/−19 and −17/−18 of the template for pre- and post-translocated complexes, respectively. The footprinting reaction is mediated by binding of $Fe^{2+}$ ions at or near the RNase H active site (Gotte et al. 1998). Oxidation of the bound metal ions leads to the generation of hydroxyl radicals responsible for the site-specific cleavage of the template. Once the template is cleaved, the fragments can be separated by denaturing polyacrylamide gel electrophoresis, and the ratio of pre- and post-translocated complexes can be examined. A metal-free method of site-specific footprinting has also been developed (Marchand and Gotte 2003). This involves treatment of RT: DNA/DNA complexes with potassium peroxynitrite (KOONO), which reacts with the sulfur of C280 in RT subunit p66 to cleave the DNA. The ability to track a single residue, C280, provides this technique with higher resolution than $Fe^{2+}$ footprinting or DNase I protection experiments that can give rise to multiple overlapping cuts at adjacent positions (Marchand and Gotte 2003; Meyer et al. 2007). In KOONO footprinting, as the enzyme moves between pre- and post-translocation conformations, a single cut is produced at template positions −8 or −7, respectively.

Site-specific footprinting experiments with stalled RT: DNA/DNA complexes in the presence of the next complementary nucleotide reveal a shift from pre- to post-translocation conformations. Comparison of $Fe^{2+}$ and KOONO techniques revealed that both the RNase H and DNA polymerase domains (C280) move together, with identical responses to dNTP binding-induced transition from the pre- to post-translocation conformation (Marchand and Gotte 2003). Similar experiments with the PPi analogue phosphonoformic acid (PFA) revealed increased stability of the pre-translocation conformation (Marchand et al. 2007a). Equivalent results have been observed for both the binding of complementary dNTP and PFA using DNase I and exonuclease protection assays (Meyer et al. 2007). In all of these experiments, incorporation was prevented by the use of chain-terminated primers; therefore, the movement of the RT: DNA/DNA complex could not be energetically linked to catalysis or the release of PPi. These observations support a passive model of translocation in HIV-1 RT driven by the binding of the next complementary dNTP.

## 9.5 Translocation and Drug Resistance

As described above, post-translocated complexes are amenable to nucleotide incorporation, while pre-translocated complexes are not. Conversely, nucleotide excision requires the pre-translocated conformation. As the process of nucleotide incorporation results in the release of PPi, the reverse reaction can use PPi as a substrate to break the phosphodiester bond and excise the 3′ terminal nucleotide. While PPi is the most efficient substrate for excision, ATP can also act as a PPi donor for the excision reaction, causing the release of the terminal nucleotide in the form of a dinucleoside tetraphosphate.

### *9.5.1 NRTIs*

All approved NRTIs inhibit DNA synthesis through chain termination following incorporation into the nascent DNA chain. NRTIs lack a 3′ OH group necessary for phosphodiester bond formation with the next nucleotide. In order for DNA synthesis to continue, the NRTI must be removed from the 3′ primer end, and this is accomplished by nucleotide excision. Resistance via the excision mechanism is associated with accumulation of a series of mutations near the polymerase active site (Kellam et al. 1992). These mutations are referred to as thymidine analogue-associated mutations (TAMs) due to their initial association with resistance to thymidine analogues AZT and d4T. Classical TAMs include M41L, D67N, K70R, L210W, T215F/Y, and K219Q/E (Kellam et al. 1992; Hooker et al. 1996; Boucher et al. 1992; Larder and Kemp 1989; Richman et al. 1991; St Clair et al. 1991). In phenotypic drug susceptibility assays, TAMs confer the highest level of resistance to AZT with lower levels seen with d4T and other NRTIs (Lafeuillade and Tardy 2003). Increased excision of AZT by TAMs-containing RT in the presence of either PPi (Arion et al. 1998) or NTPs (Meyer et al. 1998) was initially reported in 1998.

Crystal structures of AZT-terminated primers in the pre- and post-translocated conformations show a steric clash between the incorporated AZT at the 3′ primer terminus and an incoming dNTP in the post-translocation conformation (Sarafianos et al. 2002, 2003). Thus, AZT primarily mediates the effect on translocation in this context, while the selectivity for ATP as the PPi source is mediated by interactions between TAMs and the base moiety of ATP.

The effect of AZT on the translocational equilibrium has direct functional consequences for excision. If RT translocates from the pre- to the post-conformational state, it moves the scissile bond out of position, preventing the excision reaction (Sarafianos et al. 2002). Moreover, post-translocated complexes that accept binding of an incoming nucleotide form stable closed complexes (Tong et al. 1997). "Dead-end complexes" (DECs) with a chain-terminated primer and a bound nucleotide are incapable of both nucleotide incorporation and nucleotide excision reactions. Footprinting experiments have confirmed biochemically that AZT-terminated primers preferentially reside in the pre-translocation conformation in the absence of dNTP and require higher concentrations of dNTP to trap a post-translocated complex than primers terminated with ddTTP (Marchand and Gotte 2003). This finding suggests that since AZT stalls RT predominantly in the pre-translocation conformation, the susceptibility of the complex to excision is increased, promoting resistance. Decreased resistance to other NRTIs conferred by TAMs is usually explained by increased DEC formation, as these NRTIs can form stable complexes at concentrations of dNTP within the physiological range (Meyer et al. 2000).

An additional factor observed to affect both translocation and excision is the nucleic acid sequence context. Site-specific footprinting has shown that the sequence context alone can influence the translocational equilibrium, with certain sequences inducing either heavily pre- or post-translocational states, while others can induce equilibria that are more balanced (Marchand and Gotte 2003; Scarth et al. 2010). Sequences on which RT exhibits decreased access to the pre-translocated conformation are deficient at the excision reaction (Marchand and Gotte 2003). In a separate study the sequence context was seen to heavily affect rates of excision (Meyer et al. 2004). Although the authors did not correlate their results with the translocational equilibrium, it is conceivable that the effects are related. It has been shown that AZT-terminated primers are able to overcome sequence-specific effects, while d4T remained susceptible to them (Scarth and Gotte 2008). This provides further explanation for effects seen in phenotypic drug susceptibility assays in which the sequence context of the entire genome is reflected. Sequences that favor the pre-translocation conformation would be resistant to dead-end complex formation for d4T, while AZT would resist binding of the next nucleotide at all positions.

### 9.5.2 Non-obligate Chain Terminators

2′-Deoxy-4′-C-ethynyl-2-fluoroadenosine (EFdA) (Fig. 9.2) is among a group of 4′-substituted NRTIs that require a 3′OH group for activity and are therefore classified as non-obligate chain terminators. Since the first report of

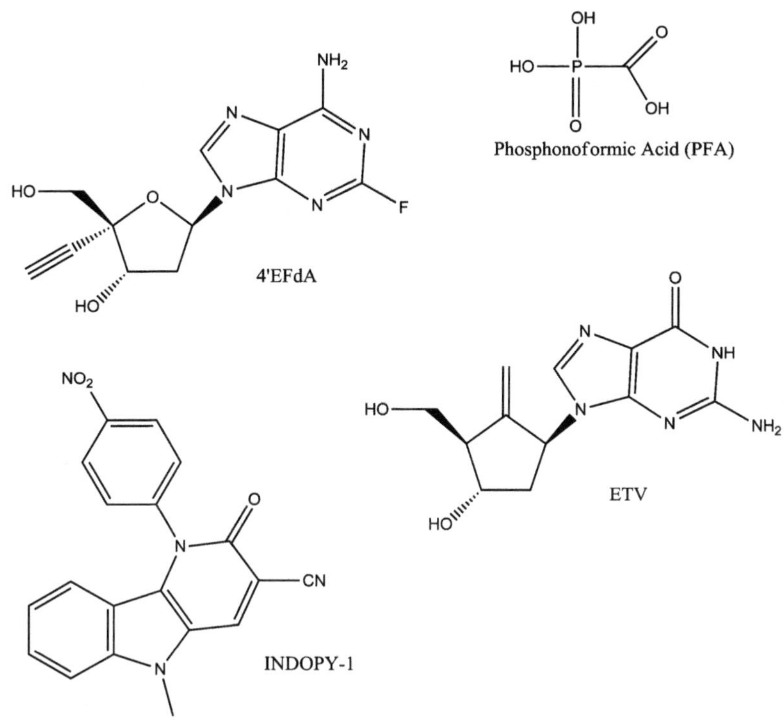


**Fig. 9.2** The chemical structures of representative compounds from different classes of RT inhibitor

4′-azido-thymidine in 1992, a wide range of molecules of this class have been synthesized (Kawamoto et al. 2008; Maag et al. 1992). EFdA stands out as the most potent of those synthesized with $EC_{50}$ values in the sub-nM range (Kawamoto et al. 2008). Once incorporated, EFdA inhibits polymerization as a chain terminator by preventing the translocation of RT (Michailidis et al. 2009). Modeling suggests that the 4’ethynyl group of EFdA fits into a hydrophobic pocket on the enzyme. This translocation-deficient mechanism of chain termination prevents binding of the next nucleotide and subsequent incorporation despite the presence of a 3′OH group at the primer terminus. The translocation block can be seen in site-specific footprinting experiments. EFdA is able to remain active against NRTI-resistant strains containing TAMs while being efficiently excised. This is a result of the extremely high level of potency and the ability of excised EFdA to be efficiently reincorporated.

### *9.5.3 NNRTIs*

Several NNRTIs, namely, nevirapine (NVP), efavirenz (EFV), etravirine (ETR), and most recently rilpivirine, are important components of combination antiretroviral therapy. Unlike NRTIs, NNRTIs are administered in their active form and are highly specific to HIV-1 RT. The class of compounds is highly diverse with over 50 families of molecules. Despite the immense diversity in structure, NNRTIs traditionally have been found to act through a common mechanism. NNRTIs inhibit RT by binding to an inducible hydrophobic pocket approximately 10 Å from the polymerase active site. NNRTI binding acts to inhibit DNA polymerization noncompetitively relative to dNTP substrates by affecting the positioning of the polymerase active site. This mechanism was first proposed based on structural data from co-crystals of unliganded HIV-1 RT complexed with primer/template and dNTPs, or with bound NNRTIs (reviewed in (Zhou et al. 2006)), and is supported by kinetic data (Sluis-Cremer et al. 2004). The structure of RT in complex with a DNA substrate together with nevirapine has elucidated the mechanism of NNRTI inhibition and the effect on RT translocation (Das et al. 2012). As NVP binds to the NNRTI-binding pocket, a structural rearrangement of the polymerase active site occurs. This has the effect of moving the 3′ primer end about 5.5 Å away from its position in the P site to a new position called P' (Das et al. 2012).

### *9.5.4 Other RT Inhibitors*

#### 9.5.4.1 PFA

PFA is a polymerase active site inhibitor that acts through a mechanism distinct from that of NRTIs (Fig. 9.2). The use of PFA is limited to salvage therapy due to problems associated with its clinical use including poor bioavailability (Noormohamed et al. 1998) and adverse effects (Gerard and Salmon-Ceron 1995; Canestri et al. 2006). The use of PFA in salvage therapy revealed a relationship to the excision of AZT by TAMs (Tachedjian et al. 1996). In these patients, treatment with PFA can produce initially favorable results with the majority of NNRTI and NRTI resistance mutations conferring no resistance to PFA. PFA resistance-conferring mutations, however, eventually develop which cause increased susceptibility to AZT (Mathiesen et al. 2007). The mechanism of re-sensitization to AZT by PFA resistance-conferring mutations has been determined to be reduction of the excision reaction (Tachedjian et al. 1996; Arion et al. 2000; Meyer et al. 2003).

The relationship between AZT sensitization and PFA resistance was linked to the translocational equilibrium by characterization of the specific mechanism of action of PFA (Marchand et al. 2007a). Inhibition with PFA on long templates was seen at "hot spots" along the template. These positions exhibit a strong bias toward the pre-translocation conformation in site-specific footprinting experiments. PFA binding at these sequences results in the formation of stable closed complexes, while sequences heavily biased to post-translocation are literally resistant to PFA binding. The binding of PFA increases in the presence of divalent metal ions which points to a PFA binding site at the active site. Binding of PFA and the natural dNTP substrate are, therefore, mutually exclusive.

In agreement with an overlapping binding site of PFA and PPi, resistance to PFA occurs in the context of mutations that directly interact with the β- and γ-phosphates of the incoming dNTP, such as K65R (Hammond et al. 2001) and R72A (Sarafianos et al. 1995). Resistance to PFA and sensitization to AZT can also occur through a decrease in the amount of pre-translocated complexes as seen with E89K. E89 is located in the palm subdomain adjacent to template position n-2, relatively removed from the putative PPi/PFA binding site (Huang et al. 1998). E89K affects the translocational equilibrium by disrupting the normal distribution between pre- and post-translocated conformations, causing the enzyme to slide beyond this typical register (Marchand et al. 2007a). Without the pre-translocated conformation, PFA cannot exert its inhibitory effect nor can RT excise AZT.

Similarly mutations F61A and A62V in the fingers subdomain of RT that decrease and enhance the stability of pre-translocation complexes, respectively, were shown to affect both binding and inhibition by PFA (Scarth et al. 2010). The substitution F61A does not occur naturally and is known to negatively affect DNA synthesis (Fisher and Prasad 2002; Fisher et al. 2003; Mandal et al. 2006; Upadhyay et al. 2010). A62V has been implicated as a compensatory mutation in the multidrug-resistant Q151M complex (Maeda et al. 1998) and can partially restore replication deficits found with K65R (Svarovskaia et al. 2008) and, as mentioned earlier, confers high-level excision activity in 69ss-containing complexes (Matamoros et al. 2004). Site-specific footprinting experiments with binary RT complexes showed that F61A strongly favors a post-translocation conformation, while A62V favors a pre-translocation conformation. Further, A62V was shown to increase the proportion of pre-translocated complexes when placed on multiple drug-resistant backgrounds including the Q151M- and 69ss-containing complexes. These effects on translocation were associated with strong resistance to PFA with F61A-containing RT and increased susceptibility with A62V-containing RT in binding and inhibition assays (Scarth et al. 2010). PFA also inhibits the HCMV polymerase (UL54), and a crystal of PFA bound to a chimeric polymerase comprising both UL54 and the homologous polymerase from the bacteriophage RB69 (gp43) shows PFA bound to an untranslocated, closed-fingers complex, despite the absence of a bound nucleotide. This confirms the notion that PFA binds at the PPi-binding site and to pre-translocated complexes (Zahn et al. 2011).

#### 9.5.4.2 NcRTIs

A novel class of RT inhibitors has recently been described (Jochmans et al. 2005, 2006). These compounds are not nucleotide analogues and are not incorporated as chain terminators. Instead these compounds bear more structural resemblance to NNRTIs. Unlike NNRTIs, however, they block polymerization by competing with dNTPs for binding at the polymerase active site of RT. Due to their unique mechanism of action, these compounds have been termed nucleotide-competing RT inhibitors (NcRTIs). Of the NcRTIs so far identified, there are two families of molecules. The first, the INDOPYs, including INDOPY-1, are polymerase active site inhibitors (Fig. 9.2). The second class of NcRTIs, the 4-dimethylamino-6-vinylpyrimidines (DAVPs), bind a novel site near the polymerase active site. Studies with INDOPY-1 revealed that this compound binds to post-translocated RT complexes, creating stable closed complexes, analogous to DEC formation seen with dNTP binding to RT containing a chain-terminated primer/template. Unlike dNTP binding, INDOPY-1 binding is not directed by the templated nucleotide. Instead the presence of a pyrimidine (preferentially TTP) at the 3' primer terminus directs binding of INDOPY-1 (Jochmans et al. 2006). Resistance-conferring mutations map to the polymerase active site and overlap with NRTI resistance-conferring mutations M184V and Y115F (Ehteshami et al. 2008). M184V and Y115F confer resistance by discrimination whereby the enzyme preferentially binds dNTP over INDOPY-1 relative to WT. In contrast, K65R has been shown to confer hypersusceptibility to INDOPY-1 through decreased binding of dNTP with increased binding of the NcRTI (Ehteshami et al. 2008). The mutation F61A, which results in a strong bias to the post-translocated conformation, also confers a large increase in susceptibility to INDOPY-1, indicating the importance of translocation conformation in the binding and inhibition by INDOPY-1.

### *9.5.5 Conclusion*

RT translocation is a necessary step two cycles of nucleotide incorporation and, as such, DNA synthesis and viral replication. Inhibitors of translocation such as PFA and INDOPY-1 can potently affect the translocational equilibrium. Thus, assays can be developed to screen for novel translocation inhibitors. In this context it is important to note that both PFA and INDOPY-1 were shown to be ineffective in affecting the translocational equilibrium on RNA-primed reactions (Auger et al. 2011). RT is able to flip between different conformations such as polymerase competent and RNase competent. In the case of RNA-primed reactions, such as the initiation of (+)-strand DNA synthesis, RT tends to flip between orientations so much that a translocational equilibrium can be difficult to establish. This is why PFA and INDOPY-1 are ineffective at this stage of reverse transcription. It is clear, therefore, that in order to screen efficiently for novel translocation inhibitors, the selection of a proper nucleic acid sequence, as well as the proper sugar backbone (DNA vs. RNA), is very important to ensure that a proper translocational equilibrium is formed. A potential difficulty to that end is the nature of the 3′ primer end. This does not make a difference in the case of PFA, but is very important in the case of INDOPY-1. Perhaps multiple screens will be necessary in order to properly canvas for novel inhibitors of RT translocation.

## References

Arion D, Kaushik N, McCormick S, Borkow G, Parniak MA (1998) Phenotypic mechanism of HIV-1 resistance to 3'-azido-3'-deoxythymidine (AZT): increased polymerization processivity and enhanced sensitivity to pyrophosphate of the mutant viral reverse transcriptase. Biochemistry 37:15908–15917

Arion D, Sluis-Cremer N, Parniak MA (2000) Mechanism by which phosphonoformic acid resistance mutations restore 3'-azido-3'-deoxythymidine (AZT) sensitivity to AZT-resistant HIV-1 reverse transcriptase. J Biol Chem 275:9251–9255

Auger A, Beilhartz GL, Zhu S, Cauchon E, Falgueyret JP, Grobler JA, Ehteshami M, Gotte M, Melnyk RA (2011) Impact of primer-induced conformational dynamics of HIV-1 reverse transcriptase on polymerase translocation and inhibition. J Biol Chem 286:29575–29583

Bar-Nahum G, Epshtein V, Ruckenstein AE, Rafikov R, Mustaev A, Nudler E (2005) A ratchet mechanism of transcription elongation and its control. Cell 120:183–193

Boucher CA, O'Sullivan E, Mulder JW, Ramautarsing C, Kellam P, Darby G, Lange JM, Goudsmit J, Larder BA (1992) Ordered appearance of zidovudine resistance mutations during treatment of 18 human immunodeficiency virus-positive subjects. J Infect Dis 165:105–110

Canestri A, Ghosn J, Wirden M, Marguet F, Ktorza N, Boubezari I, Dominguez S, Bossi P, Caumes E, Calvez V, Katlama C (2006) Foscarnet salvage therapy for patients with late-stage HIV disease and multiple drug resistance. Antivir Ther 11:561–566

Das K, Martinez SE, Bauman JD, Arnold E (2012) HIV-1 reverse transcriptase complex with DNA and nevirapine reveals non-nucleoside inhibition mechanism. Nat Struct Mol Biol 19:253–259

Ding J, Das K, Hsiou Y, Sarafianos SG, Clark AD Jr, Jacobo-Molina A, Tantillo C, Hughes SH, Arnold E (1998) Structure and functional implications of the polymerase active site region in a complex of HIV-1 RT with a double-stranded DNA template-primer and an antibody Fab fragment at 2.8 A resolution. J Mol Biol 284:1095–1111

Ehteshami M, Scarth BJ, Tchesnokov EP, Dash C, Le Grice SF, Hallenberger S, Jochmans D, Gotte M (2008) Mutations M184V and Y115F in HIV-1 reverse transcriptase discriminate against "nucleotidecompeting reverse transcriptase inhibitors". J Biol Chem 283:29904–29911

Fisher TS, Darden T, Prasad VR (2003) Substitutions at Phe61 in the beta3-beta4 hairpin of HIV-1 reverse transcriptase reveal a role for the Fingers subdomain in strand displacement DNA synthesis. J Mol Biol 325:443–459

Fisher TS, Prasad VR (2002) Substitutions of Phe61 located in the vicinity of template 5′-overhang influence polymerase fidelity and nucleoside analog sensitivity of HIV-1 reverse transcriptase. J Biol Chem 277:22345–22352

Foster JE, Holmes SF, Erie DA (2001) Allosteric binding of nucleoside triphosphates to RNA polymerase regulates transcription elongation. Cell 106:243–252

Gerard L, Salmon-Ceron D (1995) Pharmacology and clinical use of foscarnet. Int J Antimicrob Agents 5:209–217

Gnatt AL, Cramer P, Fu J, Bushnell DA, Kornberg RD (2001) Structural basis of transcription: an RNA polymerase II elongation complex at 3.3 A resolution. Science 292:1876–1882

Gong XQ, Zhang C, Feig M, Burton ZF (2005) Dynamic error correction and regulation of downstream bubble opening by human RNA polymerase II. Mol Cell 18:461–470

Gotte M, Maier G, Gross HJ, Heumann H (1998) Localization of the active site of HIV-1 reverse transcriptase-associated RNase H domain on a DNA template using site-specific generated hydroxyl radicals. J Biol Chem 273:10139–10146

Hammond JL, Koontz DL, Bazmi HZ, Beadle JR, Hostetler SE, Kini GD, Aldern KA, Richman DD, Hostetler KY, Mellors JW (2001) Alkylglycerol prodrugs of phosphonoformate are potent in vitro inhibitors of nucleoside-resistant human immunodeficiency virus type 1 and select for resistance mutations that suppress zidovudine resistance. Antimicrob Agents Chemother 45:1621–1628

Hooker DJ, Tachedjian G, Solomon AE, Gurusinghe AD, Land S, Birch C, Anderson JL, Roy BM, Arnold E, Deacon NJ (1996) An in vivo mutation from leucine to tryptophan at position 210 in human immunodeficiency virus type 1 reverse transcriptase contributes to high-level resistance to 3'-azido-3'-deoxythymidine. J Virol 70:8010–8018

Huang H, Chopra R, Verdine GL, Harrison SC (1998) Structure of a covalently trapped catalytic complex of HIV-1 reverse transcriptase: implications for drug resistance. Science 282: 1669–1675

Ignatov ME, Berdis AJ, Le Grice SF, Barkley MD (2005) Attenuation of DNA replication by HIV-1 reverse transcriptase near the central termination sequence. Biochemistry 44:5346–5356

Jochmans D, Kesteleyn B, Marchand B, Gotte M, Ivens T, Dehertogh P, Peeters A, Pauwels R, Wigerinck P, Hertogs K (2005) Identification and biochemical characterization of a new class of HIV inhibitors: nucleotide-competing reverse transcriptase inhibitors. In: Conference of retroviruses and opportunistic infections. Hynes Convention center, Boston

Jochmans D, Deval J, Kesteleyn B, Van Marck H, Bettens E, De Baere I, Dehertogh P, Ivens T, Van Ginderen M, Van Schoubroeck B, Ehteshami M, Wigerinck P, Gotte M, Hertogs K (2006) Indolopyridones inhibit human immunodeficiency virus reverse transcriptase with a novel mechanism of action. J Virol 80:12283–12292

Kati WM, Johnson KA, Jerva LF, Anderson KS (1992) Mechanism and fidelity of HIV reverse transcriptase. J Biol Chem 267:25988–25997

Kawamoto A, Kodama E, Sarafianos SG, Sakagami Y, Kohgo S, Kitano K, Ashida N, Iwai Y, Hayakawa H, Nakata H, Mitsuya H, Arnold E, Matsuoka M (2008) 2′-deoxy-4′-C-ethynyl-2-halo-adenosines active against drug-resistant human immunodeficiency virus type 1 variants. Int J Biochem Cell Biol 40:2410–2420

Kellam P, Boucher CA, Larder BA (1992) Fifth mutation in human immunodeficiency virus type 1 reverse transcriptase contributes to the development of high-level resistance to zidovudine. Proc Natl Acad Sci USA 89:1934–1938

Lafeuillade A, Tardy JC (2003) Stavudine in the face of cross-resistance between HIV-1 nucleoside reverse transcriptase inhibitors: a review. AIDS Rev 5:80–86

Larder BA, Kemp SD (1989) Multiple mutations in HIV-1 reverse transcriptase confer high-level resistance to zidovudine (AZT). Science 246:1155–1158

Li Y, Korolev S, Waksman G (1998) Crystal structures of open and closed forms of binary and ternary complexes of the large fragment of Thermus aquaticus DNA polymerase I: structural basis for nucleotide incorporation. EMBO J 17:7514–7525

Liu S, Abbondanzieri EA, Rausch JW, Le Grice SF, Zhuang X (2008) Slide into action: dynamic shuttling of HIV reverse transcriptase on nucleic acid substrates. Science 322:1092–1097

Maag H, Rydzewski RM, McRoberts MJ, Crawford-Ruth D, Verheyden JP, Prisbe EJ (1992) Synthesis and anti-HIV activity of 4'-azido- and 4'-methoxynucleosides. J Med Chem 35:1440–1451

Maeda Y, Venzon DJ, Mitsuya H (1998) Altered drug sensitivity, fitness, and evolution of human immunodeficiency virus type 1 with pol gene mutations conferring multi-dideoxynucleoside resistance. J Infect Dis 177:1207–1213

Mandal D, Dash C, Le Grice SF, Prasad VR (2006) Analysis of HIV-1 replication block due to substitutions at F61 residue of reverse transcriptase reveals additional defects involving the RNase H function. Nucleic Acids Res 34:2853–2863

Marchand B, Gotte M (2003) Site-specific footprinting reveals differences in the translocation status of HIV-1 reverse transcriptase. Implications for polymerase translocation and drug resistance. J Biol Chem 278:35362–35372

Marchand B, Tchesnokov EP, Gotte M (2007a) The pyrophosphate analogue foscarnet traps the pretranslocational state of HIV-1 reverse transcriptase in a Brownian ratchet model of polymerase translocation. J Biol Chem 282:3337–3346

Marchand B, White KL, Ly JK, Margot NA, Wang R, McDermott M, Miller MD, Gotte M (2007b) Effects of the translocation status of human immunodeficiency virus type 1 reverse transcriptase on the efficiency of excision of tenofovir. Antimicrob Agents Chemother 51:2911–2919

Matamoros T, Franco S, Vazquez-Alvarez BM, Mas A, Martinez MA, Menendez-Arias L (2004) Molecular determinants of multi-nucleoside analogue resistance in HIV-1 reverse transcriptases containing a dipeptide insertion in the fingers subdomain: effect of mutations D67N and T215Y on removal of thymidine nucleotide analogues from blocked DNA primers. J Biol Chem 279:24569–24577

Mathiesen S, Dam E, Roge B, Joergensen LB, Laursen AL, Gerstoft J, Clavel F (2007) Long-term foscarnet therapy remodels thymidine analogue mutations and alters resistance to zidovudine and lamivudine in HIV-1. Antivir Ther 12:335–343

Meyer PR, Matsuura SE, So AG, Scott WA (1998) Unblocking of chain-terminated primer by HIV-1 reverse transcriptase through a nucleotide-dependent mechanism. Proc Natl Acad Sci USA 95:13471–13476

Meyer PR, Matsuura SE, Schinazi RF, So AG, Scott WA (2000) Differential removal of thymidine nucleotide analogues from blocked DNA chains by human immunodeficiency virus reverse transcriptase in the presence of physiological concentrations of 2'-deoxynucleoside triphosphates. Antimicrob Agents Chemother 44:3465–3472

Meyer PR, Matsuura SE, Zonarich D, Chopra RR, Pendarvis E, Bazmi HZ, Mellors JW, Scott WA (2003) Relationship between 3'-azido-3'-deoxythymidine resistance and primer unblocking activity in foscarnetresistant mutants of human immunodeficiency virus type 1 reverse transcriptase. J Virol 77:6127–6137

Meyer PR, Smith AJ, Matsuura SE, Scott WA (2004) Effects of primer-template sequence on ATPdependent removal of chain-terminating nucleotide analogues by HIV-1 reverse transcriptase. J Biol Chem 279:45389–45398

Meyer PR, Rutvisuttinunt W, Matsuura SE, So AG, Scott WA (2007) Stable complexes formed by HIV-1 reverse transcriptase at distinct positions on the primer-template controlled by binding deoxynucleoside triphosphates or foscarnet. J Mol Biol 369:41–54

Michailidis E, Marchand B, Kodama EN, Singh K, Matsuoka M, Kirby KA, Ryan EM, Sawani AM, Nagy E, Ashida N, Mitsuya H, Parniak MA, Sarafianos SG (2009) Mechanism of inhibition of HIV-1 reverse transcriptase by 4'-Ethynyl-2-fluoro-2'-deoxyadenosine triphosphate, a translocation-defective reverse transcriptase inhibitor. J Biol Chem 284:35681–35691

Noormohamed FH, Youle MS, Higgs CJ, Martin-Munley S, Gazzard BG, Lant AF (1998) Pharmacokinetics and absolute bioavailability of oral foscarnet in human immunodeficiency virus-seropositive patients. Antimicrob Agents Chemother 42:293–297

Patel PH, Jacobo-Molina A, Ding J, Tantillo C, Clark AD Jr, Raag R, Nanni RG, Hughes SH, Arnold E (1995) Insights into DNA polymerization mechanisms from structure and function analysis of HIV-1 reverse transcriptase. Biochemistry 34:5351–5363

Richman DD, Guatelli JC, Grimes J, Tsiatis A, Gingeras T (1991) Detection of mutations associated with zidovudine resistance in human immunodeficiency virus by use of the polymerase chain reaction. J Infect Dis 164:1075–1081

Rothwell PJ, Berger S, Kensch O, Felekyan S, Antonik M, Wohrl BM, Restle T, Goody RS, Seidel CA (2003) Multiparameter single-molecule fluorescence spectroscopy reveals heterogeneity of HIV-1 reverse transcriptase:primer/template complexes. Proc Natl Acad Sci USA 100: 1655–1660

Sarafianos SG, Pandey VN, Kaushik N, Modak MJ (1995) Site-directed mutagenesis of arginine 72 of HIV-1 reverse transcriptase. Catalytic role and inhibitor sensitivity. J Biol Chem 270: 19729–19735

Sarafianos SG, Clark AD Jr, Das K, Tuske S, Birktoft JJ, Ilankumaran P, Ramesha AR, Sayer JM, Jerina DM, Boyer PL, Hughes SH, Arnold E (2002) Structures of HIV-1 reverse transcriptase with pre- and posttranslocation AZTMP-terminated DNA. EMBO J 21:6614–6624

Sarafianos SG, Clark AD Jr, Tuske S, Squire CJ, Das K, Sheng D, Ilankumaran P, Ramesha AR, Kroth H, Sayer JM, Jerina DM, Boyer PL, Hughes SH, Arnold E (2003) Trapping HIV-1 reverse transcriptase before and after translocation on DNA. J Biol Chem 278:16280–16288

Scarth BJ, Gotte M (2008) High phenotypic resistance to AZT correlates with an increased availability of pre-translocated complexes throughout the viral genome: Abstract 848. In: 15th conference on retroviruses and opportunistic infections, Boston

Scarth B, McCormick S, Gotte M (2010) Effects of mutations F61A and A62V in the fingers subdomain of HIV-1 reverse transcriptase on the translocational equilibrium. J Mol Biol 405: 349–360

Sluis-Cremer N, Temiz NA, Bahar I (2004) Conformational changes in HIV-1 reverse transcriptase induced by nonnucleoside reverse transcriptase inhibitor binding. Curr HIV Res 2:323–332

St Clair MH, Martin JL, Tudor-Williams G, Bach MC, Vavro CL, King DM, Kellam P, Kemp SD, Larder BA (1991) Resistance to ddI and sensitivity to AZT induced by a mutation in HIV-1 reverse transcriptase. Science 253:1557–1559

Steitz TA (1998) A mechanism for all polymerases. Nature 391:231–232

Steitz TA (1999) DNA polymerases: structural diversity and common mechanisms. J Biol Chem 274:17395–17398

Svarovskaia ES, Feng JY, Margot NA, Myrick F, Goodman D, Ly JK, White KL, Kutty N, Wang R, Borroto-Esoda K, Miller MD (2008) The A62V and S68G mutations in HIV-1 reverse transcriptase partially restore the replication defect associated with the K65R mutation. J Acquir Immune Defic Syndr 48:428–436

Tachedjian G, Mellors J, Bazmi H, Birch C, Mills J (1996) Zidovudine resistance is suppressed by mutations conferring resistance of human immunodeficiency virus type 1 to foscarnet. J Virol 70:7171–7181

Tong W, Lu CD, Sharma SK, Matsuura S, So AG, Scott WA (1997) Nucleotide-induced stable complex formation by HIV-1 reverse transcriptase. Biochemistry 36:5749–5757

Upadhyay AK, Talele TT, Pandey VN (2010) Impact of template overhang-binding region of HIV-1 RT on the binding and orientation of the duplex region of the template-primer. Mol Cell Biochem 338:19–33

Wohrl BM, Krebs R, Goody RS, Restle T (1999) Refined model for primer/template binding by HIV-1 reverse transcriptase: pre-steady-state kinetic analyses of primer/template binding and nucleotide incorporation events distinguish between different binding modes depending on the nature of the nucleic acid substrate. J Mol Biol 292:333–344

Yin YW, Steitz TA (2004) The structural mechanism of translocation and helicase activity in T7 RNA polymerase. Cell 116:393–404

Zahn KE, Tchesnokov EP, Gotte M, Doublie S (2011) Phosphonoformic acid inhibits viral replication by trapping the closed form of the DNA polymerase. J Biol Chem 286:25246–25255

Zhou Z, Lin X, Madura JD (2006) HIV-1 RT nonnucleoside inhibitors and their interaction with RT for antiviral drug development. Infect Disord Drug Targets 6:391–413

# Chapter 10
# tRNA Primer Sequestration as an Antiviral Strategy

**Christopher P. Jones and Karin Musier-Forsyth**

## Abbreviations

| | |
|---|---|
| FRET | Förster resonance energy transfer |
| HIV-1 | Human immunodeficiency virus type 1 |
| LysRS | Lysyl-tRNA synthetase |
| MA | Matrix |
| NC | Nucleocapsid |
| NMR | Nuclear magnetic resonance |
| nt | Nucleotide(s) |
| ORF | Open reading frame |
| PBS | Primer-binding site |
| RSV | Rous sarcoma virus |
| RT | Reverse transcriptase |
| TrpRS | Tryptophanyl-tRNA synthetase |
| UTR | Untranslated region |
| vRNA | Viral RNA |
| WT | Wild type |

*Conflict of interest statement*: None declared.

C.P. Jones • K. Musier-Forsyth (✉)
Department of Chemistry and Biochemistry, Center for RNA Biology, and Center for Retroviral Research, The Ohio State University, 43210 Columbus, OH, USA
e-mail: musier@chemistry.ohio-state.edu

S.F.J. LeGrice and M. Götte (eds.), *Human Immunodeficiency Virus Reverse Transcriptase: A Bench-to-Bedside Success*, DOI 10.1007/978-1-4614-7291-9_10,

## 10.1 Introduction

From retroviral initiation to eukaryotic genome replication, priming cDNA is a challenging task. An underlying problem is that the template lacks a 3′-OH substrate required for faithful initiation of RNA or DNA synthesis by the replicative enzyme, whether it is HIV-1 reverse transcriptase (RT) or eukaryotic pol α. The solutions to this problem are various – in the case of hepadnaviruses, the replicase itself uses a tyrosine residue for the substrate to mimic a primer's 3′-OH (reviewed in (Salas 1991)), and in the case of eukaryotic replication, a second enzyme, the pol α-primase domain, serves the role of synthesizing a primer for use by the replicase pol α. Retroviruses best exemplify genomic brevity by their ability to accomplish so many activities in so few nucleotides (nt). Thus, retroviruses and retrotransposons have solved the priming problem by co-opting an abundant highly conserved cellular factor to serve as the primer for reverse transcriptase and adapting their genomic viral RNA (vRNA) sequences to be complementary to their primers (Dahlberg et al. 1974; Harada et al. 1975; Kikuchi et al. 1986; Peters and Glover 1980). Although the primer is always a tRNA for retroviruses, the specific primer used by a retrovirus subgroup is unique, with all lentiviruses including HIV-1 using solely $tRNA^{Lys3}$ in vivo (reviewed in (Mak and Kleiman 1997; Marquet et al. 1995)).

Upon infection of a host cell and fusion with the plasma membrane, the HIV-1 particle releases its capsid into the cell's cytoplasm (Fig. 10.1). Contained within the capsid are components necessary for viral replication, including reverse transcriptase, the annealed $tRNA^{Lys3}$/vRNA primer/template duplex, the replication chaperone, nucleocapsid protein (NC), and other factors. Reverse transcription begins from the 3′-OH of the CCA end of $tRNA^{Lys3}$, which is annealed to the highly conserved primer-binding site (PBS) region on the vRNA. NC facilitates reverse transcription, in part, by destabilizing structured regions of the vRNA to allow RT read through (Ji et al. 1996; Wu et al. 1996). Once reverse transcription is complete, the cDNA is imported into the nucleus and integrated into cellular genomic DNA. Newly synthesized vRNA serves as the template for both viral protein synthesis (in both spliced and unspliced forms) and as the genomic material for incorporation into newly assembled virions (unspliced form only). During assembly, the Gag protein packages the unspliced genomic vRNA together with other necessary viral proteins and cellular components, such as the tRNA primer, and altogether the ~100 nm virus particle buds off from the infected cell. The tRNA content of virions is critical for viral infectivity due to the essential role the primer plays in initiating cDNA synthesis – overexpression of $tRNA^{Lys3}$ yields more infectious virions with higher $tRNA^{Lys3}$ content, and preventing $tRNA^{Lys3}$ incorporation produces less infectious virions (Guo et al. 2003). Thus, preventing $tRNA^{Lys}$ packaging and inhibiting priming are viable strategies for reducing virus infectivity (Fig. 10.1).

In this chapter, we discuss investigations into the mechanism of priming of HIV-1 reverse transcription with emphasis on how findings may impact the development of antiretroviral drugs targeting $tRNA^{Lys3}$ priming. First, we review what is known about the mechanism of tRNA annealing and the structure of the primer/template complex. As virion tRNA content is modulated by the reverse transcriptase (RT) domain of GagPol and the cellular factor lysyl-tRNA synthetase (LysRS), the

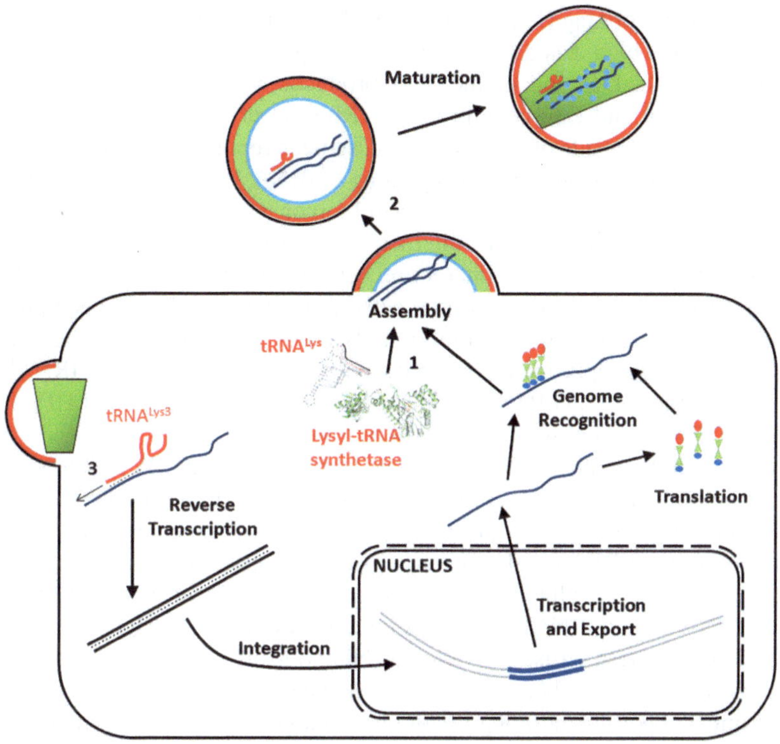


**Fig. 10.1** Simplified schematic of the HIV-1 life cycle with emphasis on roles played by the tRNA$^{Lys3}$ primer including (1) packaging in complex with human LysRS and Gag during assembly, (2) annealing by the NC domain of Gag, and (3) initiating reverse transcription via extension of annealed tRNA$^{Lys3}$ by RT. As the precise timing of annealing is unknown, only the budded virion is depicted with annealed tRNA$^{Lys3}$, although the primer may be annealed prior to or during assembly. For clarity, reverse transcription is shown in the cytoplasm after uncoating although the two processes are likely coordinated (Hulme et al. 2011), and translation of only the Gag polyprotein is shown

contribution of these accessory proteins to tRNA packaging and annealing is also discussed. Finally, we review efforts to date to directly prevent priming and to inhibit primer tRNA$^{Lys3}$ packaging, noting recent developments in understanding the complex interactions between the protein and RNA components in the initiation complex. Reverse transcription initiation (tRNA$^{Lys3}$ priming) is not the only priming event in the HIV-1 life cycle, and the subsequent polypurine tract priming events for positive strand synthesis will not be discussed here.

## 10.2 Mechanism of tRNA$^{Lys3}$ Annealing

Duplex formation between the 3′ 18 nt of tRNA$^{Lys3}$ and a complementary and absolutely conserved 18-nt PBS on the HIV-1 genome is the primary interaction that allows for initiation of DNA synthesis by RT. This annealing reaction is catalyzed by the NC domain of Gag (Feng et al. 1999; Jones et al. 2011; Roldan et al. 2005; Wu et al. 2010), which acts as a nucleic acid chaperone (Darlix et al. 2011; Levin et al. 2005, 2010). Deletion of the PBS produces noninfectious virions despite expression of Gag and Env gene products (Rhim et al. 1991). Additional tRNA-vRNA interactions outside of the PBS have also been reported, including an interaction between the anticodon stem/variable loop of $tRNA^{Lys3}$ and a sequence upstream of the PBS (Iwatani et al. 2003). An A-rich loop or bulge in the genomes of some isolates has also been reported to interact with the U-rich anticodon of $tRNA^{Lys3}$ (Isel et al. 1993). An alternative interaction between the TΨC stem and nt 123–130 of the HIV-1 genome termed the primer activation signal (PAS) has been proposed to regulate reverse transcription initiation (Beerens and Berkhout 2002; Beerens et al. 2001). The PAS appears to contribute to the virus's preference for $tRNA^{Lys3}$, as mutating both the PBS and PAS is required to switch the preference of HIV-1 from $tRNA^{Lys3}$ to $tRNA^{Lys1,2}$ (Abbink et al. 2004).

Reversion from an altered PBS sequence back to wild type (WT) can occur in a single virus replication cycle, as the PBS sequence in the viral cDNA is copied directly from the tRNA primer. Efforts to engineer HIV-1 to use alternative primers by mutation of the PBS have shown varied success – either immediate reversion to $tRNA^{Lys3}$ or replication failure is observed in most cases. However, use of $tRNA^{Lys1,2}$, $tRNA^{His}$, $tRNA^{Met}$, and $tRNA^{Glu}$ produces infectious virions although the replication characteristics of these viruses are altered (Ni and Morrow 2007; Wei et al. 2005). A similar observation has been made in avian leukosis virus where the use of other tRNA primers is possible although with observed growth defects (Whitcomb et al. 1995). Aberrant annealing has also been observed – either due to the use of the incorrect tRNA to misprime reverse transcription (Colicelli and Goff 1986a, b; Wakefield et al. 1995) or due to the priming of cellular mRNAs with short "tRNA-like" primers (Taylor and Cywinski 1984). HIV-1 RT can cope with misaligned annealing, and minor insertions and deletions to the PBS can still be extended at low levels, eventually leading to the recovery of WT PBS upon virus replication (Das and Berkhout 1995).

The precise timing of $tRNA^{Lys3}$ annealing in the HIV-1 life cycle is not known. However, protease-negative virions contain annealed $tRNA^{Lys3}$ (Fu et al. 1997; Huang et al. 1997; Stewart et al. 1990) and mature virions contain a greater amount of tRNA stably annealed to vRNA (Guo et al. 2009). Thus, annealing has been described as a multistep process that is initially catalyzed by Gag in the cell and later remodeled by the Gag cleavage product NC in the mature virion. A corresponding maturation step in genome dimerization has also been proposed (Jalalirad and Laughrea 2010), suggesting that liberation of NC from Gag leads to changes in chaperone activity as well. Highlighting a newly discovered role for NC in the HIV-1 life cycle, recent studies have observed a loss in virus infectivity in NC zinc-finger mutant viruses in which reverse transcription occurs prematurely (i.e., priming and extension in the cytoplasm prior to budding) (Houzet et al. 2008; Thomas et al. 2008). However, the loss in infectivity is not due to early reverse transcription (Thomas et al. 2011). These studies demonstrate that WT virus suppresses reverse transcription until after assembly and budding.

In vitro, tRNA annealing can be observed by mixing $tRNA^{Lys3}$ and vRNA in the presence of NC (Chan et al. 1999; De Rocquigny et al. 1992) or Gag (Jones et al. 2011; Roldan et al. 2005; Wu et al. 2010). The tRNA may also be hybridized to the template by thermal denaturation and still be competent for reverse transcription (Waters et al. 1975). Previous reports found no difference in the structure of the template between heat-annealed and NC-annealed RNAs (Brule et al. 2002; Kim et al. 2012). Alternatively, vRNA and $tRNA^{Lys3}$ gently extracted from virions can serve as substrates for in vitro extension assays in which extension via radiolabel incorporation is a measure of the amount of annealed complex (Cen et al. 2009; Huang et al. 1997). A detailed kinetic analysis of tRNA annealing in the presence and absence of NC found that NC accelerates annealing by up to ~$10^5$-fold and that the uncatalyzed reaction would proceed with a $t_{1/2}$ of ~4.5 days (Hargittai et al. 2004). Ten- to 100-fold of this rate enhancement is due to NC's destabilization of secondary structure within the PBS template, and a further $10^3$- to $10^4$-fold enhancement is achieved by NC-facilitated duplex nucleation (i.e., bringing the strands together).

In vivo, the initial annealing step is performed by Gag (Feng et al. 1999), which facilitates the annealing reaction at least tenfold more slowly in vitro when compared to mature NC (Jones et al. 2011; Wu et al. 2010). The reduced chaperone activity of Gag relative to NC is partly attributed to the matrix (MA) domain, which also interacts with nucleic acids. Inositol phosphates can stimulate Gag's tRNA annealing activity by competing with MA to prevent it from interacting with nucleic acids (Jones et al. 2011). In addition, deletion of the MA domain stimulates Gag's chaperone activity.

## 10.3 Structure of the Annealed tRNA/vRNA Complex

The structure of the HIV-1 genome has been probed using chemical and enzymatic methods (reviewed in (Lu et al. 2011b)), NMR spectroscopy (Lu et al. 2011a; Puglisi and Puglisi 2011; Tisné et al. 2004), and mass spectrometry-based approaches (Turner et al. 2009; Yu et al. 2008). Recently, the secondary structure of the entire genome has been investigated using selective 2′-hydroxyl acylation analyzed by primer extension (SHAPE) (Watts et al. 2009). The three-dimensional structure of the complete ~340-nt 5′ UTR, a highly structured region of the genome, is unknown, and the differences in the reported secondary structures likely reflect the dynamics of the vRNA, as well as sequence differences between different isolates (Goldschmidt et al. 2004). Recent work has suggested that the variety of secondary structure models is an accurate depiction of conformational heterogeneity, and it is likely that the HIV-1 genome adopts alternate conformations depending on its involvement in translation or packaging (Lu et al. 2011a) or its monomeric/dimeric state. Examination of long constructs has proven difficult; however, shorter regions of the genome have been examined via NMR, and high-resolution structural models are available for various stem loops, including SL1 (Lawrence et al. 2003), SL2 (Amarasinghe et al. 2000), SL3 (De Guzman et al. 1998), and SL4 (Kerwood et al. 2001) of the packaging signal; the end of TAR (Aboul-ela et al. 1996); and an A-rich stemloop upstream of the PBS in the MAL isolate (Puglisi and Puglisi 1998).

Understanding how the 5′ UTR structure changes as a consequence of $tRNA^{Lys3}$ primer annealing poses an even greater challenge. The crystal structure of the fully modified primer tRNA alone has been solved to 3.3 Å and adopts a canonical L-shaped tRNA fold (Benas et al. 2000). Studies to understand the mechanism of tRNA annealing have been performed to probe how NC affects tRNA structure and to determine which parts of NC mediate changes within $tRNA^{Lys3}$ and vRNA (Barraud et al. 2007; Tisné et al. 2001, 2003, 2004). NC's interactions with the primer tRNA appear to be primarily within the acceptor stem, on the inside of the L shape, and on the D stem at the point of tertiary interactions in the loop (Tisné et al. 2001). Using either NC or heat, $tRNA^{Lys3}$ was annealed to an 18-nt PBS sequence or a 65-nt RNA containing the PBS and examined by NMR. These studies found that melting initiated at the four-way junction and continued at the base of the acceptor and TΨC stems (Tisné et al. 2004). These findings are consistent with terbium probing, which also revealed that NC first disrupts the metal binding pockets of the tRNA, including those that stabilize the tRNA tertiary fold at the D/TΨC junction (Hargittai et al. 2001). An important finding from these studies and FRET-based studies (Chan et al. 1999) is that NC weakly disrupts the overall structure of $tRNA^{Lys3}$ and that complete unwinding of the acceptor stem is not observed in the absence of vRNA.

Chemical and enzymatic probing of the annealed complex (MAL isolate) has been performed (Isel et al. 1995), and SHAPE was also used to determine the secondary structural models of the free vRNA and annealed $tRNA^{Lys3}$/vRNA complex (NL4-3 isolate) (Wilkinson et al. 2008). These models are consistent in predicting both the 18-nt duplex formed between vRNA and $tRNA^{Lys3}$, as well as the previously proposed additional interactions (Isel et al. 1993; Iwatani et al. 2003) despite sequence differences between isolates. Additional footprinting assays with the $tRNA^{Lys3}$/vRNA complex (MAL isolate) in the presence of RT led to a proposed model for the initiation complex, which predicted that RT could change orientations relative to its $tRNA^{Lys3}$/vRNA substrate (Isel et al. 1999). More recently, the initiation complex of the MAL isolate has been investigated using NMR (Puglisi and Puglisi 2011), and substantial secondary structural changes were observed in both the template and primer $tRNA^{Lys3}$ in the annealed duplex compared to the free RNAs. One of these changes occurring upstream from the 18-nt PBS duplex involves the formation of a long stable hairpin, which is hypothesized to be a block to RT initiation (further discussed below) (Liu et al. 2010). Similar to the study with the NL4-3-derived PBS (Tisné et al. 2004), the interaction between the U-rich $tRNA^{Lys3}$ anticodon and an A-rich loop in the MAL vRNA was not observed by NMR even though DMS probing/foot printing showed protection of the A-rich loop in the presence of either transcribed or authentic $tRNA^{Lys3}$ (Puglisi and Puglisi 2011). Taken together, these results suggest that the structure of the annealed template is significantly different from the free vRNA template, but the precise interactions present remain to be determined.

A recent single molecule Förster resonance energy transfer (FRET) study (Liu et al. 2010) was able to observe the dynamic interactions occurring between RT and the $tRNA^{Lys3}$/vRNA duplex and demonstrate that a stemloop within the template prevents RT from extending the $tRNA^{Lys3}$ primer past 3 nt. Using RT labeled with a fluorescent donor and annealed $tRNA^{Lys3}$/PBS template labeled with a fluorescent acceptor, two FRET states were observable, corresponding to RT bound in a polymerase orientation (productive for extension) or an RNase H orientation (nonproductive for extension). In the presence of extended $tRNA^{Lys3}+3$, which mimics the $tRNA^{Lys3}$ primer after 3 nt have been added to the CCA end, RT remained primarily in the nonproductive orientation. However, in the presence of $tRNA^{Lys3}+6$, which mimics the primer after 6 nt of extension, RT was found predominantly in the productive orientation, which was attributed to destabilization of a hairpin upstream of the PBS (Liu et al. 2010). The presence of NC was also sufficient to destabilize this hairpin, and mutation of the stemloop to destabilize its secondary structure also allowed for productive RT binding. These findings support a mechanism in which RT prevents premature initiation by remaining in a nonproductive orientation on the template but binds productively and extends the tRNA primer once NC is present.

## 10.4 Accessory Factors to tRNA Packaging

Retroviruses are ribonucleoprotein complexes that package various levels of host tRNAs. In some cases, the total tRNA content of virions reflects the cellular milieu in which assembling particles coalesce (Kelly et al. 2003); however, many studies have noted that the tRNA in cells differs considerably from the tRNA in retroviruses (Jiang et al. 1993; Pavon-Eternod et al. 2010; Waters 1975; Waters et al. 1975, 1980). Since the vRNA serves as both template for Gag translation and future packaging material, it has been suggested that newly translated Gag interacts with nearby vRNA, thereby shutting off translation for that particular vRNA and eventually leading to its incorporation into assembling virions. Even though Gag binds relatively tightly to all RNAs, Gag preferentially binds the packaging signal or $\psi$ region of the genome in the 5′ UTR and beginning of the Gag open reading frame (ORF), which allows the precise selection of the full-length vRNA instead of spliced versions of vRNA or other cellular mRNAs (reviewed in (Lu et al. 2011b)). Thus, preference for $\Psi$ excludes other cellular RNAs from being packaged. tRNAs and other noncoding relatively short RNAs are exceptions: many tRNAs are packaged into each HIV-1 virion along with copies of 7SL RNA (Onafuwa-Nuga et al. 2006). MLV also packages 7SL RNA and a variety of other cellular RNAs, including Ro RNP RNAs, 5S rRNA, and U6 snRNA (Onafuwa-Nuga et al. 2005). Of the tRNAs packaged in HIV-1, $tRNA^{Lys}$ (~35 %) and $tRNA^{Asn(GUU)}$ (11 %) are the most abundant and are enriched in virions, as their relevant percentages in the cytoplasm are 4.5 % and 3.4 % of total tRNA, respectively (Pavon-Eternod et al. 2010).

In avian retroviruses, it was suggested that RT within GagPol is the major determinant for tRNA packaging and that RT binds specifically to its primer $tRNA^{Trp}$ (Cordell et al. 1979; Garret et al. 1984; Panet et al. 1975; Peters and Hu 1980). Studies of HIV-1 also initially suggested RT to be crucial for tRNA packaging via an interaction between RT and the anticodon domain of its $tRNA^{Lys3}$ primer (Baratet al. 1989; Mak et al. 1994). However, later reports excluded the anticodon sequence to be required for RT-tRNA interaction and attributed the overall tRNA fold to be the recognition motif for RT, which binds preferentially to tRNAs but otherwise shows no specificity (Arion et al. 1996; Oude Essink et al. 1995; Thrall et al. 1996). Moreover, the contribution of GagPol is essential for tRNA packaging as Gag-only virus-like particles contain >80 % less $tRNA^{Lys}$ than WT particles (Pavon-Eternod et al. 2010). Nevertheless, the specificity must be derived from another factor. The interaction between RT and the tRNA/vRNA complex is not unlike the interaction between ribosomes and internal ribosome entry sites – both occur at the 5′ UTR of the RNA in question, recognize tRNA or tRNA-like structural motifs, and signal the starting points for cDNA and protein synthesis, respectively.

The observation that all $tRNA^{Lys}$ isoacceptors, $tRNA^{Lys3}$ and $tRNA^{Lys1,2}$, are selectively packaged into HIV-1 virions and that HIV-1 RT only generally recognizes the tRNA fold implies that a common interacting factor specific for $tRNA^{Lys}$ is involved in selective primer packaging. A likely candidate, lysyl-tRNA synthetase (LysRS), is a translation factor and natural binding partner of all $tRNA^{Lys}$ isoacceptors, which are aminoacylated with lysine by LysRS for use by the ribosome in protein synthesis. Indeed, it was discovered that LysRS is packaged into HIV-1 virions in an ~1:1 correspondence with $tRNA^{Lys}$ (Cen et al. 2001, 2002) while other aminoacyl-tRNA synthetases and associated factors are undetectable in particles (Halwani et al. 2004). In RSV, which primes reverse transcription with $tRNA^{Trp}$, tryptophanyl-tRNA synthetase (TrpRS) is packaged whereas other aminoacyl-tRNA synthetases are not, suggesting that other retroviruses may also package tRNA/synthetase complexes (Cen et al. 2002). The ability of LysRS to bind its cognate tRNA influences packaging efficiency, as $tRNA^{Lys3}$ anticodon mutations reduce their incorporation (Javanbakht et al. 2002). In vitro studies and cell-based assays suggest that LysRS is packaged via an interaction between the motif 1 dimerization interface of the synthetase and the CA domain of Gag (Javanbakht et al. 2003). A recent mass spectrometry-based proteomics analysis also identified the multisynthetase complex (MSC) as an interacting partner of Gag's MA domain, although the significance of this interaction is unclear (Jager et al. 2012). LysRS is present in cells in premitochondrial, mitochondrial, and cytoplasmic forms (Cen et al. 2001, 2002; Kaminska et al. 2007). These forms are differentiated by the inclusion of a signal peptide in the premitochondrial LysRS, which is cleaved in the mitochondrial form. Cytoplasmic and mitochondrial LysRS are derived from the same gene, which undergoes alternative splicing, and are identical except for ~20 amino acids at the N-terminus (Tolkunova et al. 2000; Dias 2012).

In addition to its canonical function in tRNA aminoacylation, LysRS also has additional "moonlighting" jobs in the cell (reviewed in (Nechushtan et al. 2009)). LysRS is believed to be stored in the MSC for translation and mobilized for its other cellular roles. For example, in mast cells LysRS functions as an Ap4A-producing transcription factor in complex with microphthalmia transcription factor (MITF) and is released from the MSC via a conformational change induced by phosphorylation (Fang et al. 2011; Yannay-Cohen et al. 2009). tRNAs are channeled for translation and are rarely free in the cytoplasm in their non-aminoacylated form; however, separate pools of active synthetase/tRNA have been observed that do not participate in translation (Kyriacou and Deutscher 2008). Thus, the source of primer $tRNA^{Lys}$ and LysRS in HIV-1 is either non-MSC bound or mobilized from the MSC via a viral signal. Based upon siRNA knockdown data, in early time points, a 20–25 % decrease in total cellular LysRS correlated with an 80 % decrease in packaged LysRS, suggesting that newly synthesized LysRS might be the cellular source of LysRS in virions. LysRS is presumably bound to $tRNA^{Lys3}$ when it encounters the assembly complex, and the RT domain of GagPol may simultaneously interact with the primer tRNA as RT recognizes the features of tRNAs that are not identity elements for LysRS. Mitochondrial LysRS has also been proposed to be the source of LysRS in HIV-1 (Kaminska et al. 2007).

The state of $tRNA^{Lys}$ is critical for priming because the site of lysine attachment in aminoacylated tRNA is the CCA-3′-OH, which is also the substrate for RT (Shimada et al. 1994). In avian myeloblastosis virus, which uses $tRNA^{Trp}$ as its primer, RT was reported to have deacylation activity in the presence of Trp-$tRNA^{Trp}$ and viral RNA but not in the presence of $tRNA^{Trp}$ alone (Litvak et al. 1983; Sarih et al. 1982), suggesting that primer/template annealing facilitates $tRNA^{Trp}$ deacylation by RT. In HIV-1, aminoacylated Lys-$tRNA^{Lys3}$ cannot be extended by HIV-1 RT, and RT does not accelerate the rate of deacylation of Lys-$tRNA^{Lys3}$ (Rigourd et al. 2003). Spontaneous deacylation of charged tRNAs and the synthetase-catalyzed reverse reaction also occurs to varying extents (Bonnet and Ebel 1972). Alternatively, $tRNA^{Lys3}$ charging may be prevented by an unknown mechanism such that $tRNA^{Lys3}$ is packaged in an uncharged state.

## 10.5 Inhibitors of tRNA Priming

The most desirable drug targets in the HIV-1 genome are highly conserved sites, and many of the most highly conserved RNA sequences are found within the 5′ UTR, such as the PBS. Strategies for inhibiting tRNA priming include (1) preventing packaging of the $tRNA^{Lys3}$ primer by disruption of the $tRNA^{Lys}$/LysRS/Gag/GagPol packaging complex, (2) blocking NC's chaperone activity and thus preventing tRNA annealing, and (3) inhibiting $tRNA^{Lys3}$ extension by RT (Fig. 10.1). Disrupting the packaging complex could involve inhibition of the interaction between Gag and LysRS. Cyclic peptides selected for binding to Gag's CA domain have been recently developed and shown to effectively compete for the Gag/LysRS interaction in vitro (Dewan et al. 2012). Inhibition of NC's chaperone activity is a promising strategy, as multiple steps in reverse transcription would be inhibited using this approach.

More recent research in antiretroviral therapies has been progressing toward the goal of disrupting NC/$tRNA^{Lys3}$ interactions (Graham et al. 2011). The anticodon stemloop of $tRNA^{Lys3}$ is highly posttranscriptionally modified, containing 5-methoxycarbonylmethyl-2-thiouridine ($mcm^5s^2U_{34}$ or $S_{34}$), 2-methylthio-$N^6$-threonylcarbamoyladenosine ($ms^2t^6A_{37}$), and pseudouridine ($\Psi_{39}$) modifications.

According to a recent report, NC's affinity to the tRNA$^{Lys3}$ anticodon is enhanced by these modifications, which allow NC to better melt the anticodon stemloop and facilitate tRNA annealing (Graham et al. 2011). A more general series of NC inhibitors, which would potentially prevent priming as well as other RNA refolding activities chaperoned by NC, are the zinc-ejecting class of drugs, such as S-acyl-2-mercaptobenzamide thioesters and others (de Rocquigny et al. 2008; Miller Jenkins et al. 2007, 2010; Morcock et al. 2008; Pannecouque et al. 2010; Rice et al. 1993; Wallace et al. 2009). These inhibitors irreversibly react with the cysteines of the CCHC zinc fingers, thus preventing zinc binding and eliminating NC's duplex destabilization activity (reviewed in (Supuran et al. 2004)). As the zinc fingers also play a well established role in $\psi$ recognition, this class of drugs could also disrupt virus assembly provided that they penetrate cell membranes. A unique feature of some compounds in this drug class is inactivation of NC followed by regeneration of the active drug by cellular enzymes (Miller Jenkins et al. 2010). One recent screen for inhibitors of NC interactions with SL2, a stemloop in $\psi$ region of the vRNA, identified compounds with modest antiviral activity (Breuer et al. 2012). Another class of NC-inhibiting compounds, methylated oligonucleotides, compete for NC/vRNA interactions and inhibit viral replication (Avilov et al. 2012; Grigorov et al. 2011).

For inhibitors of RT initiation, one strategy is to use a mimic of the primer tRNA$^{Lys3}$ to block annealing such that RT initiation is prevented or leads to inert cDNA. For example, one paradigm is to provide a tRNA$^{Lys3}$ mimic that would inactivate RT. Alternatively, the annealed primer/template complex could be targeted and cleaved. Antisense phosphorothioate oligonucleotides to the Gag, Tat, and Rev ORFs and siRNAs to the 5′ UTR, Vif, Tat, Rev, and Nef have been shown to be inhibitors of HIV replication (Coburn and Cullen 2002; Jacque et al. 2002; Lisziewicz et al. 1993), and phosphorothioate oligonucleotides derived from the sequence of tRNA$^{Lys3}$ have been shown to inhibit RT (El Dirani-Diab et al. 1997). A tRNA$^{Lys3}$ mispriming mimic designed to be complementary to a region in TAR was found to modestly inhibit HIV-1 replication (10- to 50-fold) due to its ability to prime reverse transcription from both the PBS and the TAR sequence (Lu et al. 1997). This tRNA contained a modified acceptor stem that allowed for TAR priming, and complementary mutations in the acceptor stem modestly impaired PBS priming. A second tRNA$^{Lys3}$ mimicry strategy used targeted ribozyme cleavage with a tRNA$^{Lys3}$ mimic fused to a ribozyme and was shown to be active in vitro (Westaway et al. 1995). Aside from the challenges posed by drug delivery, a difficulty with these strategies is that any primer mimic that resembles the authentic tRNA$^{Lys3}$ might compete for tRNA$^{Lys3}$ in translation or disrupt charging and thus be toxic. Indirect methods targeting tRNA$^{Lys3}$ priming or the PBS have also been developed. For example, targeting the integrated PBS cDNA with an engineered zinc-finger transcription factor designed to repress HIV-1 transcription from the 5′ LTR leads to escape mutations in the PBS (e.g., single or double point mutations), which in turn reduce the rate of viral replication (Eberhardy et al. 2006).

Cellular defenses against retroviruses like HIV-1 include strategies that are often thwarted by HIV-1 accessory proteins. APOBEC3G, a cellular cytidine deaminase, has potent anti-HIV-1 activity in viruses lacking the Vif accessory protein. In addition to hypermutation of viral cDNA, APOBEC3G also appears to function in blocking RT elongation (Bishop et al. 2008; Iwatani et al. 2007) and has been reported to inhibit $tRNA^{Lys3}$ annealing (Guo et al. 2009). Moreover, the presence of cleaved primer/template complexes in HIV-1-infected cells suggests that cellular countermeasures against retroviruses by targeting annealed complexes may have already evolved (Yeung et al. 2009). RNA interference (RNAi) is utilized by the cell to silence retrotransposons (Yang and Kazazian 2006), which also have complex 5′ UTRs and prime reverse transcription using cellular tRNAs. Future anti-priming strategies derived from known host defense mechanisms may also provide new approaches to combat HIV-1.

## References

Abbink TE, Beerens N, Berkhout B (2004) Forced selection of a human immunodeficiency virus type 1 variant that uses a non-self tRNA primer for reverse transcription: involvement of viral RNA sequences and the reverse transcriptase enzyme. J Virol 78(19):10706–10714

Aboul-ela F, Karn J, Varani G (1996) Structure of HIV-1 TAR RNA in the absence of ligands reveals a novel conformation of the trinucleotide bulge. Nucleic Acids Res 24(20):3974–3981

Amarasinghe GK, De Guzman RN, Turner RB, Summers MF (2000) NMR structure of stem-loop SL2 of the HIV-1 psi RNA packaging signal reveals a novel A-U-A base-triple platform. J Mol Biol 299(1):145–156

Arion D, Harada R, Li X, Wainberg MA, Parniak MA (1996) HIV-1 reverse transcriptase shows no specificity for the binding of primer $tRNA^{Lys3}$. Biochem Biophys Res Commun 225(3):839–843

Avilov SV, Boudier C, Gottikh M, Darlix JL, Mély Y (2012) Characterization of the inhibition mechanism of HIV-1 nucleocapsid protein chaperone activities by methylated oligoribonucleotides. Antimicrob Agents Chemother 56(2):1010–1018

Barat C, Lullien V, Schatz O, Keith G, Nugeyre MT, Gruninger-Leitch F et al (1989) HIV-1 reverse transcriptase specifically interacts with the anticodon domain of its cognate primer tRNA. EMBO J 8(11):3279–3285

Barraud P, Gaudin C, Dardel F, Tisné C (2007) New insights into the formation of HIV-1 reverse transcription initiation complex. Biochimie 89(10):1204–1210

Beerens N, Berkhout B (2002) The tRNA primer activation signal in the human immunodeficiency virus type 1 genome is important for initiation and processive elongation of reverse transcription. J Virol 76(5):2329–2339

Beerens N, Groot F, Berkhout B (2001) Initiation of HIV-1 reverse transcription is regulated by a primer activation signal. J Biol Chem 276(33):31247–31256

Benas P, Bec G, Keith G, Marquet R, Ehresmann C, Ehresmann B et al (2000) The crystal structure of HIV reverse-transcription primer $tRNA^{Lys3}$ shows a canonical anticodon loop. RNA 6(10):1347–1355

Bishop KN, Verma M, Kim EY, Wolinsky SM, Malim MH (2008) APOBEC3G inhibits elongation of HIV-1 reverse transcripts. PLoS Pathog 4(12):e1000231

Bonnet J, Ebel JP (1972) Interpretation of incomplete reactions in tRNA aminoacylation. Aminoacylation of yeast $tRNA^{Val}$ II with yeast valyl-tRNA synthetase. Eur J Biochem 31(2):335–344

Breuer S, Chang MW, Yuan J, Torbett BE (2012) Identification of HIV-1 inhibitors targeting the nucleocapsid protein. J Med Chem 55(11):4968–4977

Brule F, Marquet R, Rong L, Wainberg MA, Roques BP, Le Grice SF et al (2002) Structural and functional properties of the HIV-1 RNA-$tRNA^{Lys3}$ primer complex annealed by the nucleocapsid protein: comparison with the heat-annealed complex. RNA 8(1):8–15

Cen S, Khorchid A, Javanbakht H, Gabor J, Stello T, Shiba K et al (2001) Incorporation of lysyl-tRNA synthetase into human immunodeficiency virus type 1. J Virol 75(11):5043–5048
Cen S, Javanbakht H, Kim S, Shiba K, Craven R, Rein A et al (2002) Retrovirus-specific packaging of aminoacyl-tRNA synthetases with cognate primer tRNAs. J Virol 76(24):13111–13115
Cen S, Guo F, Kleiman L (2009) Methods for analysis of incorporation and annealing methods for analysis of $tRNA^{Lys}$ in HIV-1. Methods Mol Biol 485:223–232
Chan B, Weidemaier K, Yip WT, Barbara PF, Musier-Forsyth K (1999) Intra-tRNA distance measurements for nucleocapsid protein dependent tRNA unwinding during priming of HIV reverse transcription. Proc Natl Acad Sci 96(2):459–464
Coburn GA, Cullen BR (2002) Potent and specific inhibition of human immunodeficiency virus type 1 replication by RNA interference. J Virol 76(18):9225–9231
Colicelli J, Goff SP (1986a) Isolation of a recombinant murine leukemia virus utilizing a new primer tRNA. J Virol 57(1):37–45
Colicelli J, Goff SP (1986b) Structure of a cloned circular retroviral DNA containing a tRNA sequence between the terminal repeats. J Virol 57(2):674–677
Cordell B, Swanstrom R, Goodman HM, Bishop JM (1979) $tRNA^{Trp}$ as primer for RNA-directed DNA polymerase: structural determinants of function. J Biol Chem 254(6):1866–1874
Dahlberg JE, Sawyer RC, Taylor JM, Faras AJ, Levinson WE, Goodman HM et al (1974) Transcription of DNA from the 70S RNA of Rous sarcoma virus. I. Identification of a specific 4S RNA which serves as primer. J Virol 13(5):1126–1133
Darlix JL, Godet J, Ivanyi-Nagy R, Fosse P, Mauffret O, Mély Y (2011) Flexible nature and specific functions of the HIV-1 nucleocapsid protein. J Mol Biol 410(4):565–581
Das AT, Berkhout B (1995) Efficient extension of a misaligned tRNA-primer during replication of the HIV-1 retrovirus. Nucleic Acids Res 23(8):1319–1326
De Guzman RN, Wu ZR, Stalling CC, Pappalardo L, Borer PN, Summers MF (1998) Structure of the HIV-1 nucleocapsid protein bound to the SL3 psi-RNA recognition element. Science 279(5349):384–388
De Rocquigny H, Gabus C, Vincent A, Fournie-Zaluski MC, Roques B, Darlix JL (1992) Viral RNA annealing activities of human immunodeficiency virus type 1 nucleocapsid protein require only peptide domains outside the zinc fingers. Proc Natl Acad Sci 89(14):6472–6476
de Rocquigny H, Shvadchak V, Avilov S, Dong CZ, Dietrich U, Darlix JL et al (2008) Targeting the viral nucleocapsid protein in anti-HIV-1 therapy. Mini Rev Med Chem 8(1):24–35
Dewan V, Liu T, Chen KM, Qian Z, Xiao Y, Kleiman L et al (2012) Cyclic peptide inhibitors of HIV-1 capsid-human lysyl-tRNA synthetase interaction. ACS Chem Biol 7(4):761–769
Dias J, Octobre G, Kobbi L, Comisso M, Flisiak S, Mirande M (2012) Activation of human mitochondrial lysyl-tRNA synthetase upon maturation of its premitochondrial precursor. Biochemistry 51(4):909–916
Eberhardy SR, Goncalves J, Coelho S, Segal DJ, Berkhout B, Barbas CF 3rd (2006) Inhibition of human immunodeficiency virus type 1 replication with artificial transcription factors targeting the highly conserved primer-binding site. J Virol 80(6):2873–2883
El Dirani-Diab R, Sarih-Cottin L, Delord B, Dumon B, Moreau S, Toulme JJ et al (1997) Phosphorothioate oligonucleotides derived from human immunodeficiency virus type 1 (HIV-1) primer $tRNA^{Lys3}$ are strong inhibitors of HIV-1 reverse transcriptase and arrest viral replication in infected cells. Antimicrob Agents Chemother 41(10):2141–2148
Fang P, Zhang HM, Shapiro R, Marshall AG, Schimmel P, Yang XL et al (2011) Structural context for mobilization of a human tRNA synthetase from its cytoplasmic complex. Proc Natl Acad Sci 108(20):8239–8244
Feng YX, Campbell S, Harvin D, Ehresmann B, Ehresmann C, Rein A (1999) The human immunodeficiency virus type 1 Gag polyprotein has nucleic acid chaperone activity: possible role in dimerization of genomic RNA and placement of tRNA on the primer binding site. J Virol 73(5):4251–4256
Fu W, Ortiz-Conde BA, Gorelick RJ, Hughes SH, Rein A (1997) Placement of tRNA primer on the primer-binding site requires pol gene expression in avian but not murine retroviruses. J Virol 71(9):6940–6946

Garret M, Romby P, Giegé R, Litvak S (1984) Interactions between avian myeloblastosis reverse transcriptase and $tRNA^{Trp}$. Mapping of complexed tRNA with chemicals and nucleases. Nucleic Acids Res 12(5):2259–2271

Goldschmidt V, Paillart JC, Rigourd M, Ehresmann B, Aubertin AM, Ehresmann C et al (2004) Structural variability of the initiation complex of HIV-1 reverse transcription. J Biol Chem 279(34):35923–35931

Graham WD, Barley-Maloney L, Stark CJ, Kaur A, Stolyarchuk K, Sproat B et al (2011) Functional recognition of the modified human $tRNA^{Lys3(UUU)}$ anticodon domain by HIV's nucleocapsid protein and a peptide mimic. J Mol Biol 410(4):698–715

Grigorov B, Bocquin A, Gabus C, Avilov S, Mély Y, Agopian A et al (2011) Identification of a methylated oligoribonucleotide as a potent inhibitor of HIV-1 reverse transcription complex. Nucleic Acids Res 39(13):5586–5596

Guo F, Cen S, Niu M, Javanbakht H, Kleiman L (2003) Specific inhibition of the synthesis of human lysyl-tRNA synthetase results in decreases in $tRNA^{Lys}$ incorporation, $tRNA^{Lys3}$ annealing to viral RNA, and viral infectivity in human immunodeficiency virus type 1. J Virol 77(18):9817–9822

Guo F, Saadatmand J, Niu M, Kleiman L (2009) Roles of Gag and NCp7 in facilitating $tRNA^{Lys3}$ annealing to viral RNA in human immunodeficiency virus type 1. J Virol 83(16): 8099–8107

Halwani R, Cen S, Javanbakht H, Saadatmand J, Kim S, Shiba K et al (2004) Cellular distribution of Lysyl-tRNA synthetase and its interaction with Gag during human immunodeficiency virus type 1 assembly. J Virol 78(14):7553–7564

Harada F, Sawyer RC, Dahlberg JE (1975) A primer ribonucleic acid for initiation of in vitro Rous sarcarcoma virus deoxyribonucleic acid synthesis. J Biol Chem 250(9):3487–3497

Hargittai MR, Mangla AT, Gorelick RJ, Musier-Forsyth K (2001) HIV-1 nucleocapsid protein zinc finger structures induce $tRNA^{Lys3}$ structural changes but are not critical for primer/template annealing. J Mol Biol 312(5):985–997

Hargittai MR, Gorelick RJ, Rouzina I, Musier-Forsyth K (2004) Mechanistic insights into the kinetics of HIV-1 nucleocapsid protein-facilitated tRNA annealing to the primer binding site. J Mol Biol 337(4):951–968

Houzet L, Morichaud Z, Didierlaurent L, Muriaux D, Darlix JL, Mougel M (2008) Nucleocapsid mutations turn HIV-1 into a DNA-containing virus. Nucleic Acids Res 36(7):2311–2319

Huang Y, Wang J, Shalom A, Li Z, Khorchid A, Wainberg MA et al (1997) Primer $tRNA_3^{Lys}$ on the viral genome exists in unextended and two-base extended forms within mature human immunodeficiency virus type 1. J Virol 71(1):726–728

Hulme AE, Perez O, Hope TJ (2011) Complementary assays reveal a relationship between HIV-1 uncoating and reverse transcription. Proc Natl Acad Sci 108(24):9975–9980

Isel C, Marquet R, Keith G, Ehresmann C, Ehresmann B (1993) Modified nucleotides of $tRNA^{Lys3}$ modulate primer/template loop-loop interaction in the initiation complex of HIV-1 reverse transcription. J Biol Chem 268(34):25269–25272

Isel C, Ehresmann C, Keith G, Ehresmann B, Marquet R (1995) Initiation of reverse transcription of HIV-1: secondary structure of the HIV-1 RNA/$tRNA^{Lys3}$ (template/primer). J Mol Biol 247(2):236–250

Isel C, Westhof E, Massire C, Le Grice SF, Ehresmann B, Ehresmann C et al (1999) Structural basis for the specificity of the initiation of HIV-1 reverse transcription. EMBO J 18(4): 1038–1048

Iwatani Y, Rosen AE, Guo J, Musier-Forsyth K, Levin JG (2003) Efficient initiation of HIV-1 reverse transcription in vitro. Requirement for RNA sequences downstream of the primer binding site abrogated by nucleocapsid protein-dependent primer-template interactions. J Biol Chem 278(16):14185–14195

Iwatani Y, Chan DS, Wang F, Maynard KS, Sugiura W, Gronenborn AM et al (2007) Deaminase-independent inhibition of HIV-1 reverse transcription by APOBEC3G. Nucleic Acids Res 35(21):7096–7108

Jacque JM, Triques K, Stevenson M (2002) Modulation of HIV-1 replication by RNA interference. Nature 418(6896):435–438

Jager S, Cimermancic P, Gulbahce N, Johnson JR, McGovern KE, Clarke SC et al (2012) Global landscape of HIV-human protein complexes. Nature 481(7381):365–370

Jalalirad M, Laughrea M (2010) Formation of immature and mature genomic RNA dimers in wild-type and protease-inactive HIV-1: differential roles of the Gag polyprotein, nucleocapsid proteins NCp15, NCp9, NCp7, and the dimerization initiation site. Virology 407(2):225–236

Javanbakht H, Cen S, Musier-Forsyth K, Kleiman L (2002) Correlation between $tRNA^{Lys3}$ aminoacylation and its incorporation into HIV-1. J Biol Chem 277(20):17389–17396

Javanbakht H, Halwani R, Cen S, Saadatmand J, Musier-Forsyth K, Göttlinger H et al (2003) The interaction between HIV-1 Gag and human lysyl-tRNA synthetase during viral assembly. J Biol Chem 278(30):27644–27651

Ji X, Klarmann GJ, Preston BD (1996) Effect of human immunodeficiency virus type 1 (HIV-1) nucleocapsid protein on HIV-1 reverse transcriptase activity in vitro. Biochemistry 35(1):132–143

Jiang M, Mak J, Ladha A, Cohen E, Klein M, Rovinski B et al (1993) Identification of tRNAs incorporated into wild-type and mutant human immunodeficiency virus type 1. J Virol 67(6):3246–3253

Jones CP, Datta SA, Rein A, Rouzina I, Musier-Forsyth K (2011) Matrix domain modulates HIV-1 Gag's nucleic acid chaperone activity via inositol phosphate binding. J Virol 85(4):1594–1603

Kaminska M, Shalak V, Francin M, Mirande M (2007) Viral hijacking of mitochondrial lysyl-tRNA synthetase. J Virol 81(1):68–73

Kelly NJ, Palmer MT, Morrow CD (2003) Selection of retroviral reverse transcription primer is coordinated with tRNA biogenesis. J Virol 77(16):8695–8701

Kerwood DJ, Cavaluzzi MJ, Borer PN (2001) Structure of SL4 RNA from the HIV-1 packaging signal. Biochemistry 40(48):14518–14529

Kikuchi Y, Ando Y, Shiba T (1986) Unusual priming mechanism of RNA-directed DNA synthesis in copia retrovirus-like particles of *Drosophila*. Nature 323(6091):824–826

Kim J, Roberts A, Yuan H, Xiong Y, Anderson KS (2012) Nucleocapsid protein annealing of a primer-template enhances (+)-strand DNA synthesis and fidelity by HIV-1 reverse transcriptase. J Mol Biol 415(5):866–880

Kyriacou SV, Deutscher MP (2008) An important role for the multienzyme aminoacyl-tRNA synthetase complex in mammalian translation and cell growth. Mol Cell 29(4):419–427

Lawrence DC, Stover CC, Noznitsky J, Wu Z, Summers MF (2003) Structure of the intact stem and bulge of HIV-1 Psi-RNA stem-loop SL1. J Mol Biol 326(2):529–542

Levin JG, Guo J, Rouzina I, Musier-Forsyth K (2005) Nucleic acid chaperone activity of HIV-1 nucleocapsid protein: critical role in reverse transcription and molecular mechanism. Prog Nucleic Acid Res Mol Biol 80:217–286

Levin JG, Mitra M, Mascarenhas A, Musier-Forsyth K (2010) Role of HIV-1 nucleocapsid protein in HIV-1 reverse transcription. RNA Biol 7(6):754–774

Lisziewicz J, Sun D, Metelev V, Zamecnik P, Gallo RC, Agrawal S (1993) Long-term treatment of human immunodeficiency virus-infected cells with antisense oligonucleotide phosphorothioates. Proc Natl Acad Sci 90(9):3860–3864

Litvak S, Sarih L, Fournier M, von der Haar F, Labouesse B, Araya A (1983) Involvement of tRNA in retrovirus expression: biological implications of reverse transcriptase-primer tRNA interactions. Recent Results Cancer Res 84:184–190

Liu S, Harada BT, Miller JT, Le Grice SF, Zhuang X (2010) Initiation complex dynamics direct the transitions between distinct phases of early HIV reverse transcription. Nat Struct Mol Biol 17(12):1453–1460

Lu Y, Planelles V, Li X, Palaniappan C, Day B, Challita-Eid P et al (1997) Inhibition of HIV-1 replication using a mutated $tRNA^{Lys3}$ primer. J Biol Chem 272(23):14523–14531

Lu K, Heng X, Garyu L, Monti S, Garcia EL, Kharytonchyk S et al (2011a) NMR detection of structures in the HIV-1 5'-leader RNA that regulate genome packaging. Science 334(6053):242–245

Lu K, Heng X, Summers MF (2011b) Structural determinants and mechanism of HIV-1 genome packaging. J Mol Biol 410(4):609–633

Mak J, Kleiman L (1997) Primer tRNAs for reverse transcription. J Virol 71(11):8087–8095

Mak J, Jiang M, Wainberg MA, Hammarskjold ML, Rekosh D, Kleiman L (1994) Role of Pr160gag-pol in mediating the selective incorporation of $tRNA^{Lys}$ into human immunodeficiency virus type 1 particles. J Virol 68(4):2065–2072

Marquet R, Isel C, Ehresmann C, Ehresmann B (1995) tRNAs as primer of reverse transcriptases. Biochimie 77(1–2):113–124

Miller Jenkins LM, Hara T, Durell SR, Hayashi R, Inman JK, Piquemal JP et al (2007) Specificity of acyl transfer from 2-mercaptobenzamide thioesters to the HIV-1 nucleocapsid protein. J Am Chem Soc 129(36):11067–11078

Miller Jenkins LM, Ott DE, Hayashi R, Coren LV, Wang D, Xu Q et al (2010) Small-molecule inactivation of HIV-1 NCp7 by repetitive intracellular acyl transfer. Nat Chem Biol 6(12):887–889

Morcock DR, Thomas JA, Sowder RC 2nd, Henderson LE, Crise BJ, Gorelick RJ (2008) HIV-1 inactivation by 4-vinylpyridine is enhanced by dissociating $Zn^{2+}$ from nucleocapsid protein. Virology 375(1):148–158

Nechushtan H, Kim S, Kay G, Razin E (2009) Chapter 1: The physiological role of lysyl tRNA synthetase in the immune system. Adv Immunol 103:1–27

Ni N, Morrow CD (2007) Impact of forced selection of tRNAs on HIV-1 replication and genome stability highlight preferences for selection of certain tRNAs. Virus Res 124(1–2):29–37

Onafuwa-Nuga AA, King SR, Telesnitsky A (2005) Nonrandom packaging of host RNAs in Moloney murine leukemia virus. J Virol 79(21):13528–13537

Onafuwa-Nuga AA, Telesnitsky A, King SR (2006) 7SL RNA, but not the 54-kd signal recognition particle protein, is an abundant component of both infectious HIV-1 and minimal virus-like particles. RNA 12(4):542–546

Oude Essink BB, Das AT, Berkhout B (1995) Structural requirements for the binding of $tRNA^{Lys3}$ to reverse transcriptase of the human immunodeficiency virus type 1. J Biol Chem 270(40):23867–23874

Panet A, Haseltine WA, Baltimore D, Peters G, Harada F, Dahlberg JE (1975) Specific binding of tryptophan transfer RNA to avian myeloblastosis virus RNA-dependent DNA polymerase (reverse transcriptase). Proc Natl Acad Sci USA 72(7):2535–2539

Pannecouque C, Szafarowicz B, Volkova N, Bakulev V, Dehaen W, Mély Y et al (2010) Inhibition of HIV-1 replication by a bis-thiadiazolbenzene-1, 2-diamine that chelates zinc ions from retroviral nucleocapsid zinc fingers. Antimicrob Agents Chemother 54(4):1461–1468

Pavon-Eternod M, Wei M, Pan T, Kleiman L (2010) Profiling non-lysyl tRNAs in HIV-1. RNA 16(2):267–273

Peters G, Glover C (1980) tRNA's and priming of RNA-directed DNA synthesis in mouse mammary tumor virus. J Virol 35(1):31–40

Peters GG, Hu J (1980) Reverse transcriptase as the major determinant for selective packaging of tRNA's into Avian sarcoma virus particles. J Virol 36(3):692–700

Puglisi EV, Puglisi JD (1998) HIV-1 A-rich RNA loop mimics the tRNA anticodon structure. Nat Struct Biol 5(12):1033–1036

Puglisi EV, Puglisi JD (2011) Secondary structure of the HIV reverse transcription initiation complex by NMR. J Mol Biol 410(5):863–874

Rhim H, Park J, Morrow CD (1991) Deletions in the $tRNA^{Lys}$ primer-binding site of human immunodeficiency virus type 1 identify essential regions for reverse transcription. J Virol 65(9):4555–4564

Rice WG, Schaeffer CA, Harten B, Villinger F, South TL, Summers MF et al (1993) Inhibition of HIV-1 infectivity by zinc-ejecting aromatic C-nitroso compounds. Nature 361(6411):473–475

Rigourd M, Bec G, Benas P, Le Grice SF, Ehresmann B, Ehresmann C et al (2003) Effects of tRNA$^{Lys3}$ aminoacylation on the initiation of HIV-1 reverse transcription. Biochimie 85(5):521–525

Roldan A, Warren OU, Russell RS, Liang C, Wainberg MA (2005) A HIV-1 minimal Gag protein is superior to nucleocapsid at in vitro annealing and exhibits multimerization-induced inhibition of reverse transcription. J Biol Chem 280(17):17488–17496

Salas M (1991) Protein-priming of DNA replication. Annu Rev Biochem 60:39–71

Sarih L, Fournier M, von der Haar F, Labouesse J, Litvak S (1982) Avian myeloblastosis reverse transcriptase deacylates tryptophanyl-tRNA. Nucleic Acids Res 10(22):7387–7393

Shimada M, Hosaka H, Takaku H, Smith JS, Roth MJ, Inouye S et al (1994) Specificity of priming reaction of HIV-1 reverse transcriptase, 2'-OH or 3'-OH. J Biol Chem 269(6):3925–3927

Stewart L, Schatz G, Vogt VM (1990) Properties of avian retrovirus particles defective in viral protease. J Virol 64(10):5076–5092

Supuran CT, Innocenti A, Mastrolorenzo A, Scozzafava A (2004) Antiviral sulfonamide derivatives. Mini Rev Med Chem 4(2):189–200

Taylor JM, Cywinski A (1984) A defective retrovirus particle (SE21Q1b) packages and reverse transcribes cellular RNA, utilizing tRNA-like primers. J Virol 51(2):267–271

Thomas JA, Bosche WJ, Shatzer TL, Johnson DG, Gorelick RJ (2008) Mutations in human immunodeficiency virus type 1 nucleocapsid protein zinc fingers cause premature reverse transcription. J Virol 82(19):9318–9328

Thomas JA, Shatzer TL, Gorelick RJ (2011) Blocking premature reverse transcription fails to rescue the HIV-1 nucleocapsid-mutant replication defect. Retrovirology 8:46

Thrall SH, Reinstein J, Wöhrl BM, Goody RS (1996) Evaluation of human immunodeficiency virus type 1 reverse transcriptase primer tRNA binding by fluorescence spectroscopy: specificity and comparison to primer/template binding. Biochemistry 35(14):4609–4618

Tisné C, Roques BP, Dardel F (2001) Heteronuclear NMR studies of the interaction of tRNA$^{Lys3}$ with HIV-1 nucleocapsid protein. J Mol Biol 306(3):443–454

Tisné C, Roques BP, Dardel F (2003) Specific recognition of primer tRNA$^{Lys3}$ by HIV-1 nucleocapsid protein: involvement of the zinc fingers and the N-terminal basic extension. Biochimie 85(5):557–561

Tisné C, Roques BP, Dardel F (2004) The annealing mechanism of HIV-1 reverse transcription primer onto the viral genome. J Biol Chem 279(5):3588–3595

Tolkunova E, Park H, Xia J, King MP, Davidson E (2000) The human lysyl-tRNA synthetase gene encodes both the cytoplasmic and mitochondrial enzymes by means of an unusual alternative splicing of the primary transcript. J Biol Chem 275(45):35063–35069

Turner KB, Yi-Brunozzi HY, Brinson RG, Marino JP, Fabris D, Le Grice SF (2009) SHAMS: combining chemical modification of RNA with mass spectrometry to examine polypurine tract-containing RNA/DNA hybrids. RNA 15(8):1605–1613

Wakefield JK, Wolf AG, Morrow CD (1995) Human immunodeficiency virus type 1 can use different tRNAs as primers for reverse transcription but selectively maintains a primer binding site complementary to tRNA$^{Lys3}$. J Virol 69(10):6021–6029

Wallace GS, Cheng-Mayer C, Schito ML, Fletcher P, Miller Jenkins LM, Hayashi R et al (2009) Human immunodeficiency virus type 1 nucleocapsid inhibitors impede trans infection in cellular and explant models and protect nonhuman primates from infection. J Virol 83(18): 9175–9182

Waters LC (1975) Transfer RNAs associated with the 70S RNA of AKR murine leukemia virus. Biochem Biophys Res Commun 65(3):1130–1136

Waters LC, Mullin BC, Ho T, Yang WK (1975) Ability of tryptophan tRNA to hybridize with 35S RNA of avian myeloblastosis virus and to prime reverse transcription in vitro. Proc Natl Acad Sci 72(6):2155–2159

Waters LC, Mullin BC, Bailiff EG, Popp RA (1980) Differential association of transfer RNAs with the genomes of murine, feline and primate retroviruses. Biochim Biophys Acta 608(1):112–126

Watts JM, Dang KK, Gorelick RJ, Leonard CW, Bess JW Jr, Swanstrom R et al (2009) Architecture and secondary structure of an entire HIV-1 RNA genome. Nature 460(7256):711–716

Wei M, Cen S, Niu M, Guo F, Kleiman L (2005) Defective replication in human immunodeficiency virus type 1 when non-primers are used for reverse transcription. J Virol 79(14):9081–9087

Westaway SK, Larson GP, Li S, Zaia JA, Rossi JJ (1995) A chimeric tRNA(Lys3)-ribozyme inhibits HIV replication following virion assembly. Nucleic Acids Symp Ser 33:194–199

Whitcomb JM, Ortiz-Conde BA, Hughes SH (1995) Replication of avian leukosis viruses with mutations at the primer binding site: use of alternative tRNAs as primers. J Virol 69(10): 6228–6238

Wilkinson KA, Gorelick RJ, Vasa SM, Guex N, Rein A, Mathews DH et al (2008) High-throughput SHAPE analysis reveals structures in HIV-1 genomic RNA strongly conserved across distinct biological states. PLoS Biol 6(4):e96

Wu W, Henderson LE, Copeland TD, Gorelick RJ, Bosche WJ, Rein A et al (1996) Human immunodeficiency virus type 1 nucleocapsid protein reduces reverse transcriptase pausing at a secondary structure near the murine leukemia virus polypurine tract. J Virol 70(10):7132–7142

Wu T, Datta SA, Mitra M, Gorelick RJ, Rein A, Levin JG (2010) Fundamental differences between the nucleic acid chaperone activities of HIV-1 nucleocapsid protein and Gag or Gag-derived proteins: biological implications. Virology 405(2):556–567

Yang N, Kazazian HH Jr (2006) L1 retrotransposition is suppressed by endogenously encoded small interfering RNAs in human cultured cells. Nat Struct Mol Biol 13(9):763–771

Yannay-Cohen N, Carmi-Levy I, Kay G, Yang CM, Han JM, Kemeny DM et al (2009) LysRS serves as a key signaling molecule in the immune response by regulating gene expression. Mol Cell 34(5):603–611

Yeung ML, Bennasser Y, Watashi K, Le SY, Houzet L, Jeang KT (2009) Pyrosequencing of small non-coding RNAs in HIV-1 infected cells: evidence for the processing of a viral-cellular double-stranded RNA hybrid. Nucleic Acids Res 37(19):6575–6586

Yu ET, Hawkins A, Eaton J, Fabris D (2008) MS3D structural elucidation of the HIV-1 packaging signal. Proc Natl Acad Sci 105(34):12248–12253

# Part IV
# HIV Genetic Variability and the Problem of Drug Resistance

# Chapter 11
# HIV Reverse Transcriptase Fidelity, Clade Diversity, and Acquisition of Drug Resistance

**Luis Menéndez-Arias**

## 11.1 Introduction

The human immunodeficiency virus (HIV) is a lentivirus (a member of the family Retroviridae) that causes acquired immunodeficiency syndrome (AIDS) in humans. Two types of HIV have been identified, HIV-1 and HIV-2. Genetic variability is one of the hallmarks of HIV. Its genetic diversity is manifested worldwide through the identification of numerous clades as well as within individuals, where pairwise comparisons of viral sequences revealed differences of up to 2.5 %. HIV-1 variants have been classified into four major phylogenetic groups: M (main), O (outlier), N (non-M/non-O), and P. The M group includes at least nine genetically distinct clades, designated as subtypes A, B, C, D, F, G, H, J, and K (Buonaguro et al. 2007; Ramirez et al. 2008; Plantier et al. 2009) (Fig. 11.1). Subtype B (group M) includes the first HIV isolates identified in the early 80s and is the most prevalent clade in Western Europe and the Americas (Hemelaar et al. 2011). Therefore, most virological studies and characterizations of HIV-1 proteins have been carried out with reference laboratory strains belonging to group M subtype B.

Genetic variation in retroviral populations results from the influence of several factors: (1) high mutation and recombination rates, (2) the size of the viral population, (3) the high multiplication rate of the virus, and (4) selective forces (e.g., immune pressure, host factors). Historically, the first evidence of retrovirus variation was reported by Rous and Murphy (Rous and Murphy 1913) who described different types of tumors originating from Rous sarcoma virus (RSV) infection. Thirty years later, Duran-Reynals reported variations in Rous sarcomas that

L. Menéndez-Arias (✉)
Centro de Biología Molecular "Severo Ochoa" (Consejo Superior de Investigaciones Científicas & Universidad Autónoma de Madrid), c/Nicolás Cabrera 1; Campus de Cantoblanco, 28049 Madrid, Spain
e-mail: lmenendez@cbm.uam.es

S.F.J. LeGrice and M. Götte (eds.), *Human Immunodeficiency Virus Reverse Transcriptase: A Bench-to-Bedside Success*, DOI 10.1007/978-1-4614-7291-9_11, 

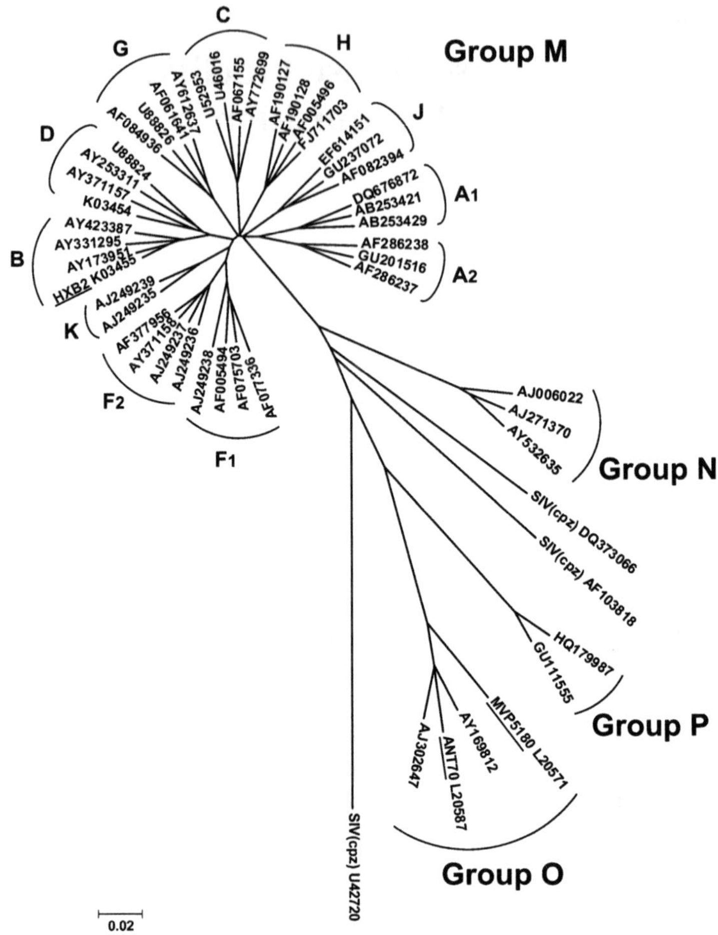


**Fig. 11.1** Phylogenetic tree analysis of the *pol* gene of HIV-1 reference strains. Evolutionary relationships were inferred using the neighbor-joining method and a bootstrap of 1,000 replicates. The tree is drawn to scale, with branch lengths in the same units as those of the evolutionary distances used to infer the phylogenetic tree. The evolutionary distances were computed with the maximum composite likelihood method (Tamura et al. 2004) and are in the units of the number of base substitutions per site. All positions containing gaps and missing data were eliminated from the dataset, and a total of 2,743 nucleotides were used in the analysis. Phylogenetic analyses were conducted in MEGA4 (Tamura et al. 2007), and the reference sequences were retrieved from the Los Alamos HIV Sequence Database (http://www.hiv.lanl.gov/content/sequence/NEWALIGN/align.html). HIV sequences are identified by their GenBank accession codes. *SIV (cpz)* simian immunodeficiency virus infecting chimpanzees. Well-characterized reference strains are underlined

appeared when the virus was passaged in ducklings (Duran-Reynals 1942). In those experiments, there were tumors that developed early (within 30 days) and others appeared several months after RSV inoculation. Early tumors were similar to the original Rous sarcoma and contained spindle-shaped cells. Cell-free extracts of these early tumors produced sarcoma in chicken but not in ducklings. However, extracts from late lesions were effective in producing tumors in ducklings, but not in chickens. These results were the first evidence demonstrating distinct biological properties associated with different retrovirus variants. Using a focus assay for transformation by RSV, Howard Temin later demonstrated that genetic differences in RSV could lead to different morphological phenotypes (Temin 1960).

The genomes of all retroviruses consist of two molecules of RNA which are single-stranded, positive sense and have a 5′ cap and a 3′-polyadenine tail (similar to messenger RNA). The conversion of genomic RNA into integration-competent double-stranded DNA is a critical step in the retroviral life cycle. This process is catalyzed by the virally encoded reverse transcriptase (RT) whose discovery in 1970 (Baltimore 1970; Temin and Mizutani 1970) represents one of the biggest breakthroughs in modern biology. Unlike cellular DNA polymerases, retroviral RTs are devoid of $3' \rightarrow 5'$ exonucleolytic proofreading activity and are therefore less accurate while synthesizing DNA. The RT uses the viral genomic RNA and a tRNA as template and primer, respectively, to synthesize the minus-strand DNA (cDNA). Then, its complementary strand (i.e., plus-strand DNA) is generated by the DNA-dependent DNA polymerase activity of the enzyme, in a reaction requiring the newly synthesized minus-strand DNA as template and the polypurine tract (PPT) as primer (for a review, see Herschhorn and Hizi 2010).

Although retroviral variation is largely a result of errors made by the RT during reverse transcription, cellular RNA polymerase II as well as viral and host factors also contribute to the retroviral mutation rate (O'Neil et al. 2002). In principle, proviral DNA replication that involves the participation of proofreading-proficient DNA polymerases (i.e., DNA polymerases δ, γ, and/or ε) should have a relatively minor effect on the retroviral mutation rate due to the high fidelity of those enzymes (Kunkel and Bebenek 2000; McCulloch and Kunkel 2008).

## 11.2 Intrinsic Fidelity of HIV-1 RT in Comparison with Other Retroviral RTs: Overall Error Rates and Mutational Spectra

The accuracy of DNA synthesis catalyzed by purified retroviral RTs can be estimated by measuring mutant frequencies in a reporter gene (e.g., *lacZα*) or in enzymological assays (also known as gel-based assays) (for a review, see Menéndez-Arias 2002). Among the first group, reversion assays were initially used to determine the base substitution fidelity of DNA polymerases (Weymouth and Loeb 1978). In these assays, the single-strand phage DNA used as template for DNA synthesis contained nonsense codons (*opal* or *amber*) in essential phage genes (e.g., ΦX174 *am*3 DNA or M13mp2 DNA with an *opal* codon in the *lacZ* gene). Revertant plaques were scored after transfection of host bacterial cells with copied DNA (Preston et al. 1988; Roberts et al. 1989). However, these methods focus on only a few sites, and estimates are restricted to specific base substitution errors (mostly, $G \rightarrow A$).

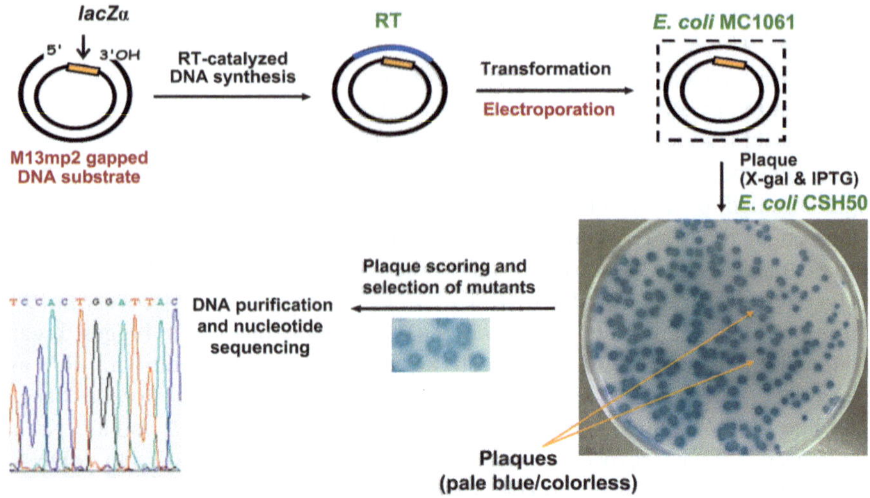


**Fig. 11.2** Experimental outline of the M13mp2 *lacZα*-based forward fidelity assay. Adapted from Bebenek and Kunkel (1995)

Forward mutation assays using the M13mp2 *lacZα* gene as a mutational target have been widely used to determine the fidelity of many DNA polymerases including HIV-1 RT. In these assays, the substrate is a gapped double-stranded M13mp2 DNA duplex, where the *lacZα* sequence of one of the DNA strands has been deleted. The DNA-dependent DNA synthesis reaction is carried out in the presence of RT and saturating concentrations of all dNTPs. The reaction product is then used to transform *Escherichia coli*, and mutants are scored in X-Gal (5-bromo-4-chloro-3-indolyl-β-D-galactoside) indicator plates by the number of plaques with an altered color phenotype (pale blue or colorless), in a lawn of *E. coli* α-complementation host cells (CSH50 strain). Sequence analysis of DNA extracted from mutants can be performed to identify the introduced mutations (for a detailed description of the procedure, see Bebenek and Kunkel 1995) (Fig. 11.2). Synonymous mutations are not detectable using this method. However, forward mutation assays allow for the detection of nucleotide substitutions at 125 different sites, as well as nucleotide additions or deletions at 199 different positions. Therefore, it provides a fidelity assessment based on a relatively large number of mutational sites. Related assays based on the usage of different vectors [e.g., pBluescript or Litmus 29 (Ji and Loeb 1992; Boyer and Hughes 2000)] or mutational targets [e.g., the HIV-1 *env* variable region 1 (Ji and Loeb 1994)] have also been described.

Early studies showed that the RTs of avian myeloblastosis virus (AMV) and Moloney murine leukemia virus (MLV) were 10–15 times more accurate than the HIV-1 RT in forward mutation assays (Roberts et al. 1988, 1989). However, reported error rates for HIV-1 RT using the M13-based forward mutation assay range from $6.0\times10^{-5}$ to $6.7\times10^{-4}$ substitutions per nucleotide (Roberts et al. 1988; Eckert

**Table 11.1** Estimated error rates of retroviral RTs in comparison with wild-type HIV-1 RT (group M subtype B clade) determined with the forward mutation assay and using M13mp2 *lacZα* as reporter gene

| RT | Mutation rate | Accuracy[a] (fold change) | HIV-1 RT | References |
|---|---|---|---|---|
| HIV-1 group O | $5.5\times10^{-5}$ | 2.5 | BH10 (p66/p51) | Álvarez et al. (2009) |
| $SIV_{agm}$ | $2.9\times10^{-5}$ | 1.8 | HXB2 (p66/p51) | Stuke et al. (1997) |
| $SIV_{mne}$ | $1.2\times10^{-4}$ (CL8)[b] | 1.3 | BH10 (p66/p51)[c] | Diamond et al. (2003) |
| | $1.6\times10^{-5}$ (170) | 9.7 | | |
| FIV | $6.2\times10^{-5}$ | 2.8 | BH10 (p66/p51)[c] | Operario et al. (2005) |
| MLV | $3.3\times10^{-5}$ | 14.3 | (p66/p61)[d] | Roberts et al. (1988, 1989) |
| | $1.2\times10^{-5}$ | 11.7 | BH10 (p66/p51) | Barrioluengo et al. (2012) |
| XMRV | $1.2\times10^{-5}$ | 11.7 | BH10 (p66/p51) | Barrioluengo et al. (2012) |
| FeLV | $5.8\times10^{-6}$ | 29 | BH10 (p66/p51)[c] | Operario et al. (2005) |
| AMV | $5.9\times10^{-5}$ | 9.5 | (p66/p61)[d] | Roberts et al. (1988, 1989) |
| PFV | $1.7\times10^{-4}$ | 0.4 | BH10 (p66/p51) | Boyer et al. (2007) |
| R2 retrotransposon | $1.2\times10^{-4}$ | 1.2 | BH10 (p66/p66) | Jamburuthugoda and Eickbush (2011) |

*SIV* simian immunodeficiency virus (agm, African green monkey; mne, pig-tailed macaque), *FIV* feline immunodeficiency virus, *XMRV* xenotropic murine leukemia virus-related virus, *FeLV* feline leukemia virus, *PFV* prototype primate foamy virus

[a]Increase relative to the wild-type HIV-1 RT enzyme

[b]CL8 and 170 are two variants of SIVmne that infected the same macaque. The strain 170 was obtained earlier in the asymptomatic phase, while CL8 was isolated in the late symptomatic phase of the infection

[c]Data for HIV-1 RT were not provided in the referenced papers. For comparative purposes, mutation rates for HIV-1 RT obtained in the same lab were taken from (Weiss et al. 2004)

[d]Purified from HIV-1 group M subtype B virions (unreported strain) (Roberts et al. 1988)

and Kunkel 1993; Rezende et al. 1998a; Weiss et al. 2004; Matamoros et al. 2008). This variability can be attributed to the different assay conditions (buffer, ionic strength, pH, etc.) and to the specific RT form used in these experiments. The HIV-1 RT is a heterodimer composed of subunits of 66 and 51 kDa, designated as p66 and p51, respectively. Since the catalytic site of its DNA polymerase activity resides within p66, a number of studies have been carried out with p66/p66 homodimers that retain wild-type activity. Furthermore, RTs used in those experiments are mostly derived from HIV-1 group M subtype B isolates, but their sequences might not be identical since different strains of HIV-1 have been used (e.g., BH10, NL4-3, and NY5). Not surprisingly, published mutation rates for a given retroviral RT show large variations even when obtained by using the same method and reporter gene. In this context, relative fidelities are more relevant for a proper comparison between different RTs. Table 11.1 summarizes data obtained from side-by-side comparisons between various retroviral RTs and the HIV-1 RT using the M13mp2 *lacZα* forward mutation assay. In general, oncoretroviral RTs (i.e., those of AMV, MLV, XMRV, and FeLV) are around 5- to 15-fold more accurate than the RTs of lentiviruses (i.e., HIV-1, SIV, and FIV), foamy viruses (PFV), or retrotransposons (R2).

Methods based on the use of reporter genes allow for the identification of errors by nucleotide sequencing. The mutational spectrum has been obtained only for a few retroviral RTs, such as AMV RT (Roberts et al. 1989), MLV RT (Barrioluengo et al. 2011), XMRV RT (Barrioluengo et al. 2012), PFV RT (Boyer et al. 2007), the non-LTR R2 retrotransposon RT (Jamburuthugoda and Eickbush 2011), $SIV_{agm}$ RT (Stuke et al. 1997), and HIV-1 RTs (Roberts et al. 1988; Eckert and Kunkel 1993; Rezende et al. 1998a; Weiss et al. 2004; Matamoros et al. 2008; Álvarez et al. 2009). These and other studies show that RTs tend to accumulate errors at specific sites (hot spots), sometimes at very high rates (Roberts et al. 1988; Bebenek et al. 1989; Stuke et al. 1997; Rezende et al. 1998a; Álvarez et al. 2009). In assays carried out with HIV-1 RT, hot spots are usually located at the boundaries of homopolymeric runs and could be explained by invoking a dislocation mechanism. In this model, base substitution errors as well as insertions or deletions of one nucleotide could initiate by template-primer slippage (Bebenek et al. 1993; reviewed in Menéndez-Arias 2009). A comparison of mutational spectra of HIV-$1_{BH10}$ (group M subtype B) RT and MLV RT is given in Fig. 11.3. The MLV RT shows an eightfold higher error rate for frameshifts (i.e., one-nucleotide insertions or deletions) in comparison with the wild-type HIV-$1_{BH10}$ RT and a higher tendency to produce large deletions.

Most of the frameshifts produced by the HIV-1 RT occur at homopolymeric runs and could originate from misalignment and template/primer slippage (Bebenek et al. 1989, 1993). In addition, hot spots are frequently observed at sequences with a higher frequency of termination of processive DNA synthesis (Bebenek et al. 1989), although no correlation has been found between RT errors and termination of DNA polymerization. The $SIV_{agm}$ RT also shows a high tendency to produce frameshifts (Stuke et al. 1997). However, the frameshift error rate of the wild-type HIV-1 group O RT was found to be very low (Álvarez et al. 2009). The RTs of AMV, XMRV, and PFV showed a higher frameshift error rate than the HIV-$1_{BH10}$ RT (Roberts et al. 1989; Boyer et al. 2007; Barrioluengo et al. 2012). MLV and AMV RTs showed a similar tendency to introduce frameshifts in nucleotide runs as compared with non-runs, while the XMRV RT was prone to introduce frameshifts at heteropolymeric sequences (Barrioluengo et al. 2012).

## 11.3 HIV-1 RT: Base Substitution Error Rates and Nucleotide Incorporation Kinetics

Several laboratories have reported base substitution error rates for the wild-type HIV-1 (group M subtype B) RT, obtained by using forward mutation assays. These estimates were rather consistent and in the range of $6.3 \times 10^{-5}$ to $1.9 \times 10^{-4}$ base substitutions per nucleotide (Bebenek et al. 1989; Drosopoulos and Prasad 1998; Rezende et al. 1998a; Lewis et al. 1999; Boyer et al. 2007; Álvarez et al. 2009). Similar values were obtained for the HIV-1 group O RT ($1.1 \times 10^{-4}$) (Álvarez et al. 2009) and for the PFV RT ($5.8 \times 10^{-5}$) (Boyer et al. 2007). In contrast, AMV and MLV RTs were 2.5 and 8 times more accurate than the HIV-1 RT in M13mp2 *lacZα*

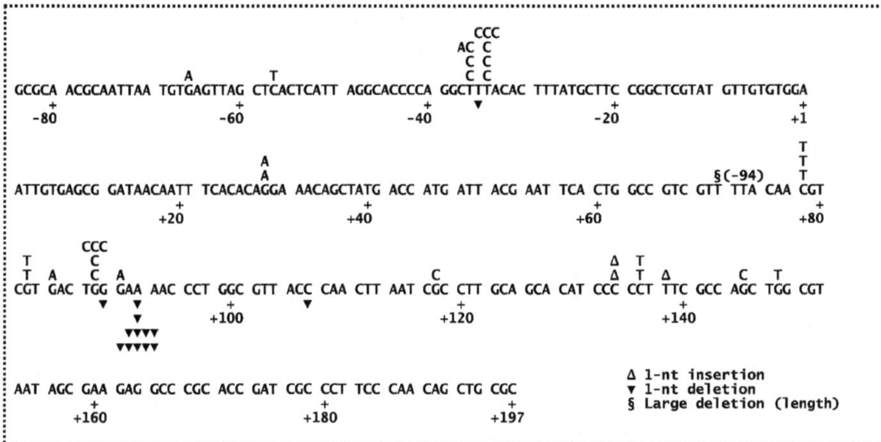


| Enzyme | Total plaques | Mutant plaques | Mutant frequency |
|---|---|---|---|
| HIV-$1_{BH10}$ RT | 6736 | 139 | 206.3 x $10^{-4}$ |

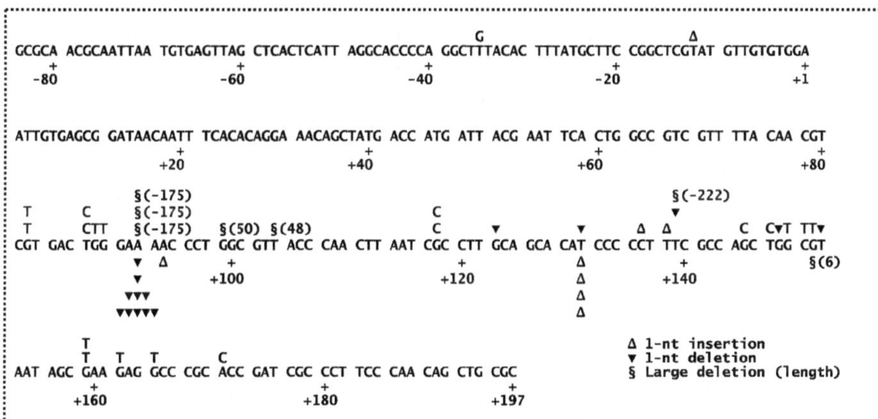


| Enzyme | Total plaques | Mutant plaques | Mutant frequency |
|---|---|---|---|
| MLV RT | 29648 | 40 | 13.5 x $10^{-4}$ |

**(15.3)**

**Fig. 11.3** Distribution and types of mutations generated by HIV-$1_{BH10}$ RT and MLV RT in M13mp2 *lacZα*-based forward mutation assays. The data shown were taken from Álvarez et al. (2009) and Barrioluengo et al. (2011)

forward mutation assays. Reported base substitution error rates obtained with AMV RT and with MLV RT were $2.7\times10^{-5}$ and $8.5\times10^{-6}$, respectively (Roberts et al. 1989; Barrioluengo et al. 2011).

The fixation of a base substitution involves the incorporation of an incorrect nucleotide, followed by the extension of the mismatched primer. These events can be analyzed by enzyme kinetics using quantitative gel electrophoresis. The DNA polymerization reactions should be preferably performed under pre-steady-state conditions to avoid distortions in the calculation of the polymerization rate and nucleotide affinity due to the contribution of the template-primer dissociation rate (Kati et al. 1992; Kellinger and Johnson 2010). Relevant kinetic parameters in these assays are $k_{pol}$ (nucleotide incorporation rate) and $K_d$ (equilibrium dissociation constant), while the catalytic efficiency of the polymerase reaction can be determined by the $k_{pol}/K_d$ ratio. Although reported values can vary depending on the assay conditions and template-primer used in the assays, $k_{pol}$ values for nucleotide incorporations catalyzed by HIV-1 RT are usually within 10–80 $s^{-1}$, while the $K_d$ ranges from 5 to 35 μM (Kati et al. 1992; Zinnen et al. 1994; Kerr and Anderson 1997; Mendieta et al. 2008; Kisic et al. 2011).

An estimate of the misinsertion fidelity can be obtained from the ratio between the catalytic efficiencies of incorporation of incorrect nucleotides relative to correct ones on a specific template-primer (reviewed in Echols and Goodman 1991; Johnson 1993). Similar assays carried out with template-primers having a mismatch at the 3′-end of the DNA primer are used to determine the fidelity of the RT in mispair extension reactions. In these assays, the incorporated nucleotide is a correct one, but the template-primer substrates can have either matched or mismatched termini. Reported mispair extension efficiencies for HIV-1 RT are usually within the range of $10^{-2}$ to $10^{-4}$, while misinsertion ratios range from $10^{-3}$ to $10^{-6}$. Similar values have been reported for assays carried out under steady-state conditions, where the relevant kinetic parameters are $k_{cat}$ (instead of $k_{pol}$) and the apparent $K_m$ (instead of $K_d$) (for a review, see Menéndez-Arias 2002). The efficient extension of mismatched 3′ termini of DNA has been recognized as a major determinant of the low accuracy of HIV-1 RT and other retroviral RTs (Perrino et al. 1989; Yu and Goodman 1992).

In general, retroviral RTs form transversions with lower efficiencies than transitions. Misinsertion ratios are very low for A:G, G:G, and C:C pairs but higher for G:T and T:G mismatches (Mendelman et al. 1989; Yu and Goodman 1992; Sala et al. 1995). The highest mispair extensions are frequently obtained for G:T or T:G mispairs (Matamoros et al. 2008; Álvarez et al. 2009; Barrioluengo et al. 2012). These observations are broadly consistent with the higher frequencies of T→C and G→A substitutions identified in forward mutation assays (Bebenek et al. 1989; Ji and Loeb 1992; Eckert and Kunkel 1993).

Extensive comparisons of misincorporation and mispair extension fidelities of HIV-1, MLV, and AMV RTs based on nucleotide incorporation kinetics failed to provide consistent results. Steady-state kinetic studies carried out with model heteropolymeric template-primers have shown differences between HIV-1 and AMV RTs that affect specific misincorporations (Yu and Goodman 1992). For example, HIV-1 RT has a stronger tendency to incorporate T opposite T. However, in the same sequence context, G:T mispairs are extended 20 times more efficiently by the AMV RT than by the wild-type HIV-1 enzyme. In the case of MLV RT, it has been shown that A:C, A:A, and A:G mispairs are more efficiently extended by HIV-1 and HIV-2 RTs than by the MLV enzyme (Bakhanashvili and Hizi 1992, 1993), in agreement with their lower fidelity in *lacZ*-based assays. However, a large study involving a systematic evaluation of misincorporation frequencies at 46 different positions in the *gag* sequence revealed great variability (Ricchetti and Buc 1990). Although differences were observed at specific base pairs, contrary to expectation, the HIV-1 RT did not always emerge as the less faithful of all three enzymes.

## 11.4 RNA- Versus DNA-Dependent Fidelity of DNA Synthesis

RTs utilize the viral genomic RNA as template to obtain a cDNA that is later used as template to synthesize the complementary DNA chain and then complete the synthesis of double-stranded proviral DNA. In general, misinsertion and mispair extension fidelities do not seem to be significantly affected by the type of nucleic acid (RNA or DNA) used as template, as shown for the RTs of HIV-1, MLV, and AMV in assays carried out under steady-state conditions (Yu and Goodman 1992; Bakhanashvili and Hizi 1993). However, some relevant differences have been reported in pre-steady-state kinetic assays carried out with specific template-primers. Thus, Kerr and Anderson (1997) showed that the misinsertion fidelity of C, A, or G opposite A was 13–24 times more accurate with an RNA template than with a DNA template, in assays carried out with HIV-1 RT and 45/22mer duplexes bearing the same nucleotide sequence (except for the presence of U instead of T in the RNA template).

The comparison of error rates obtained with HIV-1 RT in reporter gene-based assays using RNA versus DNA templates failed to provide consistent results. Thus, overall error rates using M13 and *env* V1 as vector and reporter gene, respectively, were $1.9 \times 10^{-4}$ for DNA-dependent DNA synthesis and $2.0 \times 10^{-4}$ for RNA-dependent DNA synthesis (Ji and Loeb 1994). Similar values were obtained using pBluescript and *lacZ* ($1.7 \times 10^{-4}$ using a DNA template and $1.4 \times 10^{-4}$ using an RNA template) (Ji and Loeb 1992). These values were reduced in the case of Moloney MLV RT, but still differences between RNA and DNA were very small ($2.7 \times 10^{-5}$ and $3.4 \times 10^{-5}$ for RNA and DNA, respectively) (Ji and Loeb 1992).

However, other studies have shown that the RNA-dependent DNA synthesis is about two- to threefold more accurate than the DNA-dependent DNA synthesis, in reactions catalyzed by the HIV-1 RT (Boyer et al. 1992). In any case, the largest differences were observed when comparing error patterns generated while copying RNA or DNA. Fidelity with RNA was >10-fold higher for single nucleotide substitutions and minus-one nucleotide errors at five different homopolymeric positions (Boyer et al. 1992), suggesting that misaligned intermediates are formed or used less frequently with an RNA template-primer than with a DNA template-primer.

In RNA-dependent DNA polymerization reactions, both MLV and HIV-1 RTs showed similar misinsertion fidelities when analyzed by pre-steady-state kinetics (Skasko et al. 2005). However, the MLV enzyme showed about threefold lower mispair extension fidelity, compared with the HIV-1 RT. These differences were attributed to the lower affinity of MLV RT for mismatched template-primers.

## 11.5 Structural Role of Amino Acid Residues in the Fidelity of HIV-1 RT

As a major target of antiretroviral therapy, the HIV-1 RT has been the subject of a large number of biochemical studies aimed towards understanding the role of specific amino acids in drug resistance and nucleotide specificity (for a review, see Menéndez-Arias 2010, 2013). Crystal structures of HIV-1 RT bound to antiretroviral drugs (Kohlstaedt et al. 1992), double-stranded DNA or RNA/DNA complexes (Jacobo-Molina et al. 1993; Sarafianos et al. 2001), or associated with a template-primer and dNTP (Huang et al. 1998) revealed the presence of four subdomains (fingers, thumb, palm, and connection) in both subunits p66 and p51, and the RNase H domain occupying the C-terminal region of p66 and absent from p51. DNA polymerase catalytic residues ($Asp^{110}$, $Asp^{185}$, and $Asp^{186}$) locate at the palm subdomain of p66. Residues in the vicinity of the catalytic triad form the dNTP-binding pocket. These amino acids are $Lys^{65}$, $Arg^{72}$, $Asp^{113}$, $Ala^{114}$, $Tyr^{115}$, and $Gln^{151}$. Other neighboring residues, such as $Leu^{74}$, $Pro^{157}$, $Phe^{160}$, $Tyr^{183}$, and $Met^{184}$, could also indirectly affect dNTP binding (Huang et al. 1998).

The effects of many amino acid substitutions on the fidelity of DNA synthesis have been studied by using *lacZ*-based forward mutation and gel-based nucleotide incorporation assays (for recent reviews, see Menéndez-Arias 2009; Matamoros et al. 2011).

It has been proposed that polymerase fidelity may be a function of the flexibility of the dNTP-binding pocket (Harris et al. 1998). However, further studies have shown that molecular determinants of fidelity are not confined to the dNTP-binding site or its vicinity. Residues interacting with the nucleic acid template and/or the DNA primer, including minor groove-binding track residues or amino acids in the RNase H primer grip, may also contribute to the fidelity of the RT. A summary of the effects of single amino acid substitutions on the accuracy of HIV-1 RT, as determined by using the forward mutation assay, is given in Fig. 11.4. Several residues in the dNTP-binding site play a key role in nucleotide selectivity. Thus, $Tyr^{115}$ acts as a "steric gate," preventing the incorporation of ribonucleotides (or analogues having a 2′ hydroxyl group) (Gao et al. 1997; Cases-González et al. 2000). In addition, nonconservative substitutions affecting this position cause a significant reduction in the fidelity of the RT (Martín-Hernández et al. 1996, 1997; Boyer and Hughes 2000; Jonckheere et al. 2000). In contrast, replacing $Lys^{65}$ by Arg leads to a large increase of fidelity in different sequence contexts, without affecting the catalytic efficiency of the RT (Shah et al. 2000; Garforth et al. 2010; Barrioluengo et al. 2011). Nonconservative substitutions affecting neighboring residue $Lys^{66}$ (e.g. K66A, K66N or K66T) increase misinsertion and mispair extension fidelity, although the K66R mutant was akin to wild-type HIV-1 RT (Lwatula et al. 2012).

The drug resistance-associated mutations K65R and M184V (or M184I) decrease the viral RT processivity (Back et al. 1996; Sharma et al. 2009). Substituting Val or Ile for $Met^{184}$ has a relatively minor effect on fidelity, although M184I has been shown to decrease the overall error rate by fourfold in forward mutation assays (Wainberg et al. 1996; Drosopoulos and Prasad 1998; Rezende et al. 1998a;

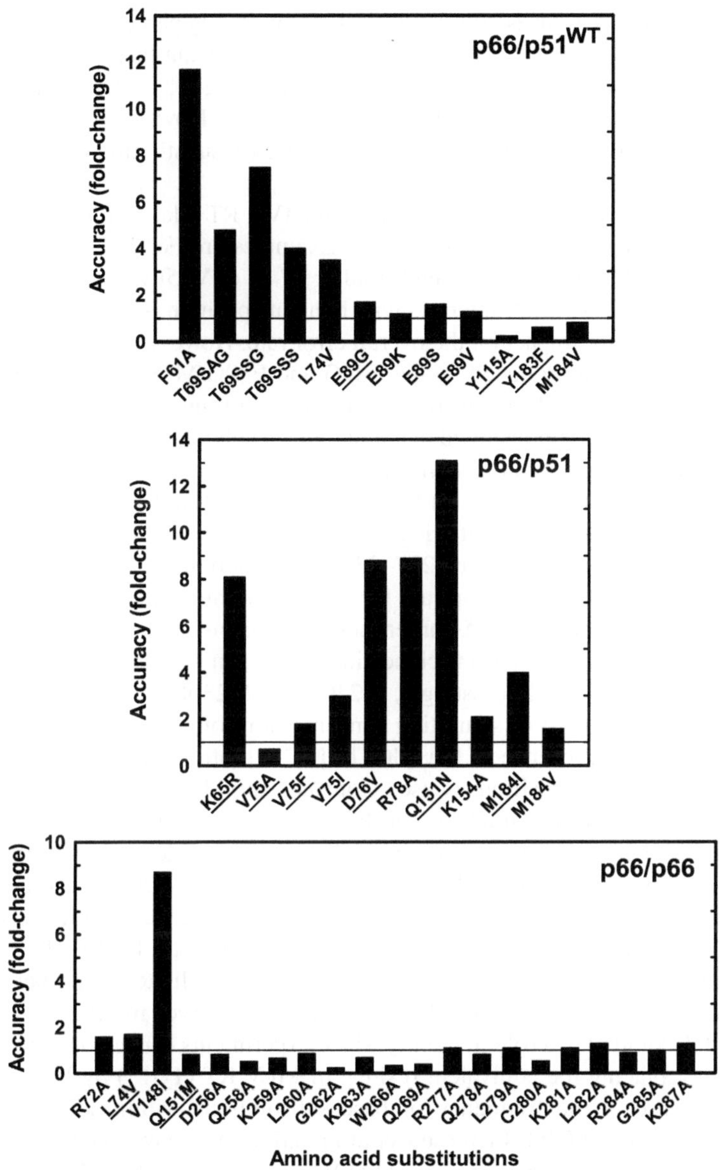


**Fig. 11.4** Effects of single amino acid substitutions on the fidelity of HIV-1 RT, as estimated by using the M13mp2 *lacZα*-based forward mutation assay. Data are shown in three different groups based on the reference wild-type RT used in the assays. The indicated mutations were introduced in the p66 subunit only (*upper panel*) or in both subunits (*middle* and *lower panels*). The increase in accuracy is given as fold change relative to the wild-type enzyme. The effects of underlined substitutions have been confirmed in misinsertion and mispair extension fidelity assays. For references, see reviews by Menéndez-Arias (2009) and Matamoros et al. (2011)

reviewed in Menéndez-Arias 2002). $Gln^{151}$ is also probably part of the fidelity center of the RT, since its substitution by Asn leads to a significant increase in fidelity, as demonstrated in forward mutation assays, as well as in gel-based enzymatic assays (Weiss et al. 2000, 2002; Jamburuthugoda et al. 2005). However, as in the case of the nonconservative substitutions of $Tyr^{115}$, the Q151N substitution reduces the catalytic efficiency of the RT.

Other amino acid changes that influence the HIV-1 RT fidelity affect residues that interact with the template-primer overhang. Examples are F61A, D76V, and R78A that decrease the RT error rate by >eightfold or L74V or V75I whose effects are less pronounced (Fig. 11.4). $Glu^{89}$ interacts with template position −2, and its substitution has a relatively minor impact on the overall error rate. However, detailed analysis of the mutational spectra produced by the E89G and E89G/M184V RTs, followed by further characterization of mutant RTs with other substitutions at this position (i.e., E89A, E89D, E89N), demonstrated that mutations involving $Glu^{89}$ have an impact on −1 frameshifting. Interestingly, similar effects were also observed with mutant RTs having the amino acid substitutions K154A or K154R (Hamburgh et al. 2006) and could be related to a potential salt bridge between the side chains of $Glu^{89}$ and $Lys^{154}$.

Primer grip residues (positions 227–235) are important to maintain the primer terminus in an orientation appropriate for the nucleophilic attack. The amino acid substitutions F227A and W229A confer increased misinsertion fidelity in gel-based assays but show lower catalytic efficiency in comparison with the wild-type enzyme (Wisniewski et al. 1999). Interestingly, M230L and M230I are mutations that have a minor impact on fidelity at saturating concentrations of dNTP but show relatively low accuracy in the presence of low [dCTP]/[dTTP] ratios, thereby facilitating the appearance of G → A transitions (Cases-González and Menéndez-Arias 2004).

Several residues in the nucleic acid-binding cleft are known to contribute to template-primer binding and processivity. Their role in fidelity has been demonstrated with mutant RTs bearing the substitutions G262A or W266A. These enzymes showed decreased fidelity and a higher tendency to produce frameshifts in comparison with the wild-type RT (Beard et al. 1994; Bebenek et al. 1995). The RNase H primer grip interacts with the DNA primer strand and facilitates its proper position relative to the RNase H active site. $Tyr^{501}$ within the conserved motif Asp-Ser-X-Tyr of HIV-1 RT could play a role in fidelity, since experiments carried out ex vivo have shown that replacing the equivalent residue of MLV RT ($Tyr^{586}$) by Phe leads to a 17-fold increase of the frequency of base substitutions within adenine-thymine tracts (Zhang et al. 2002a). Increased viral mutation rates have been also obtained after introducing mutations S557A, A558V, or Q559L in the RNase H-coding region of an MLV vector (Mbisa et al. 2005). However, none of those mutations were tested in cell-free assays.

## 11.6 HIV-1 RT as a Biotechnological Tool

RTs are widely used in recombinant DNA technology to synthesize cDNA from messenger RNA (mRNA). AMV and MLV RTs are the most frequently used RTs in many research applications, including diagnostic detection of gene products, cDNA cloning, end-point and real-time reverse transcription-polymerase chain reaction (RT-PCR), and gene expression analysis (e.g., microarrays). HIV-1 (group M subtype B) RT variants show higher activity than MLV and AMV RTs at 50–60 °C (Gerard et al. 2002; Yasukawa et al. 2008; Álvarez et al. 2009). However, the reduced accuracy of the HIV-1 RT constitutes a disadvantage that limits an extensive use in RNA amplification.

Wild-type RTs from different HIV-1 clades (e.g., subtype B and subtype C) differ in about 10–15 % of their amino acid sequence but share similar biochemical properties (Xu et al. 2010). However, the phylogenetically distant HIV-1 group O contains an RT that differs from HIV-1 group M subtype B RTs in about 21 % of its amino acid sequence (Quiñones-Mateu et al. 1997). Compared to HIV-$1_{BH10}$ RT, the group O polymerase showed increased stability in the presence of urea (Menéndez-Arias et al. 2001) and higher efficiency in RT-PCR reactions when the reverse transcription step was carried out at temperatures as high as 68 °C (Álvarez et al. 2009). Its remarkable thermal stability was similar to that shown by the AffinityScript RT (Agilent Technologies), an engineered MLV RT that contains five mutations (i.e., E69K/E302R/W313F/L435G/N454K) that promote tighter binding to the template-primer (Arezi and Hogrefe 2009).

The wild-type HIV-1 group O RT is more faithful than the HIV-$1_{BH10}$ enzyme (Álvarez et al. 2009), but it is about sixfold less accurate than the wild-type MLV RT (Barrioluengo et al. 2011). The fidelity of HIV-1 group O RT was increased above the levels of MLV RT by introducing the substitution K65R or the combination K65R/V75I, without a significant loss of its thermal stability (Barrioluengo et al. 2011). The effect of increasing the temperature on the fidelity of RNA-dependent or DNA-dependent DNA synthesis catalyzed by RTs has not been studied in detail. Qualitative assessments based on primer extension reactions carried out in the presence of three dNTPs and misinsertion fidelity assays showed that the MLV RT was more accurate at 55 °C than at 37 °C (Malboeuf et al. 2001). However, the significance of these data is limited by the fact that at 55 °C, the enzyme retains only 6 % of the catalytic efficiency observed at 37 °C.

## 11.7 HIV Mutation Rates: The Discrepancies Between Ex Vivo and In Vitro Estimates of Replication Fidelity

Accurate estimates of mutation rates are important to understand viral evolution. In an extensive analysis, Sanjuán et al. (2010) reported mutation rates (as substitutions per nucleotide per cell infection, s/n/c) for a number of viruses to show an inverse relationship between genome size and the mutation rate per nucleotide, which can be further strengthened by extending the analysis to viroids and bacteria (Fig. 11.5). Mutation rates for several retroviruses have been determined ex vivo by using a retroviral vector containing a mutation reporter gene (often *lacZ* or its truncated derivative *lacZα*), a packaging cell line, and target cells in conditions that allow the vector to complete one round of replication (for reviews, see Yu et al. 1997;

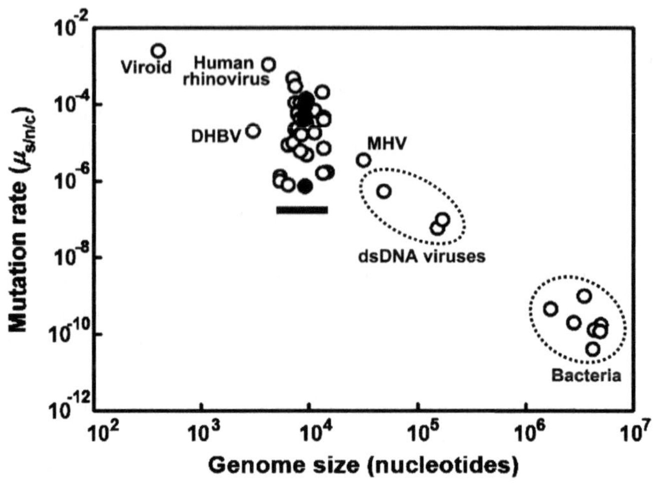


**Fig. 11.5** Relationship between genome size and mutation rate. The mutation rates for several types of viruses including poliovirus, rhinovirus, duck hepatitis B virus (DHBV), mouse hepatitis virus (MHV), retrovirus, and double-stranded DNA viruses (e.g., herpes simplex virus type 1 and bacteriophages-λ and T2) are reported as substitutions per nucleotide per cell infection (Sanjuán et al. 2010). Reported mutation rates for HIV-1 are represented with filled circles (Mansky and Temin 1995; Mansky 1996; Mansky and Bernard 2000; Mansky et al. 2000; Gao et al. 2004; Huang and Wooley 2005; Laakso and Sutton 2006; Abram et al. 2010). The *horizontal bar* indicates the genome size span of single-stranded RNA viruses, including retroviruses. The viroid mutation rate was taken from Gago et al. (2000). Bacterial mutation rates were obtained from Lynch (2006), and genome sizes of *Bacillus anthracis*, *Deinococcus radiodurans*, *Escherichia coli*, *Helicobacter pylori*, *Mycobacterium tuberculosis*, *Salmonella enterica*, and *Sulfolobus acidocaldarius* were collected from http://www.genomesize.com/prokaryotes

Svarovskaia et al. 2003). As in the case of forward mutation assays used to evaluate the intrinsic fidelity of RTs, not all mutations result in a detectable phenotypic change in the gene product. Mutation rates are estimated on the base of a specific reporter gene and may not reflect an average of the complete genome. However, these techniques provide data in the context of a replicating complex inside the cell and in the presence of viral and host factors that could modulate the mutation rate.

Mutation rates for HIV-1 have been estimated at around $3.4 \times 10^{-5}$ per nucleotide and replication cycle in HeLa and CEM cells (Mansky and Temin 1995; Mansky 1996) and $1.4 \times 10^{-5}$ in HOS cells (Abram et al. 2010). These values are about ten times lower than anticipated from the error rates obtained in forward mutation assays carried out with the RT alone. This discrepancy appears to be specific for HIV-1 since reported estimates of the mutation rate of MLV ranged from $2 \times 10^{-6}$ to $2 \times 10^{-5}$ (Monk et al. 1992; Varela-Echavarría et al. 1992), close to the values obtained with the purified enzyme (Table 11.1). A similar mutation rate is predicted for the closely related gammaretrovirus XMRV, based on the reported frequencies of *lacZ* inactivation in single cycle of replication assays (Ndongwe et al. 2012).

Apart from MLV, there is good agreement between error rates determined in vitro and ex vivo with avian retroviruses and foamy viruses. Although the method used to determine the RSV mutation rate was based on a denaturing gel electrophoresis and rather different from those used for other viruses, the obtained value of $1.4 \times 10^{-4}$ (Leider et al. 1988) was only twice as high as the one obtained in vitro with the AMV RT (Table 11.1). In the case of foamy viruses, an error rate of $4 \times 10^{-4}$ has been reported for PFV cultured in HEK 293T cells after a single round of reverse transcription (Gärtner et al. 2009). This estimate was roughly similar to the one obtained in vitro with the recombinant RT.

Mutational spectra generated in experiments carried out ex vivo were rather similar for all tested retroviruses (i.e., spleen necrosis virus, MLV, PFV, HIV-1), with a large predominance of transitions (most frequently, $G \rightarrow A$) and a significant proportion of frameshifts, usually occurring at nucleotide runs (Svarovskaia et al. 2003; and references therein). Identified base substitutions and insertions/deletions accumulate at hot spots in the *lacZα* gene. The most common nucleotide substitutions (i.e., 5′-GG to -GA) have been attributed to the deamination of cytidines to uridines in the viral minus-strand DNA by the host APOBEC enzymes (for a review, see Vartanian et al. 2003). A recent study has demonstrated that the orientation of the *lacZα* reporter gene had a minimal impact on the HIV-1 error rate (estimated as $1.4 \times 10^{-5}$ mutations/bp/cycle), but influenced the distribution of hot spots (Abram et al. 2010). In this analysis, carried out in 293T producer cells that contain very low levels of APOBEC3G, authors found a small proportion of $G \rightarrow A$ mutations, but relatively high levels of $A \rightarrow G$ transitions. $A \rightarrow G$ hypermutations have been previously found in proviral sequences of Rous-associated virus type 1 (RAV-1), avian leukosis virus, and in spleen necrosis virus and have been attributed to the activity of cellular adenosine deaminases that act on the viral RNA (ADAR enzymes) (Kim et al. 1996).

### 11.7.1 The Influence of Viral and Host Factors on the HIV-1 Mutation Rate

Viral and host factors intervene at different steps of the viral life cycle and could modulate HIV-1 variation in vivo. Apart from the viral RT, the cellular transcriptional machinery (RNA polymerase II) and replicative host DNA polymerases (e.g., DNA polymerases $\alpha$, $\delta$, $\gamma$, $\varepsilon$) and mismatch repair mechanisms also participate in the replication of the viral genome. The contribution to mutation rates of host DNA polymerases, involved in the replication of integrated proviral DNA is expected to be minimal due to their high fidelity (Fig. 11.6). However, the host-cell RNA polymerase II could have a significant effect on the viral mutation rate, as it participates in the synthesis of viral RNAs using the proviral DNA as template. The contributions of the viral RT and the RNA polymerase II have been evaluated in a single cycle of replication, by analyzing mutations in the HIV-1 long terminal repeats (LTRs) (O'Neil et al. 2002). Ten out of twenty-one mutations were found in both

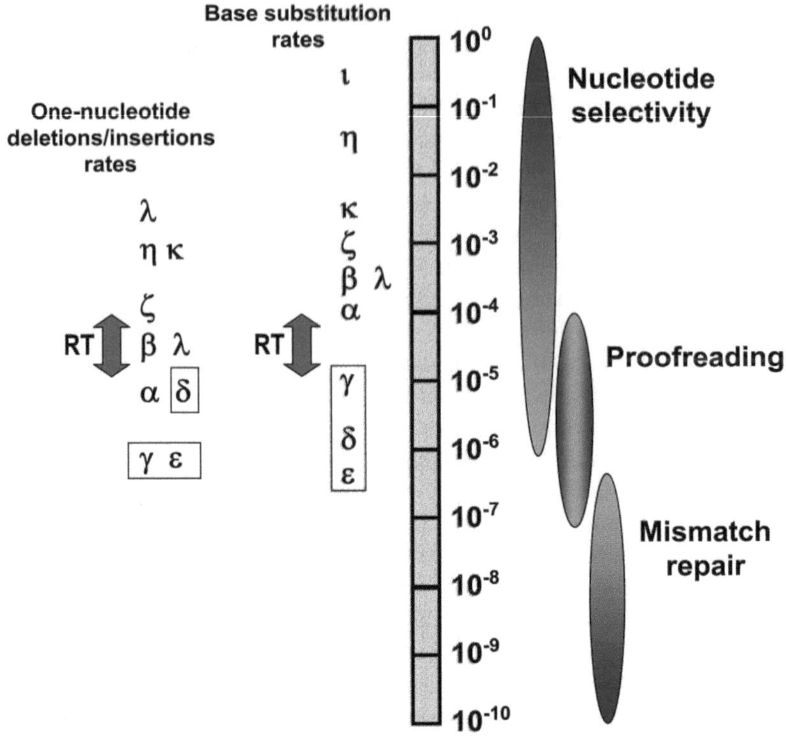


**Fig. 11.6** Polymerase error rates and mechanisms contributing to fidelity in eukaryotic DNA replication. Base substitution and frameshift error rates of selected DNA polymerases belonging to families A (γ), B (α, δ, ε, ζ), and Y (η, κ, ι) are shown in comparison with those obtained with retroviral RTs, using the forward mutation assay (Matsuda et al. 2000; Kunkel 2004; McCulloch and Kunkel 2008; and references therein). Exonuclease proficient polymerases are boxed

LTRs, while the remaining eleven appeared in only one of them. These 11 mutations would have been incorporated during the later stages of minus-strand DNA synthesis by the HIV-1 RT, thereby indicating that this enzyme has an important influence upon generation of viral diversity (O'Neil et al. 2002).

Imbalanced nucleotide levels reduce genomic stability even in the presence of proofreading and mismatch repair mechanisms (Kumar et al. 2011). As demonstrated for spleen necrosis virus and MLV, treatment with thymidine analogues or hydroxyurea that leads to low [dCTP]/[dTTP] ratios promotes an increase of the viral mutation rate (Julias and Pathak 1998). Reverse transcription in the presence of biased dNTP pools (i.e., low [dCTP]/[dTTP] ratios) can result in G→A hypermutagenesis (Martinez et al. 1994), and the magnitude of these effects could be modulated by RT mutations, as demonstrated with mutants affecting Met$^{230}$ (Cases-González and Menéndez-Arias 2004).

The viral mutation rate can be modulated by HIV proteins. One of them is Vpr, an accessory polypeptide of 96 amino acids. Vpr facilitates the incorporation of the nuclear form of uracil DNA glycosylase (UNG2) into virions. The deletion of the *vpr* gene in HIV-1 produces a fourfold increase in the viral mutation rate, resulting from a higher frequency of G→A and C→T transition mutations (Mansky et al. 2000). These nucleotide substitutions could originate from increased uracil content in the DNA, due to cytosine deamination occurring during minus-strand DNA synthesis or simply by misincorporation of U instead of T. Several retroviruses contain dUTPases in their viral genomes. These enzymes maintain low levels of UTP to avoid uridine incorporation into DNA. Examples of dUTPase-encoding retroviruses are FIV, equine infectious anemia virus, mouse mammary tumor virus, and Mason-Pfizer monkey virus (Elder et al. 1992; Köppe et al. 1994).

Apolipoprotein B mRNA-editing, catalytic polypeptide-like enzymes (APOBEC) are host-cell cytidine deaminases that can be incorporated into HIV particles. The viral protein Vif (viral infectivity factor) binds and triggers the degradation of human APOBEC3 proteins (typically APOBEC3F and APOBEC3G) (reviewed in Aguiar and Peterlin 2008). In the absence of Vif, APOBEC3 enzymes induce deamination of C→U residues in the negative strand of the retroviral DNA. The modified nucleic acid can be degraded after formation of abasic lesions (due to uracil DNA glycosylation) and further degradation by endonucleases or could be used as template leading to G→A hypermutation on the plus-strand DNA. Moderate levels of APOBEC3G-mediated G→A hypermutation could have a potential effect on HIV-1 evolution (Jern et al. 2009; Sadler et al. 2010). Since APOBEC3-driven mutagenesis occurs in a GG/GA nucleotide context, this mechanism could facilitate the emergence of specific drug resistance-associated mutations such as D67N or M184I in the viral RT, thereby inducing the appearance of drug-resistant progenies (Mulder et al. 2008).

Recent studies suggest that the nucleocapsid protein (NC) could have a critical role in mediating highly specific and efficient reverse transcription (for a review, see Levin et al. 2010). The HIV-1 NC protein is a nucleic acid chaperone that allows the viral RNA to reach the most thermodynamically stable arrangement. NC can destabilize secondary structure and alleviate pausing, thereby facilitating reverse transcription (Zhang et al. 2002b). The NC protein may have an effect on the fidelity of plus-strand priming by selecting the specific polypurine tract (Post et al. 2009; Godet et al. 2011). However, available data on its effects on misinsertion and mispair extension fidelity of DNA synthesis are limited. A qualitative analysis of reverse transcription products suggested that NC could facilitate repair of mispair termini, although the molecular mechanism involved has not been elucidated (Bampi et al. 2006). NC enhances the stability of the RT-nucleic acid substrate complex and increases the percentage of catalytically competent polymerase. In the presence of NC, there is a reduction in the nucleotide incorporation rate, although its impact on misinsertion fidelity has not been evaluated (Grohmann et al. 2008). However, in experiments carried out with oligonucleotides mimicking the HIV-1 primer binding site of plus-strand DNA synthesis, it has been shown that the rate of nucleotide incorporation is substantially increased (three- to sevenfold) by NCp7-annealed template-primer as compared to heat-annealed template-primers (Kim et al. 2012). NCp7-annealed template-primer also produces a threefold increase in the fidelity of HIV-1 RT by slowing the rate of incorporation of an incorrect nucleotide, as determined by pre-steady-state kinetics (Kim et al. 2012).

### *11.7.2 Impact of RT Mutations on HIV Variability*

Host and cells factors described in the previous section seem to mitigate the effects of HIV-1 RT on the retroviral mutation rates. Therefore, there are justifiable concerns about the relevance of error rates obtained in vitro with the purified RT. However, the determination of mutant frequencies in one round of replication, using a panel of HIV-1 RT variants, reveals meaningful correlations between error rates obtained in vitro and ex vivo referred to the impact of specific amino acid substitutions on HIV variability (Mansky et al. 2003). When introduced in recombinant HIV-1 clones, single amino acid substitutions in the RT such as K65R, D76V, R78A, Q151N, or W229A rendered viruses that showed lower mutant frequencies in comparison with the wild type. Four out of five of those amino acid changes were previously characterized as antimutators in enzymatic and forward mutation assays (Fig. 11.3), and the fifth one (W229A) displayed higher mispair extension fidelity in gel-based assays (Wisniewski et al. 1999). Interestingly, single-mutants Y115A and Y501W increased the HIV-1 mutation frequency in one round of replication assays, while other mutations such as D67N, K70R, L74V, E89G, K154A, M184V, F227A, and I505A had no significant effect on the viral error rates.

## 11.8 Emergence of HIV Drug-Resistant Variants and Their Impact on Virus Evolution

Although several studies have shown a correlation between AIDS progression and the rate of HIV diversification, these effects are largely dependent on the kinetics of the immune cell levels (Lee et al. 2008). The evolution of the viral polymerase is hardly affected by disease progression, although there is at least one example in which authors identified a more faithful RT in the late stage of SIV infection (Diamond et al. 2001). Theoretically, more faithful polymerases could limit genomic diversity in viral populations. By reducing the complexity of the mutant spectrum, viral populations would be less adaptable to a complex environment. This concept has been demonstrated for poliovirus after infecting mice with a mutant virus carrying a high-fidelity polymerase (Vignuzzi et al. 2006) and could have potential applications for obtaining more stable attenuated virus vaccines (Vignuzzi et al. 2008). On the other hand, studies using chikungunya virus as a model indicate that the reduced genetic diversity derived from increased replication fidelity may have a negative impact on viral fitness (Coffey et al. 2011).

There has been discussion on the effects of drug resistance mutations in the diversification of HIV populations. Based on the increased misinsertion and mispair extension fidelity of mutant HIV-1 RTs bearing the amino acid substitution M184V (Wainberg et al. 1996), authors suggested that this mutation appearing during lamivudine treatment could prevent or delay the emergence of resistance mutations associated with further antiretroviral therapies. This prediction was not confirmed in clinical studies (Jonckheere et al. 1998; Keulen et al. 1999). This was due in part to the modest effect of M184V on the fidelity of the RT but also to the influence of a number of factors including the viral population size, the number of replication cycles, and the viral replication rate. In addition, antiretroviral treatment with nucleoside analogues (e.g., zidovudine or lamivudine) could have a mutagenic effect in vivo, as demonstrated in cell cultures (Mansky and Bernard 2000). Moreover, evolution of antiretroviral drug resistance is characterized by significant fitness loss followed by repair strategies that include compensatory mutations in the target gene (for a review, see Menéndez-Arias et al. 2003).

Available data indicate that multidrug-resistant HIV-1 RTs show reduced error rates in comparison with the wild-type enzyme, as determined in vitro with the M13-based forward mutation assay. The increase in accuracy was estimated at around twofold for the A62V/V75I/F77L/F116Y/Q151M RT, eightfold for the A62V/T69SSS/L210W/R211K/L214F/T215Y RT, and >15-fold for M41L/T69SSG/L210W/R211K/L214F/T215Y RT (Rezende et al. 1998b; Curr et al. 2006). In contrast, a three- to fourfold increase in the mutation rates was obtained ex vivo with HIV-1 bearing in the RT the drug resistance-associated mutations M41L/T215Y or M41L/D67N/K70R/T215Y (Mansky and Bernard 2000). In any case, the evolutionary potential of any of those combinations of mutations has not been studied in a clinical context.

## 11.9 Lethal Mutagenesis

Increasing the viral error rate above a critical threshold could lead to the loss of genetic information. RNA viruses such as polio, hepatitis C, or retroviruses replicate very close to the critical mutation rate. A relatively small increase in the mutation rate could drive them into extinction ("error catastrophe"). Lethal mutagenesis (or viral decay acceleration) is an investigational therapeutic strategy based on the use of mutagens that could drive viral populations to extinction (for reviews, see Anderson et al. 2004; Domingo et al. 2005). Laboratory studies have demonstrated that therapeutics capable of increasing the mutation frequency as little as twofold results in collapse of the viral population (Loeb et al. 1999). Nucleoside analogues (e.g., 3′-azidothymidine, 5-azacytidine, and 5-hydroxydeoxycytidine; Fig. 11.7) have been shown to effectively increase mutation rates of various retroviruses, including spleen necrosis virus, MLV, FIV, and HIV-1 (Pathak and Temin 1992; LaCasse et al. 1996; Julias et al. 1997; Mansky and Bernard 2000; Dapp et al. 2009). These effects are consistent with an observed increase in the frequencies of G→C transversions (5-azacytidine) or G→A transitions (3′-azidothymidine). Mechanisms of action for these nucleosides involve misincorporation and altered base pairing with or without previous modification by cellular enzymes (e.g., ribonucleotide reductase, kinases). In addition, a mutagenic side effect results from alterations in the dNTP pools as shown, for example, in the case of zidovudine (Julias and Pathak 1998), or through the administration of 2′-deoxycytidine and tetrahydrouridine that can reverse the characteristic G→A mutational bias found in HIV-1 (Balzarini et al. 2001).

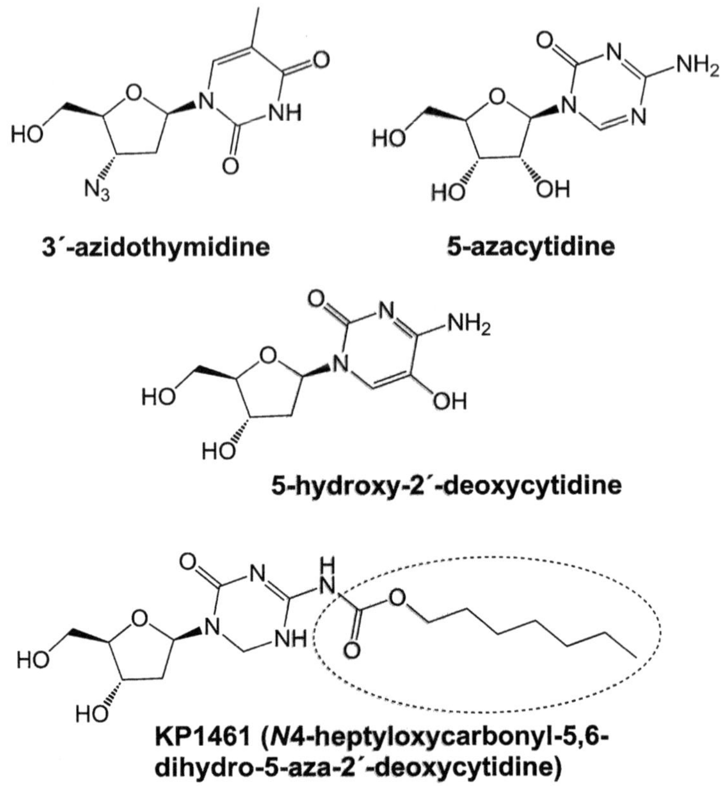


**Fig. 11.7** Chemical structures of mutagenic nucleosides. KP-1461 is a pro-drug of KP-1212 (5,6-dihydro-5-aza-2′-deoxycytidine)

Despite concerns on the toxicity and carcinogenicity of mutagenic nucleosides, a novel analogue for treatment of HIV infection (KP1212/KP1461) has completed Phase 2a clinical trials, showing a statistically significant increase in viral mutation frequency (Mullins et al. 2011). KP1461 is a cytidine analogue (Fig. 11.7) that promotes G→A and A→G substitutions due to its ability to form base pairs with both guanosine and adenine. The mutagen could act synergistically with APOBEC3 proteins by increasing the frequency of G→A transitions.

## 11.10 Conclusions

Almost 25 years after the first reliable estimates of the fidelity of retroviral RTs, we have learned a lot about nucleotide specificity that ultimately has led to the development of potent antiretroviral agents. However, evidence showing that the fidelity center of HIV-1 RT expands beyond the nucleotide-binding site, together with recent studies revealing the role of connection subdomain and RNase H domain mutations in nucleoside analogue resistance (for review, see Menéndez-Arias 2008, 2013), warrants further RT structure-activity studies. More faithful RTs are deemed necessary in deep sequencing and other massive sequencing techniques to decrease the error background resulting from the conversion of RNA to cDNA. On the other hand, further studies will be necessary to gain a deeper knowledge of the mechanisms that modulate the viral mutation rate in vivo, including how novel or previously known viral and host-cell factors participate in reverse transcription and in the generation of HIV genomes. This information may open novel possibilities for the control of the AIDS pandemic.

**Acknowledgments** I thank past and present members of our group and collaborators elsewhere for their contribution to RT fidelity studies over the years. I also thank Verónica Barrioluengo for her assistance in the preparation of Fig. 11.2. This work was supported in part by the Ministry of Science and Innovation of Spain (grant BIO2010/15542) and an institutional grant from the Fundación Ramón Areces.

# References

Abram ME, Ferris AL, Shao W, Alvord WG, Hughes SH (2010) Nature, position, and frequency of mutations made in a single cycle of HIV-1 replication. J Virol 84:9864–9878

Aguiar RS, Peterlin BM (2008) APOBEC3 proteins and reverse transcription. Virus Res 134:74–85

Álvarez M, Matamoros T, Menéndez-Arias L (2009) Increased thermostability and fidelity of DNA synthesis of wild-type and mutant HIV-1 group O reverse transcriptases. J Mol Biol 392:872–884

Anderson JP, Daifuku R, Loeb LA (2004) Viral error catastrophe by mutagenic nucleosides. Annu Rev Microbiol 58:183–205

Arezi B, Hogrefe H (2009) Novel mutations in Moloney murine leukemia virus reverse transcriptase increase thermostability through tighter binding to template-primer. Nucleic Acids Res 37:473–481

Back NKT, Nijhuis M, Keulen W, Boucher CAB, Oude Essink BB, van Kuilenburg ABP, van Gennip AH, Berkhout B (1996) Reduced replication of 3TC-resistant HIV-1 variants in primary cells due to a processivity defect of the reverse transcriptase enzyme. EMBO J 15:4040–4049

Bakhanashvili M, Hizi A (1992) Fidelity of the RNA-dependent DNA synthesis exhibited by the reverse transcriptases of human immunodeficiency virus types 1 and 2 and of murine leukemia virus: mispair extension frequencies. Biochemistry 31:9393–9398

Bakhanashvili M, Hizi A (1993) The fidelity of the reverse transcriptases of human immunodeficiency viruses and murine leukemia virus, exhibited by the mispair extension frequencies, is sequence dependent and enzyme related. FEBS Lett 319:201–205

Baltimore D (1970) RNA-dependent DNA polymerase in virions of RNA tumour viruses. Nature 226:1209–1211

Balzarini J, Camarasa M-J, Pérez-Pérez M-J, San-Félix A, Velázquez S, Perno C-F, De Clercq E, Anderson JN, Karlsson A (2001) Exploitation of the low fidelity of human immunodeficiency virus type 1 (HIV-1) reverse transcriptase and the nucleotide composition bias in the HIV-1 genome to alter the drug resistance development of HIV. J Virol 75:5772–5777

Bampi C, Bibillo A, Wendeler M, Divita G, Gorelick RJ, Le Grice SFJ, Darlix J-L (2006) Nucleotide excision repair and template-independent addition by HIV-1 reverse transcriptase in the presence of nucleocapsid protein. J Biol Chem 281:11736–11743

Barrioluengo V, Álvarez M, Barbieri D, Menéndez-Arias L (2011) Thermostable HIV-1 group O reverse transcriptase variants with the same fidelity as murine leukaemia virus reverse transcriptase. Biochem J 436:599–607

Barrioluengo V, Wang Y, Le Grice SFJ, Menéndez-Arias L (2012) Intrinsic DNA synthesis fidelity of xenotropic murine leukemia virus-related virus reverse transcriptase. FEBS J 279:1433–1444

Beard WA, Stahl SJ, Kim H-R, Bebenek K, Kumar A, Strub M-P, Becerra SP, Kunkel TA, Wilson SH (1994) Structure/function studies of human immunodeficiency virus type 1 reverse transcriptase – alanine scanning mutagenesis of an α-helix in the thumb subdomain. J Biol Chem 269:28091–28097

Bebenek K, Kunkel TA (1995) Analyzing fidelity of DNA polymerases. Methods Enzymol 262:217–232

Bebenek K, Abbotts J, Roberts JD, Wilson SH, Kunkel TA (1989) Specificity and mechanism of error-prone replication by human immunodeficiency virus-1 reverse transcriptase. J Biol Chem 264:16948–16956

Bebenek K, Abbotts J, Wilson SH, Kunkel TA (1993) Error-prone polymerization by HIV-1 reverse transcriptase – contribution of template-primer misalignment, miscoding, and termination probability to mutational hot spots. J Biol Chem 268:10324–10334

Bebenek K, Beard WA, Casas-Finet JR, Kim H-R, Darden TA, Wilson SH, Kunkel TA (1995) Reduced frameshift fidelity and processivity of HIV-1 reverse transcriptase mutants containing alanine substitutions in helix H of the thumb subdomain. J Biol Chem 270:19516–19523

Boyer PL, Hughes SH (2000) Effects of amino acid substitutions at position 115 on the fidelity of human immunodeficiency virus type 1 reverse transcriptase. J Virol 74:6494–6500

Boyer JC, Bebenek K, Kunkel TA (1992) Unequal human immunodeficiency virus type 1 reverse transcriptase error rates with RNA and DNA templates. Proc Natl Acad Sci USA 89:6919–6923

Boyer PL, Stenbak CR, Hoberman D, Linial ML, Hughes SH (2007) In vitro fidelity of the prototype primate foamy virus (PFV) RT compared to HIV-1 RT. Virology 367:253–264

Buonaguro L, Tornesello ML, Buonaguro FM (2007) Human immunodeficiency virus type 1 subtype distribution in the worldwide epidemic: pathogenetic and therapeutic implications. J Virol 81:10209–10219

Cases-González CE, Menéndez-Arias L (2004) Increased G → A transition frequencies displayed by primer grip mutants of human immunodeficiency virus type 1 reverse transcriptase. J Virol 78:1012–1019

Cases-González CE, Gutiérrez-Rivas M, Menéndez-Arias L (2000) Coupling ribose selection to fidelity of DNA synthesis: the role of Tyr-115 of human immunodeficiency virus type 1 reverse transcriptase. J Biol Chem 275:19759–19767

Coffey LL, Beeharry Y, Bordería AV, Blanc H, Vignuzzi M (2011) Arbovirus high fidelity variant loses fitness in mosquitoes and mice. Proc Natl Acad Sci USA 108:16038–16043

Curr K, Tripathi S, Lennerstrand J, Larder BA, Prasad VR (2006) Influence of naturally occurring insertions in the fingers subdomain of human immunodeficiency virus type 1 reverse transcriptase on polymerase fidelity and mutation frequencies in vitro. J Gen Virol 87:419–428

Dapp MJ, Clouser CL, Patterson S, Mansky LM (2009) 5-Azacytidine can induce lethal mutagenesis in human immunodeficiency virus type 1. J Virol 83:11950–11958

Diamond TL, Kimata J, Kim B (2001) Identification of a simian immunodeficiency virus reverse transcriptase variant with enhanced replicational fidelity in the late stage of viral infection. J Biol Chem 276:23624–23631

Diamond TL, Souroullas G, Weiss KK, Lee KY, Bambara RA, Dewhurst S, Kim B (2003) Mechanistic understanding of an altered fidelity simian immunodeficiency virus reverse transcriptase mutation, V148I, identified in a pig-tailed macaque. J Biol Chem 278:29913–29924

Domingo E, Escarmís C, Lázaro E, Manrubia SC (2005) Quasispecies dynamics and RNA virus extinction. Virus Res 107:129–139

Drosopoulos WC, Prasad VR (1998) Increased misincorporation fidelity observed for nucleoside analog resistance mutations M184V and E89G in human immunodeficiency virus type 1 reverse transcriptase does not correlate with the overall error rate measured in vitro. J Virol 72:4224–4230

Duran-Reynals F (1942) The reciprocal infection of ducks and chickens with tumor-inducing viruses. Cancer Res 2:343–369

Echols H, Goodman MF (1991) Fidelity mechanisms in DNA replication. Annu Rev Biochem 60:477–511

Eckert KA, Kunkel TA (1993) Fidelity of DNA synthesis catalyzed by human DNA polymerase α and HIV-1 reverse transcriptase: effect of reaction pH. Nucleic Acids Res 21:5212–5220

Elder JH, Lerner DL, Hasselkus-Light CS, Fontenot DJ, Hunter E, Luciw PA, Montelaro RC, Phillips TR (1992) Distinct subsets of retroviruses encode dUTPase. J Virol 66:1791–1794

Gago S, Elena SF, Flores R, Sanjuán R (2009) Extremely high mutation rate of a hammerhead viroid. Science 323:1308

Gao G, Orlova M, Georgiadis MM, Hendrickson WA, Goff SP (1997) Conferring RNA polymerase activity to a DNA polymerase: a single residue in reverse transcriptase controls substrate selection. Proc Natl Acad Sci USA 94:407–411

Gao F, Chen Y, Levy DN, Conway JA, Kepler TB, Hui H (2004) Unselected mutations in the human immunodeficiency virus type 1 genome are mostly nonsynonymous and often deleterious. J Virol 78:2426–2433

Garforth SJ, Domaoal RA, Lwatula C, Landau MJ, Meyer AJ, Anderson KS, Prasad VR (2010) K65R and K65A substitutions in HIV-1 reverse transcriptase enhance polymerase fidelity by decreasing both dNTP misinsertion and mispaired primer extension efficiencies. J Mol Biol 401:33–44

Gärtner K, Wiktorowicz T, Park J, Mergia A, Rethwilm A, Scheller C (2009) Accuracy estimation of foamy virus genome copying. Retrovirology 6:32

Gerard GF, Potter RJ, Smith MD, Rosenthal K, Dhariwal G, Lee J, Chatterjee DK (2002) The role of template-primer in protection of reverse transcriptase from thermal inactivation. Nucleic Acids Res 30:3118–3129

Godet J, Ramalanjaona N, Sharma KK, Richert L, de Rocquigny H, Darlix J-L, Duportail G, Mély Y (2011) Specific implications of the HIV-1 nucleocapsid zinc fingers in the annealing of the primer binding site complementary sequences during the obligatory plus strand transfer. Nucleic Acids Res 39:6633–6645

Grohmann D, Godet J, Mély Y, Darlix J-L, Restle T (2008) HIV-1 nucleocapsid traps reverse transcriptase on nucleic acid substrates. Biochemistry 47:12230–12240

Hamburgh ME, Curr KA, Monaghan M, Rao VR, Tripathi S, Preston BD, Sarafianos S, Arnold E, Darden T, Prasad VR (2006) Structural determinants of slippage-mediated mutations by human immunodeficiency virus type 1 reverse transcriptase. J Biol Chem 281:7421–7428

Harris D, Kaushik N, Pandey PK, Yadav PN, Pandey VN (1998) Functional analysis of amino acid residues constituting the dNTP binding pocket of HIV-1 reverse transcriptase. J Biol Chem 273:33624–33634

Hemelaar J, Gouws E, Ghys PD, Osmanov S, WHO-UNAIDS network for HIV Isolation and Characterisation (2011) Global trends in molecular epidemiology of HIV-1 during 2000–2007. AIDS 25:679–689

Herschhorn A, Hizi A (2010) Retroviral reverse transcriptases. Cell Mol Life Sci 67:2717–2747

Huang KJ, Wooley DP (2005) A new cell-based assay for measuring the forward mutation rate of HIV-1. J Virol Methods 124:95–104

Huang H, Chopra R, Verdine GL, Harrison SC (1998) Structure of a covalently trapped catalytic complex of HIV-1 reverse transcriptase: implications for drug resistance. Science 282:1669–1675

Jacobo-Molina A, Ding J, Nanni RG, Clark AD Jr, Lu X, Tantillo C, Williams RL, Kamer G, Ferris AL, Clark P, Hizi A, Hughes SH, Arnold E (1993) Crystal structure of human immunodeficiency virus type 1 reverse transcriptase complexed with double-stranded DNA at 3.0 Å resolution shows bent DNA. Proc Natl Acad Sci USA 90:6320–6324

Jamburuthugoda VK, Eickbush TH (2011) The reverse transcriptase encoded by the non-LTR retrotransposon R2 is as error-prone as that encoded by HIV-1. J Mol Biol 407:661–672

Jamburuthugoda VK, Guo D, Wedekind JE, Kim B (2005) Kinetic evidence for interaction of human immunodeficiency virus type 1 reverse transcriptase with the 3′-OH of the incoming dTTP substrate. Biochemistry 44:10635–10643

Jern P, Russell RA, Pathak VK, Coffin JM (2009) Likely role of APOBEC3G-mediated G-to-A mutations in HIV-1 evolution and drug resistance. PLoS Pathog 5:e1000367

Ji J, Loeb LA (1992) Fidelity of HIV-1 reverse transcriptase copying RNA in vitro. Biochemistry 31:954–958

Ji J, Loeb LA (1994) Fidelity of HIV-1 reverse transcriptase copying a hypervariable region of the HIV-1 *env* gene. Virology 199:323–330

Johnson KA (1993) Conformational coupling in DNA polymerase fidelity. Annu Rev Biochem 62:685–713

Jonckheere H, Witvrouw M, De Clercq E, Anné J (1998) Lamivudine resistance of HIV type 1 does not delay development of resistance to nonnucleoside HIV type 1-specific reverse transcriptase inhibitors as compared with wild-type HIV type 1. AIDS Res Hum Retroviruses 14:249–253

Jonckheere H, De Clercq E, Anné J (2000) Fidelity analysis of HIV-1 reverse transcriptase mutants with an altered amino-acid sequence at residues Leu74, Glu89, Tyr115, Tyr183 and Met184. Eur J Biochem 267:2658–2665

Julias JG, Pathak VK (1998) Deoxyribonucleoside triphosphate pool imbalances in vivo are associated with an increased retroviral mutation rate. J Virol 72:7941–7949

Julias JG, Kim T, Arnold G, Pathak VK (1997) The antiretrovirus drug 3′-azido-3′-deoxythymidine increases the retrovirus mutation rate. J Virol 71:4254–4263

Kati WM, Johnson KA, Jerva LF, Anderson KS (1992) Mechanism and fidelity of HIV reverse transcriptase. J Biol Chem 267:25988–25997

Kellinger MW, Johnson KA (2010) Nucleotide-dependent conformational change governs specificity and analog discrimination by HIV reverse transcriptase. Proc Natl Acad Sci USA 107:7734–7739

Kerr SG, Anderson KS (1997) RNA dependent DNA replication fidelity of HIV-1 reverse transcriptase: evidence of discrimination between DNA and RNA substrates. Biochemistry 36:14056–14063

Keulen W, van Wijk A, Schuurman R, Berkhout B, Boucher CAB (1999) Increased polymerase fidelity of lamivudine-resistant HIV-1 variants does not limit their evolutionary potential. AIDS 13:1343–1349

Kim T, Mudry RA Jr, Rexrode CA 2nd, Pathak VK (1996) Retroviral mutation rates and A-to-G hypermutations during different stages of retroviral replication. J Virol 70:7594–7602

Kim J, Roberts A, Yuan H, Xiong Y, Anderson KS (2012) Nucleocapsid protein annealing of a primer-template enhances (+)-strand DNA synthesis and fidelity by HIV-1 reverse transcriptase. J Mol Biol 415:866–880

Kisic M, Matamoros T, Nevot M, Mendieta J, Martinez-Picado J, Martínez MA, Menéndez-Arias L (2011) Thymidine analogue excision and discrimination modulated by mutational complexes including single amino acid deletions of Asp-67 or Thr-69 in HIV-1 reverse transcriptase. J Biol Chem 286:20615–20624

Kohlstaedt LA, Wang J, Friedman JM, Rice PA, Steitz TA (1992) Crystal structure at 3.5 Å resolution of HIV-1 reverse transcriptase complexed with an inhibitor. Science 256:1783–1790

Köppe B, Menéndez-Arias L, Oroszlan S (1994) Expression and purification of the mouse mammary tumor virus *gag-pro* transframe protein p30 and characterization of its dUTPase activity. J Virol 68:2313–2319

Kumar D, Abdulovic AL, Viberg J, Nilsson AK, Kunkel TA, Chabes A (2011) Mechanisms of mutagenesis in vivo due to imbalanced dNTP pools. Nucleic Acids Res 39:1360–1371

Kunkel TA (2004) DNA replication fidelity. J Biol Chem 279:16895–16898

Kunkel TA, Bebenek K (2000) DNA replication fidelity. Annu Rev Biochem 69:497–529

Laakso MM, Sutton RE (2006) Replicative fidelity of lentiviral vectors produced by transient infection. Virology 348:406–417

LaCasse RA, Remington KM, North TW (1996) The mutation frequency of feline immunodeficiency virus enhanced by 3′-azido-3′-deoxythymidine. J Acquir Immune Defic Syndr Hum Retrovirol 12:26–32

Lee HY, Perelson AS, Park S-C, Leitner T (2008) Dynamic correlation between intrahost HIV-1 quasispecies evolution and disease progression. PLoS Comput Biol 4:e1000240
Leider JM, Palese P, Smith FI (1988) Determination of the mutation rate of a retrovirus. J Virol 62:3084–3091
Levin JG, Mitra M, Mascarenhas A, Musier-Forsyth K (2010) Role of HIV-1 nucleocapsid protein in HIV-1 reverse transcription. RNA Biol 7:754–774
Lewis DA, Bebenek K, Beard WA, Wilson SH, Kunkel TA (1999) Uniquely altered DNA replication fidelity conferred by an amino acid change in the nucleotide binding pocket of human immunodeficiency virus type 1 reverse transcriptase. J Biol Chem 274:32924–32930
Loeb LA, Essigmann JM, Kazazi F, Zhang J, Rose KD, Mullins JI (1999) Lethal mutagenesis of HIV with mutagenic nucleoside analogs. Proc Natl Acad Sci USA 96:1492–1497
Lwatula C, Garforth SJ, Prasad VR (2012) Lys66 residue as a determinant of high mismatch extension and misinsertion rates of HIV-1 reverse transcriptase. FEBS J 279:4010–4024
Lynch M (2006) The origins of eukaryotic gene structure. Mol Biol Evol 23:450–468
Malboeuf CM, Isaacs SJ, Tran NH, Kim B (2001) Thermal effects on reverse transcription: improvement of accuracy and processivity in cDNA synthesis. Biotechniques 30:1074–1084
Mansky LM (1996) Forward mutation rate of human immunodeficiency virus type 1 in a T lymphoid cell line. AIDS Res Hum Retroviruses 12:307–314
Mansky LM, Bernard LC (2000) 3′-azido-3′-deoxythymidine (AZT) and AZT-resistant reverse transcriptase can increase the in vivo mutation rate of human immunodeficiency virus type 1. J Virol 74:9532–9539
Mansky LM, Temin HM (1995) Lower in vivo mutation rate of human immunodeficiency virus type 1 than that predicted from the fidelity of purified reverse transcriptase. J Virol 69:5087–5094
Mansky LM, Preveral S, Selig L, Benarous R, Benichou S (2000) The interaction of Vpr with uracil DNA glycosylase modulates the human immunodeficiency virus type 1 in vivo mutation rate. J Virol 74:7039–7047
Mansky LM, Le Rouzic E, Benichou S, Gajary LC (2003) Influence of reverse transcriptase variants, drugs, and Vpr on human immunodeficiency virus type 1 mutant frequencies. J Virol 77:2071–2080
Martinez MA, Vartanian JP, Wain-Hobson S (1994) Hypermutagenesis of RNA using human immunodeficiency virus type 1 reverse transcriptase and biased dNTP concentrations. Proc Natl Acad Sci USA 91:11787–11791
Martín-Hernández AM, Domingo E, Menéndez-Arias L (1996) Human immunodeficiency virus type 1 reverse transcriptase: role of Tyr115 in deoxynucleotide binding and misinsertion fidelity of DNA synthesis. EMBO J 15:4434–4442
Martín-Hernández AM, Gutiérrez-Rivas M, Domingo E, Menéndez-Arias L (1997) Mispair extension fidelity of human immunodeficiency virus type 1 reverse transcriptases with amino acid substitutions affecting Tyr115. Nucleic Acids Res 25:1383–1389
Matamoros T, Kim B, Menéndez-Arias L (2008) Mechanistic insights into the role of Val75 of HIV-1 reverse transcriptase in misinsertion and mispair extension fidelity of DNA synthesis. J Mol Biol 375:1234–1248
Matamoros T, Álvarez M, Barrioluengo V, Betancor G, Menéndez-Arias L (2011) Reverse transcriptase and retroviral replication. In: Kušić-Tišma J (ed) DNA replication and related cellular process. InTech, Rijeka, p. 111–142. Available from http://www.intechopen.com/articles/show/title/reverse-transcriptase-and-retroviral-replication
Matsuda T, Bebenek K, Masutani C, Hanaoka F, Kunkel TA (2000) Low fidelity DNA synthesis by human DNA polymerase-η. Nature 404:1011–1013
Mbisa JL, Nikolenko GN, Pathak VK (2005) Mutations in the RNase H primer grip domain of murine leukemia virus reverse transcriptase decrease efficiency and accuracy of plus-strand DNA transfer. J Virol 79:419–427
McCulloch SD, Kunkel TA (2008) The fidelity of DNA synthesis by eukaryotic replicative and translation synthesis polymerases. Cell Res 18:148–161

Mendelman LV, Boosalis MS, Petruska J, Goodman MF (1989) Nearest neighbor influences on DNA polymerase insertion fidelity. J Biol Chem 264:14415–14423

Mendieta J, Cases-González CE, Matamoros T, Ramírez G, Menéndez-Arias L (2008) A $Mg^{2+}$-induced conformational switch rendering a competent DNA polymerase catalytic complex. Proteins 71:565–574

Menéndez-Arias L (2002) Molecular basis of fidelity of DNA synthesis and nucleotide specificity of retroviral reverse transcriptases. Prog Nucleic Acid Res Mol Biol 71:91–147

Menéndez-Arias L (2008) Mechanisms of resistance to nucleoside analogue inhibitors of HIV-1 reverse transcriptase. Virus Res 134:124–146

Menéndez-Arias L (2009) Mutation rates and intrinsic fidelity of retroviral reverse transcriptases. Viruses 1:1137–1165

Menéndez-Arias L (2010) Molecular basis of human immunodeficiency virus drug resistance: an update. Antiviral Res 85:210–231

Menéndez-Arias L (2013) Molecular basis of human immunodeficiency virus type 1 drug resistance: overview and recent developments. Antiviral Res 98:93–120

Menéndez-Arias L, Abraha A, Quiñones-Mateu ME, Mas A, Camarasa M-J, Arts EJ (2001) Functional characterization of chimeric reverse transcriptases with polypeptide subunits of highly divergent HIV-1 group M and O strains. J Biol Chem 276:27470–27479

Menéndez-Arias L, Martínez MA, Quiñones-Mateu ME, Martinez-Picado J (2003) Fitness variations and their impact on the evolution of antiretroviral drug resistance. Curr Drug Targets Infect Disord 3:355–371

Monk RJ, Malik FG, Stokesberry D, Evans LH (1992) Direct determination of the point mutation rate of a murine retrovirus. J Virol 66:3683–3689

Mulder LCF, Harari A, Simon V (2008) Cytidine deamination induced HIV-1 drug resistance. Proc Natl Acad Sci USA 105:5501–5506

Mullins JI, Heath L, Hughes JP, Kicha J, Styrchak S, Wong KG, Rao U, Hansen A, Harris KS, Laurent JP, Li D, Simpson JH, Essigmann JM, Loeb LA, Parkins J (2011) Mutation of HIV-1 genomes in a clinical population treated with the mutagenic nucleoside KP1461. PLoS One 6:e15135

Ndongwe TP, Adedeji AO, Michailidis E, Ong YT, Hachiya A, Marchand B, Ryan EM, Rai DK, Kirby KA, Whatley AS, Burke DH, Johnson M, Ding S, Zheng Y-M, Liu S-L, Kodama E-I, Delviks-Frankenberry KA, Pathak VK, Mitsuya H, Parniak MA, Singh K, Sarafianos SG (2012) Biochemical, inhibition and inhibitor resistance studies of xenotropic murine leukemia virus-related virus reverse transcriptase. Nucleic Acids Res 40:345–359

O'Neil PK, Sun G, Yu H, Ron Y, Dougherty JP, Preston BD (2002) Mutational analysis of HIV-1 long terminal repeats to explore the relative contribution of reverse transcriptase and RNA polymerase II to viral mutagenesis. J Biol Chem 277:38053–38061

Operario DJ, Reynolds HM, Kim B (2005) Comparison of DNA polymerase activities between recombinant feline immunodeficiency and leukemia virus reverse transcriptases. Virology 335:106–121

Pathak VK, Temin HM (1992) 5-Azacytidine and RNA secondary structure increase the retrovirus mutation rate. J Virol 66:3093–3100

Perrino FW, Preston BD, Sandell LL, Loeb LA (1989) Extension of mismatched 3′ termini of DNA is a major determinant of the infidelity of human immunodeficiency virus type 1 reverse transcriptase. Proc Natl Acad Sci USA 86:8343–8347

Plantier JC, Leoz M, Dickerson JE, De Oliveira F, Cordonnier F, Lemée V, Damond F, Robertson DL, Simon F (2009) A new human immunodeficiency virus derived from gorillas. Nat Med 15:871–872

Post K, Kankia B, Gopalakrishnan S, Yang V, Cramer E, Saladores P, Gorelick RJ, Guo J, Musier-Forsyth K, Levin JG (2009) Fidelity of plus-strand priming requires the nucleic acid chaperone activity of HIV-1 nucleocapsid protein. Nucleic Acids Res 37:1755–1766

Preston BD, Poiesz BJ, Loeb LA (1988) Fidelity of HIV-1 reverse transcriptase. Science 242:1168–1171

Quiñones-Mateu ME, Soriano V, Domingo E, Menéndez-Arias L (1997) Characterization of the reverse transcriptase of a human immunodeficiency virus type 1 group O isolate. Virology 236:364–373

Ramirez BC, Simon-Loriere E, Galetto R, Negroni M (2008) Implications of recombination for HIV diversity. Virus Res 134:64–73

Rezende LF, Drosopoulos WC, Prasad VR (1998a) The influence of 3TC resistance mutation M184I on the fidelity and error specificity of human immunodeficiency virus type 1 reverse transcriptase. Nucleic Acids Res 26:3066–3072

Rezende LF, Curr K, Ueno T, Mitsuya H, Prasad VR (1998b) The impact of multidideoxynucleoside resistance-conferring mutations in human immunodeficiency virus type 1 reverse transcriptase on polymerase fidelity and error specificity. J Virol 72:2890–2895

Ricchetti M, Buc H (1990) Reverse transcriptases and genomic variability: the accuracy of DNA replication is enzyme specific and sequence dependent. EMBO J 9:1583–1593

Roberts JD, Bebenek K, Kunkel TA (1988) The accuracy of reverse transcriptase from HIV-1. Science 242:1171–1173

Roberts JD, Preston BD, Johnston LA, Soni A, Loeb LA, Kunkel TA (1989) Fidelity of two retroviral reverse transcriptases during DNA-dependent DNA synthesis in vitro. Mol Cell Biol 9:469–476

Rous P, Murphy JB (1913) Variations in a chicken sarcoma caused by a filterable agent. J Exp Med 17:219–231

Sadler HA, Stenglein MD, Harris RS, Mansky LM (2010) APOBEC3G contributes to HIV-1 variation through sublethal mutagenesis. J Virol 84:7396–7404

Sala M, Wain-Hobson S, Schaeffer F (1995) Human immunodeficiency virus type 1 reverse transcriptase ${}_{t}$G:T mispair formation on RNA and DNA templates with mismatched primers: a kinetic and thermodynamic study. EMBO J 14:4622–4627

Sanjuán R, Nebot MR, Chirico N, Mansky LM, Belshaw R (2010) Viral mutation rates. J Virol 84:9733–9748

Sarafianos SG, Das K, Tantillo C, Clark AD Jr, Ding J, Whitcomb JM, Boyer PL, Hughes SH, Arnold E (2001) Crystal structure of HIV-1 reverse transcriptase in complex with a polypurine tract RNA: DNA. EMBO J 20:1449–1461

Shah FS, Curr KA, Hamburgh ME, Parniak M, Mitsuya H, Arnez JG, Prasad VR (2000) Differential influence of nucleoside analog-resistance mutations K65R and L74V on the overall mutation rate and error specificity of human immunodeficiency virus type 1 reverse transcriptase. J Biol Chem 275:27037–27044

Sharma PL, Nettles JH, Feldman A, Rapp K, Schinazi RF (2009) Comparative analysis of in vitro processivity of HIV-1 reverse transcriptases containing mutations 65R, 74V, 184V and 65R+74V. Antiviral Res 83:317–323

Skasko M, Weiss KK, Reynolds HM, Jamburuthugoda V, Lee K, Kim B (2005) Mechanistic differences in RNA-dependent DNA polymerization and fidelity between murine leukemia virus and HIV-1 reverse transcriptases. J Biol Chem 280:12190–12200

Stuke AW, Ahmad-Omar O, Hoefer K, Hunsmann G, Jentsch KD (1997) Mutations in the SIV *env* and the M13 *lacZα* gene generated in vitro by reverse transcriptases and DNA polymerases. Arch Virol 142:1139–1154

Svarovskaia ES, Cheslock SR, Zhang W-H, Hu W-S, Pathak VK (2003) Retroviral mutation rates and reverse transcriptase fidelity. Front Biosci 8:D117–D134

Tamura K, Nei M, Kumar S (2004) Prospects for inferring very large phylogenies by using the neighbor-joining method. Proc Natl Acad Sci USA 101:11030–11035

Tamura K, Dudley J, Nei M, Kumar S (2007) MEGA4: molecular evolutionary genetics analysis (MEGA) software version 4.0. Mol Biol Evol 24:1596–1599

Temin H (1960) The control of cellular morphology in embryonic cells infected with Rous sarcoma virus in vitro. Virology 10:182–197

Temin HM, Mizutani S (1970) RNA-dependent DNA polymerase in virions of Rous sarcoma virus. Nature 226:1211–1213

Varela-Echavarría A, Garvey N, Preston BD, Dougherty JP (1992) Comparison of Moloney murine leukemia virus mutation rate with the fidelity of its reverse transcriptase in vitro. J Biol Chem 267:24681–24688

Vartanian JP, Sommer P, Wain-Hobson S (2003) Death and the retrovirus. Trends Mol Med 9:409–413

Vignuzzi M, Stone JK, Arnold JJ, Cameron CE, Andino R (2006) Quasispecies diversity determines pathogenesis through cooperative interactions in a viral population. Nature 439:344–348

Vignuzzi M, Wendt E, Andino R (2008) Engineering attenuated virus vaccines by controlling replication fidelity. Nat Med 14:154–161

Wainberg MA, Drosopoulos WC, Salomon H, Hsu M, Borkow G, Parniak MA, Gu Z, Song Q, Manne J, Islam S, Castriota G, Prasad VR (1996) Enhanced fidelity of 3TC-selected mutant HIV-1 reverse transcriptase. Science 271:1282–1285

Weiss KK, Isaacs SJ, Tran NH, Adman ET, Kim B (2000) Molecular architecture of the mutagenic active site of human immunodeficiency virus type 1 reverse transcriptase: roles of the β8-αE loop in fidelity, processivity, and substrate interactions. Biochemistry 39:10684–10694

Weiss KK, Bambara RA, Kim B (2002) Mechanistic role of residue Gln[151] in error prone DNA synthesis by human immunodeficiency virus type 1 (HIV-1) reverse transcriptase (RT) – pre-steady state kinetic study of the Q151N HIV-1 RT mutant with increased fidelity. J Biol Chem 277:22662–22669

Weiss KK, Chen R, Skasko M, Reynolds HM, Lee K, Bambara RA, Mansky LM, Kim B (2004) A role for dNTP binding of human immunodeficiency virus type 1 reverse transcriptase in viral mutagenesis. Biochemistry 43:4490–4500

Weymouth LA, Loeb LA (1978) Mutagenesis during in vitro DNA synthesis. Proc Natl Acad Sci USA 75:1924–1928

Wisniewski M, Palaniappan C, Fu Z, Le Grice SFJ, Fay P, Bambara RA (1999) Mutations in the primer grip region of HIV reverse transcriptase can increase replication fidelity. J Biol Chem 274:28175–28184

Xu H-T, Quan Y, Asahchop E, Oliveira M, Moisi D, Wainberg MA (2010) Comparative biochemical analysis of recombinant reverse transcriptase enzymes of HIV-1 subtype B and subtype C. Retrovirology 7:80

Yasukawa K, Nemoto D, Inouye K (2008) Comparison of the thermal stabilities of reverse transcriptases from avian myeloblastosis virus and Moloney murine leukaemia virus. J Biochem 143:261–268

Yu H, Goodman MF (1992) Comparison of HIV-1 and avian myeloblastosis virus reverse transcriptase fidelity on RNA and DNA templates. J Biol Chem 267:10888–10896

Yu H, Jetzt AE, Dougherty JP (1997) Use of single-cycle analysis to study rates and mechanisms of retroviral mutation. Methods 12:325–336

Zhang W-H, Svarovskaia ES, Barr R, Pathak VK (2002a) Y586F mutation in murine leukemia virus reverse transcriptase decreases fidelity of DNA synthesis in regions associated with adenine-thymine tracts. Proc Natl Acad Sci USA 99:10090–10095

Zhang W-H, Hwang CK, Hu W-S, Gorelick RJ, Pathak VK (2002b) Zinc finger domain of murine leukemia virus nucleocapsid protein enhances the rate of viral DNA synthesis in vivo. J Virol 76:7473–7484

Zinnen S, Hsieh J-C, Modrich P (1994) Misincorporation and mispaired primer extension by human immunodeficiency virus reverse transcriptase. J Biol Chem 269:24195–24202

# Chapter 12
# APOBECs and Their Role in Proviral DNA Synthesis

**John S. Albin and Reuben S. Harris**

## 12.1 Introduction to APOBEC3 Proteins

The human genome has a total of seven *APOBEC3* DNA cytosine deaminase genes arrayed on chromosome 22 (*A3A*, *A3B*, *A3C*, *A3D*, *A3F*, *A3G,* and *A3H*), the result of a series of tandem duplications that have expanded the locus throughout primate evolution (Conticello 2008; LaRue et al. 2008, 2009). This expansion, combined with evidence for strong positive selection among most of these *A3* genes, suggests that a primary function of their protein products is defense of the genome against any number of retroelements encountered throughout evolution (Duggal et al. 2011; OhAinle et al. 2006; Sawyer et al. 2004). The list of retroelements for which direct evidence of A3 restriction exists currently includes endogenous LTR and non-LTR retrotransposons as well as exogenous members of the *Hepadnaviridae* and *Retroviridae* families [e.g., (Bogerd et al. 2006; Derse et al. 2007; Esnault et al. 2005, 2006; Harris et al. 2003a; Löchelt et al. 2005; Mangeat et al. 2003; Muckenfuss et al. 2006; Navarro et al. 2005; Okeoma et al. 2007; Russell et al. 2005; Sheehy et al. 2002; Stenglein and Harris 2006; Turelli et al. 2004)].

Chief among the examples of viruses restricted by A3 proteins is human immunodeficiency virus type 1 (HIV-1, hereafter abbreviated HIV). In the absence of the viral accessory protein virion infectivity factor (Vif), HIV infectivity is dramatically reduced when produced from certain cell lines [e.g., (Gabuzda et al. 1992, 1994; Strebel et al. 1987)]. Cell fusion experiments demonstrated that this restriction in so-called nonpermissive cells is due to the presence of a dominant-acting cellular factor suppressible by Vif function (Madani and Kabat 1998; Simon et al. 1998), and subsequent analysis of differentially expressed genes in permissive versus nonpermissive cells established A3G as a source of restriction in nonpermissive

J.S. Albin • R.S. Harris (✉)
Department of Biochemistry, Molecular Biology and Biophysics, Institute for Molecular Virology, Center for Genome Engineering, Masonic Cancer Center, University of Minnesota, Minneapolis, MN 55455, USA
e-mail: rsh@umn.edu

S.F.J. LeGrice and M. Götte (eds.), *Human Immunodeficiency Virus Reverse Transcriptase: A Bench-to-Bedside Success*, DOI 10.1007/978-1-4614-7291-9_12,

cells (Sheehy et al. 2002). Additional work revealed that Vif nucleates an E3 ligase complex consisting of CUL5, ELOB/C, RBX2, and cofactor CBFβ for A3G polyubiquitination and subsequent proteasomal degradation, thus obviating A3G antiviral functions (Conticello et al. 2003; Hultquist et al. 2012; Jäger et al. 2011; Kao et al. 2003; Marin et al. 2003; Mehle et al. 2004; Sheehy et al. 2002, 2003; Stopak et al. 2003; Yu et al. 2003; Zhang et al. 2011).

Work toward understanding the mechanism of A3G's antiviral action was facilitated by the coincident discovery that A3G and its RNA-editing homolog APOBEC1 can each deaminate DNA cytosines to yield uracils (Harris et al. 2002). Combined with the well-established phenomenon of retroviral hypermutation in which HIV and other retroviruses had been noted to accumulate large numbers of plus-strand G-to-A mutations (Janini et al. 2001; Pathak and Temin 1990; Vartanian et al. 1991, 1994), particularly in the absence of Vif (Lecossier et al. 2003), this led several groups to hypothesize and demonstrate that A3G can restrict HIV by introducing extensive C-to-U mutations into minus-strand cDNA during reverse transcription prior to synthesis of the complementary viral plus-strand (Harris et al. 2003a; Mangeat et al. 2003; Zhang et al. 2003). This minus-strand C-to-U transition subsequently templates the insertion of A during plus-strand synthesis, providing a clear explanation for the G-to-A hypermutation phenomenon (Harris et al. 2003a; Mangeat et al. 2003; Yu et al. 2004; Zhang et al. 2003). From the beginning, restriction was thereby inextricably linked to the ability of A3G to catalyze mutations in the viral genome.

## 12.2 Deaminase-Dependent Restriction Mechanisms

Restriction by A3 proteins is dependent on their expression in virus-producing cells and on their encapsidation during virus assembly (Harris et al. 2003a; Mangeat et al. 2003; Zhang et al. 2003). This is thought to occur through an association of A3 proteins with the viral nucleocapsid protein via an RNA bridge (Alce and Popik 2004; Bogerd and Cullen 2008; Cen et al. 2004; Luo et al. 2004; Schafer et al. 2004; Svarovskaia et al. 2004; Zennou et al. 2004). In particular, this interaction may be facilitated by the cellular 7SL RNA, which is selectively packaged into HIV-1 virions (Tian et al. 2007; Wang et al. 2007, 2008). The importance of RNA in facilitating the encapsidation of A3 proteins is further supported by the observation that mutations in the N-terminus of A3G that ablate RNA binding also ablate Gag interaction, encapsidation, and, consequently, restriction (Friew et al. 2009; Huthoff et al. 2009; Huthoff and Malim 2007; Navarro et al. 2005). In total, it is thought that $7 \pm 4$ molecules of A3G are packaged per particle and that as few as one may be measurably antiviral (Browne et al. 2009; Xu et al. 2007).

Once associated with the viral core, A3 proteins are positioned to exert their restrictive activities upon nascent reverse-transcribed cDNA in target cells. Interpretations of the relative importance of different putative mechanisms of A3G inhibition of HIV may vary, but a broad point of unity in the field is that deamination is a major component of the restriction mechanism, as initially proposed [(Harris et al. 2003a; Mangeat et al. 2003; Zhang et al. 2003) and see Sect. 12.3 below]. Thus, we begin a consideration of the impact of A3 proteins on proviral DNA synthesis with a more detailed rendering of the deamination process, focusing on the prototypical A3 protein A3G, but with the understanding that A3D, A3F, and A3H are thought to work similarly.

### 12.2.1 Mechanism of A3G Deamination and Hypermutation

The polynucleotide intermediates that occur in the course of HIV reverse transcription include the single-stranded RNA genome (ssRNA), a DNA:RNA heteroduplex following synthesis of the minus-strand cDNA templated by the plus-strand RNA genome, single-stranded DNA (ssDNA) resulting from RNaseH degradation of the plus-strand ssRNA template after minus-strand synthesis, and double-stranded DNA (dsDNA) after completion of plus-strand cDNA (Fig. 12.1a–d). Among these, A3G is capable of binding to both ssRNA and ssDNA (Iwatani et al. 2006; Yu et al. 2004). Binding to ssRNA is thought to segregate A3G into enzymatically inactive high molecular mass complexes (Chelico et al. 2006; Wedekind et al. 2006), and this may also be true of association between A3G and viral genomic ssRNA (Soros et al. 2007). In its low molecular mass form, however, A3G is capable of deaminating ssDNA substrates (Harris et al. 2003a; Iwatani et al. 2006; Suspène et al. 2004; Yu et al. 2004). This ssDNA specificity is reinforced by the observation that there are two gradients of 5′ to 3′ increasing levels of hypermutation in the viral genome, one from the primer binding site to the central polypurine tract and another from the central polypurine tract to the 3′ polypurine tract (Suspène et al. 2006; Yu et al. 2004). Thus, there is a correlation between the most heavily deaminated regions of the viral genome and those regions of the minus-strand that remain single-stranded for the longest duration during reverse transcription.

Some authors have noted an increased C-to-T transition rate in the viral plus-strand as well, particularly with A3F (Albin et al. 2010a; Bishop et al. 2004; Liddament et al. 2004; Yu et al. 2004). It is not clear whether these represent a modest ability of A3F to deaminate RNA, RNA:DNA heteroduplex, or dsDNA templates or a more mundane phenomenon such as transient ssDNA states within the predominantly double-stranded context in which the plus-strand generally exists. Regardless, there is no evidence for A3G activity on any non-ssDNA substrate, including free nucleoside (Iwatani et al. 2006; Suspène et al. 2004).

While A3G is termed a "deaminase," it actually consists of two conserved deaminase motifs, N- and C-terminal. Between these, only the C-terminus is capable of deaminating ssDNA cytosine to uracil (Haché et al. 2005; Iwatani et al. 2006; Langlois et al. 2005; Navarro et al. 2005; Newman et al. 2005). The catalytically inactive N-terminus, in contrast, is more important for binding to single-stranded nucleic acid substrates, an activity that mediates RNA-dependent oligomerization and encapsidation and that may also help orient A3G for processive deamination

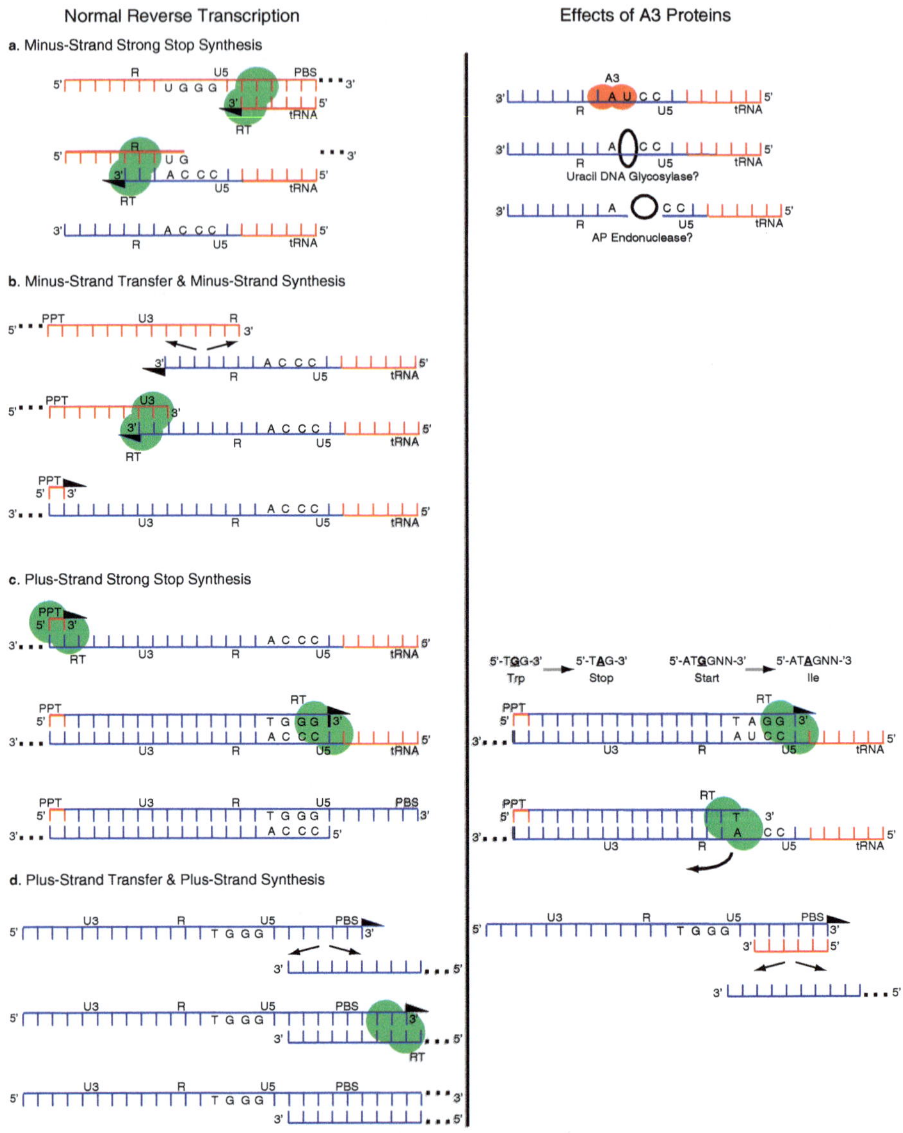


**Fig. 12.1** Normal reverse transcription is depicted in the *left* column, while points at which A3 proteins may impede reverse transcription by deaminase-dependent mechanisms are depicted at the *right*. References are distributed throughout the main text. (**a**) A3 deamination of a target cytosine to uracil creates a substrate on which a uracil DNA glycosylase may act (*blank oval*), creating an abasic site that may be further processed by an unspecified AP endonuclease to degrade viral cDNA (*blank circle*). (**b**) Steps unaffected per published mechanisms of A3 deaminase-dependent restriction. *Paired arrows* indicate annealing of the viral minus-strand to complementary viral genomic RNA. (**c**) Persistence of uracil in the viral minus-strand is a mutagenic event that, among many possibilities, may introduce premature stop codons or ablate initiating methionine codons. Alternatively, persistence of an abasic site subsequent to A3-dependent deamination and

(Huthoff et al. 2009 #1974; Huthoff and Malim 2007 #1965; Friew et al. 2009 #2128; Gooch and Cullen 2008 #2129; Li et al. 2011 #2127; Navarro et al. 2005 #528; Chelico et al. 2010 #2152). Although it has been proposed that monomeric A3G may also be capable of encapsidation, deamination, and restriction (Opi et al. 2006), recent evidence suggests the particular mutant employed in this study, while monomeric, may not actually encapsidate, raising the question of whether its apparent ability to restrict represents a specific effect on HIV or a nonspecific poisoning of producer cells [(Friew et al. 2009; Shlyakhtenko et al. 2011) and Mueller and Harris Labs unpublished data].

Deamination by A3G occurs in the $3' \rightarrow 5'$ direction (Chelico et al. 2006; Furukawa et al. 2009) and can proceed from 3 to at least 100 nucleotides along ssDNA substrates in the absence of any associated cofactors (Chelico et al. 2006). Although, as stated above, the C-terminal domain is required for actual catalysis, the inactive N-terminal deaminase domain is important for regulating the directionality and processivity of deamination (Chelico et al. 2010). Movement of the enzyme from one region of ssDNA to another may be achieved through some combination of random "jumping" motions to different segments (Chelico et al. 2006) and/or via intersegmental "transfers" in which the simultaneous binding of A3G to distinct ssDNA segments may facilitate its movement from one region to another (Nowarski et al. 2008).

The oligomeric form of A3G that mediates deamination in the context of infection is not entirely clear. Some in vitro data have implicated A3G monomers as the enzymatically active form (Nowarski et al. 2008), others dimers (Chelico et al. 2006; Shlyakhtenko et al. 2011). Atomic force microscopy studies that visualize A3G bound to ssDNA substrates have found a range of oligomeric forms, with a predominance of monomers and higher-order oligomers (Chelico et al. 2008, 2010), although more recent evidence derived from a refinement of this technique suggests a model in which unbound A3G, in agreement with the aforementioned prior studies, is predominantly monomeric, while bound A3G can take on a dimeric form with substantial components of monomers and higher-order oligomers also apparent (Shlyakhtenko et al. 2011). Oligomerization of A3G does not appear to be required for deaminase activity (Chelico et al. 2010), but the existence of these oligomeric forms may promote enzyme processivity (Chelico et al. 2008) or the ability of A3G to "jump" or "transfer" to new substrates (Feng and Chelico 2011; Nowarski et al. 2008).

**Fig. 12.1** (continued) processing by a uracil DNA glycosylase enzyme may impede synthesis of the plus-strand (loss of RT processivity indicated by a *curved arrow*) or, alternatively, result in the insertion of adenine across from the abasic site, yielding the same mutagenic event as minus-strand uracil persistence. (**d**) A3 deaminase-dependent aberrant RNaseH processing of the tRNA primer may prevent liberation of the viral plus-strand for transfer to the 5′ end of the genome and subsequent priming of additional plus-strand synthesis. *Paired arrows* indicate annealing of the viral plus-strand to complementary minus-strand cDNA

Different A3 proteins exhibit different dinucleotide preferences for deamination, with A3G preferring 5′-CC-3′ dinucleotides (deaminated nucleotide underlined, 5′-GG-3′ on the plus-strand), often within a broader 5′-CCCA-3′ context (5′-TGGG on the plus-strand) [e.g., (Beale et al. 2004; Harris et al. 2002, 2003a; Mangeat et al. 2003; Yu et al. 2004; Zhang et al. 2003) and Fig. 12.1c], and other A3 proteins including A3F typically showing varying degrees of preference for 5′-TC-3′ motifs (5′-GA-3′ on the plus-strand) [e.g., (Bishop et al. 2004; Hultquist et al. 2011; Liddament et al. 2004; Zheng et al. 2004) and reviewed in (Albin and Harris 2010)]. This complicates the attribution of hypermutation patterns to different A3 proteins. That is, while 5′-GG-3′ context mutations are most likely attributable to A3G since this is the sole A3 protein that displays a strong preference for this dinucleotide context, the 5′-GA-3′ context mutations could be introduced by any of several A3 proteins present and potentially active in primary T cells as well as by the myeloid-specific editing activity of A3A (Koning et al. 2009, 2011; Refsland et al. 2010; Thielen et al. 2010). Since diverse authors have described substantial amounts of both 5′-GG-3′ and 5′-GA-3′ context hypermutation in patient-derived viral sequences [reviewed in (Albin and Harris 2010)], the question of which A3 proteins contribute to the sequence diversification of HIV remains an outstanding and subtly complex issue in the field. Emergent data derived from gene targeting in naturally nonpermissive cells suggest, however, that A3F and A3D likely contribute to 5′-GA-3′ context hypermutation on top of the 5′-GG-3′ context hypermutation characteristic of A3G activity (Refsland et al. 2012).

The structural components of A3F and A3G that guide these context preferences have been mapped to loop sequences between the β4 strand and the α4 helix of active deaminase domains, which is thought to interact with ssDNA sequences 5′ of the target cytosine (Carpenter et al. 2010; Conticello 2008; Kohli et al. 2009, 2010; Langlois et al. 2005). The effect of these loop sequences may further be enhanced by a more N-terminal region corresponding to arginine residues along one of the predicted paths of ssDNA bound to the A3G C-terminus, although the exact placement of ssDNA on A3G remains controversial and, fundamentally, unknown (Carpenter et al. 2010; Chen et al. 2008; Holden et al. 2008). An additional study focusing on A3F and A3C further narrowed these determinants primarily to a single amino acid change within this same region (Langlois et al. 2005). Interestingly, these putative sequence recognition motifs are found adjacent to or extending into the conserved α4 helix where most known determinants of A3 susceptibility to Vif occur, with the most prominent single amino acid determinant of target sequence specificity in A3F also corresponding to the equivalent of A3G N-terminal amino acid D128, the best known of these determinants of Vif susceptibility (Albin et al. 2010b; Bogerd et al. 2004; Mangeat et al. 2004; Schröfelbauer et al. 2004; Smith and Pathak 2010; Xu et al. 2004; Zhen et al. 2010). Thus, while the ability of different deaminase domains among diverse A3 proteins to mediate distinct processes – Vif binding, RNA binding, sequence recognition, etc. – implies that the structure of each will be somewhat different, determining the structure of the C-terminus of A3F may be particularly informative for understanding how this critical region influences the antiviral properties of A3 proteins and may explain why some authors have noted a direct effect of Vif on the deaminase activity of A3G (Britan-Rosich et al. 2011; Santa-Marta et al. 2005).

On the other side of this interaction, a study utilizing nucleoside analogs as molecular probes for the DNA determinants recognized by A3G during the deamination process has suggested several important features of the target deamination hotspot itself beyond the aforementioned sequence specificity observations (Rausch et al. 2009). Insertion of an abasic site and alteration of the 2′ deoxyribose within a deamination hotspot are both potent inhibitors of deamination at a target cytosine, and ring positions 3 or 4 of the nucleotide base 5′ of the target cytosine may be particularly important for hydrogen bonding with A3G residues involved in substrate recognition. These data are consistent with the exquisite specificity of A3G for DNA substrates as well as the importance of active site aromatic residues for enzyme function (see below).

To date, seven structures of the catalytically active C-terminus of A3G have been solved by either NMR or X-ray crystallography (Chen et al. 2008; Furukawa et al. 2009; Harjes et al. 2009; Holden et al. 2008; Li et al. 2012; Shandilya et al. 2010). These structures show that A3G has a concave active site with a floor containing critical hydrophobic residues likely involved in interaction with substrate bases, as also suggested by the above nucleoside analog studies. Surrounding the outer edges above this concave active site are a number of positively charged residues that may orient ssDNA within the active site, although the exact path taken by ssDNA remains a point of controversy with competing "brim" (Chen et al. 2008) and "kink" (Holden et al. 2008) interaction models proposing somewhat different DNA-binding paths. Barring a co-crystal structure, ideally including the full-length enzyme, this debate may be difficult to resolve, since molecular dynamics simulations have suggested that the DNA-binding loops are flexible and that substrate binding may induce a conformational change in A3G (Autore et al. 2010).

### *12.2.2 cDNA Degradation*

In addition to the mutagenic implications of A3G action, early models of restriction proposed that A3G might trigger the degradation of viral cDNA through a base excision repair pathway [(Harris et al. 2003a, b; Mangeat et al. 2003; Zhang et al. 2003) and Fig. 12.1a]. Potential degradation was considered an important component of any explanatory model because prior observations had indicated that decreases in cDNA accumulation to roughly half of wild type among early reverse transcription products with still more potent effects on late reverse transcription products were the primary phenotypic differences between Vif-deficient and Vif-proficient viruses (Sova and Volsky 1993; von Schwedler et al. 1993). In this model, the presence of DNA uracils introduced by A3G would cause a uracil DNA glycosylase enzyme to excise the offending uracil, yielding an abasic site that would then be a target for cleavage by an AP endonuclease (Harris et al. 2003a, b; Mangeat et al. 2003; Zhang et al. 2003). The first critical step in this proposed degradative pathway, uracil excision, is easily envisaged given the fact that UNG2 is packaged into HIV virions, and there is some evidence that this is important for the ability of HIV to repair uracil-containing cDNA (Chen et al. 2004; Mansky et al. 2000; Priet et al. 2005). Indeed, sequencing of proviruses with identical integration sites but different subsets of G-to-A mutations suggests that some uracilated minus-strand cDNAs do undergo repair, although the timing of this is unknown (Yu et al. 2004).

In contrast, others have found that the presence of UNG2 in virions may be detrimental to virus infectivity, presumably via cDNA degradation at the same abasic sites thought to lead to DNA repair in the aforementioned studies (Schröfelbauer et al. 2005; Yang et al. 2007a). This view is not universally held, as others have reported that the presence of UNG2 or another uracil DNA glycosylase, SMUG1, is simply irrelevant to viral infectivity (Kaiser and Emerman 2006; Langlois and Neuberger 2008; Mbisa et al. 2007; Schumacher et al. 2008). It has even been proposed, somewhat counterintuitively, that heavy uracilation of viral cDNA may actually represent an adaptive strategy by which HIV can prevent autointegration (Yan et al. 2011).

Setting aside the role of UNG2 in determining the fate of uracilated cDNA for a moment, a separate but related question is what endonuclease(s) might cause the ultimate degradation of cDNA containing abasic sites as proposed. APE1 has been found to be packaged in virions along with UNG2 in a scenario perfectly supporting the base excision repair model of cDNA degradation as originally proposed, but we are unaware of any additional reports confirming this effect (Yang et al. 2007a). It is not clear that an active degradation mechanism would be necessary to result in decreased cDNA accumulation, however, as the mere existence of a number of abasic sites in the minus-strand strand may simply impede the ability of reverse transcriptase to synthesize the plus-strand (Fig. 12.1c). Reverse transcriptase is capable of inserting dAMP across from an abasic site, a phenomenon that may further be stimulated by Vif (Cancio et al. 2004), but this process appears to also reduce enzyme processivity (Cai et al. 1993). Thus, while it is clear that a fundamental aspect of A3-dependent restriction is the decreased accumulation of cDNA, whether this is the result of cDNA degradation, deaminase-dependent decreases in reverse transcript synthesis, or an alternative mechanism such as one of the deaminase-independent models to be discussed below remains an open and interesting question.

### *12.2.3 Summary of Deaminase-Dependent Reverse Transcription Mechanisms*

The deaminase-dependent mutation of the HIV genome is a widely accepted mechanism by which A3G can inhibit the virus. The picture surrounding cDNA degradation, however, is somewhat less clear. It seems the crux of the argument is whether an abasic site is generated during reverse transcription since reverse transcriptase should not be able to distinguish cDNA uracils from thymines (Klarmann et al. 2003).

The ssDNA specificity of A3 proteins has been taken to imply that the substrate for uracil DNA glycosylase activity should also be ssDNA. This ssDNA limitation, in turn, would narrow the available uracil DNA glycosylase enzymes to those with specificity for single-stranded DNA, namely, UNG1/2 and SMUG1 in human cells, and as discussed above, there is little consensus on the involvement of these two enzymes in retroviral restriction. It is important to remember, however, that ssDNA, like ssRNA, is capable of taking on secondary and tertiary structures, sometimes with thoroughly impressive effects [e.g., (Breaker 2004)]. Such cDNA folded back on itself, based on the extensive structure of the ssRNA regions of the viral genome first reverse transcribed [e.g., (Watts et al. 2009)], would likely represent an appropriate template for alternative uracil DNA glycosylases with double-stranded specificity such as TDG or MBD4 (Krokan et al. 2002) without invoking the specter of unidentified enzymes, a particularly intriguing possibility given the interaction of MBD4 with relatives of the A3 enzymes (Rai et al. 2008). The assumption of single-strandedness as depicted in Fig. 12.1a and in many similar figures in prior reviews may therefore need to be revisited.

If the abasic site forms, it is a short leap to cleavage of the phosphodiester backbone, if such cleavage would in fact be necessary versus simple stalling of reverse transcriptase (Fig. 12.1c). This could occur via enzymatic means as initially proposed, via APE1 or an alternative enzyme. Spontaneous cleavage may also be a viable scenario, but it is, to date, an untested alternative hypothesis. For example, the deamination reaction itself releases ammonia, which may hypothetically create a locally basic environment in the viral core favoring destabilization of an abasic site. One could also imagine the apparent cDNA degradation observed experimentally as an artifact caused by sensitivity of an abasic site to the heat denaturation typical of the PCR procedures that have been utilized universally, to our knowledge, to amplify viral reverse transcripts for further analysis [e.g., (Borman et al. 1995; Courcoul et al. 1995; Goncalves et al. 1996; Simon and Malim 1996; Sova and Volsky 1993; von Schwedler et al. 1993)]. While we acknowledge that alternative routes to cDNA degradation are highly speculative, we propose that, given the lack of consensus on the exact mechanism by which decreased cDNA accumulation occurs, it may be time to start looking a bit farther outside the box for an answer to this important mechanistic question.

## 12.3 Deaminase-Independent Restriction Mechanisms

While the first papers describing the mechanism of A3G action focused on the implications of deamination for HIV restriction, multiple labs soon thereafter began to note that deaminase-independent mechanisms of restriction might also be at play. This was based on the simple observation that overexpression of A3G containing mutations of conserved residues critical for deaminase activity, such as the zinc coordinating residues and the catalytic glutamate of the conserved HxE/PCxxC motif, might not fully ablate restriction (Bishop et al. 2006; Newman et al. 2005; Shindo et al. 2003). Although at odds with identical experiments published in the initial descriptions of A3G's deaminase-dependent restriction activities

(Mangeat et al. 2003; Zhang et al. 2003), this has grown to become a point of contention within the field. The details surrounding these proposed deaminase-independent mechanisms fall into several categories, but these studies are united in the suggestion that the decreased accumulation of reverse transcripts characteristic of the nonpermissive phenotype is not due to cDNA degradation but rather to the inhibition of cDNA synthesis in a manner not dependent on the introduction of uracils or abasic sites into viral cDNA.

### *12.3.1 Physical Obstruction of Reverse Transcriptase*

One mechanism by which deaminase-independent restriction may occur is the simple obstruction of reverse transcriptase during cDNA synthesis as depicted in Fig. 12.2a (Bishop et al. 2008; Holmes et al. 2007; Iwatani et al. 2007). Per this model, A3G accumulates on viral genomic RNA and impedes the processivity of reverse transcriptase such that the reaction eventually falters. Several lines of evidence support this mechanism. First, A3G has RNA-binding activity (Chelico et al. 2006; Iwatani et al. 2006; Yu et al. 2004). Second, in vitro reverse transcription reactions using A3G or a catalytically inactive A3G mutant inhibit the elongation of reverse transcriptase products (Iwatani et al. 2007). Finally, the length of minus-strand strong-stop cDNA derived from endogenous reverse transcription reactions in particles – and thus presumably independent of any downstream catabolic pathway – inversely correlates with the amount of A3G or A3G catalytic mutant present (Bishop et al. 2008).

### *12.3.2 Inhibition of tRNA Priming of Reverse Transcription*

Another potential deaminase-independent mechanism of inhibition proposed is the inhibition of tRNA priming during reverse transcription [(Guo et al. 2006; Yang et al. 2007b) and Fig. 12.2a]. Most of these observations were made using wild-type A3G or A3F. To extend the observation of decreased reverse transcription to the lack of a deamination requirement, however, these authors employed deletion mutants as their inactive deaminases of choice. Despite the obvious implication that this priming deficiency may be due to A3G's RNA-binding activity, either via competition for the primer binding site or perhaps sequestration of the primer itself, this effect has specifically been proposed to be a result of the interaction between A3G and the HIV nucleocapsid protein (Guo et al. 2007). One potential problem with this particular study is that seeing the in vitro reverse transcription effect requires A3G:NC ratios 10 times greater than those found in naturally restricted virions, but the authors propose that the actual A3G:NC ratios present during the critical tRNA annealing step may be more in line with the enhanced relative levels of A3G they used.

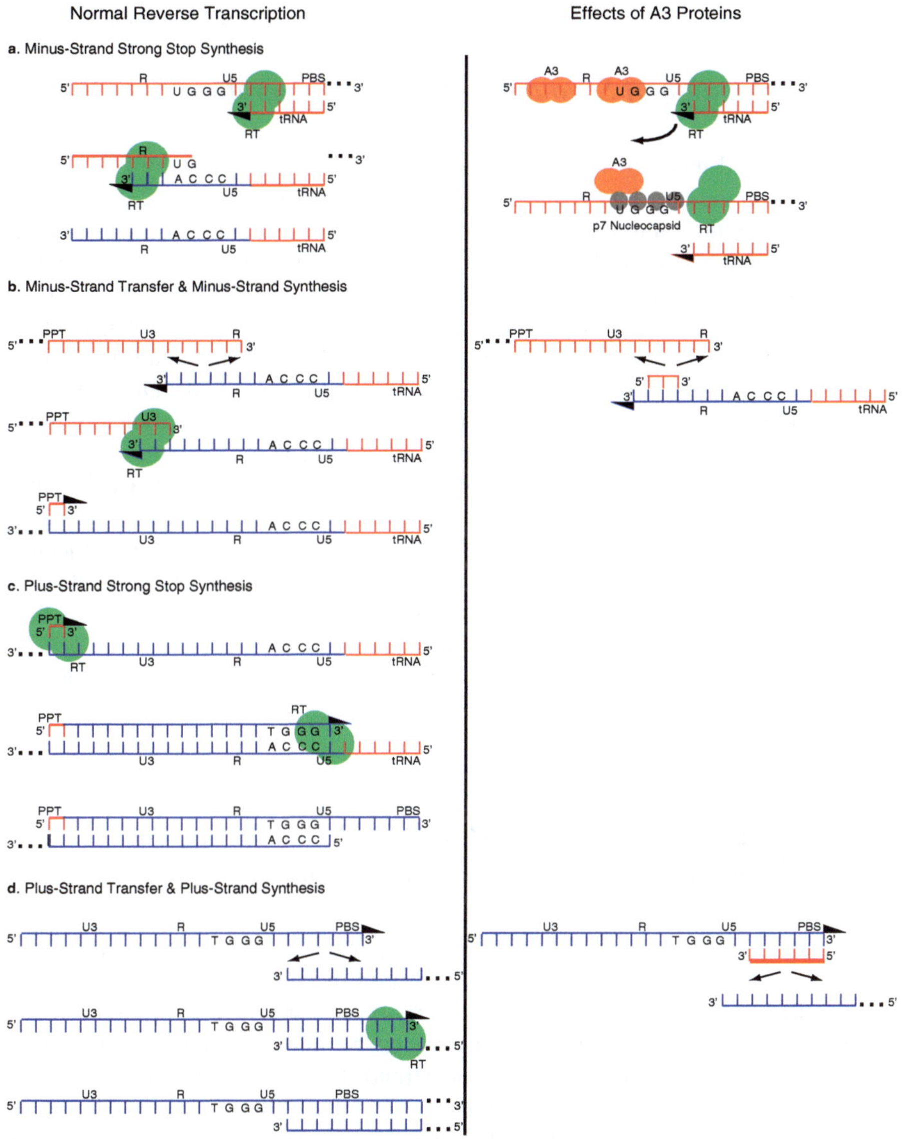


**Fig. 12.2** Normal reverse transcription is depicted in the *left* column, while points at which A3 proteins may impede reverse transcription by deaminase-independent mechanisms are depicted at the *right*. References are distributed throughout the main text. (**a**) Binding of A3 proteins to genomic RNA may impede the processivity of reverse transcriptase, preventing the completion of cDNA synthesis (loss of RT processivity indicated by a *curved arrow*). Alternatively, the reverse transcription reaction may fail to prime in a manner dependent on A3 interaction with the viral nucleocapsid protein (*gray circles*). (**b**) A3 deaminase-independent aberrant RNaseH processing of the template genomic RNA may fail to liberate the minus-strand strong-stop cDNA to transfer to the 3′ end of the genome and prime additional minus-strand synthesis. *Paired arrows* indicate annealing of the viral minus-strand to complementary viral genomic RNA. (**c**) Steps unaffected per published mechanisms of A3 deaminase-independent restriction. (**d**) A3-dependent aberrant RNaseH processing of the tRNA primer may fail to liberate the plus-strand for transfer to the 5′ end of the genome and subsequent priming of additional plus-strand synthesis. *Paired arrows* indicate annealing of the viral plus-strand to complementary minus-strand cDNA

### *12.3.3 Inhibition of Strand Transfer and Integration*

A third proposed mechanism of deaminase-independent inhibition by A3G is the inhibition of strand transfer events during reverse transcription (Fig. 12.2b, d). One paper describes the inhibition of both minus- and plus-strand transfers due to a deaminase-independent mechanism of A3G (Li et al. 2007), perhaps caused by the inhibition of RNaseH processing via A3G binding to the DNA:RNA heteroduplex, although A3G appears to bind such heteroduplexes relatively weakly (Iwatani et al. 2006; Yu et al. 2004). Another finds an effect primarily on plus-strand transfer in a deaminase-dependent fashion [(Mbisa et al. 2007) and Fig. 12.1d]. Discrepancies may be due in part to the different mutants used to render A3G catalytically inactive – truncation mutants in the former and point mutants in the latter. Similarly, there have been reports of an effect of A3G on HIV integration, but whether this process is deaminase-dependent (Mbisa et al. 2007) or, at least partially, deaminase-independent (Luo et al. 2007; Mbisa et al. 2010) is unclear. One interesting mechanistic note contained within one of these reports, however, is the finding that A3G appears to interfere with cleavage of the tRNA primer, creating aberrant ends that may contribute to the downstream inhibition of integration (Mbisa et al. 2007). Exactly how this may occur is unknown, but it echoes speculation on protection of the DNA:RNA heteroduplex (Li et al. 2007).

### *12.3.4 Discrepancies Among Proposed Deaminase-Independent Mechanisms*

Interpretations of deaminase-independent mechanisms of reverse transcription are as complicated as those of the cDNA degradation mechanisms above. One of the primary drawbacks of models focused on early reverse transcription products such as a physical blockage of reverse transcriptase or the inhibition of reverse transcriptase priming is that the inhibition of minus-strand strong-stop cDNA is not universally observed under nonpermissive conditions (Li et al. 2004; von Schwedler et al. 1993) and when it is observed, this effect may be overcome by priming with DNA rather than the natural $tRNA_{lys}3$ or by removing nucleocapsid from the reaction (Guo et al. 2006, 2007; Li et al. 2007; Yang et al. 2007b). Moreover, even if one finds a defect in early RT products, a specific effect on tRNA priming may not be apparent either (Bishop et al. 2008; Iwatani et al. 2007).

Rather, all authors agree that, whether early reverse transcription is impeded or not, late reverse transcription products are decreased. This may be consistent with those studies proposing effects on strand transfer. Alternatively, this could reflect a cumulative effect of deaminase-independent restriction mechanisms – at multiple steps, in multiple combinations with each other – or any mix of the above. It has even been proposed that A3G may interact directly with reverse transcriptase to mediate the inhibition of proviral cDNA synthesis (Wang et al. 2012) and, similarly, that Vif itself is a component of reverse transcription complexes that directly facilitates cDNA synthesis (Carr et al. 2006, 2008).

### *12.3.5 Summary of Deaminase-Independent Restriction Mechanisms*

A large body of evidence now supports a variety of deaminase-independent mechanisms of proviral DNA synthesis inhibition. That said, significant discrepancies remain in this literature as in the putative degradation of viral cDNA. The sheer diversity of the deaminase-independent restriction mechanisms described above could merely reflect the fact that RNA binding may be a critical component in all reactions. Alternatively, caution is generally appropriate when interpreting broadly positive results, as they may be explained by a pervasive artifact.

Apropos of pervasive artifacts, several papers now suggest that deaminase-independent mechanisms may be overexpression artifacts (Browne et al. 2009; Miyagi et al. 2007, 2010; Schumacher et al. 2008). That is, proponents of deaminase-independent restriction mechanisms universally rely on the transient overexpression of catalytic mutants to make their core functional arguments. When one stably expresses these mutants or titrates back the quantities transiently expressed, however, deaminase-independent restriction is largely absent. As of this writing, there are no published reports of the use of a stably expressed catalytic mutant in combination with a more reductionist approach such as the endogenous reverse transcription reaction.

The questionable status of deaminase-independent restriction by A3G does not, however, negate the possibility that deaminase-independent mechanisms may be more prominent for other A3 proteins such as A3F (Holmes et al. 2007) or that the induction of A3 proteins might theoretically raise expression levels to the range of deaminase-independent relevance (Hultquist et al. 2011; Koning et al. 2009; Refsland et al. 2010). Moreover, despite skepticism about the veracity of deaminase-independent restriction claims, there is arguably more evidence supporting the mechanistic details of how this may occur than there is for the mechanistic details surrounding cDNA degradation. As posited above, however, this could merely be a consequence of faulty assumptions about how decreased cDNA accumulation may occur.

## 12.4 G-to-A Mutations In Vitro and In Vivo

One of the most compelling pieces of circumstantial evidence supporting deamination as a primary determinant of A3G action is that it has been clear since long before the discovery of A3G that retroviruses, including HIV, can at times undergo a curious process referred to as hypermutation. In these situations, large numbers of G-to-A plus-strand transitions are apparent in a subpopulation of viral sequences (Janini et al. 2001; Pathak and Temin 1990; Vartanian et al. 1991, 1994). Thus, controversies pertaining to the potential role of deamination in reduced cDNA accumulation aside, it is widely accepted that the deaminase activity of A3 proteins does affect the HIV mutation rate.

### 12.4.1 A3 Proteins and In Vivo Hypermutation

While perhaps not entirely appropriate due to the clear in vitro effect of A3 proteins on cDNA accumulation, hypermutation is often taken as a surrogate for A3 activity in the evaluation of the potential clinical impact of A3 proteins with an eye toward their use as therapeutics. This approach has yielded a mixed picture for the potential efficacy of A3 proteins in vivo. A number of publications support a positive correlation between hypermutation levels in patient sequences and reduced viremia and/or increased CD4+ T-cell counts (Land et al. 2008; Pace et al. 2006; Vazquez-Perez et al. 2009), but an equal number fail to find any association (Gandhi et al. 2008; Piantadosi et al. 2009; Ulenga et al. 2008a). Aside from the cDNA accumulation confounder, it is possible that this lack of clarity may further suffer from the fact that Vif is presumably largely functional in these patient-derived isolates. Thus, its ongoing suppression of A3 proteins may be sufficient to lower their apparent activity to a level difficult to assess. Unfortunately, the situation has turned out much the same when assessing A3 mRNA levels [(Amoedo et al. 2011; Biasin et al. 2007; Cho et al. 2006; Jin et al. 2005; Reddy et al. 2010; Ulenga et al. 2008b; Vazquez-Perez et al. 2009) and reviewed in more detail in (Albin and Harris 2010)]. Among recent positive data, however, an analysis of the effects of interferon induction in HIV/HCV coinfected human patients has shown a modest correlation between decreased viral load and the upregulation of A3F, but not A3G (Pillai et al. 2012).

The search for A3 polymorphisms of functional significance has achieved similarly modest success. For example, polymorphisms in A3H determine stability and restriction activity (OhAinle et al. 2008; Tan et al. 2009). Another common variation in the *A3* locus is deletion of A3B; homozygosity for this deletion has previously been associated with enhanced susceptibility to HIV infection and a higher viral set point (An et al. 2009), but an additional study failed to find this effect in a separate cohort (Itaya et al. 2010). As with other areas of A3 research, however, the most intensive efforts to date have been directed toward the search for polymorphisms in A3G, and although many have been found, the potential clinical implications are generally unclear [reviewed in (Albin and Harris 2010)]. It has been suggested that one variant, A3G H186R, may be associated with an unfavorable course of infection (An et al. 2004; Reddy et al. 2010), but a smaller study focused on a population where the allele is less prevalent failed to observe these effects (Do et al. 2005). Nevertheless, the H186R variant of A3G is worthy of further in vitro characterization in light of the recent finding that this change results in decreased processivity of the enzyme along ssDNA substrates and consequently decreased deaminase activity (Feng and Chelico 2011).

In summary, the effect of A3 proteins on overall viral replication in vivo is a mixed picture. While it would be traditional to simply state the need for further research at this point, we would instead advocate for this line of inquiry a change in approach. Using hypermutation exclusively as a readout, for example, is clearly confounded. The use of mRNA rather than protein levels as a marker of A3 expression is a necessary evil, but this too is likely limited further by the failure to consider the full range of A3 proteins that may be involved in restriction besides A3G, such as A3F, A3H, A3D, and perhaps others [e.g., (Berger et al. 2010, 2011b; Bishop et al. 2004; Dang et al. 2006; Hultquist et al. 2011; Liddament et al. 2004; OhAinle et al. 2008; Peng et al. 2007; Tan et al. 2009; Wiegand et al. 2004; Zheng et al. 2004)]. Admittedly, doing a comprehensive clinical study that considers the *A3* locus as a whole, either via polymorphic content or overall expression levels, would be a massive undertaking, and the current lack of antibodies sufficient for protein expression analysis of the entire human A3 repertoire leaves any analysis of A3 expression wanting. With respect to the potential effects of *A3* polymorphisms, however, advances in deep sequencing technology have made genotyping of the full *A3* locus in populations of interest such as highly exposed seronegative individuals feasible. Such a study may be the best way forward in addressing whether, which, and how A3 proteins act in vivo.

### *12.4.2 Hypermutation as an Adaptive Strategy*

A hallmark of HIV is its high mutation rate (Mansky and Temin 1995), yet this mutation rate likely represents an equilibrium selected by the virus through evolution to optimize host immune evasion without collapsing into error catastrophe (Harris 2008; Loeb et al. 1999). In this respect, A3 proteins are of interest not only for their ability to damage viral genomic integrity but also for their potential ability to change the pace at which the virus may evolve. For example, several authors have written on the theoretical ability of A3 proteins to introduce mutations into the viral genome that may confer resistance to antiviral drugs, promote immune evasion, or even facilitate coreceptor switching (Berkhout and de Ronde 2004; Bewick et al. 2011; Haché et al. 2006; Pillai et al. 2008).

The levels of mutation characteristic of hypermutation, on the face of it, would seem far too high to be of benefit to HIV. Yet current evidence suggests that there are multiple mechanisms by which HIV may harness the mutagenic power of A3 proteins without suffering fatal consequences. The first of these centers on the concept of the fitness of a viral population versus that of any individual genome. If a particular viral genome encodes a Vif protein that is hypofunctional in its ability to counteract A3 proteins, an event that appears to happen at least 20 % of the time

(Albin et al. 2010b; Simon et al. 2005), processive deamination will introduce a large number of G-to-A mutations into the progeny of that viral genome. Frequently, these may be premature stop codons or mutations ablating the start codon of a given gene – consider the A3G target sequence preference and the codons for tryptophan (5′-UGG-3′) and methionine (5′-AUG-3′), for example (Fig. 12.1c). Yet recombination can allow the virus to resurrect hypermutated sequences into a functional genomic background (Mulder et al. 2008). Thus, even when hypermutation is at full strength, the virus can utilize the expanded sequence space afforded by exposure to A3 proteins to acquire adaptive mutations.

Even when heavily hit by hypermutation, the virus retains at least two additional ways of averting certain death. One is purely stochastic in that it is possible for a gene to suffer heavy levels of hypermutation but still remain functional (Harris et al. 2003a). A second is so-called purifying selection, a process by which the viral genomes represented at each stage of replication from integrated provirus through packaged genome are observed to be progressively less mutated, presumably due to loss of function at various stages of replication – for example, mutations in the viral U3 promoter may select against the transcription of heavily mutated proviruses versus relatively intact counterparts (Russell et al. 2009).

Given the diverse mechanisms by which HIV may turn hypermutation to its own advantage, it may under certain conditions actually be beneficial for the virus to acquire mutations causing defects in Vif. For example, the presence of the hypofunctional Vif mutant K22H has been correlated with patients experiencing antiretroviral treatment failure, and these patients further display an enrichment for drug resistance mutations in a context consistent with A3G action (Fourati et al. 2010). While this would seemingly suggest that the potentiation of A3 proteins as a therapeutic strategy may be dangerous, others have found that, despite the potential contribution of A3G and other A3 proteins to drug resistance mutations, this effect is relatively modest and not necessarily a reason to avoid the therapeutic use of A3 proteins (Jern et al. 2009).

It may also be possible for HIV to acquire sublethal levels of mutation from A3 proteins (Sadler et al. 2010), presumably through some residual encapsidation despite the presence of Vif, whether by chance or design. This is echoed by a similar story in which cytotoxic T-cell escape mutations consistent with A3G action were found early in infection, implying a role for A3G in providing the sequence diversity on which selection may act from even the earliest stages of infection (Wood et al. 2009). Indeed, even a Vif-proficient virus passaged in the presence of A3G can acquire a drug resistance mutation in a favorable dinucleotide context much more rapidly than the virus itself (Kim et al. 2010). It is thus at least theoretically possible that the virus could benefit from the presence of A3G at some level. Here, it is also important to reiterate the distinction between a viral population and an individual viral genome, as debilitated viruses may suffer mutation rates several times greater than their wild-type counterparts without necessarily impacting the health of the overall titer of the population as a whole [e.g., (Haché et al. 2008)].

## 12.5 Summary and Emerging Themes

Targeting proviral synthesis has been a successful strategy for the pharmaceutical inhibition of viral spread, as the very existence of this compendium would suggest. It is perhaps not all that surprising, then, that evolution should have found the same viral vulnerability long before science.

A3 proteins are, to date, the best characterized natural mechanism by which cells can inhibit proviral DNA synthesis. It is clear, however, that our understanding of this process is far from complete. Although not all authors are in agreement on the exact stage of replication inhibited, there are several stories outside of those described above that bear some similarity to the story of Vif and the discovery of A3 proteins, i.e., some particular cellular environment prevents successful reverse transcription. Wild-type HIV suffers incomplete reverse transcription and/or an integration block in resting T cells, which is suppressible by target cell activation (Stevenson et al. 1990; Zack et al. 1990). Certain cell types display a Nef-suppressible defect in reverse transcription (Aiken and Trono 1995; Schwartz et al. 1995). Capsid protein variation and capsid binding of cyclophilin A protect HIV from restriction factor TRIM5α, which otherwise causes a postentry restriction occurring before or during early reverse transcription (Sayah et al. 2004; Stremlau et al. 2004; Towers et al. 2003). Finally, the viral accessory protein Vpx in HIV-2 and certain SIV species facilitates successful reverse transcription and/or nuclear import of the preintegration complex (Fletcher et al. 1996; Goujon et al. 2007; Kaushik et al. 2009).

The last of these, the Vpx-suppressible myeloid block to infection, has become a hot topic of late due to the discovery of SAMHD1 as a Vpx-suppressible factor in myeloid cells that restricts lentiviral infection (Berger et al. 2011a; Hrecka et al. 2011; Laguette et al. 2011). It has been postulated that the apparent ability of SAMHD1 to suppress reverse transcription is related to its triphosphohydrolase activity, which may decrease dNTP concentrations to levels incompatible with successful reverse transcription (Goldstone et al. 2011; Powell et al. 2011).

Despite this recent progress, a number of questions remain as to the exact mechanism by which SAMHD1 may act to restrict lentiviral infection in myeloid cells. For example, core-associated Vpx (Kewalramani and Emerman 1996) must somehow encounter nuclear SAMHD1 (Rice et al. 2009) in order to degrade the restriction factor and allow reverse transcription in the cytoplasmic viral core to proceed. Moreover, this must happen quite quickly to allow dNTP levels to reequilibrate, since the late reverse transcription products of a Vpx-encoding virus in macrophages peak between 6 and 24 h post-infection [e.g., (Gramberg et al. 2010; Sunseri et al. 2011)]. This degradation mechanism must also be exceedingly robust if a limited quantity of core-associated Vpx is to counteract continuously synthesized, interferon-regulable, and enzymatically active SAMHD1. Since the dNTP levels that permit efficient reverse transcription by HIV-1 reverse transcriptase in macrophages are essentially equivalent to those found in "nonpermissive" THP-1 cells (Diamond et al. 2004; Goldstone et al. 2011), one must also ask how low intracellular dNTP levels must drop to define "permissive" versus "nonpermissive" conditions and whether SAMHD1 can actually achieve these levels given its allosteric substrate regulation by dGTP (Goldstone et al. 2011; Powell et al. 2011). At a more fundamental level, there is no universal agreement on the more basic question of when the myeloid cell block to restriction occurs, with some authors implicating nuclear import over reverse transcription (e.g., (Fletcher et al. 1996; Neil et al. 2001)). It remains possible that other factors may also contribute to the myeloid cell block, with several studies implicating A3A as the effector arm of this phenomenon (Berger et al. 2010, 2011b; Peng et al. 2007), consistent with its role in the clearance of foreign DNA (Stenglein et al. 2010) as well as its ability to mutate incoming viral cDNA in myeloid cells (Koning et al. 2011; Thielen et al. 2010).

One might also consider the possibility that these discrepancies as well as the recurrence of the theme of incomplete reverse transcription under diverse conditions of infection might point to a deeper, undiscovered truth. To make an analogy to the prototypical cellular regulator of viral replication, students are not taught to think of eIF-2 and RNaseL as lone gunmen of viral restriction. Rather, these are the end points of a vast cellular conspiracy that starts at interferon and cascades through diverse intermediates to activate the manifestation of restriction by the cell's weapons of choice. Recently, it has been found that the restriction factor TRIM5α acts to stimulate just such an alteration of the broader cellular environment (Pertel et al. 2011). We speculate that further study will eventually point not to the diversity of cellular restrictions of exogenous viral reverse transcription but rather to a broader network regulating endogenous viral reverse transcription in the cell. Given that the most commonly encountered retrovirus in the cell is the endogenous retroelement, it is likely that constitutive systems act to control reverse transcription for the sake of genome stability and of whatever purpose is apparently served by having nearly half of one's genome consist of viral information. Such a function has been attributed, for example, to TREX1, the absence of which, like SAMHD1, can cause Aicardi-Goutières syndrome (Stetson et al. 2008). We would propose, therefore, that the next frontier in the study of A3 proteins and of other restriction factors lies at the hypothetical upstream points of regulation that determine whether a cell is ultimately permissive or nonpermissive for proviral DNA synthesis.

**Acknowledgments** This work was funded by the National Institutes of Health (R01 AI064046 and P01 GM091743 to RSH). JSA was supported in part by the National Institute on Drug Abuse and by the University of Minnesota Medical Scientist Training Program (F30 DA026310 and T32 GM008244, respectively).

## References

Aiken C, Trono D (1995) Nef stimulates human immunodeficiency virus type 1 proviral DNA synthesis. J Virol 69(8):5048–5056

Albin JS, Harris RS (2010) Interactions of host APOBEC3 restriction factors with HIV-1 in vivo: implications for therapeutics. Expert Rev Mol Med 12(e4):1–26

Albin JS, Haché G, Hultquist JF, Brown WL, Harris RS (2010a) Long-term restriction by APOBEC3F selects human immunodeficiency virus type 1 variants with restored Vif function. J Virol 84(19):10209–10219

Albin JS, LaRue RS, Weaver JA, Brown WL, Shindo K, Harjes E, Harris RS (2010b) A single amino acid in human APOBEC3F alters susceptibility to HIV-1 Vif. J Biol Chem 285(52): 40785–40792

Alce TM, Popik W (2004) APOBEC3G is incorporated into virus-like particles by a direct interaction with HIV-1 Gag nucleocapsid protein. J Biol Chem 279(33):34083–34086

Amoedo ND, Afonso AO, Cunha SM, Oliveira RH, Machado ES, Soares MA (2011) Expression of APOBEC3G/3F and G-to-A hypermutation levels in HIV-1-Infected children with different profiles of disease progression. PLoS One 6(8):e24118

An P, Bleiber G, Duggal P, Nelson G, May M, Mangeat B et al (2004) APOBEC3G genetic variants and their influence on the progression to AIDS. J Virol 78(20):11070–11076

An P, Johnson R, Phair J, Kirk GD, Yu XF, Donfield S, Winkler CA (2009) APOBEC3B deletion and risk of HIV-1 acquisition. J Infect Dis 200(7):1054–1058

Autore F, Bergeron JR, Malim MH, Fraternali F, Huthoff H (2010) Rationalisation of the differences between APOBEC3G structures from crystallography and NMR studies by molecular dynamics simulations. PLoS One 5(7):e11515

Beale RC, Petersen-Mahrt SK, Watt IN, Harris RS, Rada C, Neuberger MS (2004) Comparison of the differential context-dependence of DNA deamination by APOBEC enzymes: correlation with mutation spectra in vivo. J Mol Biol 337(3):585–596

Berger A, Munk C, Schweizer M, Cichutek K, Schule S, Flory E (2010) Interaction of Vpx and apolipoprotein B mRNA-editing catalytic polypeptide 3 family member A (APOBEC3A) correlates with efficient lentivirus infection of monocytes. J Biol Chem 285(16): 12248–12254

Berger A, Sommer AF, Zwarg J, Hamdorf M, Welzel K, Esly N et al (2011a) SAMHD1-Deficient CD14+ Cells from individuals with Aicardi-Goutières syndrome are highly susceptible to HIV-1 infection. PLoS Pathog 7(12):e1002425

Berger G, Durand S, Fargier G, Nguyen XN, Cordeil S, Bouaziz S et al (2011b) APOBEC3A is a specific inhibitor of the early phases of HIV-1 infection in myeloid cells. PLoS Pathog 7(9):e1002221

Berkhout B, de Ronde A (2004) APOBEC3G versus reverse transcriptase in the generation of HIV-1 drug-resistance mutations. AIDS 18(13):1861–1863

Bewick S, Wu J, Lenaghan SC, Yang R, Zhang M, Hamel W (2011) The R5 to X4 coreceptor switch: a dead-end path, or a strategic maneuver? Lessons from a game theoretic analysis. Bull Math Biol 73(10):2339–2356

Biasin M, Piacentini L, Lo Caputo S, Kanari Y, Magri G, Trabattoni D et al (2007) Apolipoprotein B mRNA-editing enzyme, catalytic polypeptide-like 3G: a possible role in the resistance to HIV of HIV-exposed seronegative individuals. J Infect Dis 195(7):960–964

Bishop KN, Holmes RK, Sheehy AM, Davidson NO, Cho SJ, Malim MH (2004) Cytidine deamination of retroviral DNA by diverse APOBEC proteins. Curr Biol 14(15):1392–1396

Bishop KN, Holmes RK, Malim MH (2006) Antiviral potency of APOBEC proteins does not correlate with cytidine deamination. J Virol 80(17):8450–8458

Bishop KN, Verma M, Kim EY, Wolinsky SM, Malim MH (2008) APOBEC3G inhibits elongation of HIV-1 reverse transcripts. PLoS Pathog 4(12):e1000231

Bogerd HP, Cullen BR (2008) Single-stranded RNA facilitates nucleocapsid: APOBEC3G complex formation. RNA 14(6):1228–1236

Bogerd HP, Doehle BP, Wiegand HL, Cullen BR (2004) A single amino acid difference in the host APOBEC3G protein controls the primate species specificity of HIV type 1 virion infectivity factor. Proc Natl Acad Sci USA 101(11):3770–3774

Bogerd HP, Wiegand HL, Hulme AE, Garcia-Perez JL, O'Shea KS, Moran JV, Cullen BR (2006) Cellular inhibitors of long interspersed element 1 and Alu retrotransposition. Proc Natl Acad Sci USA 103(23):8780–8785

Borman AM, Quillent C, Charneau P, Dauguet C, Clavel F (1995) Human immunodeficiency virus type 1 Vif- mutant particles from restrictive cells: role of Vif in correct particle assembly and infectivity. J Virol 69(4):2058–2067

Breaker RR (2004) Natural and engineered nucleic acids as tools to explore biology. Nature 432(7019):838–845

Britan-Rosich E, Nowarski R, Kotler M (2011) Multifaceted counter-APOBEC3G mechanisms employed by HIV-1 Vif. J Mol Biol 410(5):1065–1076
Browne EP, Allers C, Landau NR (2009) Restriction of HIV-1 by APOBEC3G is cytidine deaminase-dependent. Virology 387(2):313–321
Cai H, Bloom LB, Eritja R, Goodman MF (1993) Kinetics of deoxyribonucleotide insertion and extension at abasic template lesions in different sequence contexts using HIV-1 reverse transcriptase. J Biol Chem 268(31):23567–23572
Cancio R, Spadari S, Maga G (2004) Vif is an auxiliary factor of the HIV-1 reverse transcriptase and facilitates abasic site bypass. Biochem J 383(Pt. 3):475–482
Carpenter MA, Rajagurubandara E, Wijesinghe P, Bhagwat AS (2010) Determinants of sequence-specificity within human AID and APOBEC3G. DNA Repair 9(5):579–587
Carr JM, Davis AJ, Coolen C, Cheney K, Burrell CJ, Li P (2006) Vif-deficient HIV reverse transcription complexes (RTCs) are subject to structural changes and mutation of RTC-associated reverse transcription products. Virology 351(1):80–91
Carr JM, Coolen C, Davis AJ, Burrell CJ, Li P (2008) Human immunodeficiency virus 1 (HIV-1) virion infectivity factor (Vif) is part of reverse transcription complexes and acts as an accessory factor for reverse transcription. Virology 372(1):147–156
Cen S, Guo F, Niu M, Saadatmand J, Deflassieux J, Kleiman L (2004) The interaction between HIV-1 Gag and APOBEC3G. J Biol Chem 279(32):33177–33184
Chelico L, Pham P, Calabrese P, Goodman MF (2006) APOBEC3G DNA deaminase acts processively $3' \rightarrow 5'$ on single-stranded DNA. Nat Struct Mol Biol 13(5):392–399
Chelico L, Sacho EJ, Erie DA, Goodman MF (2008) A model for oligomeric regulation of APOBEC3G cytosine deaminase-dependent restriction of HIV. J Biol Chem 283(20): 13780–13791
Chelico L, Prochnow C, Erie DA, Chen XS, Goodman MF (2010) Structural model for deoxycytidine deamination mechanisms of the HIV-1 inactivation enzyme APOBEC3G. J Biol Chem 285(21):16195–16205
Chen R, Le Rouzic E, Kearney JA, Mansky LM, Benichou S (2004) Vpr-mediated incorporation of UNG2 into HIV-1 particles is required to modulate the virus mutation rate and for replication in macrophages. J Biol Chem 279(27):28419–28425
Chen KM, Harjes E, Gross PJ, Fahmy A, Lu Y, Shindo K et al (2008) Structure of the DNA deaminase domain of the HIV-1 restriction factor APOBEC3G. Nature 452(7183):116–119
Cho SJ, Drechsler H, Burke RC, Arens MQ, Powderly W, Davidson NO (2006) APOBEC3F and APOBEC3G mRNA levels do not correlate with human immunodeficiency virus type 1 plasma viremia or CD4+ T-cell count. J Virol 80(4):2069–2072
Conticello SG (2008) The AID/APOBEC family of nucleic acid mutators. Genome Biol 9(6):229
Conticello SG, Harris RS, Neuberger MS (2003) The Vif protein of HIV triggers degradation of the human antiretroviral DNA deaminase APOBEC3G. Curr Biol 13(22):2009–2013
Courcoul M, Patience C, Rey F, Blanc D, Harmache A, Sire J et al (1995) Peripheral blood mononuclear cells produce normal amounts of defective Vif- human immunodeficiency virus type 1 particles which are restricted for the preretrotranscription steps. J Virol 69(4):2068–2074
Dang Y, Wang X, Esselman WJ, Zheng YH (2006) Identification of APOBEC3DE as another antiretroviral factor from the human APOBEC family. J Virol 80(21):10522–10533
Derse D, Hill SA, Princler G, Lloyd P, Heidecker G (2007) Resistance of human T cell leukemia virus type 1 to APOBEC3G restriction is mediated by elements in nucleocapsid. Proc Natl Acad Sci USA 104(8):2915–2920
Diamond TL, Roshal M, Jamburuthugoda VK, Reynolds HM, Merriam AR, Lee KY et al (2004) Macrophage tropism of HIV-1 depends on efficient cellular dNTP utilization by reverse transcriptase. J Biol Chem 279(49):51545–51553
Do H, Vasilescu A, Diop G, Hirtzig T, Heath SC, Coulonges C et al (2005) Exhaustive genotyping of the CEM15 (APOBEC3G) gene and absence of association with AIDS progression in a French cohort. J Infect Dis 191(2):159–163
Duggal NK, Malik HS, Emerman M (2011) Positive selection of Apobec3DE in chimpanzees has driven breadth in anti-viral activity. J Virol 85(21):11361–11371

Esnault C, Heidmann O, Delebecque F, Dewannieux M, Ribet D, Hance AJ et al (2005) APOBEC3G cytidine deaminase inhibits retrotransposition of endogenous retroviruses. Nature 433(7024):430–433

Esnault C, Millet J, Schwartz O, Heidmann T (2006) Dual inhibitory effects of APOBEC family proteins on retrotransposition of mammalian endogenous retroviruses. Nucleic Acids Res 34(5):1522–1531

Feng Y, Chelico L (2011) Intensity of deoxycytidine deamination of HIV-1 proviral DNA by the retroviral restriction factor APOBEC3G is mediated by the noncatalytic domain. J Biol Chem 286(13):11415–11426

Fletcher TM 3rd, Brichacek B, Sharova N, Newman MA, Stivahtis G, Sharp PM et al (1996) Nuclear import and cell cycle arrest functions of the HIV-1 Vpr protein are encoded by two separate genes in HIV-2/SIV(SM). EMBO J 15(22):6155–6165

Fourati S, Malet I, Binka M, Boukobza S, Wirden M, Sayon S et al (2010) Partially active HIV-1 Vif alleles facilitate viral escape from specific antiretrovirals. AIDS 24(15):2313–2321

Friew YN, Boyko V, Hu WS, Pathak VK (2009) Intracellular interactions between APOBEC3G, RNA, and HIV-1 Gag: APOBEC3G multimerization is dependent on its association with RNA. Retrovirology 6:56

Furukawa A, Nagata T, Matsugami A, Habu Y, Sugiyama R, Hayashi F et al (2009) Structure, interaction and real-time monitoring of the enzymatic reaction of wild-type APOBEC3G. EMBO J 28(4):440–451

Gabuzda DH, Lawrence K, Langhoff E, Terwilliger E, Dorfman T, Haseltine WA, Sodroski J (1992) Role of Vif in replication of human immunodeficiency virus type 1 in CD4+ T lymphocytes. J Virol 66(11):6489–6495

Gabuzda DH, Li H, Lawrence K, Vasir BS, Crawford K, Langhoff E (1994) Essential role of vif in establishing productive HIV-1 infection in peripheral blood T lymphocytes and monocyte/macrophages. J Acquir Immune Defic Syndr 7(9):908–915

Gandhi SK, Siliciano JD, Bailey JR, Siliciano RF, Blankson JN (2008) Role of APOBEC3G/F-mediated hypermutation in the control of human immunodeficiency virus type 1 in elite suppressors. J Virol 82(6):3125–3130

Goldstone DC, Ennis-Adeniran V, Hedden JJ, Groom HC, Rice GI, Christodoulou E et al (2011) HIV-1 restriction factor SAMHD1 is a deoxynucleoside triphosphate triphosphohydrolase. Nature 480(7377):379–382

Goncalves J, Korin Y, Zack J, Gabuzda D (1996) Role of Vif in human immunodeficiency virus type 1 reverse transcription. J Virol 70(12):8701–8709

Gooch BD, Cullen BR (2008) Functional domain organization of human APOBEC3G. Virology 379(1):118–124

Goujon C, Riviere L, Jarrosson-Wuilleme L, Bernaud J, Rigal D, Darlix JL, Cimarelli A (2007) SIVSM/HIV-2 Vpx proteins promote retroviral escape from a proteasome-dependent restriction pathway present in human dendritic cells. Retrovirology 4:2

Gramberg T, Sunseri N, Landau NR (2010) Evidence for an activation domain at the amino terminus of simian immunodeficiency virus Vpx. J Virol 84(3):1387–1396

Guo F, Cen S, Niu M, Saadatmand J, Kleiman L (2006) Inhibition of formula-primed reverse transcription by human APOBEC3G during human immunodeficiency virus type 1 replication. J Virol 80(23):11710–11722

Guo F, Cen S, Niu M, Yang Y, Gorelick RJ, Kleiman L (2007) The interaction of APOBEC3G with human immunodeficiency virus type 1 nucleocapsid inhibits tRNA3Lys annealing to viral RNA. J Virol 81(20):11322–11331

Haché G, Liddament MT, Harris RS (2005) The retroviral hypermutation specificity of APOBEC3F and APOBEC3G is governed by the C-terminal DNA cytosine deaminase domain. J Biol Chem 280(12):10920–10924

Haché G, Mansky LM, Harris RS (2006) Human APOBEC3 proteins, retrovirus restriction, and HIV drug resistance. AIDS Rev 8(3):148–157

Haché G, Shindo K, Albin JS, Harris RS (2008) Evolution of HIV-1 isolates that use a novel Vif-independent mechanism to resist restriction by human APOBEC3G. Curr Biol 18(11):819–824

Harjes E, Gross PJ, Chen KM, Lu Y, Shindo K, Nowarski R et al (2009) An extended structure of the APOBEC3G catalytic domain suggests a unique holoenzyme model. J Mol Biol 389(5):819–832

Harris RS (2008) Enhancing immunity to HIV through APOBEC. Nat Biotechnol 26(10): 1089–1090

Harris RS, Petersen-Mahrt SK, Neuberger MS (2002) RNA editing enzyme APOBEC1 and some of its homologs can act as DNA mutators. Mol Cell 10(5):1247–1253

Harris RS, Bishop KN, Sheehy AM, Craig HM, Petersen-Mahrt SK, Watt IN et al (2003a) DNA deamination mediates innate immunity to retroviral infection. Cell 113(6):803–809

Harris RS, Sheehy AM, Craig HM, Malim MH, Neuberger MS (2003b) DNA deamination: not just a trigger for antibody diversification but also a mechanism for defense against retroviruses. Nat Immunol 4(7):641–643

Holden LG, Prochnow C, Chang YP, Bransteitter R, Chelico L, Sen U et al (2008) Crystal structure of the anti-viral APOBEC3G catalytic domain and functional implications. Nature 456(7218):121–124

Holmes RK, Koning FA, Bishop KN, Malim MH (2007) APOBEC3F can inhibit the accumulation of HIV-1 reverse transcription products in the absence of hypermutation. Comparisons with APOBEC3G. J Biol Chem 282(4):2587–2595

Hrecka K, Hao C, Gierszewska M, Swanson SK, Kesik-Brodacka M, Srivastava S et al (2011) Vpx relieves inhibition of HIV-1 infection of macrophages mediated by the SAMHD1 protein. Nature 474(7353):658–661

Hultquist JF, Lengyel JA, Refsland EW, Larue RS, Lackey L, Brown WL, Harris RS (2011) Human and rhesus APOBEC3D, APOBEC3F, APOBEC3G, and APOBEC3H Demonstrate a conserved capacity to restrict Vif-deficient HIV-1. J Virol 85(21):11220–11234

Hultquist JF, Binka M, LaRue RS, Simon V, Harris RS (2012) Vif proteins of human and simian immunodeficiency viruses require cellular CBFβ to degrade APOBEC3 restriction factors. J Virol 68(5):2874–2877

Huthoff H, Malim MH (2007) Identification of amino acid residues in APOBEC3G required for regulation by human immunodeficiency virus type 1 Vif and Virion encapsidation. J Virol 81(8):3807–3815

Huthoff H, Autore F, Gallois-Montbrun S, Fraternali F, Malim MH (2009) RNA-dependent oligomerization of APOBEC3G is required for restriction of HIV-1. PLoS Pathog 5(3):e1000330

Itaya S, Nakajima T, Kaur G, Terunuma H, Ohtani H, Mehra N, Kimura A (2010) No evidence of an association between the APOBEC3B deletion polymorphism and susceptibility to HIV infection and AIDS in Japanese and Indian populations. J Infect Dis 202(5):815–816, author reply 816–817

Iwatani Y, Takeuchi H, Strebel K, Levin JG (2006) Biochemical activities of highly purified, catalytically active human APOBEC3G: correlation with antiviral effect. J Virol 80(12):5992–6002

Iwatani Y, Chan DS, Wang F, Maynard KS, Sugiura W, Gronenborn AM et al (2007) Deaminase-independent inhibition of HIV-1 reverse transcription by APOBEC3G. Nucleic Acids Res 35(21):7096–7108

Jäger S, Kim DY, Hultquist JF, Shindo K, Larue RS, Kwon E et al (2011) Vif hijacks CBFβ to degrade APOBEC3G and promote HIV-1 infection. Nature 481(7381):371–375

Janini M, Rogers M, Birx DR, McCutchan FE (2001) Human immunodeficiency virus type 1 DNA sequences genetically damaged by hypermutation are often abundant in patient peripheral blood mononuclear cells and may be generated during near-simultaneous infection and activation of CD4(+) T cells. J Virol 75(17):7973–7986

Jern P, Russell RA, Pathak VK, Coffin JM (2009) Likely role of APOBEC3G-mediated G-to-A mutations in HIV-1 evolution and drug resistance. PLoS Pathog 5(4):e1000367

Jin X, Brooks A, Chen H, Bennett R, Reichman R, Smith H (2005) APOBEC3G/CEM15 (hA3G) mRNA levels associate inversely with human immunodeficiency virus viremia. J Virol 79(17):11513–11516

Kaiser SM, Emerman M (2006) Uracil DNA glycosylase is dispensable for human immunodeficiency virus type 1 replication and does not contribute to the antiviral effects of the cytidine deaminase APOBEC3G. J Virol 80(2):875–882

Kao S, Khan MA, Miyagi E, Plishka R, Buckler-White A, Strebel K (2003) The human immunodeficiency virus type 1 Vif protein reduces intracellular expression and inhibits packaging of APOBEC3G (CEM15), a cellular inhibitor of virus infectivity. J Virol 77(21):11398–11407

Kaushik R, Zhu X, Stranska R, Wu Y, Stevenson M (2009) A cellular restriction dictates the permissivity of nondividing monocytes/macrophages to lentivirus and gammaretrovirus infection. Cell Host Microbe 6(1):68–80

Kewalramani VN, Emerman M (1996) Vpx association with mature core structures of HIV-2. Virology 218(1):159–168

Kim EY, Bhattacharya T, Kunstman K, Swantek P, Koning FA, Malim MH (2010) Human APOBEC3G-mediated editing can promote HIV-1 sequence diversification and accelerate adaptation to selective pressure. J Virol 84(19):10402–10405

Klarmann GJ, Chen X, North TW, Preston BD (2003) Incorporation of uracil into minus strand DNA affects the specificity of plus strand synthesis initiation during lentiviral reverse transcription. J Biol Chem 278(10):7902–7909

Kohli RM, Abrams SR, Gajula KS, Maul RW, Gearhart PJ, Stivers JT (2009) A portable hot spot recognition loop transfers sequence preferences from APOBEC family members to activation-induced cytidine deaminase. J Biol Chem 284(34):22898–22904

Kohli RM, Maul RW, Guminski AF, McClure RL, Gajula KS, Saribasak H et al (2010) Local sequence targeting in the AID/APOBEC family differentially impacts retroviral restriction and antibody diversification. J Biol Chem 285(52):40956–40964

Koning FA, Newman EN, Kim EY, Kunstman KJ, Wolinsky SM, Malim MH (2009) Defining APOBEC3 expression patterns in human tissues and hematopoietic cell subsets. J Virol 83:9474–9485

Koning FA, Goujon C, Bauby H, Malim MH (2011) Target cell-mediated editing of HIV-1 cDNA by APOBEC3 proteins in human macrophages. J Virol 85(24):13448–13452

Krokan HE, Drablos F, Slupphaug G (2002) Uracil in DNA–occurrence, consequences and repair. Oncogene 21(58):8935–8948

Laguette N, Sobhian B, Casartelli N, Ringeard M, Chable-Bessia C, Segeral E et al (2011) SAMHD1 is the dendritic- and myeloid-cell-specific HIV-1 restriction factor counteracted by Vpx. Nature 474(7353):654–657

Land AM, Ball TB, Luo M, Pilon R, Sandstrom P, Embree JE et al (2008) Human immunodeficiency virus (HIV) type 1 proviral hypermutation correlates with CD4 count in HIV-infected women from Kenya. J Virol 82(16):8172–8182

Langlois MA, Neuberger MS (2008) Human APOBEC3G can restrict retroviral infection in avian cells and acts independently of both UNG and SMUG1. J Virol 82(9):4660–4664

Langlois MA, Beale RC, Conticello SG, Neuberger MS (2005) Mutational comparison of the single-domained APOBEC3C and double-domained APOBEC3F/G anti-retroviral cytidine deaminases provides insight into their DNA target site specificities. Nucleic Acids Res 33(6):1913–1923

LaRue RS, Jónsson SR, Silverstein KAT, Lajoie M, Bertrand D, El-Mabrouk N et al (2008) The artiodactyl APOBEC3 innate immune repertoire shows evidence for a multi-functional domain organization that existed in the ancestor of placental mammals. BMC Mol Biol 9(1):104, 120 pages

LaRue RS, Andrésdóttir V, Blanchard Y, Conticello SG, Derse D, Emerman M et al (2009) Guidelines for naming nonprimate APOBEC3 genes and proteins. J Virol 83(2):494–497

Lecossier D, Bouchonnet F, Clavel F, Hance AJ (2003) Hypermutation of HIV-1 DNA in the absence of the Vif protein. Science 300(5622):1112

Li J, Potash MJ, Volsky DJ (2004) Functional domains of APOBEC3G required for antiviral activity. J Cell Biochem 92(3):560–572

Li XY, Guo F, Zhang L, Kleiman L, Cen S (2007) APOBEC3G inhibits DNA strand transfer during HIV-1 reverse transcription. J Biol Chem 282(44):32065–32074

Li X, Ma J, Zhang Q, Zhou J, Yin X, Zhai C et al (2011) Functional analysis of the two cytidine deaminase domains in APOBEC3G. Virology 414(2):130–136

Li M, Shandilya SM, Carpenter MA, Rathore A, Brown WL, Perkins AL et al (2012) First-in-class small molecule inhibitors of the single-strand DNA cytosine deaminase APOBEC3G. ACS Chem Biol 7(3):506–517

Liddament MT, Brown WL, Schumacher AJ, Harris RS (2004) APOBEC3F properties and hypermutation preferences indicate activity against HIV-1 in vivo. Curr Biol 14(15):1385–1391
Löchelt M, Romen F, Bastone P, Muckenfuss H, Kirchner N, Kim YB et al (2005) The antiretroviral activity of APOBEC3 is inhibited by the foamy virus accessory Bet protein. Proc Natl Acad Sci USA 102(22):7982–7987
Loeb LA, Essigmann JM, Kazazi F, Zhang J, Rose KD, Mullins JI (1999) Lethal mutagenesis of HIV with mutagenic nucleoside analogs. Proc Natl Acad Sci USA 96(4):1492–1497
Luo K, Liu B, Xiao Z, Yu Y, Yu X, Gorelick R, Yu XF (2004) Amino-terminal region of the human immunodeficiency virus type 1 nucleocapsid is required for human APOBEC3G packaging. J Virol 78(21):11841–11852
Luo K, Wang T, Liu B, Tian C, Xiao Z, Kappes J, Yu XF (2007) Cytidine deaminases APOBEC3G and APOBEC3F interact with human immunodeficiency virus type 1 integrase and inhibit proviral DNA formation. J Virol 81(13):7238–7248
Madani N, Kabat D (1998) An endogenous inhibitor of human immunodeficiency virus in human lymphocytes is overcome by the viral Vif protein. J Virol 72(12):10251–10255
Mangeat B, Turelli P, Caron G, Friedli M, Perrin L, Trono D (2003) Broad antiretroviral defence by human APOBEC3G through lethal editing of nascent reverse transcripts. Nature 424(6944):99–103
Mangeat B, Turelli P, Liao S, Trono D (2004) A single amino acid determinant governs the species-specific sensitivity of APOBEC3G to Vif action. J Biol Chem 279(15):14481–14483
Mansky LM, Temin HM (1995) Lower in vivo mutation rate of human immunodeficiency virus type 1 than that predicted from the fidelity of purified reverse transcriptase. J Virol 69(8): 5087–5094
Mansky LM, Preveral S, Selig L, Benarous R, Benichou S (2000) The interaction of Vpr with uracil DNA glycosylase modulates the human immunodeficiency virus type 1 In vivo mutation rate. J Virol 74(15):7039–7047
Marin M, Rose KM, Kozak SL, Kabat D (2003) HIV-1 Vif protein binds the editing enzyme APOBEC3G and induces its degradation. Nat Med 9(11):1398–1403
Mbisa JL, Barr R, Thomas JA, Vandegraaff N, Dorweiler IJ, Svarovskaia ES et al (2007) Human immunodeficiency virus type 1 cDNAs produced in the presence of APOBEC3G exhibit defects in plus-strand DNA transfer and integration. J Virol 81(13):7099–7110
Mbisa JL, Bu W, Pathak VK (2010) APOBEC3F and APOBEC3G inhibit HIV-1 DNA integration by different mechanisms. J Virol 84(10):5250–5259
Mehle A, Strack B, Ancuta P, Zhang C, McPike M, Gabuzda D (2004) Vif overcomes the innate antiviral activity of APOBEC3G by promoting its degradation in the ubiquitin-proteasome pathway. J Biol Chem 279(9):7792–7798
Miyagi E, Opi S, Takeuchi H, Khan M, Goila-Gaur R, Kao S, Strebel K (2007) Enzymatically active APOBEC3G is required for efficient inhibition of human immunodeficiency virus type 1. J Virol 81(24):13346–13353
Miyagi E, Brown CR, Opi S, Khan M, Goila-Gaur R, Kao S et al (2010) Stably expressed APOBEC3F has negligible antiviral activity. J Virol 84(21):11067–11075
Muckenfuss H, Hamdorf M, Held U, Perkovic M, Lower J, Cichutek K et al (2006) APOBEC3 proteins inhibit human LINE-1 retrotransposition. J Biol Chem 281(31):22161–22172
Mulder LC, Harari A, Simon V (2008) Cytidine deamination induced HIV-1 drug resistance. Proc Natl Acad Sci USA 105(14):5501–5506
Navarro F, Bollman B, Chen H, Konig R, Yu Q, Chiles K, Landau NR (2005) Complementary function of the two catalytic domains of APOBEC3G. Virology 333(2):374–386
Neil S, Martin F, Ikeda Y, Collins M (2001) Postentry restriction to human immunodeficiency virus-based vector transduction in human monocytes. J Virol 75(12):5448–5456
Newman EN, Holmes RK, Craig HM, Klein KC, Lingappa JR, Malim MH, Sheehy AM (2005) Antiviral function of APOBEC3G can be dissociated from cytidine deaminase activity. Curr Biol 15(2):166–170
Nowarski R, Britan-Rosich E, Shiloach T, Kotler M (2008) Hypermutation by intersegmental transfer of APOBEC3G cytidine deaminase. Nat Struct Mol Biol 15(10):1059–1066

OhAinle M, Kerns JA, Malik HS, Emerman M (2006) Adaptive evolution and antiviral activity of the conserved mammalian cytidine deaminase APOBEC3H. J Virol 80(8):3853–3862

OhAinle M, Kerns JA, Li MM, Malik HS, Emerman M (2008) Antiretroelement activity of APOBEC3H was lost twice in recent human evolution. Cell Host Microbe 4(3): 249–259

Okeoma CM, Lovsin N, Peterlin BM, Ross SR (2007) APOBEC3 inhibits mouse mammary tumour virus replication in vivo. Nature 445(7130):927–930

Opi S, Takeuchi H, Kao S, Khan MA, Miyagi E, Goila-Gaur R et al (2006) Monomeric APOBEC3G is catalytically active and has antiviral activity. J Virol 80(10):4673–4682

Pace C, Keller J, Nolan D, James I, Gaudieri S, Moore C, Mallal S (2006) Population level analysis of human immunodeficiency virus type 1 hypermutation and its relationship with APOBEC3G and *vif* genetic variation. J Virol 80(18):9259–9269

Pathak VK, Temin HM (1990) Broad spectrum of in vivo forward mutations, hypermutations, and mutational hotspots in a retroviral shuttle vector after a single replication cycle: substitutions, frameshifts, and hypermutations. Proc Natl Acad Sci USA 87(16):6019–6023

Peng G, Greenwell-Wild T, Nares S, Jin W, Lei KJ, Rangel ZG et al (2007) Myeloid differentiation and susceptibility to HIV-1 are linked to APOBEC3 expression. Blood 110(1):393–400

Pertel T, Hausmann S, Morger D, Zuger S, Guerra J, Lascano J et al (2011) TRIM5 is an innate immune sensor for the retrovirus capsid lattice. Nature 472(7343):361–365

Piantadosi A, Humes D, Chohan B, McClelland RS, Overbaugh J (2009) Analysis of the percentage of human immunodeficiency virus type 1 sequences that are hypermutated and markers of disease progression in a longitudinal cohort, including one individual with a partially defective Vif. J Virol 83(16):7805–7814

Pillai SK, Wong JK, Barbour JD (2008) Turning up the volume on mutational pressure: is more of a good thing always better? (A case study of HIV-1 Vif and APOBEC3). Retrovirology 5:26

Pillai SK, Abdel-Mohsen M, Guatelli J, Skasko M, Monto A, Fujimoto K et al (2012) Role of retroviral restriction factors in the interferon-alpha-mediated suppression of HIV-1 in vivo. Proc Natl Acad Sci USA 109(8):3035–3040

Powell RD, Holland PJ, Hollis T, Perrino FW (2011) Aicardi-Goutières syndrome gene and HIV-1 restriction factor SAMHD1 Is a dGTP-regulated deoxynucleotide triphosphohydrolase. J Biol Chem 286(51):43596–43600

Priet S, Gros N, Navarro JM, Boretto J, Canard B, Querat G, Sire J (2005) HIV-1-associated uracil DNA glycosylase activity controls dUTP misincorporation in viral DNA and is essential to the HIV-1 life cycle. Mol Cell 17(4):479–490

Rai K, Huggins IJ, James SR, Karpf AR, Jones DA, Cairns BR (2008) DNA demethylation in zebrafish involves the coupling of a deaminase, a glycosylase, and gadd45. Cell 135(7): 1201–1212

Rausch JW, Chelico L, Goodman MF, Le Grice SF (2009) Dissecting APOBEC3G substrate specificity by nucleoside analog interference. J Biol Chem 284(11):7047–7058

Reddy K, Winkler CA, Werner L, Mlisana K, Abdool Karim SS, Ndung'u T (2010) APOBEC3G expression is dysregulated in primary HIV-1 infection and polymorphic variants influence CD4+ T-cell counts and plasma viral load. AIDS 24(2):195–204

Refsland EW, Stenglein MD, Shindo K, Albin JS, Brown WL, Harris RS (2010) Quantitative profiling of the full APOBEC3 mRNA repertoire in lymphocytes and tissues: implications for HIV-1 restriction. Nucleic Acids Res 38(13):4274–4284

Refsland EW, Hultquist JF, Harris RS (2012) Endogenous origins of HIV-1 G-to-A hypermutation. PLoS Pathog 8(7):e1002800

Rice GI, Bond J, Asipu A, Brunette RL, Manfield IW, Carr IM et al (2009) Mutations involved in Aicardi-Goutières syndrome implicate SAMHD1 as regulator of the innate immune response. Nat Genet 41(7):829–832

Russell RA, Wiegand HL, Moore MD, Schafer A, McClure MO, Cullen BR (2005) Foamy virus Bet proteins function as novel inhibitors of the APOBEC3 family of innate antiretroviral defense factors. J Virol 79(14):8724–8731

Russell RA, Moore MD, Hu WS, Pathak VK (2009) APOBEC3G induces a hypermutation gradient: purifying selection at multiple steps during HIV-1 replication results in levels of G-to-A mutations that are high in DNA, intermediate in cellular viral RNA, and low in virion RNA. Retrovirology 6:16

Sadler HA, Stenglein MD, Harris RS, Mansky LM (2010) APOBEC3G contributes to HIV-1 variation through sublethal mutagenesis. J Virol 84(14):7396–7404

Santa-Marta M, da Silva FA, Fonseca AM, Goncalves J (2005) HIV-1 Vif can directly inhibit apolipoprotein B mRNA-editing enzyme catalytic polypeptide-like 3G-mediated cytidine deamination by using a single amino acid interaction and without protein degradation. J Biol Chem 280(10):8765–8775

Sawyer SL, Emerman M, Malik HS (2004) Ancient adaptive evolution of the primate antiviral DNA-editing enzyme APOBEC3G. PLoS Biol 2(9):E275

Sayah DM, Sokolskaja E, Berthoux L, Luban J (2004) Cyclophilin A retrotransposition into TRIM5 explains owl monkey resistance to HIV-1. Nature 430(6999):569–573

Schafer A, Bogerd HP, Cullen BR (2004) Specific packaging of APOBEC3G into HIV-1 virions is mediated by the nucleocapsid domain of the gag polyprotein precursor. Virology 328(2): 163–168

Schröfelbauer B, Chen D, Landau NR (2004) A single amino acid of APOBEC3G controls its species-specific interaction with virion infectivity factor (Vif). Proc Natl Acad Sci USA 101(11):3927–3932

Schröfelbauer B, Yu Q, Zeitlin SG, Landau NR (2005) Human immunodeficiency virus type 1 Vpr induces the degradation of the UNG and SMUG uracil-DNA glycosylases. J Virol 79(17):10978–10987

Schumacher AJ, Haché G, MacDuff DA, Brown WL, Harris RS (2008) The DNA deaminase activity of human APOBEC3G is required for Ty1, MusD, and human immunodeficiency virus type 1 restriction. J Virol 82(6):2652–2660

Schwartz O, Marechal V, Danos O, Heard JM (1995) Human immunodeficiency virus type 1 Nef increases the efficiency of reverse transcription in the infected cell. J Virol 69(7):4053–4059

Shandilya SM, Nalam MN, Nalivaika EA, Gross PJ, Valesano JC, Shindo K et al (2010) Crystal structure of the APOBEC3G catalytic domain reveals potential oligomerization interfaces. Structure 18(1):28–38

Sheehy AM, Gaddis NC, Choi JD, Malim MH (2002) Isolation of a human gene that inhibits HIV-1 infection and is suppressed by the viral Vif protein. Nature 418(6898):646–650

Sheehy AM, Gaddis NC, Malim MH (2003) The antiretroviral enzyme APOBEC3G is degraded by the proteasome in response to HIV-1 Vif. Nat Med 9(11):1404–1407

Shindo K, Takaori-Kondo A, Kobayashi M, Abudu A, Fukunaga K, Uchiyama T (2003) The enzymatic activity of CEM15/Apobec-3G is essential for the regulation of the infectivity of HIV-1 virion but not a sole determinant of its antiviral activity. J Biol Chem 278(45):44412–44416

Shlyakhtenko LS, Lushnikov AY, Li M, Lackey L, Harris RS, Lyubchenko YL (2011) Atomic force microscopy studies provide direct evidence for dimerization of the HIV restriction factor APOBEC3G. J Biol Chem 286(5):3387–3395

Simon JH, Malim MH (1996) The human immunodeficiency virus type 1 Vif protein modulates the postpenetration stability of viral nucleoprotein complexes. J Virol 70(8):5297–5305

Simon JH, Gaddis NC, Fouchier RA, Malim MH (1998) Evidence for a newly discovered cellular anti-HIV-1 phenotype. Nat Med 4(12):1397–1400

Simon V, Zennou V, Murray D, Huang Y, Ho DD, Bieniasz PD (2005) Natural variation in Vif: differential impact on APOBEC3G/3F and a potential role in HIV-1 diversification. PLoS Pathog 1(1):e6

Smith JL, Pathak VK (2010) Identification of specific determinants of human APOBEC3F, APOBEC3C, and APOBEC3DE and African green monkey APOBEC3F that interact with HIV-1 Vif. J Virol 84(24):12599–12608

Soros VB, Yonemoto W, Greene WC (2007) Newly synthesized APOBEC3G is incorporated into HIV virions, inhibited by HIV RNA, and subsequently activated by RNase H. PLoS Pathog 3(2):e15

Sova P, Volsky DJ (1993) Efficiency of viral DNA synthesis during infection of permissive and nonpermissive cells with vif-negative human immunodeficiency virus type 1. J Virol 67(10):6322–6326

Stenglein MD, Harris RS (2006) APOBEC3B and APOBEC3F inhibit L1 retrotransposition by a DNA deamination-independent mechanism. J Biol Chem 281(25):16837–16841

Stenglein MD, Burns MB, Li M, Lengyel J, Harris RS (2010) APOBEC3 proteins mediate the clearance of foreign DNA from human cells. Nat Struct Mol Biol 17(2):222–229

Stetson DB, Ko JS, Heidmann T, Medzhitov R (2008) Trex1 prevents cell-intrinsic initiation of autoimmunity. Cell 134(4):587–598

Stevenson M, Stanwick TL, Dempsey MP, Lamonica CA (1990) HIV-1 replication is controlled at the level of T cell activation and proviral integration. EMBO J 9(5):1551–1560

Stopak K, de Noronha C, Yonemoto W, Greene WC (2003) HIV-1 Vif blocks the antiviral activity of APOBEC3G by impairing both its translation and intracellular stability. Mol Cell 12(3):591–601

Strebel K, Daugherty D, Clouse K, Cohen D, Folks T, Martin MA (1987) The HIV 'A' (sor) gene product is essential for virus infectivity. Nature 328(6132):728–730

Stremlau M, Owens CM, Perron MJ, Kiessling M, Autissier P, Sodroski J (2004) The cytoplasmic body component TRIM5alpha restricts HIV-1 infection in Old World monkeys. Nature 427(6977):848–853

Sunseri N, O'Brien M, Bhardwaj N, Landau NR (2011) Human immunodeficiency virus type 1 modified to package Simian immunodeficiency virus Vpx efficiently infects macrophages and dendritic cells. J Virol 85(13):6263–6274

Suspène R, Sommer P, Henry M, Ferris S, Guetard D, Pochet S et al (2004) APOBEC3G is a single-stranded DNA cytidine deaminase and functions independently of HIV reverse transcriptase. Nucleic Acids Res 32(8):2421–2429

Suspène R, Rusniok C, Vartanian JP, Wain-Hobson S (2006) Twin gradients in APOBEC3 edited HIV-1 DNA reflect the dynamics of lentiviral replication. Nucleic Acids Res 34(17):4677–4684

Svarovskaia ES, Xu H, Mbisa JL, Barr R, Gorelick RJ, Ono A et al (2004) Human apolipoprotein B mRNA-editing enzyme-catalytic polypeptide-like 3G (APOBEC3G) is incorporated into HIV-1 virions through interactions with viral and nonviral RNAs. J Biol Chem 279(34):35822–35828

Tan L, Sarkis PT, Wang T, Tian C, Yu XF (2009) Sole copy of Z2-type human cytidine deaminase APOBEC3H has inhibitory activity against retrotransposons and HIV-1. FASEB J 23(1): 279–287

Thielen BK, McNevin JP, McElrath MJ, Hunt BV, Klein KC, Lingappa JR (2010) Innate immune signaling induces high levels of TC-specific deaminase activity in primary monocyte-derived cells through expression of APOBEC3A isoforms. J Biol Chem 285(36):27753–27766

Tian C, Wang T, Zhang W, Yu XF (2007) Virion packaging determinants and reverse transcription of SRP RNA in HIV-1 particles. Nucleic Acids Res 35(21):7288–7302

Towers GJ, Hatziioannou T, Cowan S, Goff SP, Luban J, Bieniasz PD (2003) Cyclophilin A modulates the sensitivity of HIV-1 to host restriction factors. Nat Med 9(9):1138–1143

Turelli P, Mangeat B, Jost S, Vianin S, Trono D (2004) Inhibition of hepatitis B virus replication by APOBEC3G. Science 303(5665):1829

Ulenga NK, Sarr AD, Hamel D, Sankale JL, Mboup S, Kanki PJ (2008a) The level of APOBEC3G (hA3G)-related G-to-A mutations does not correlate with viral load in HIV type 1-infected individuals. AIDS Res Hum Retroviruses 24(10):1285–1290

Ulenga NK, Sarr AD, Thakore-Meloni S, Sankale JL, Eisen G, Kanki PJ (2008b) Relationship between human immunodeficiency type 1 infection and expression of human APOBEC3G and APOBEC3F. J Infect Dis 198(4):486–492

Vartanian JP, Meyerhans A, Asjo B, Wain-Hobson S (1991) Selection, recombination, and G→A hypermutation of human immunodeficiency virus type 1 genomes. J Virol 65(4):1779–1788

Vartanian JP, Meyerhans A, Sala M, Wain-Hobson S (1994) G→A hypermutation of the human immunodeficiency virus type 1 genome: evidence for dCTP pool imbalance during reverse transcription. Proc Natl Acad Sci USA 91(8):3092–3096

Vazquez-Perez JA, Ormsby CE, Hernandez-Juan R, Torres KJ, Reyes-Teran G (2009) APOBEC3G mRNA expression in exposed seronegative and early stage HIV infected individuals decreases with removal of exposure and with disease progression. Retrovirology 6:23

von Schwedler U, Song J, Aiken C, Trono D (1993) Vif is crucial for human immunodeficiency virus type 1 proviral DNA synthesis in infected cells. J Virol 67(8):4945–4955

Wang T, Tian C, Zhang W, Luo K, Sarkis PT, Yu L (2007) 7SL RNA mediates virion packaging of the antiviral cytidine deaminase APOBEC3G. J Virol 81(23):13112–13124

Wang T, Tian C, Zhang W, Sarkis PT, Yu XF (2008) Interaction with 7SL RNA but not with HIV-1 genomic RNA or P bodies is required for APOBEC3F virion packaging. J Mol Biol 375(4):1098–1112

Wang X, Ao Z, Chen L, Kobinger G, Peng J, Yao X (2012) The cellular antiviral protein APOBEC3G interacts with HIV-1 reverse transcriptase and inhibits its function during viral replication. J Virol 86(7):3777–3786

Watts JM, Dang KK, Gorelick RJ, Leonard CW, Bess JW Jr, Swanstrom R et al (2009) Architecture and secondary structure of an entire HIV-1 RNA genome. Nature 460(7256):711–716

Wedekind JE, Gillilan R, Janda A, Krucinska J, Salter JD, Bennett RP et al (2006) Nanostructures of APOBEC3G support a hierarchical assembly model of high molecular mass ribonucleoprotein particles from dimeric subunits. J Biol Chem 281(50):38122–38126

Wiegand HL, Doehle BP, Bogerd HP, Cullen BR (2004) A second human antiretroviral factor, APOBEC3F, is suppressed by the HIV-1 and HIV-2 Vif proteins. EMBO J 23(12):2451–2458

Wood N, Bhattacharya T, Keele BF, Giorgi E, Liu M, Gaschen B et al (2009) HIV evolution in early infection: selection pressures, patterns of insertion and deletion, and the impact of APOBEC. PLoS Pathog 5(5):e1000414

Xu H, Svarovskaia ES, Barr R, Zhang Y, Khan MA, Strebel K, Pathak VK (2004) A single amino acid substitution in human APOBEC3G antiretroviral enzyme confers resistance to HIV-1 virion infectivity factor-induced depletion. Proc Natl Acad Sci USA 101(15):5652–5657

Xu H, Chertova E, Chen J, Ott DE, Roser JD, Hu WS, Pathak VK (2007) Stoichiometry of the antiviral protein APOBEC3G in HIV-1 virions. Virology 360(2):247–256

Yan N, O'Day E, Wheeler LA, Engelman A, Lieberman J (2011) HIV DNA is heavily uracilated, which protects it from autointegration. Proc Natl Acad Sci USA 108(22):9244–9249

Yang B, Chen K, Zhang C, Huang S, Zhang H (2007a) Virion-associated uracil DNA glycosylase-2 and apurinic/apyrimidinic endonuclease are involved in the degradation of APOBEC3G-edited nascent HIV-1 DNA. J Biol Chem 282(16):11667–11675

Yang Y, Guo F, Cen S, Kleiman L (2007b) Inhibition of initiation of reverse transcription in HIV-1 by human APOBEC3F. Virology 365(1):92–100

Yu X, Yu Y, Liu B, Luo K, Kong W, Mao P, Yu XF (2003) Induction of APOBEC3G ubiquitination and degradation by an HIV-1 Vif-Cul5-SCF complex. Science 302(5647):1056–1060

Yu Q, Konig R, Pillai S, Chiles K, Kearney M, Palmer S et al (2004) Single-strand specificity of APOBEC3G accounts for minus-strand deamination of the HIV genome. Nat Struct Mol Biol 11(5):435–442

Zack JA, Arrigo SJ, Weitsman SR, Go AS, Haislip A, Chen IS (1990) HIV-1 entry into quiescent primary lymphocytes: molecular analysis reveals a labile, latent viral structure. Cell 61(2): 213–222

Zennou V, Perez-Caballero D, Gottlinger H, Bieniasz PD (2004) APOBEC3G incorporation into human immunodeficiency virus type 1 particles. J Virol 78(21):12058–12061

Zhang H, Yang B, Pomerantz RJ, Zhang C, Arunachalam SC, Gao L (2003) The cytidine deaminase CEM15 induces hypermutation in newly synthesized HIV-1 DNA. Nature 424(6944):94–98

Zhang W, Du J, Evans SL, Yu Y, Yu XF (2011) T-cell differentiation factor CBFβ regulates HIV-1 Vif-mediated evasion of host restriction. Nature 481(7381):376–379

Zhen A, Wang T, Zhao K, Xiong Y, Yu XF (2010) A single amino acid difference in human APOBEC3H variants determines HIV-1 Vif sensitivity. J Virol 84(4):1902–1911

Zheng YH, Irwin D, Kurosu T, Tokunaga K, Sata T, Peterlin BM (2004) Human APOBEC3F is another host factor that blocks human immunodeficiency virus type 1 replication. J Virol 78(11):6073–6076

# Chapter 13
# Role of RNase H Activity in NRTI/NNRTI Drug Resistance

**Gilda Tachedjian and Nicolas Sluis-Cremer**

## 13.1 Introduction

HIV-1 reverse transcriptase (RT) is responsible for conversion of the viral single-stranded (+)RNA into double-stranded DNA. The enzyme is multifunctional and possesses two distinct activities: (i) a DNA polymerase activity that can use either RNA or DNA as a template and (ii) a ribonuclease H (RNase H) activity that degrades the RNA strand in RNA/DNA duplexes.

HIV-1 RT consists of a 560-residue 66kDa subunit (p66) and a p66-derived 440-residue 51kDa (p51) subunit (Fig. 13.1). The p66 subunit can be divided into DNA polymerase (residues 1-315), connection (residues 316-437), and RNase H (residues 438-560) domains (Kohlstaedt et al. 1992; Jacobo-Molina et al. 1993). In comparison, p51 comprises only the polymerase and connection domains. Each p66/p51 RT molecule has one binding cleft for the template/primer (T/P) nucleic acid substrate, one DNA polymerization active site, and one RNase H active site. Whereas the DNA polymerase and RNase H active sites reside in the p66 subunit only, the T/P nucleic acid substrate-binding cleft is formed by residues from the polymerase domain of p66, the connection domains of p66 and p51, the thumb domain of p51, and the RNase H domain of p66 (Sarafianos et al. 2009).

HIV-1 RT is a key target for the development of antiretroviral drugs. To date, the United States (US) Food and Drug Administration (FDA) has approved 13 reverse transcriptase inhibitors (RTIs) that can be classified into two distinct groups

G. Tachedjian, B.Sc. (Hons), Ph.D. (✉)
Retroviral Biology and Antivirals Laboratory, Centre for Virology, Burnet Institute,
85 Commercial Rd, GPO Box 2284, Melbourne, VIC 3004, Australia
e-mail: gildat@burnet.edu.au

N. Sluis-Cremer, B.Sc. (Hons), Ph.D.
Division of Infectious Diseases, University of Pittsburgh School of Medicine,
Scaife Hall, Suite 817, 3550 Terrace Street, Pittsburgh, PA 15261, USA
e-mail: nps2@pitt.edu

S.F.J. LeGrice and M. Götte (eds.), *Human Immunodeficiency Virus Reverse Transcriptase: A Bench-to-Bedside Success*, DOI 10.1007/978-1-4614-7291-9_13,

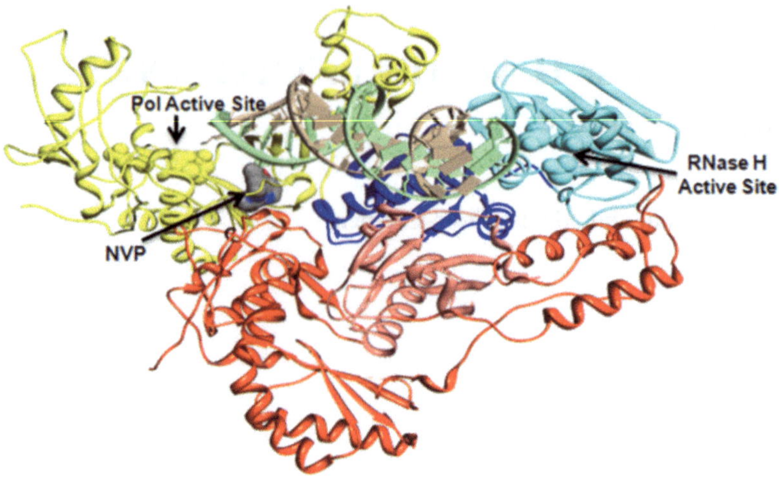


**Fig. 13.1** Three dimensional structure of HIV-1 RT. The polymerase, connection and RNase H domains are colored yellow, blue and cyan in the p66 subunit, respectively. The polymerase and connection domains are colored red and pink in the p51 subunit, respectively. Key residues in the DNA polymerase (D110, D185, D186) and RNase H (D443, E478, D498) active sites are shown in space fill. Nucleic acid T/P and the NNRTI NVP are also depicted. The figure was generated using Chimera and the PDB coordinates 3V81 Das et al. 2012

(Fig. 13.2). These include the nucleoside and nucleotide RT inhibitors (NRTIs), which are analogues of naturally occurring deoxyribonucleoside triphosphates (dNTPs) that lack a 3′-hydroxyl group on the ribose sugar. The triphosphate forms of NRTI (NRTI-TP) inhibit HIV RT DNA polymerization by acting as chain terminators of nucleic acid synthesis (Furman et al. 1986). FDA-approved NRTIs include zidovudine (AZT), stavudine (d4T), didanosine (ddI), zalcitabine (ddC), lamivudine (3TC), emtricitabine (FTC), abacavir (ABC), and tenofovir (TFV) disoproxil fumarate (TDF) (Fig. 13.2a). By contrast, the nonnucleoside RT inhibitors (NNRTIs) are a group of amphiphilic compounds that bind to a hydrophobic pocket in HIV-1 RT – termed the NNRTI-binding pocket (BP) – that is near to but distinct from the polymerase active site. NNRTIs are allosteric inhibitors of HIV-1 RT DNA polymerization reactions (Rittinger et al. 1995; Spence et al. 1995; Das et al. 2012). The FDA-approved NNRTIs include nevirapine (NVP), delavirdine (DLV), efavirenz (EFV), etravirine (ETV), and rilpivirine (RIL) (Fig. 13.2b).

Although combinations of RTIs have profoundly reduced morbidity and mortality from HIV-1 infection, their long-term efficacy is limited by the selection of drug-resistant variants of HIV-1 and cross-resistance among RTIs (Clavel and Hance 2004). Mutations known to confer resistance to both NRTIs and NNRTIs have been identified by in vitro passage experiments and from sequences of viral isolates from patients experiencing virologic failure on RTI therapy. All the RTI mutations included in the most widely used resistance tables, such as that from the International AIDS Society-USA (IAS-USA) expert panel, map to the DNA polymerase domain of HIV-1 RT (Johnson et al. 2011). However, most genotypic assays do not analyze the connection and RNase H domains of RT despite the fact that these regions

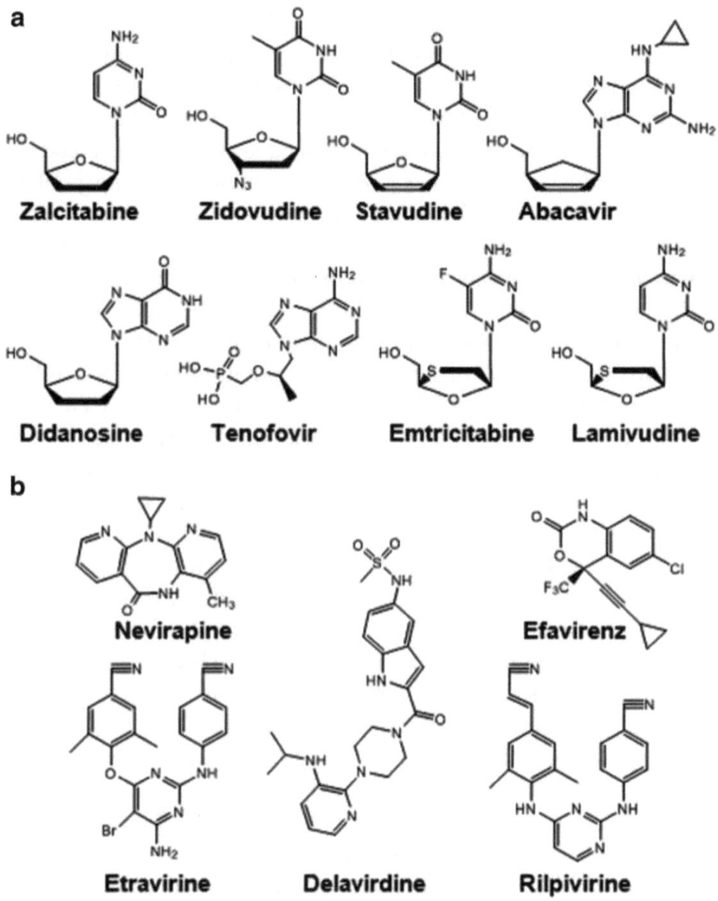


**Fig. 13.2** Chemical structures of the NRTI (**a**) and NNRTI (**b**)

contain key residues that are essential for RT structure-function (Sarafianos et al. 2009). Importantly, in the last few years, compelling evidence has emerged that implicates mutations outside the DNA polymerase domain of RT in RTI resistance. As such, this chapter reviews the discovery of drug resistance mutations outside of the polymerase domain and highlights the mechanisms by which these mutations confer NRTI and NNRTI resistance.

## 13.2 Historical Perspective: Key Paradigm-Shifting Discoveries

AZT was the first NRTI approved by the US FDA for the treatment of HIV infection. HIV-1 resistance to AZT was first reported in 1989 and was mapped to mutations in the polymerase domain of RT (Larder et al. 1989; Larder and Kemp 1989).

However, as described above, the general assumption at that time was that based on their mechanism of action, NRTIs would only select resistance mutations at or near to the DNA polymerase active site. Similarly, by the time NVP was approved in 1996, its binding site had been identified by crystallography as being adjacent to the DNA polymerase active site of RT and NVP resistance (both in vitro and in patients) mapped to key residues that formed part of the NNRTI-binding pocket (Esnouf et al. 1995; Ding et al. 1995; Richman et al. 1991). As such, there appeared to be little impetus to look for drug resistance-related mutations outside of the DNA polymerase domain of RT. There was, however, one obvious exception: residue Y318 in the connection domain of HIV-1 RT that forms part of the NNRTI-binding pocket. In 1998, Pelemans et al. performed a mutational analysis study of Y318 and demonstrated in vitro that the Y318F and Y318W mutations in HIV-1 RT conferred decreased susceptibility to several NNRTIs, including NVP and DLV (Pelemans et al. 1998). In 2002, Harrigan et al. reported that the Y318F mutation was significantly associated with the use of DLV ($p=10^{-11}$) and NVP ($p=10^{-6}$) in the clinic, but not EFV ($p=0.3$)(Harrigan et al. 2002). Site-directed HIV-1 Y318F mutants in an HXB2 background displayed 42-fold-decreased susceptibility to DLV but < threefold-decreased susceptibility to NVP or EFV. However, combinations of Y318F with K103N, Y181C, or both resulted in large decreases in susceptibility to both EFV and NVP. Taken together, these studies highlighted the importance of the Y318F substitution in the connection domain of RT in HIV-1 NNRTI resistance.

In 1995, dual NRTI therapy with AZT and 3TC was shown to have significant virological and clinical benefits compared with AZT or 3TC monotherapy (Larder et al. 1995). The addition of 3TC to AZT delayed the development of AZT resistance in therapy-naïve patients and restored AZT susceptibility in patients who previously received AZT monotherapy. However, in some AZT-experienced patients, the virological response to AZT and 3TC therapy was not sustained, and virus resistant to both drugs could be identified. Although substitutions at RT codons 44, 118, 207, and 208 are associated with increased AZT resistance in viruses that carry both thymidine analogue mutations (TAMs) and M184V (Romano et al. 2002; Stoeckli et al. 2002), in some instances, these mutations were absent, and a G333D/E polymorphism in the connection domain of RT was found to facilitate dual AZT-3TC resistance (Larder et al. 1995). The role of G333D/E in dual AZT/3TC resistance was demonstrated in two ways. First, conversion of 333E to G333 in dually resistant virus reversed resistance to AZT. Second, mutation of the WT G333 residue to E in a recombinant virus that contained M41L/L210W/T215Y and M184V resulted in AZT resistance despite the presence of the M184V mutation.

In 2004, both Kagan et al. and Galli et al. analyzed HIV-1 RT mutations spanning codons 1–400 in the Quest Diagnostic clinical sample database in the US and British Columbia (BC) Centre for Excellence in HIV/AIDS genotypic drug resistance databases, respectively (Kagan et al. 2004; Galli et al. 2004). Both studies identified several mutations that appeared more frequently in samples from antiretroviral-experienced patients than in antiretroviral-naïve patients. These included the N348I, G359S, and A371V mutations. At this time, our research groups became actively involved in trying to understand the in vivo significance of these mutations and the possible mechanisms by which they conferred RTI resistance. At first, progress was slowed by technical difficulties and a critical lack of mechanistic insight. However, in 2007, we provided the first in vivo evidence that treatment with RTIs can select a mutation (i.e., N348I) outside the polymerase domain of the HIV-1 RT that confers dual-class RTI resistance (Yap et al. 2007). The N348I mutation is discussed in more depth below.

Perhaps the greatest impetus for evaluating the role of the RNase H domain in NRTI resistance resulted from key studies from the laboratory of Vinay Pathak (NCI, Frederick). In 2005, Nikolenko et al. showed that mutations that reduce the RNase H activity of RT (such as D549N and H539N) increase AZT and d4T resistance alone and further in combination with TAMs (Nikolenko et al. 2005). Based on these observations, the authors hypothesized that once AZT is incorporated into the nascent DNA chain, there was competition between primer repair (excision) and RNase H activity. RNase H cleavage of the RNA template could lead to premature dissociation of the newly synthesized chain-terminated DNA from the RNA template, resulting in irreversible termination. However, the clinical significance of this finding was unknown given that the D549N and H539N mutations are not present in clinical samples. In 2007, Nikolenko et al. performed genotypic and phenotypic analyses from 8 NRTI-experienced and 7 treatment-naïve individuals (Nikolenko et al. 2007). This study showed that the C-terminal domains obtained from the 7 treatment-naïve patients did not increase resistance to AZT. By contrast, the same domains obtained from 6 treatment-experienced patients increased resistance to AZT (2.4- to 5.7-fold) in the context of a wild-type polymerase domain. Remarkably, 5 of the 8 C-terminal domains from the treatment-experienced patients substantially increased resistance to AZT (100- to 800-fold) in the context of a polymerase domain encoding TAMs. The authors further narrowed down the region that increased AZT resistance to amino acids 293-414 in the RT connection domain. Mutational analysis demonstrated that the substitutions G335C, N348I, and A360I conferred 30-, 35-, and 30-fold increases in AZT resistance, respectively.

## 13.3 Clinically Relevant Connection and RNase H Domain Mutations

Since 2007, there have been numerous studies that have identified resistance mutations in the connection and RNase H domains of RT (Tables 13.1 and 13.2) (Fig. 13.3). Unfortunately this review cannot highlight each of these mutations. Importantly, many of them are polymorphic across different HIV-1 subtypes (Table 13.3), thus making it difficult to assess clinical significance. Therefore, in the following paragraphs, we highlight those mutations in the connection domain of subtype B and non-subtype B HIV-1 RT that impact drug susceptibility and may be of clinical significance. These include the N348I, A360V, and A376S mutations.

**Table 13.1** Drug resistance phenotype of mutations in the RT connection domain selected during antiretroviral therapy

| Mutation | Subtype[a] | Decreased RTI susceptibility in wild-type backbone | Decreased NRTI susceptibility in concert with classical mutations | Decreased NNRTI susceptibility in concert with classical mutations | Associated with virologic failure | References |
|---|---|---|---|---|---|---|
| E312Q | B | – | AZT with TAMs | – | | Nikolenko et al. (2007), Hachiya et al. (2009) |
| Y318F | B | DLV, NVP | – | DLV, NVP, EFV resistance with NNRTI-BP mutations | | Pelemans et al. (1998), Harrigan et al. (2002) |
| G333D/E | B | – | AZT and 3TC with TAMs and M184V | – | | Kemp et al. (1998), Gallego et al. (2002), Hachiya et al. (2009), Gupta et al. (2011) |
| G335C/D | B | – | AZT with TAMs | NVP (G335C) in concert with N348I | | Nikolenko et al. (2007), Hachiya et al. (2009), Nikolenko et al. (2010) |
| N348I | B | AZT, ddI[b], NVP, DLV, ETR (low) | AZT with TAM-1 and TAM-2 | DLV, N, EFV, ETR with NNRTI-BP mutations | Associated with increase in viral load at least as large as the appearance of each of the TAMs (subtype B) | Yap et al. (2007), Nikolenko et al. (2007), Hachiya et al. (2008), Gupta et al. (2010), Sluis-Cremer et al. (2010), von Wyl et al. (2010a), Lengruber et al. (2011), McCormick et al. (2011), Gupta et al. (2011), Brehm et al. (2012b), Saeng-aroon et al. (2010) |
| | C | NVP, ETV in clade C | 3TC with TAMs | | Associated with failure of first-line NVP+ d4T+3TC treatment | |
| | | | TDF with TAMs | | (CRF01_AE)<br>Associated with virologic failure of first-line ART (subtype C) | |
| K358R | B | AZT<br>NVP hypersusceptibility[c] | AZT with TAMs | | | Lengruber et al. (2011) |
| G359S | B | AZT hypersusceptibility[c] | AZT hypersusceptibility with TAM-1 and TAM-2 | | | Lengruber et al. (2011) |

| | | | | | | |
|---|---|---|---|---|---|---|
| A360I/V | B | AZT (A360V)[d] | AZT with TAMs (A360V) | NVP (A360V) with NNRTI-BP mutations<br>EFV (A360I) with NNRTI-BP mutations | | Nikolenko et al. (2007), Ehteshami et al. (2008), Santos et al. (2008), Hachiya et al. (2009)[e], von Wyl et al. (2010b), Nikolenko et al. (2010), Lengruber et al. (2011)[c], Brehm et al. (2012a) |
| V365I | B | | AZT with TAMs | | | Nikolenko et al. (2007), Hachiya et al. (2009) |
| T369I/V | B | AZT, NVP, DLV, EFV | AZT with TAM-1 TAM-2 and N348I | EFV, NVP, and DLV (T369I) with N348I<br>EFV, NVP, and DLV with NNRTI-BP mutations<br>NVP (T369V) in context of TAMs<br>ETR (T369I) with NNRTI-BP mutations | | Gupta et al. (2010), Lengruber et al. (2011) |
| A371V | B<br>CRF01_AE | | AZT with TAM-1 and TAM-2 (B and CRF01)<br>3TC, ABC with TAMs and Q509L (B)<br>D4T and ddI with G335D and TAM-1/TAM-2 (CRF01)<br>TDF with G335D and TAM-2 (CRF01) | NVP in context of TAMs | Associated with lack of virologic response | Brehm et al. (2007)[f], Lengruber et al. (2011), Tanuma et al. (2010), Dau et al. (2010) |
| A376S | B | NVP | AZT with TAMs | NVP and EFV with Pol domain from RTI-experienced patients | Associated with tenfold increased risk of virologic failure to NVP but not EFV-based ART | Nikolenko et al. (2007), Hachiya et al. (2009), Nikolenko et al. (2010), Paredes et al. (2011) |
| T386A | B | | | EFV with NNRTI-BP mutations | | Vingerhoets et al. (2005)[f] |

(continued)

**Table 13.1** (continued)

| Mutation | Subtype[a] | Decreased RTI susceptibility in wild-type backbone | Decreased NRTI susceptibility in concert with classical mutations | Decreased NNRTI susceptibility in concert with classical mutations | Associated with virologic failure | References |
|---|---|---|---|---|---|---|
| E399G/D | B | | AZT (E399D) with TAMs[g] | EFV with NNRTI-BP mutations<br>ETR with NNRTI-BP mutations | CRF01_AE – associated with failure of first-line NVP + d4T + 3TC treatment | Gupta et al. (2006), Poveda et al. (2008), Gupta et al. (2011) |
| A400T | B | | AZT with TAMs | | | Santos et al. (2008), Delviks-Frankenberry et al. (2009), Tanuma et al. (2010) |

[a]Genetic background used to determine phenotype in site-directed mutagenesis studies
[b]Reported in one study (Hachiya et al. 2008)
[c]Not significant
[d]Resistance not reported in all studies
[e]Significant decreased susceptibility to AZT, NVP, EFV not observed for A360V or A360I in WT backbone in this study
[f]In vitro selection
[g]In context of patients, degree of reduction depends on background mutations

**Table 13.2** Drug resistance phenotype of mutations in the RNase H domain

| Mutation | Subtype | Selected in vivo[a] | Decreased drug susceptibility alone[b] | Decreased drug susceptibility in concert with Pol mutations[b] | References |
|---|---|---|---|---|---|
| K451R | B | Yes | No | No with TAMs | Santos et al. (2008), Waters et al. (2009), Lengruber et al. (2011) |
| Q475A | B | No | NVP, DLV, EFV | Not done | Santos et al. (2008), Nikolenko et al. (2010) |
| D488E | B | Yes | No | AZT with TAM-2 | Santos et al. (2008), Lengruber et al. (2011), Santos et al. (2011) |
| Y501A | B | No | NVP, DLV, EFV | Not done | Santos et al. (2008), Nikolenko et al. (2010) |
| Q509L | B | No[c] | No | AZT with TAMs or A371V and TAMs | Brehm et al. (2007), Ntemgwa et al. (2007) |
| I524M | CRF01_AE | Yes Associated with failure of first-line NVP+3TC+d4T | Not done | Not done | Saeng-aroon et al. (2010) |
| P537S | CRF01_AE | Yes Associated with failure of first-line NVP+3TC+d4T | Not done | Not done | Saeng-aroon et al. (2010) |
| Q547K | B | Yes | No | AZT with TAM-1 | Lengruber et al. (2011), Santos et al. (2011) |
| H549N | B | No | NVP, DLV | NVP, DLV, EFV, ETR with NNRTI-BP mutations AZT with TAMs | Nikolenko et al. (2005), Ntemgwa et al. (2007), Nikolenko et al. (2010) |

[a]Selected in the presence of ART
[b]Confirmed by site-directed mutagenesis studies
[c]In vitro selection

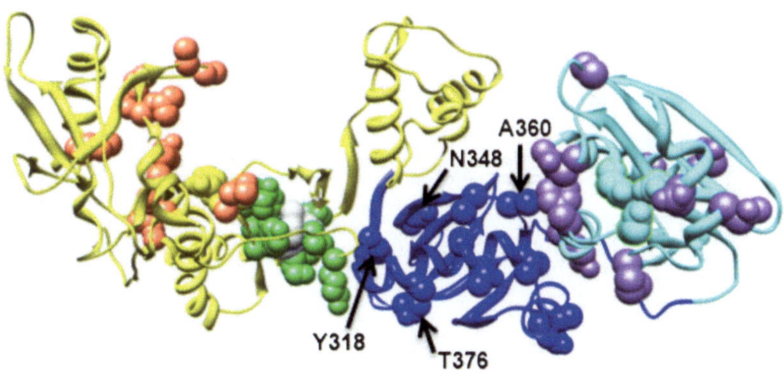


**Fig. 13.3** Three dimensional structure of the p66 subunit of HIV-1 RT showing locations of mutations that confer RTI resistance. The DNA polymerase domain is colored *yellow* and the locations of the NRTI and NNRTI resistance mutations are shown in *orange* and *green*, respectively. The connection domain is shown in *blue*. Key residues described in this review, including Y318, N348, A360 and T376, are highlighted. The RNase H domain is shown in *cyan*, and mutations in this domain are colored purple. The figure was generated using Chimera and the PDB coordinates 3HVT

**Table 13.3** Prevalence of mutation in naive and RTI-experienced individuals in subtypes. Data are adapted from the Stanford HIV Drug Resistance Database (http://hivdb.stanford.edu/)

| | Prevalence of mutation in naive and RTI-experienced individuals in subtypes | | | | | | | | | | | | | | | |
|---|---|---|---|---|---|---|---|---|---|---|---|---|---|---|---|---|
| | A | | B | | C | | D | | F | | G | | AE | | AG | |
| Mutation | Naive | RTI | Naive | RTI | Naive | RTI | Naive | RTI | Naive | RTI | Naive | RTI | Naive | RTI | Naive | RTI |
| E312Q | 0 | 0 | 1.1 | 1.3 | 0.6 | 1.0 | 0.8 | 1.1 | 0 | 0 | 0 | 0.6 | 0 | 0 | 0.9 | 0 |
| Y318F | 0 | 0 | 0 | 1.6 | 0 | 1.3 | 0 | 0 | 0 | 0.9 | 0 | 0.9 | 0 | 0.7 | 0 | 0.7 |
| G333D | 0 | 0 | 0.8 | 0.8 | 0.1 | 0.3 | 0 | 0 | 0 | 0 | 0 | 0 | 0 | 0 | 0 | 0 |
| G333E | 0.5 | 0.9 | 7.9 | 11 | 1.1 | 2.2 | 3.3 | 4.4 | 1.8 | 5.7 | 1.3 | 1.0 | 0.3 | 0 | 0.1 | 0 |
| G335D | 87 | 86 | 2.2 | 2.5 | 80 | 69 | 53 | 41 | 18 | 1.0 | 46 | 14 | 92 | 97 | 86 | 74 |
| *N348I* | *0* | *3.9* | *0* | *9.2* | *0* | *14* | *0.8* | *9.4* | *0* | *10* | *0* | *6.7* | *0.5* | *6.2* | *0* | *5.6* |
| R358K | 12 | 12 | 5.6 | 14 | 4.2 | 10 | 5.3 | 13 | 15 | 5.0 | 3.2 | 0 | 0 | 0 | 11 | 11 |
| G359S | 91 | 98 | 7.2 | 28 | 6.5 | 6.5 | 14 | 23 | 89 | 95 | 94 | 93 | 100 | 100 | 95 | 96 |
| A360V | 0 | 3.9 | 1.2 | 5.6 | 0.5 | 0 | 1.8 | 3.3 | 0 | 0 | 0.8 | 0 | 0 | 0.9 | 0 | 1.4 |
| V365I | 0.7 | 2.0 | 3.8 | 6.7 | 3.3 | 3.9 | 1.9 | 10.0 | 6.5 | 0 | 4.0 | 6.9 | 0.5 | 0.9 | 2.2 | 1.4 |
| T369I | 0 | 2.0 | 0 | 0.5 | 0 | 0 | 0 | 0 | 0 | 0 | 0 | 0 | 0 | 0 | 0 | 0 |
| T369V | 13 | 24 | 1.5 | 5.0 | 1.1 | 1.3 | 0 | 3.3 | 2.2 | 5.0 | 1.6 | 0 | 2.8 | 3.6 | 9.9 | 15 |
| A371V | 82 | 86 | 4.7 | 23 | 3.1 | 21 | 3.8 | 10 | 30 | 25 | 56 | 93 | 96 | 99 | 86 | 96 |
| A376S | 4.7 | 14 | 5.9 | 12 | 4.7 | 4.0 | 15 | 10 | 11 | 20 | 19 | 27 | 3.9 | 1.8 | 13 | 21 |
| T386A | 6.2 | 6.0 | 1.1 | 4.5 | 0.7 | 1.4 | 1.0 | 6.7 | 0 | 5.0 | 0 | 13 | 1.5 | 0.9 | 2.8 | 4.2 |
| E399G | 0 | 0 | 0 | 1.1 | 1.3 | 0 | 1.0 | 0 | 0 | 0 | 0.9 | 0 | 0.5 | 0 | 0 | 0 |
| E399D | 9.3 | 4.0 | 14 | 16 | 3.1 | 1.5 | 7.1 | 14 | 28 | 60 | 11 | 33 | 2.9 | 9.7 | 6.9 | 12 |
| A400T | 76 | 71 | 42 | 28 | 62 | 68 | 66 | 44 | 85 | 17 | 16 | 0 | 95 | 88 | 29 | 39 |

## 13.4 Subtype B HIV-1

*N348I:* We reported that in the BC Centre for Excellence in HIV/AIDS genotypic drug resistance database, N348I was found to increase in prevalence from below 1 % in 368 treatment-naïve individuals to 12.1 % in 1,009 treatment-experienced patients ($p = 7.7 \times 10^{-12}$)(Yap et al. 2007). N348I appeared early in therapy and was highly associated with the TAMs M41L and T215Y/F ($p < 0.001$), the 3TC resistance mutations M184V/I ($p < 0.001$), and NNRTI resistance mutations K103N and Y181C/I ($p < 0.001$). The association with TAMs and NNRTI resistance mutations was consistent with the selection of N348I in patients treated with regimens that included both AZT and NVP (odds ratio 2.62, 95 % confidence interval 1.43–4.81). The appearance of N348I was associated with a significant increase in viral load ($p < 0.001$), which was as large as the viral load increases observed for any of the TAMs. However, this analysis did not account for the simultaneous selection of other RT or protease inhibitor resistance mutations on viral load. To delineate the role of this mutation in RT inhibitor resistance, N348I was introduced into HIV-1 molecular clones containing different genetic backbones. N348I decreased AZT susceptibility 2- to 4-fold in the context of wild-type HIV-1 or when combined with TAMs. N348I also decreased susceptibility to NVP (7.4-fold) and EFV (2.5-fold) and significantly potentiated resistance to these drugs when combined with K103N. Subsequently, several other studies have confirmed the role of N348I in RTI resistance; however, the association with drugs and mutations varies (Nikolenko et al. 2007; Hachiya et al. 2008; Gupta et al. 2010; von Wyl et al. 2010a). For example, Hachiya et al. reported that the acquisition of N348I was frequently observed in AZT- and/or ddI-containing therapy (12.5 %; $n = 48$; $p < 0.0001$) and was accompanied with TAMs and M184V ($n = 1/6$) in a Japanese cohort (Hachiya et al. 2008). They also showed that N348I conferred ddI and DLV resistance in addition to AZT and NVP. In this regard, we showed that N348I by itself confers only a 1.6-fold ($p = 0.019$, $n = 4$) decrease in ETV susceptibility and does not confer TDF resistance but confers resistance to both drugs when combined with TAMs or Y181C, respectively (Sluis-Cremer et al. 2010). Of interest, Price et al. provided some clinical evidence to suggest that EFV and d4T, in addition to NVP and AZT, select for the N348I mutation, whereas TDF might protect against it (Price et al. 2010).

Several NNRTI (e.g., Y181C) and NRTI (e.g., L74V and M184V) resistance mutations in HIV-1 RT are antagonistic toward TAMs that confer AZT resistance (Gotte et al. 2000; Boyer et al. 2002; Miranda et al. 2005; Larder 1992; Selmi et al. 2003). In this regard, the N348I mutation may also provide a simple genetic pathway that allows the virus to select both TAMs and mutations that are antagonistic toward TAMs. For example, von Wyl et al. reported that the rate of N348I emergence was increased by fivefold in patients who received AZT/3TC compared to those that only received AZT (von Wyl et al. 2010a). They went on to show that N348I can partially compensate for the diminution in processive DNA synthesis and the reduction in AZT excision associated with M184V. Additionally, we demonstrated that N348I compensates for the antagonism of TAMs by Y181C, L74V, and M184V (Radzio et al. 2010).

*A360V:* Ehteshami et al. reported that in the BC Centre for Excellence in HIV/AIDS database, the A360V mutation in the connection domain of HIV-1 RT was observed in less than 1 % of the untreated population ($n=15$ of 1,606) and increased to 8.5 % after treatment ($n=205$ of 2,422) (Ehteshami et al. 2008). They also noted that A360V was highly correlated with various mutations considered to be part of the TAMs cluster; for example, of the samples with A360V, 35 % were associated with M41L, 21 % with D67N, 30 % with K70R, 17 % with L210W, 42 % with T215Y/F, and 17 % with K219Q/E. Interestingly, von Wyl et al. also reported an increase of this mutation in 334 patients receiving monotherapy or dual therapy with thymidine analogues at the time of the genotypic resistance test (von Wyl et al. 2010b). However, they found that A360V was not associated with specific drug combinations and was found to emerge later than M184V or TAMs. Recently, Brehm et al. showed that the A360V mutation was selected in HIV-infected individuals that received AZT monotherapy and contributed to AZT resistance (Brehm et al. 2012a). In this study, full-length RT sequences in plasma obtained pre- and post-therapy were compared in 23 participants who received AZT monotherapy from the AIDS Clinical Trials Group study 175. Mutations significantly associated with AZT monotherapy included K70R ($p=0.003$) and T215Y ($p=0.013$) in the polymerase domain of HIV-1 RT and A360V ($p=0.041$) in the connection domain of HIV-1 RT. HIV-1 drug susceptibility assays demonstrated that A360V, either alone or in combination with TAMs, decreased AZT susceptibility in recombinant viruses containing participant-derived full-length RT sequences or site-directed mutant RT.

*A376S:* Nikolenko et al. were the first to report that the A376S mutation was strongly associated with an observed increase in AZT resistance (Nikolenko et al. 2007). Later studies reported on the selection of this mutation during drug therapy. Paredes et al. reported that preexisting A376S was associated with an increased risk of virological failure to NVP (relative hazard [RH] = 10.4; 95 % confidence interval [CI], 2.0–54.7) but did not affect EFV outcome the same way (RH = 0.5; 95 % CI, 0.1–2.2) ($p=0.013$)(Paredes et al. 2011). A376S conferred selective low-level NVP resistance in vitro. Interestingly Gupta et al. reported that virologic responses to an ETR-containing regimen were slightly diminished when A376S was present (Gupta et al. 2011).

## 13.5 N348I in Non-Subtype B HIV-1

Studies by McCormick et al. reported that N348I alone reduced the susceptibility to all NNRTIs across HIV-1 subtypes A, B, C, and D, demonstrating that the phenotype conferred by N348I is not restricted to subtype B (McCormick et al. 2011). N348I has been reported in drug-treated individuals on first-line regimens that include the combination of d4T, 3TC, and NVP used in resource-constrained countries. Tanuma et al. determined the frequencies of drug resistance mutations of the entire RT in patients with treatment failure from a cohort of circulating recombinant form (CRF) 01_AE HIV-1-infected patients in Hanoi, Vietnam (Tanuma et al. 2010). In 49 patients with treatment failure, resistance mutations to NRTIs in the N-terminal half of RT were observed in 89.8 % and N348I in 36.8 % of patients. N348I is not considered a polymorphism of CRF01_AE (Table 13.3).

Seang-aroon et al. determined resistance to the generic combination tablet of d4T, 3TC, and NVP (GPOvir) in HIV/AIDS patients treated in Thailand's national antiretroviral treatment program (Saeng-aroon et al. 2010). HIV-1 CRF01_AE sequences of 59 GPOvir-failure cases from the Lampang cohort were compared with sequences from 76 randomly selected ART-naive cases. The GPOvir-failure cases, in addition to recognized d4T-, 3TC-, and NVP-resistant mutations, also had significantly higher frequency of N348I in the connection domain.

Brehm et al. also examined whether the N348I mutation is associated with failure of first-line d4T, 3TC, and NVP or EFV combination therapy in 63 participants with subtype C HIV-1 infection enrolled in the Comprehensive International Program of Research on AIDS in South Africa (CIPRA-SA) (Brehm et al. 2012b). Y181C and M184V mutations in the RT polymerase domain were associated with failure of d4T/3TC/NVP ($p<0.01$) and K103N, V106M, and M184V with failure of d4T/3TC/EFV ($p<0.01$). N348I in the RT connection domain emerged in 45 % ($p=0.002$) and 12 % ($p=0.06$) of participants on failing regimens containing NVP or EFV, respectively. Longitudinal analyses revealed that NNRTI resistance mutations in the polymerase domain generally appeared first. N348I emerged at the same time or after M184V. N348I reduced susceptibility to NVP (8.9- to 13-fold), EFV (4- to 56-fold), and ETR (1.9- to 4.7-fold) and decreased hypersusceptibility to AZT (1.4- to 2.2-fold). This study demonstrated that N348I emerges frequently with virologic failure of first-line ART in subtype C HIV-1 infection and reduces susceptibility to NVP, EFV, ETR, and AZT. Collectively these studies provide compelling evidence that N348I is associated with virologic failure in individuals infected with non-subtype B HIV-1 which have implications for cross-resistance to subsequent ART regimens in resource-limited settings.

## 13.6 RNase H Mutations

Brehm et al. were the first to report that the Q509L mutation in the RNase H domain was selected in vitro by passaging wild-type HIV-1 in the presence of increasing concentrations of AZT (Brehm et al. 2007). The first resistance mutations to appear in HIV-1 (LAI) were two polymerase domain TAMs, D67N and K70R, and two novel mutations, A371V in the connection domain and Q509L in the RNase H domain, that together conferred up to 90-fold AZT resistance. A371V and Q509L together increased AZT resistance approximately 10- to 50-fold in combination with TAMs (M41L/L210W/T215Y or D67N/K70R/T215F) but had a minimal effect without TAMs (1.7-fold). A371V and Q509L also increased cross-resistance with TAMs to 3TC and ABC, but not d4T or ddI. Despite the ability of Q509L to potentate AZT resistance, its prevalence is low (0.2 %) in drug-treated individuals (Ntemgwa et al. 2007).

Several mutations in the RNase H domain have been described that are present at higher frequency in drug-treated compared to drug-naïve individuals (Table 13.2). The only two that have been associated with virologic failure are I524M and P537S in CRF01_AE-infected individuals (Saeng-aroon et al. 2010). However, the impact of these mutations on decreased drug susceptibility is unknown. Thus, while the original impetus for studying mutations beyond the polymerase domain came from the observation that mutations decreasing RNase H can increase AZT resistance (Nikolenko et al. 2005), most of the mutations associated with virologic failure are selected in the connection domain.

## 13.7 Impact of N348I on Viral Fitness

N348I disappears relatively rapidly upon cessation of treatment (Hachiya et al. 2008). Consistent with this observation, subtype B HIV-1 harboring N348I demonstrates decreased replication kinetics compared to wild type (Hachiya et al. 2008). However, this effect is cell-type dependent with dramatic reductions observed in MT-2 and SupT1 cells, moderate reductions in PM1 and PHA-stimulated peripheral blood mononuclear cells, and little reduction observed in H9 cells. The cell-type dependence is postulated to be due to the differences in the availability of dNTPs or a cellular factor that may compensate the N348I effect (Hachiya et al. 2008). Decreased replication mediated by N348I was confirmed in subsequent studies where the replication capacity of N348I mutant HIV was reported as 83 % of wild-type NL4.3 in 239 cells (Gupta et al. 2010) with significant 27–44 %, 33–46 %, and 18–37 % reductions in a subtype B viral background compared to wild type in experiments performed in 239T, Jurkat, and SupT1 cells, respectively (McCormick et al. 2011). N348I also significantly decreases viral replication capacity when present in subtype A, C, and D genetic backbones (McCormick et al. 2011).

## 13.8 Mechanisms by which Mutations in the Connection and RNase H Domains of HIV-1 RT Decrease NRTI Susceptibility

Two phenotypic mechanisms of NRTI resistance have been proposed, namely, NRTI-triphosphate (TP) discrimination and NRTI-monophosphate (MP) excision (Menendez-Arias 2010; Sarafianos et al. 2009). The NRTI-TP discrimination phenotype involves the acquisition of one or more resistance mutations in RT that improve the enzyme's ability to discriminate between the natural dNTP substrate and the NRTI-TP. Resistance by this mechanism is typically associated with decreased catalytic efficiency of NRTI-TP incorporation. NRTI-TP (and dNTP) catalytic efficiency is driven by two kinetic parameters: (i) the affinity of the nucleotide for the RT polymerase active site ($K_d$) and (ii) the maximum rate of nucleotide incorporation ($k_{pol}$), both of which can be determined using pre-steady-state kinetic analyses (Reardon 1992). In general, NRTI-TP discrimination is achieved by the resistance mutation affecting only one of these kinetic parameters. For example, pre-steady-state kinetic analyses have demonstrated that K65R confers resistance to ddATP (the active metabolite of ddI), 3TCTP, carbovir-TP (CBVTP, active metabolite of ABC), and TFV-diphosphate by selectively reducing $k_{pol}$ without affecting $K_d$ (Selmi et al. 2001; Deval et al. 2004; Sluis-Cremer et al. 2007). By contrast, the M184V exerts a profound effect on the $K_d$ for 3TCTP, without impacting $k_{pol}$ (Deval et al. 2004; Feng and Anderson 1999). For the excision mechanism of NRTI resistance, the mutant HIV-1 RT does not discriminate between the natural dNTP substrate and the NRTI-TP at the nucleotide incorporation step (Kerr and Anderson 1997). Instead, RT containing "excision" facilitating mutations shows an increased capacity to unblock NRTI-MP-terminated primers in the presence of physiological concentrations of ATP (Meyer et al. 1999; Boyer et al. 2001). NRTI resistance mutations associated with the excision mechanism include TAMs. The efficiency of the NRTI-MP excision reaction by RT containing TAMs depends on the chain-terminating NRTI-MP residing in the nucleotide-binding site (N-site) of the RT active site. Under physiological conditions, the binding of the next correct dNTP can drive the terminating nucleotide into the primer-binding site (P-site) resulting in the formation of a dead-end complex (DEC) (Meyer et al. 1999; Boyer et al. 2001). Biochemical studies have proposed several mechanisms by which TAMs can increase the efficiency of RT to excise NRTI-MP from chain-terminated primers. These mechanisms include (i) an increase in the binding affinity of ATP for RT (Boyer et al. 2001), (ii) an increase in the kinetic rate of ATP- or PPi-mediated NRTI-MP excision (Meyer et al. 1999; Arion et al. 1998; Ray et al. 2003), (iii) a decrease in sensitivity of RT to DEC formation (Boyer et al. 2001), and (iv) a shift in the translocation equilibrium of the primer terminus between the N- and P-sites such that the N-site is favored (Marchand and Gotte 2003).

As described above, in 2005, Nikolenko et al. proposed that mutations which reduce RNase H cleavage can contribute to the NRTI-resistant phenotype by providing more time for RT to carry out nucleotide excision and resume productive DNA synthesis (Nikolenko et al. 2005). In this regard, Yap et al. were the first to demonstrate that the N348I mutation alone or in combination with TAMs decreased the frequency of secondary RNase H cleavages that reduce the RNA/DNA duplex length of the T/P and diminish the efficiency of AZT-MP excision (Yap et al. 2007). The net result of this diminished RNase H cleavage phenotype is that enzymes that contain the N348I mutation exhibit an increased ability to excise AZT-MP from RNA/DNA T/Ps, but not DNA/DNA T/Ps. In addition, efficiency of AZT-MP excision in the presence of NVP is greater for N348I HIV-1 RT compared to wild-type enzyme, which is consistent with the ability of this mutation to reverse the synergistic inhibition of HIV-1 by AZT and NVP (Huey Yap et al. 2012). AZT-MP excision can only occur when the chain-terminating moiety resides in the N-site of the RT polymerase active site. When bound in this configuration, RT can carry out a primary RNase H cleavage event that reduces the RNA/DNA duplex to ~18 nucleotides in length (Fig. 13.4). Thereafter the enzyme has to either translocate or dissociate and rebind the T/P in a polymerase (and excision) incompetent mode that will allow for the secondary RNase H cleavages to occur that further decrease the length of the RNA:

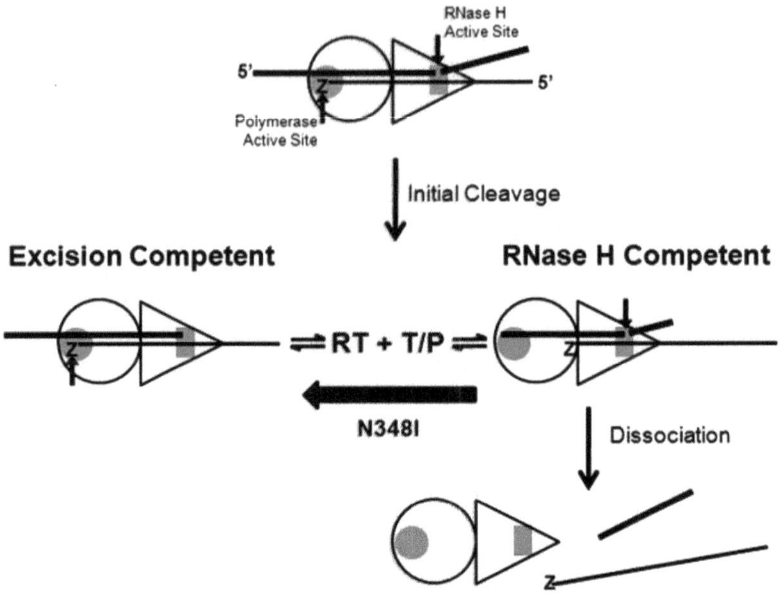


**Fig. 13.4** Proposed model for how N348I in RT affects the balance between AZT-MP excision and RNase H activity on RNA/DNA T/P. The initial RNase H cleavages reduce the RNA/DNA duplex to 15 to 18 nts in length. Because the distance between the DNA polymerase and RNase H active sites is 18 nts, RT has to bind the resultant T/P products in one of two distinct conformations to carry-out either excision or RNase H activity. Therefore, after the primary RNase H cleavages and T/P dissociation event, an equilibrium forms in which RT binds the T/P in both of these configurations. N348I selectively decreases the affinity of RT to bind the T/P in an RNase H competent mode but not an excision competent mode

DNA duplex. In this regard, Etheshami et al. elegantly demonstrated that the N348I mutation in HIV-1 RT promotes rapid dissociation from the RNase H-competent complexes that allow for secondary RNase H cleavage, whereas binding in the polymerase (and excision)-competent modes is not affected (Ehteshami et al. 2008). A similar phenotypic mechanism of AZT resistance was proposed for the A360V and Q509L mutations in HIV-1 RT (Brehm et al. 2008; Ehteshami et al. 2008; Brehm et al. 2012a). Interestingly, recent studies have suggested that the decreased RNase H phenotype of N348I RT is due to the mutation in the p51 subunit of the enzyme (Schuckmann et al. 2010; Radzio and Sluis-Cremer 2011).

Interestingly, connection domain mutations may also exert an RNase H-independent NRTI-resistance phenotype. In this regard, Ehteshami et al. found that N348I (and A360V) in combination with TAMs enhanced AZT-MP excision even in the presence of an RNase H-inactivating E478Q mutation, indicating that an RNase H-independent mechanism also contributes to the increase in AZT resistance (Ehteshami et al. 2008). They also reported that these mutations increased the processivity of RT, which could account for more efficient AZT-MP excision. Of note, Zelina et al. found that in the presence of TAMs and M184V, the G333D polymorphism increased the ability of the RT to bind to the T/P and increased ATP-mediated excision (Zelina et al. 2008). In addition, G333D increased discrimination against 3TC incorporation in the presence of M184V. The G333D mutation did not exhibit reductions in RNase H cleavage, and it was concluded that in the context of TAMs and M184V, G333D directly affects the polymerase active site, presumably as a result of long-range interactions and conformational changes in the connection domain (Zelina et al. 2008).

## 13.9 Mechanisms by which Mutations in the Connection and RNase H Domains of HIV-1 RT Decrease NNRTI Susceptibility

Typically, HIV-1 resistance to NNRTIs correlates directly with mutations of one or more RT residues in the NNRTI-BP. Mutations associated with resistance to NNRTIs include L100I, K101E, K103N, V106A/M, V108I, E138A/K, V179D, Y181C/I/V, Y188C/L/H, G190A/E/S, M230L, P236L, and Y318F (Johnson et al. 2011). These mutations can affect inhibitor binding in a number of ways. (1) They can remove one or more favorable interactions between the inhibitor and NNRTI-binding pocket. For example, the Y181C mutation eliminates $\pi$-stacking interactions between this residue and the aromatic ring of the NNRTI pharmacophore (Ren et al. 2001). (2) They can introduce steric barriers to NNRTI binding. For example, the G190E mutation introduces a bulky side-chain which may prevent NNRTI binding by sterically interfering with functional groups, such as the cyclopropyl ring of NVP (Huang et al. 2003; Sluis-Cremer et al. 2000). (3) The mutations may introduce or eliminate inter-residue contacts in the NNRTI-binding pocket, which interfere with the ability of other residues in the pocket to fold down over the NNRTI (Sluis-Cremer et al. 2000). The mechanisms by which mutations in the connection domain of HIV-1 RT confers NNRTI resistance is not entirely clear, and several different mechanisms have been proposed as described below.

The initiation of (+)-strand DNA synthesis is exquisitely sensitive to NNRTI inhibition (Grobler et al. 2007). During the initiation of (+)-strand DNA synthesis, RT can bind its T/P substrate in one of two orientations (Gotte et al. 1999; Fuentes et al. 1995; Abbondanzieri et al. 2008). The polymerase-competent mode is characterized through interaction between the polymerase active site and the 3′-end of the primer. Conversely, in the RNase H-competent mode, the RT enzyme binds with the polymerase active site in the vicinity of the DNA template so that the RNase H active site is positioned over the chimeric RNA · DNA strand that is cleaved at the junction. Thus, the efficient NNRTI-mediated inhibition during initiation of (+)-strand DNA synthesis may be explained by changes in the ratio of polymerase- to RNase H-competent complexes (Abbondanzieri et al. 2008). NNRTIs appear to promote the latter orientation, which inhibits DNA synthesis and facilitates the primer removal (Grobler et al. 2007; Abbondanzieri et al. 2008). In this regard, Biondi et al. demonstrated that N348I counteracts this effect, particularly for NVP, thus providing a novel mechanism for resistance (Biondi et al. 2010).

Schuckmann et al. demonstrated that the N348I mutation directly decreased NVP binding by primarily decreasing the association rate (i.e., $k_{on}$) of the inhibitor (Schuckmann et al. 2010). Interestingly, the authors demonstrated that N348I in both the p66 and p51 subunit of RT caused NVP resistance during RNA-dependent and DNA-dependent DNA polymerization. In this regard, the mutation in p66 alone caused NVP resistance without significantly affecting RNase H activity, whereas the mutation in p51 caused NVP resistance and impaired RNase H. As such, the authors proposed that NVP resistance may occur independently from defects in RNase H function. By contrast, Radzio and Sluis-Cremer reported that that N348I-mediated NVP resistance mapped to the mutation in the p51 subunit of RT and correlated with a decreased RNase H phenotype (Radzio and Sluis-Cremer 2011). In this regard, Nikolenko et al. proposed that connection domain mutations, by reducing RNase H cleavage, may provide more time for the NNRTI to dissociate from RT, resulting in the resumption of DNA synthesis and enhanced NNRTI resistance (Nikolenko et al. 2010).

Interestingly, Paredes et al. reported that the A376S decreases NVP affinity ($K_d$) by twofold (Paredes et al. 2011). The authors suggested that the observed decreased affinity for NVP could result from the higher affinity for double-stranded DNA of the mutant A376S RT relative to the wild-type enzyme or could be due to the contribution of Ala/Ser-376 in p66 to the dimerization interface in the RT heterodimer. In regard to the latter, however, Figueiredo et al. found no obvious correlation between NNRTI resistance and heterodimer stabilization, suggesting that the stability of RT heterodimers is unlikely to be a key player in NNRTI antiviral activity or NNRTI resistance (Figueiredo et al. 2008).

## 13.10 Conclusion

There is substantial evidence that RTIs select for resistance mutations beyond the DNA polymerase domain of HIV-1 RT. The N348I mutation is perhaps the most important and well studied of these mutations and has been observed in multiple resistance databases and across different HIV clades. Additional studies are required to determine whether N348I (and other connection domain mutations) significantly impacts the durability of second- and third-line RTI-based regimens, particularly in third world countries that use D4T/3TC or AZT/3TC in combination with NVP or EFV as roll-out therapy.

**Acknowledgments** The authors would like to thank Drs. Richard Harrigan and John Mellors for their contribution to many of the studies described above. Additionally, we are indebted to our colleagues who have made significant contributions to understanding the impact of connection and RNase H domain mutations on RTI resistance. NSC and GT gratefully acknowledge support from the National Institutes of Health grants AI081571 and GM068406 and the National Health and Medical Research (NHMRC) Project Grant 603704. GT also gratefully acknowledges support from NHMRC Senior Research Fellowship 543105 and the Victorian Operational Infrastructure Support Program received by the Burnet Institute.

## References

Abbondanzieri EA, Bokinsky G, Rausch JW, Zhang JX, Le Grice SF, Zhuang X (2008) Dynamic binding orientations direct activity of HIV reverse transcriptase. Nature 453(7192):184–189
Arion D, Kaushik N, McCormick S, Borkow G, Parniak MA (1998) Phenotypic mechanism of HIV-1 resistance to 3′-azido-3′-deoxythymidine (AZT): increased polymerization processivity and enhanced sensitivity to pyrophosphate of the mutant viral reverse transcriptase. Biochemistry 37(45):15908–15917
Biondi MJ, Beilhartz GL, McCormick S, Gotte M (2010) N348I in HIV-1 reverse transcriptase can counteract the nevirapine-mediated bias toward RNase H cleavage during plus-strand initiation. J Biol Chem 285(35):26966–26975
Boyer PL, Sarafianos SG, Arnold E, Hughes SH (2001) Selective excision of AZTMP by drug-resistant human immunodeficiency virus reverse transcriptase. J Virol 75(10):4832–4842
Boyer PL, Sarafianos SG, Arnold E, Hughes SH (2002) The M184V mutation reduces the selective excision of zidovudine 5′-monophosphate (AZTMP) by the reverse transcriptase of human immunodeficiency virus type 1. J Virol 76(7):3248–3256
Brehm JH, Koontz D, Meteer JD, Pathak V, Sluis-Cremer N, Mellors JW (2007) Selection of mutations in the connection and RNase H domains of human immunodeficiency virus type 1 reverse transcriptase that increase resistance to 3′-azido-3′-dideoxythymidine. J Virol 81(15):7852–7859
Brehm JH, Mellors JW, Sluis-Cremer N (2008) Mechanism by which a glutamine to leucine substitution at residue 509 in the ribonuclease H domain of HIV-1 reverse transcriptase confers zidovudine resistance. Biochemistry 47(52):14020–14027
Brehm JH, Scott Y, Koontz DL, Perry S, Hammer S, Katzenstein D, Mellors JW, Sluis-Cremer N (2012a) Zidovudine (AZT) monotherapy selects for the A360V mutation in the connection domain of HIV-1 reverse transcriptase. PLoS One 7(2):e31558
Brehm JH, Koontz DL, Wallis CL, Shutt KA, Sanne I, Wood R, McIntyre JA, Stevens WS, Sluis-Cremer N, Mellors JW (2012b) Frequent emergence of N348I in HIV-1 subtype C reverse transcriptase with failure of initial therapy reduces susceptibility to reverse-transcriptase inhibitors. Clin Infect Dis 55(5):737–745
Clavel F, Hance AJ (2004) HIV drug resistance. N Engl J Med 350(10):1023–1035
Das K, Martinez SE, Bauman JD, Arnold E (2012) HIV-1 reverse transcriptase complex with DNA and nevirapine reveals non-nucleoside inhibition mechanism. Nat Struct Mol Biol 19(2):253–259
Dau B, Ayers D, Singer J, Harrigan PR, Brown S, Kyriakides T, Cameron DW, Angus B, Holodniy M (2010) Connection domain mutations in treatment-experienced patients in the OPTIMA trial. J Acquir Immune Defic Syndr 54(2):160–166
Delviks-Frankenberry KA, Nikolenko GN, Maldarelli F, Hase S, Takebe Y, Pathak VK (2009) Subtype-specific differences in the human immunodeficiency virus type 1 reverse transcriptase connection subdomain of CRF01_AE are associated with higher levels of resistance to 3′-azido-3′-deoxythymidine. J Virol 83(17):8502–8513
Deval J, White KL, Miller MD, Parkin NT, Courcambeck J, Halfon P, Selmi B, Boretto J, Canard B (2004) Mechanistic basis for reduced viral and enzymatic fitness of HIV-1 reverse transcriptase containing both K65R and M184V mutations. J Biol Chem 279(1):509–516
Ding J, Das K, Moereels H, Koymans L, Andries K, Janssen PA, Hughes SH, Arnold E (1995) Structure of HIV-1 RT/TIBO R 86183 complex reveals similarity in the binding of diverse non-nucleoside inhibitors. Nat Struct Biol 2(5):407–415
Ehteshami M, Beilhartz GL, Scarth BJ, Tchesnokov EP, McCormick S, Wynhoven B, Harrigan PR, Gotte M (2008) Connection domain mutations N348I and A360V in HIV-1 reverse transcriptase enhance resistance to 3′-azido-3′-deoxythymidine through both RNase H-dependent and -independent mechanisms. J Biol Chem 283(32):22222–22232
Esnouf R, Ren J, Ross C, Jones Y, Stammers D, Stuart D (1995) Mechanism of inhibition of HIV-1 reverse transcriptase by non-nucleoside inhibitors. Nat Struct Biol 2(4):303–308

Feng JY, Anderson KS (1999) Mechanistic studies examining the efficiency and fidelity of DNA synthesis by the 3TC-resistant mutant (184V) of HIV-1 reverse transcriptase. Biochemistry 38(29):9440–9448

Figueiredo A, Zelina S, Sluis-Cremer N, Tachedjian G (2008) Impact of residues in the nonnucleoside reverse transcriptase inhibitor binding pocket on HIV-1 reverse transcriptase heterodimer stability. Curr HIV Res 6(2):130–137

Fuentes GM, Rodriguez-Rodriguez L, Fay PJ, Bambara RA (1995) Use of an oligoribonucleotide containing the polypurine tract sequence as a primer by HIV reverse transcriptase. J Biol Chem 270(47):28169–28176

Furman PA, Fyfe JA, St Clair MH, Weinhold K, Rideout JL, Freeman GA, Lehrman SN, Bolognesi DP, Broder S, Mitsuya H et al (1986) Phosphorylation of 3′-azido-3′-deoxythymidine and selective interaction of the 5′-triphosphate with human immunodeficiency virus reverse transcriptase. Proc Natl Acad Sci USA 83(21):8333–8337

Gallego O, Corral A, de Mendoza C, Rodes B, Soriano V (2002) Prevalence of G333D/E in naive and pretreated HIV-infected patients. AIDS Res Hum Retroviruses 18(12):857–860

Galli R, Wynhoven B, Sattha B, Tachedjian G, Harrigan PR (2004) Beyond codon 240: mutations in the HIV-1 reverse transcriptase selected after exposure to antiretrovirals. Paper presented at The XV International AIDS Conference, Bangkok, 11–16 July 2004

Gotte M, Maier G, Onori AM, Cellai L, Wainberg MA, Heumann H (1999) Temporal coordination between initiation of HIV (+)-strand DNA synthesis and primer removal. J Biol Chem 274(16):11159–11169

Gotte M, Arion D, Parniak MA, Wainberg MA (2000) The M184V mutation in the reverse transcriptase of human immunodeficiency virus type 1 impairs rescue of chain-terminated DNA synthesis. J Virol 74(8):3579–3585

Grobler JA, Dornadula G, Rice MR, Simcoe AL, Hazuda DJ, Miller MD (2007) HIV-1 reverse transcriptase plus-strand initiation exhibits preferential sensitivity to non-nucleoside reverse transcriptase inhibitors in vitro. J Biol Chem 282(11):8005–8010

Gupta S, Fransen S, Paxinos EE, Huang W, Stawiski E, Petropoulos CJ, Parkin NT (2006) Infrequent occurrence of mutations in the C-terminal region of reverse transcriptase modulates susceptibility to RT inhibitors. Antivir Ther 11:S143

Gupta S, Fransen S, Paxinos EE, Stawiski E, Huang W, Petropoulos CJ (2010) Combinations of mutations in the connection domain of human immunodeficiency virus type 1 reverse transcriptase: assessing the impact on nucleoside and nonnucleoside reverse transcriptase inhibitor resistance. Antimicrob Agents Chemother 54(5):1973–1980

Gupta S, Vingerhoets J, Fransen S, Tambuyzer L, Azijn H, Frantzell A, Paredes R, Coakley E, Nijs S, Clotet B, Petropoulos CJ, Schapiro J, Huang W, Picchio G (2011) Connection domain mutations in HIV-1 reverse transcriptase do not impact etravirine susceptibility and virologic responses to etravirine-containing regimens. Antimicrob Agents Chemother 55(6):2872–2879

Hachiya A, Kodama EN, Sarafianos SG, Schuckmann MM, Sakagami Y, Matsuoka M, Takiguchi M, Gatanaga H, Oka S (2008) Amino acid mutation N348I in the connection subdomain of human immunodeficiency virus type 1 reverse transcriptase confers multiclass resistance to nucleoside and nonnucleoside reverse transcriptase inhibitors. J Virol 82(7):3261–3270

Hachiya A, Shimane K, Sarafianos SG, Kodama EN, Sakagami Y, Negishi F, Koizumi H, Gatanaga H, Matsuoka M, Takiguchi M, Oka S (2009) Clinical relevance of substitutions in the connection subdomain and RNase H domain of HIV-1 reverse transcriptase from a cohort of antiretroviral treatment-naive patients. Antiviral Res 82(3):115–121

Harrigan PR, Salim M, Stammers DK, Wynhoven B, Brumme ZL, McKenna P, Larder B, Kemp SD (2002) A mutation in the 3′ region of the human immunodeficiency virus type 1 reverse transcriptase (Y318F) associated with nonnucleoside reverse transcriptase inhibitor resistance. J Virol 76(13):6836–6840

Huang W, Gamarnik A, Limoli K, Petropoulos CJ, Whitcomb JM (2003) Amino acid substitutions at position 190 of human immunodeficiency virus type 1 reverse transcriptase increase susceptibility to delavirdine and impair virus replication. J Virol 77(2):1512–1523

Huey Yap S, Herman BD, Radzio J, Sluis-Cremer N, Tachedjian G (2012) N348I in HIV-1 reverse transcriptase counteracts the synergy between zidovudine and nevirapine. J Acquir Immune Defic Syndr 61(2):153–157. doi:10.1097/QAI.0b013e3182657990

Jacobo-Molina A, Ding J, Nanni RG, Clark AD Jr, Lu X, Tantillo C, Williams RL, Kamer G, Ferris AL, Clark P et al (1993) Crystal structure of human immunodeficiency virus type 1 reverse transcriptase complexed with double-stranded DNA at 3.0 A resolution shows bent DNA. Proc Natl Acad Sci USA 90(13):6320–6324

Johnson VA, Calvez V, Gunthard HF, Paredes R, Pillay D, Shafer R, Wensing AM, Richman DD (2011) 2011 update of the drug resistance mutations in HIV-1. Top Antivir Med 19(4):156–164

Kagan R, Winters M, Merigan T, Heseltine P (2004) Expanded range of HIV-1 reverse transcriptase mutations detected through long range sequencing. Paper presented at the 11th conference on retroviruses and opportunistic infections, San Francisco, 8–11 Feb 2004

Kemp SD, Shi C, Bloor S, Harrigan PR, Mellors JW, Larder BA (1998) A novel polymorphism at codon 333 of human immunodeficiency virus type 1 reverse transcriptase can facilitate dual resistance to zidovudine and L-2′,3′-dideoxy-3′-thiacytidine. J Virol 72(6):5093–5098

Kerr SG, Anderson KS (1997) Pre-steady-state kinetic characterization of wild type and 3′-azido-3′-deoxythymidine (AZT) resistant human immunodeficiency virus type 1 reverse transcriptase: implication of RNA directed DNA polymerization in the mechanism of AZT resistance. Biochemistry 36(46):14064–14070

Kohlstaedt LA, Wang J, Friedman JM, Rice PA, Steitz TA (1992) Crystal structure at 3.5 A resolution of HIV-1 reverse transcriptase complexed with an inhibitor. Science 256(5065):1783–1790

Larder BA (1992) 3′-Azido-3′-deoxythymidine resistance suppressed by a mutation conferring human immunodeficiency virus type 1 resistance to nonnucleoside reverse transcriptase inhibitors. Antimicrob Agents Chemother 36(12):2664–2669

Larder BA, Kemp SD (1989) Multiple mutations in HIV-1 reverse transcriptase confer high-level resistance to zidovudine (AZT). Science 246(4934):1155–1158

Larder BA, Darby G, Richman DD (1989) HIV with reduced sensitivity to zidovudine (AZT) isolated during prolonged therapy. Science 243(4899):1731–1734

Larder BA, Kemp SD, Harrigan PR (1995) Potential mechanism for sustained antiretroviral efficacy of AZT-3TC combination therapy. Science 269(5224):696–699

Lengruber RB, Delviks-Frankenberry KA, Nikolenko GN, Baumann J, Santos AF, Pathak VK, Soares MA (2011) Phenotypic characterization of drug resistance-associated mutations in HIV-1 RT connection and RNase H domains and their correlation with thymidine analogue mutations. J Antimicrob Chemother 66(4):702–708

Marchand B, Gotte M (2003) Site-specific footprinting reveals differences in the translocation status of HIV-1 reverse transcriptase. Implications for polymerase translocation and drug resistance. J Biol Chem 278(37):35362–35372

McCormick AL, Parry CM, Crombe A, Goodall RL, Gupta RK, Kaleebu P, Kityo C, Chirara M, Towers GJ, Pillay D (2011) Impact of the N348I mutation in HIV-1 reverse transcriptase on nonnucleoside reverse transcriptase inhibitor resistance in non-subtype B HIV-1. Antimicrob Agents Chemother 55(4):1806–1809

Menendez-Arias L (2010) Molecular basis of human immunodeficiency virus drug resistance: an update. Antiviral Res 85(1):210–231

Meyer PR, Matsuura SE, Mian AM, So AG, Scott WA (1999) A mechanism of AZT resistance: an increase in nucleotide-dependent primer unblocking by mutant HIV-1 reverse transcriptase. Mol Cell 4(1):35–43

Miranda LR, Gotte M, Liang F, Kuritzkes DR (2005) The L74V mutation in human immunodeficiency virus type 1 reverse transcriptase counteracts enhanced excision of zidovudine monophosphate associated with thymidine analog resistance mutations. Antimicrob Agents Chemother 49(7):2648–2656

Nikolenko GN, Palmer S, Maldarelli F, Mellors JW, Coffin JM, Pathak VK (2005) Mechanism for nucleoside analog-mediated abrogation of HIV-1 replication: balance between RNase H activity and nucleotide excision. Proc Natl Acad Sci USA 102(6):2093–2098

Nikolenko GN, Delviks-Frankenberry KA, Palmer S, Maldarelli F, Fivash MJ Jr, Coffin JM, Pathak VK (2007) Mutations in the connection domain of HIV-1 reverse transcriptase increase 3′-azido-3′-deoxythymidine resistance. Proc Natl Acad Sci USA 104(1):317–322

Nikolenko GN, Delviks-Frankenberry KA, Pathak VK (2010) A novel molecular mechanism of dual resistance to nucleoside and nonnucleoside reverse transcriptase inhibitors. J Virol 84(10):5238–5249

Ntemgwa M, Wainberg MA, Oliveira M, Moisi D, Lalonde R, Micheli V, Brenner BG (2007) Variations in reverse transcriptase and RNase H domain mutations in human immunodeficiency virus type 1 clinical isolates are associated with divergent phenotypic resistance to zidovudine. Antimicrob Agents Chemother 51(11):3861–3869

Paredes R, Puertas MC, Bannister W, Kisic M, Cozzi-Lepri A, Pou C, Bellido R, Betancor G, Bogner J, Gargalianos P, Banhegyi D, Clotet B, Lundgren J, Menendez-Arias L, Martinez-Picado J (2011) A376S in the connection subdomain of HIV-1 reverse transcriptase confers increased risk of virological failure to nevirapine therapy. J Infect Dis 204(5):741–752

Pelemans H, Esnouf RM, Jonckheere H, De Clercq E, Balzarini J (1998) Mutational analysis of Tyr-318 within the non-nucleoside reverse transcriptase inhibitor binding pocket of human immunodeficiency virus type I reverse transcriptase. J Biol Chem 273(51):34234–34239

Poveda E, de Mendoza C, Pattery T, Gonzalez Mdel M, Villacian J, Soriano V (2008) Phenotypic impact of resistance mutations on etravirine susceptibility in HIV patients with prior failure to nonnucleoside analogues. AIDS 22(17):2395–2398

Price H, Asboe D, Pozniak A, Gazzard B, Fearnhill E, Pillay D, Dunn D (2010) Positive and negative drug selection pressures on the N348I connection domain mutation: new insights from in vivo data. Antivir Ther 15(2):203–211

Radzio J, Sluis-Cremer N (2011) Subunit-specific mutational analysis of residue N348 in HIV-1 reverse transcriptase. Retrovirology 8:69

Radzio J, Yap SH, Tachedjian G, Sluis-Cremer N (2010) N348I in reverse transcriptase provides a genetic pathway for HIV-1 to select thymidine analogue mutations and mutations antagonistic to thymidine analogue mutations. AIDS 24(5):659–667

Ray AS, Murakami E, Basavapathruni A, Vaccaro JA, Ulrich D, Chu CK, Schinazi RF, Anderson KS (2003) Probing the molecular mechanisms of AZT drug resistance mediated by HIV-1 reverse transcriptase using a transient kinetic analysis. Biochemistry 42(29):8831–8841

Reardon JE (1992) Human immunodeficiency virus reverse transcriptase: steady-state and pre-steady-state kinetics of nucleotide incorporation. Biochemistry 31(18):4473–4479

Ren J, Nichols C, Bird L, Chamberlain P, Weaver K, Short S, Stuart DI, Stammers DK (2001) Structural mechanisms of drug resistance for mutations at codons 181 and 188 in HIV-1 reverse transcriptase and the improved resilience of second generation non-nucleoside inhibitors. J Mol Biol 312(4):795–805

Richman D, Shih CK, Lowy I, Rose J, Prodanovich P, Goff S, Griffin J (1991) Human immunodeficiency virus type 1 mutants resistant to nonnucleoside inhibitors of reverse transcriptase arise in tissue culture. Proc Natl Acad Sci USA 88(24):11241–11245

Rittinger K, Divita G, Goody RS (1995) Human immunodeficiency virus reverse transcriptase substrate-induced conformational changes and the mechanism of inhibition by nonnucleoside inhibitors. Proc Natl Acad Sci USA 92(17):8046–8049

Romano L, Venturi G, Bloor S, Harrigan R, Larder BA, Major JC, Zazzi M (2002) Broad nucleoside-analogue resistance implications for human immunodeficiency virus type 1 reverse-transcriptase mutations at codons 44 and 118. J Infect Dis 185(7):898–904

Saeng-aroon S, Tsuchiya N, Auwanit W, Ayuthaya PI, Pathipvanich P, Sawanpanyalert P, Rojanawiwat A, Kannagi M, Ariyoshi K, Sugiura W (2010) Drug-resistant mutation patterns in CRF01_AE cases that failed d4T+3TC+nevirapine fixed-dosed, combination treatment: follow-up study from the Lampang cohort. Antiviral Res 87(1):22–29

Santos AF, Lengruber RB, Soares EA, Jere A, Sprinz E, Martinez AM, Silveira J, Sion FS, Pathak VK, Soares MA (2008) Conservation patterns of HIV-1 RT connection and RNase H domains: identification of new mutations in NRTI-treated patients. PLoS One 3(3):e1781

Santos AF, Silveira J, Muniz CP, Tornatore M, Goes LR, Mendoza-Sassi RA, Martinez AM, Tupinambas U, Greco DB, Soares MA (2011) Primary HIV-1 drug resistance in the C-terminal domains of viral reverse transcriptase among drug-naive patients from Southern Brazil. J Clin Virol 52(4):373–376

Sarafianos SG, Marchand B, Das K, Himmel DM, Parniak MA, Hughes SH, Arnold E (2009) Structure and function of HIV-1 reverse transcriptase: molecular mechanisms of polymerization and inhibition. J Mol Biol 385(3):693–713

Schuckmann MM, Marchand B, Hachiya A, Kodama EN, Kirby KA, Singh K, Sarafianos SG (2010) The N348I mutation at the connection subdomain of HIV-1 reverse transcriptase decreases binding to nevirapine. J Biol Chem 285(49):38700–38709

Selmi B, Boretto J, Sarfati SR, Guerreiro C, Canard B (2001) Mechanism-based suppression of dideoxynucleotide resistance by K65R human immunodeficiency virus reverse transcriptase using an alpha-boranophosphate nucleoside analogue. J Biol Chem 276(51):48466–48472

Selmi B, Deval J, Alvarez K, Boretto J, Sarfati S, Guerreiro C, Canard B (2003) The Y181C substitution in 3′-azido-3′-deoxythymidine-resistant human immunodeficiency virus, type 1, reverse transcriptase suppresses the ATP-mediated repair of the 3′-azido-3′-deoxythymidine 5′-monophosphate-terminated primer. J Biol Chem 278(42):40464–40472

Sluis-Cremer N, Arion D, Parniak MA (2000) Molecular mechanisms of HIV-1 resistance to nucleoside reverse transcriptase inhibitors (NRTIs). Cell Mol Life Sci 57(10):1408–1422

Sluis-Cremer N, Sheen CW, Zelina S, Torres PS, Parikh UM, Mellors JW (2007) Molecular mechanism by which the K70E mutation in human immunodeficiency virus type 1 reverse transcriptase confers resistance to nucleoside reverse transcriptase inhibitors. Antimicrob Agents Chemother 51(1):48–53

Sluis-Cremer N, Moore K, Radzio J, Sonza S, Tachedjian G (2010) N348I in HIV-1 reverse transcriptase decreases susceptibility to tenofovir and etravirine in combination with other resistance mutations. AIDS 24(2):317–319

Spence RA, Kati WM, Anderson KS, Johnson KA (1995) Mechanism of inhibition of HIV-1 reverse transcriptase by nonnucleoside inhibitors. Science 267(5200):988–993

Stoeckli TC, MaWhinney S, Uy J, Duan C, Lu J, Shugarts D, Kuritzkes DR (2002) Phenotypic and genotypic analysis of biologically cloned human immunodeficiency virus type 1 isolates from patients treated with zidovudine and lamivudine. Antimicrob Agents Chemother 46(12):4000–4003

Tanuma J, Hachiya A, Ishigaki K, Gatanaga H, Lien TT, Hien ND, Kinh NV, Kaku M, Oka S (2010) Impact of CRF01_AE-specific polymorphic mutations G335D and A371V in the connection subdomain of human immunodeficiency virus type 1 (HIV-1) reverse transcriptase (RT) on susceptibility to nucleoside RT inhibitors. Microbes Infect 12(14–15):1170–1177

Vingerhoets J, Azijn H, Fransen E, De Baere I, Smeulders L, Jochmans D, Andries K, Pauwels R, de Bethune MP (2005) TMC125 displays a high genetic barrier to the development of resistance: evidence from in vitro selection experiments. J Virol 79(20):12773–12782

von Wyl V, Ehteshami M, Symons J, Burgisser P, Nijhuis M, Demeter LM, Yerly S, Boni J, Klimkait T, Schuurman R, Ledergerber B, Gotte M, Gunthard HF (2010a) Epidemiological and biological evidence for a compensatory effect of connection domain mutation N348I on M184V in HIV-1 reverse transcriptase. J Infect Dis 201(7):1054–1062

von Wyl V, Ehteshami M, Demeter LM, Burgisser P, Nijhuis M, Symons J, Yerly S, Boni J, Klimkait T, Schuurman R, Ledergerber B, Gotte M, Gunthard HF (2010b) HIV-1 reverse transcriptase connection domain mutations: dynamics of emergence and implications for success of combination antiretroviral therapy. Clin Infect Dis 51(5):620–628

Waters JM, O'Neal W, White KL, Wakeford C, Lansdon EB, Harris J, Svarovskaia ES, Miller MD, Borroto-Esoda K (2009) Mutations in the thumb-connection and RNase H domain of HIV type-1 reverse transcriptase of antiretroviral treatment-experienced patients. Antivir Ther 14(2):231–239

Yap SH, Sheen CW, Fahey J, Zanin M, Tyssen D, Lima VD, Wynhoven B, Kuiper M, Sluis-Cremer N, Harrigan PR, Tachedjian G (2007) N348I in the connection domain of HIV-1 reverse transcriptase confers zidovudine and nevirapine resistance. PLoS Med 4(12):e335

Zelina S, Sheen CW, Radzio J, Mellors JW, Sluis-Cremer N (2008) Mechanisms by which the G333D mutation in human immunodeficiency virus type 1 reverse transcriptase facilitates dual resistance to zidovudine and lamivudine. Antimicrob Agents Chemother 52(1):157–163

# Chapter 14
# HIV Population Dynamics

**Mary Kearney and John Coffin**

## 14.1 Introduction

HIV-1 infection usually starts with the transmission of a single virion that expands and diversifies over time resulting in a chronic infection that eventually leads to the loss of immune function. Treatment with combinations of antiretroviral compounds nearly or completely blocks the infection of new cells by inhibiting the function of viral enzymes, such as reverse transcriptase, allowing the immune system to rebound. Studying HIV-1 population dynamics, such as ongoing replication, diversity, and evolution, provides insight into viral transmission, pathogenesis, persistence during antiretroviral therapy (ART), and strategies for eradication. In this chapter, we will introduce and discuss the dynamics and the genetics of HIV-1 populations in both treated and untreated individuals.

### *14.1.1 Origins and Global Diversity of HIV*

HIV is the only known infectious human lentivirus and is caused by two species that vary in their genetic composition and pathogenicity, HIV-1 and HIV-2. HIV-2 is more closely related to SIV (simian immunodeficiency virus) than to HIV-1. HIV-1 is more pathogenic and readily transmitted and is therefore largely responsible for the worldwide pandemic. HIV-1 is divided into four main groups, M (major), N (new), O (outlier), and P, each likely to have been established in humans by independent transmission events from different species of primates (Keele et al. 2006;

M. Kearney (✉)
National Cancer Institute, Frederick, MD, USA
e-mail: kearneym@mail.nih.gov

J. Coffin
Tufts University, Boston, MA, USA

S.F.J. LeGrice and M. Götte (eds.), *Human Immunodeficiency Virus Reverse Transcriptase: A Bench-to-Bedside Success*, DOI 10.1007/978-1-4614-7291-9_14,

Van Heuverswyn et al. 2006). The M group viruses were most likely transmitted from chimpanzees, our closest primate relative, and are well adapted to replication in human hosts (Keele et al. 2006). Consequently, group M viruses compose more than 90 % of current HIV infections worldwide. Group M viruses have been further subdivided into subtypes A–D, F–H, J, K, AE, and CRF. Subtype B is most prevalent in the United States, while all strains are present in Africa, leading to the well-supported theory that HIV originated in Africa and spread to other countries as intercontinental travel became more prevalent in the twentieth century.

### *14.1.2 Host Immune Response to HIV*

A single virion of HIV-1 usually enters a new host (Keele et al. 2008; Kearney et al. 2009) through the mucosa, first infecting dendritic cells, and then is rapidly transported to lymphatic tissues where it spreads to lymphocytes and macrophages. Virus replication increases in the lymphatic tissue, and virus is disseminated throughout the body. Within 2–3 weeks after exposure, an immune response is launched against HIV resulting in the proliferation of cytotoxic (CD8+) T lymphocyte (CTLs), meant to kill infected cells, and B cells meant to produce antibodies for neutralization of cell-free antigens. Briefly, dendritic cells and macrophages engulf HIV particles and stimulate cells of the immune system through a process called antigen presentation. Antigen presentation occurs by digestion of HIV particles into small peptide fragments that are transported to the cell surface of antigen presenting cells in conjunction with MHC II receptors (major histocompatibility complex II). The peptides bound to MHC II are recognized by T cell receptors on CD4+ T-helper cells. This recognition process stimulates the proliferation of CTLs and B cells of the immune system in an attempt to clear the infection. Infected cells also display antigenic peptides (called epitopes) on their cell surfaces but in conjunction with MHC class I receptors. CTLs recognize peptides bound to MHC class I on the surface of infected cells and induce cytolysis.

Targeting of the CTL response against a foreign antigen is determined by what epitopes are displayed on MHC class I molecules. Extensive polymorphism exists at the MHC genetic loci ensuring diverse recognition of a large number of antigens across a species. MHC molecules in humans are also called human leukocyte antigens (HLA). The specific alleles present at the MHC loci define the patient's HLA type. For example, A11 A68 B44 B37 describe the alleles present in the MHC class I gene (one A allele and one B allele from the mother and one A and one B from the father). The inherited alleles determine what epitopes can be successfully presented on one's cell surfaces. Although there are six allele types of MHC class I, HLA-A, -B, -C, -E, -F, and -G, there are many different HLA-A, -B, and -C alleles and only very few HLA-E, -F, and -G alleles. Consequently, HLA types containing E, F, and G alleles are rare. HLA-A and -B alleles are classified into supertypes according to epitope sequences that they recognize in common. HLA types result in different MHC structures capable of binding different peptide sequences. Specific residues within epitopes, called anchor residues, bind directly to the MHC molecules in the antigen presenting process. Anchor residues positions vary for each HLA supertype but are typically found within the 3 proximal amino acids of the peptide. For example, for the A3 supertype, anchor residues are found at position 2 and 10 within the ten-amino acid peptide fragment. Genetic mutations leading to amino acid changes at anchor residues typically abolish antigen presentation of the epitope as the MHC molecule loses its ability to bind the peptide. Mutations at codons within or adjacent to epitopes (other than anchor residues) can also affect antigen presentation either by impairing the ability of the peptide to be displayed, by impairing the processing of the peptide (i.e., proteolysis), or by impairing the recognition of the peptide by the T cell receptor. Impaired epitope presentation may ultimately lead to a loss of immune control in HIV infection.

High viral diversity and rapid evolution of HIV populations within infected individuals result in escape of immune recognition and the loss of host immune control due to the accumulation of mutations within CTL epitopes. The loss of immune control occurs in conjunction with a decline in CD4+ T-helper cells due to viral lysis since HIV infects these cells. When HIV has critically depleted the CD4+ T cell population, the individual becomes susceptible to opportunistic infections. Full-blown AIDS typically occurs in patients an average of 10 years after infection in the absence of treatment. However, its onset varies from 1–2 years in some individuals to 20 or more years in others. The variation in the time to AIDS among HIV-infected individuals reflects the diverse and complex interaction between HIV and the host immune system.

### *14.1.3 HIV Populations and Genetics*

As HIV replicates and spreads to new cells, its genome mutates creating genetic diversity in the population. The HIV RT enzyme contains no exonuclease activity or proofreading mechanism of any kind resulting in a per site mutation rate of about $3–5 \times 10^{-5}$ (Mansky and Temin 1995). In addition to its high mutation rate, HIV also has a high rate of replication. It has been estimated that $10^9$ virions are produced in the plasma in an infected individual per day with a $t_{1/2}$ of about 1–2 days (Coffin 1995). It has also been argued that every mutation at every position in the HIV genome occurs numerous times each day in an infected patient, building a dynamic swarm of genetically diverse virions called quasispecies (Coffin 1995). When a mutation arises that increases viral fitness, this mutation will be subjected to positive selection. In contrast, when a mutation impedes viral replication, it will be subjected to negative selection. In general, when the frequency of nonsynonymous (amino acid) substitutions is greater than that of synonymous (silent) substitutions (dN > dS), that protein or codon is said to be under positive selection. Conversely, when the frequency of synonymous mutations is greater than that of nonsynonymous mutations, that protein or codon is said to be under negative selection. Viral evolution is also impacted by the effective population size (Ne). Ne is defined as the number of individuals in a population actively participating in reproduction. It is important to bear in mind that the viral population size is determined by the number of cells infected at each round of replication (not by the number of virions produced by them), and the Ne is the proportion of those that are capable of infecting other target cells. If the Ne is very small then random genetic drift will be frequent, whereas if the Ne is large then natural selection will be the dominant evolutionary force. At very large Ne, evolution is generally deterministic, meaning that its future frequency can be predicted from its present frequency, mutation rate, and a coefficient for its selection pressure. At smaller Ne, evolution becomes more stochastic, meaning that it is more susceptible to random factors, and its future frequency cannot be exactly determined. While transition from deterministic to stochastic evolution increases with decreasing population size, an approximate crossover point is at a population size that equals the inverse of the mutation rate ($3–5 \times 10^5$) when all possible single mutations will likely arise with each round of replication (Rouzine and Coffin 2005). In chronic, untreated HIV infection the Ne is thought by some to be small ($10^3$–$10^4$) (Frost et al. 2000; Balagam et al. 2011; Brown 1997) suggesting that genetic drift plays an important role in HIV evolution.

A significant problem resulting from the high mutation rate and diversity of HIV is resistance to antiviral compounds. Resistance-engendering mutations are likely to arise spontaneously in a viral population and are favored for growth under the selection pressure of the drug (Coffin 1995). Resistance to drug combinations does not occur as readily as to monotherapy since mutations that convey resistance to each drug must arise on the same viral genomes in order for the virus to persist. However, failed therapy and patient noncompliance still result in the emergence of antiretroviral resistant variants despite the use of combination drugs. Understanding how viral evolution affects response to therapy and viral persistence is important for designing strategies toward the development of preventive vaccines and toward viral eradication.

## 14.2 Biology of HIV-1 Infection

### *14.2.1 Time Course of Infection*

HIV-1 infection is usually initiated by the transmission of a single virus particle and, without treatment, typically follows a course of infection that includes four phases (Fig. 14.1) (Fauci and Desrosiers 1997a):

1. *The eclipse phase* (1–2 weeks after transmission): The virus is freely replicating and spreading from the initial site of infection to the tissues that contain target cells for replication. Infection is asymptomatic, no immune response is present, and plasma viremia is undetectable by standard assays.
2. *Acute (or primary) infection phase* (~2–4 weeks): Plasma viremia is detectable, typically at high levels (up to $10^7$ copies of viral RNA per ml of blood), and the immune response begins to appear both in the form of antibodies and CTL response against HIV-1 antigens. At the end of the acute phase, the level of

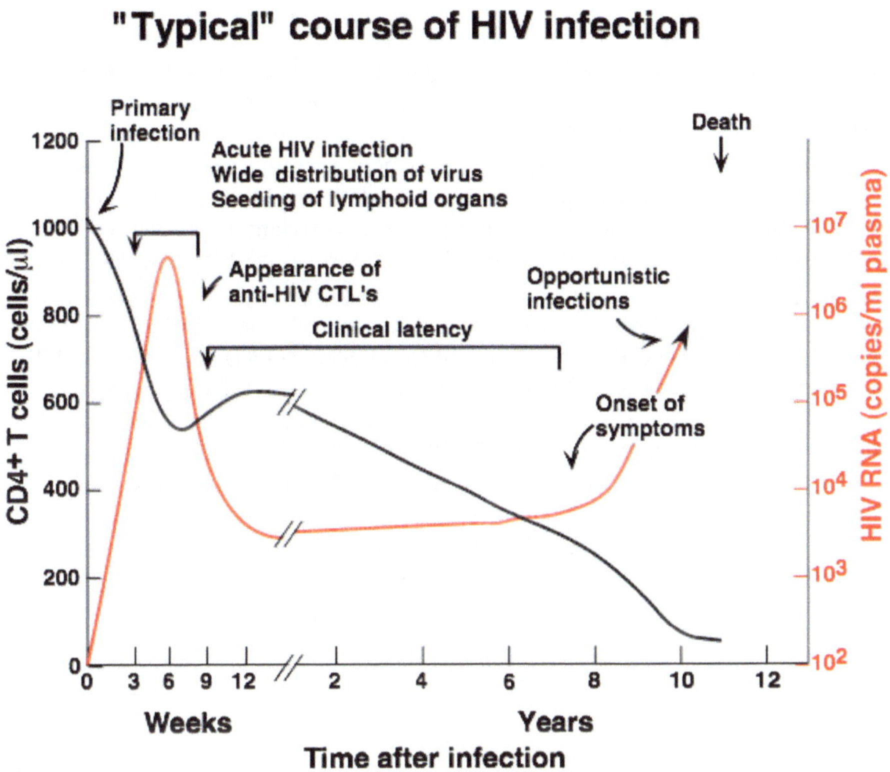


**Fig. 14.1** Time course of typical HIV infection. Patterns of CD4+ cell decline and viremia vary greatly from one patient to another (Adapted from Fauci and Desrosiers 1997a)

viremia declines sharply (100-fold or more) as CD4+ T cells die from infection and from immune control. The patient often, but not always, experiences "flu-like" symptoms during this phase.

3. *Chronic infection or "clinical latency" phase* (~1–20 years): Patients are asymptomatic as the virus infects and kills approximately $10^6$–$10^9$ CD4+ T cells daily. Viremia persists at a steady state then slowly increases as CD4+ T cells gradually decline.
4. *AIDS phase* : The number of CD4+ T cells declines to less than 200 cells/ml, immune control of viremia is lost, and opportunistic infections begin to appear. Ultimately the patient succumbs to death by wasting or infection with opportunistic agents.

### *14.2.2 HIV-1 Levels in Blood*

A wide range of HIV-1 plasma RNA levels (viral loads) are observed across infected individuals (<50 copies/ml to >$10^6$ copies/ml), but levels are remarkably steady in a single, untreated individual during the clinical latency phase, known as the *steady state*. The steady state is determined by a combination of the number of available target cells, the infectivity of the virus, the number of virions produced per cell, the productive lifetime of the infected cell, and the clearance rate of virus in the plasma. The level of the steady state is the most reliable prognostic marker for progression. The first recognition of the prognostic importance of viremia came from analysis of a large prospective study of gay men at risk for AIDS, the Multicenter AIDS Cohort Study (MACS), which concluded a strong inverse correlation between the level of viremia and the time of progression to AIDS (Mellors et al. 1996). The data showed that the group with the highest level of viremia progressed to AIDS on average about fourfold faster than the lowest group. The correlation between viremia and disease pathogenesis is straightforward. Since infected cells are targeted and killed by the immune system or die from cytopathic effects from infection, and since the replication cycle time for HIV-1 is about 1–2 days, a steady number of new cells (the number being related to the steady state) are infected and killed daily in vivo. CD4+ T cells of the immune system are a primary target for HIV-1 and consequently are killed in large numbers in infected patients. The effect is the depletion of immune cells and loss of immune function, and eventual AIDS, occurring more rapidly in patients with a higher steady state than those with lower levels of viremia.

### *14.2.3 HIV-1 Replicating Population Size*

The HIV-1 replicating population reflects the numbers of viral variants that contribute to HIV evolution. The size of the replicating population is a matter of debate, but new ultrasensitive PCR-based methods are providing further information by monitoring the frequency and selection of preexisting mutations. One such method is the ultrasensitive allele-specific PCR assay (Us-ASP) that can detect very low-frequency mutants down to 0.001 % of the virus population (Boltz et al. 2012a). This assay is currently being used to investigate the levels of drug-resistance mutations that exist before drug therapy in pigtail macaques infected with a simian immunodeficiency virus ($SIV_{mne}$) containing HIV-1 reverse transcriptase (RT-SHIV) and treated with a short course of efavirenz (EFV) monotherapy (Boltz et al. 2012a). EFV resistance mutations became readily detectable (20 % of the virus population in 7 days) after brief exposure to EFV indicating their presence in the population prior to treatment. Using Us-ASP, the EFV-resistance mutations (AAA>AAC and AAA>AAT at codon 103 in RT) were found to be present in less than 0.001 % of the plasma virus population in two infected macaques. A frequency of less than 1 in 100,000 viral genomes suggests that the number of productively HIV-1-infected cells is greater than about $10^6$ infectious cells per replication cycle. If a smaller replicating virus population is assumed, as previous studies have predicted (Frost et al. 2000; Balagam et al. 2011; Brown 1997), the probability of EFV resistance would be much lower. For example, using Poisson statistics, if the replicating population is $10^4$ infectious cells per generation, the probability that one of the two resistance alleles would be present in one animal would be <1/10. The expected probability for both mutations propagating the infection in one animal would be less than about 1 %, and both mutations in both animals would be <0.01 %. Even with a replicating population size of $10^5$, the chance of both mutations growing out in both animals is still <16 %. However, with a replicating population of >$10^6$ infectious cells per replication cycle, the probability of both mutations in both animals replicating in the population approaches 100 %. The effective population size of HIV-1 has been greatly debated with many elegant studies and complex mathematical models proposed (Coffin 1995; Balagam et al. 2011; Brown 1997; Batorsky et al. 2011; Rouzine and Coffin 1999). The use of highly sensitive allele detection methods will provide tools for the study of drug resistance that uses fundamental probability to precisely investigate the error rate of RT and the replicating population size of HIV in vivo.

### *14.2.4 Cellular Tropism and Compartmentalization*

HIV-1 infects a variety of CD4+ cell types including T cells, monocytes, macrophages, and dendritic cells. A typical HIV-1 infection begins with a variant that uses the CD4 receptor and the CCR5 coreceptor for entry primarily into dendritic cells then disseminating into macrophages and T cells. Without treatment, infection progresses and variants can emerge that are capable of using the CXCR4 coreceptor for efficient spread into an expanding repertoire of T cells types. As HIV spreads, the different cell compositions across tissues also result in the selection of distinct virus populations in different anatomical compartments (such as the gut, the semen, and the cerebral spinal fluid) contributing to the genetic diversity of HIV infection (Schnell et al. 2010). When CXCR4 variants emerge, they infect and lyse T cells aggressively and coincide with depletion of immune function, especially in HIV-1 subtype B infection.

It is thought that a small fraction of infected CD4+ T cells may return to a resting state ($G_0$) prior to viral or immune-mediated lysis, resulting in latent HIV infection (Chun et al. 1997a, b). A truly latently infected cell is one which harbors an integrated provirus but does not express viral RNA or proteins. However, it is important to note that latency is only biologically important if the provirus is able to become activated to produce virus in the future. It is likely that many cells are infected with replication-incompetent virus due to lethal mutations in the viral genome or to integration into an unfavorable site in the host genome. These cells infected with "dead" proviruses are distinct from latently infected cells and are not biologically relevant unless they become coinfected and recombine to generate a new, replication-competent virus. Latently infected cells may play an important role in HIV persistence during combination antiretroviral therapy (cART) as it is theorized that these cells become activated from time to time resulting in the low-level viremia detected in the plasma of cART-treated patients (Chun et al. 1997a, b; Cameron et al. 2010). Since HIV-infected cells that produce virus are killed by viral lysis or CTLs, it is also likely that rebound viremia after stopping cART is due to virus produced from latently infected cells that have become activated. However, the sources of persistent and rebound viremia are only recently the focus of intense research. Discovering the source of the viral reservoir during therapy will likely prove critical to curing HIV infection.

### 14.2.5 Frequency of Infected Cells

The frequency of HIV infection across different cells types is an important focus of current research. It has long been known that cells can be multiply infected with HIV (harbor more than one provirus in a single cell) since such an event is necessary for HIV recombination. However, the frequency of either singly or multiply infected cells is unknown. A new technique was developed to quantify and genetically characterize HIV-1 DNA from infected cells in vivo that can be applied to cellular subsets of blood and tissues to fully characterize the frequency of HIV-1 infection before and during suppressive therapy (Josefsson et al. 2011). This new method, called the single-cell sequencing (SCS) assay, was recently used to detect and sequence single HIV-1 proviruses in individual patient cells derived from peripheral blood (Josefsson et al. 2011). The investigators found that the frequency of HIV-1 infection of untreated patients is 1 in 1,750 CD4+ blood cells (0.06 %), with considerable patient-to-patient variation, and that the majority of infected cells contain only 1 copy of HIV-1 proviral DNA. They did not find any proviruses in 2,000–28,000 CD14+ cells and in 73,000–200,000 CD16+ cells, suggesting a very low rate of infection in monocytes. Cells containing more than 1 HIV-1 DNA molecule were rare, and they estimated the most likely frequency of CD4+ cells containing multiple HIV-1 DNA molecules to be between 2.6 % and 6.3 %. These findings agree with mathematical modeling analyses by Neher et al. and Batorsky et al., who estimated the rate of coinfection to be <10 % (Neher and Leitner 2010), and with Simmonds et al., who found evidence for 1 provirus per infected PBMC (Simmonds et al. 1990). However, these findings are far below earlier estimates based on analysis of infected spleen cells where the majority of cells appeared to contain multiple proviruses. Spleen cells may encounter HIV-1 virions more frequently than cells in the peripheral blood and may have more frequent cell-to-cell transmission, increasing their opportunity for multiple HIV-1 infection. Future studies will use SCS methods on patients initiating cART to determine the relationship between proviruses in cells from peripheral blood to those in tissue compartments.

### 14.2.6 HIV-1 Pathogenesis

The question of HIV-1 pathogenesis appears simple: HIV-1 infects and kills CD4+ T cells, resulting in immune dysfunction and the onset of opportunistic infections, marking the transition into AIDS. However, it is still not known exactly how infected cells die in vivo, the paradigm being that the virus and cell turnover rates in many SIV infections (such as $SIV_{sm}$ in sooty mangabeys) are the same as in HIV infection in humans, but monkeys do not progress to AIDS (Chakrabarti 2004; Hirsch 2004; Demma et al. 2005). This observation implies that cell killing alone cannot account for AIDS pathogenesis. SIV has evolved to replicate in a population of cells in its natural host that is replenished constantly (T cells) and in excess over what is needed for optimal health. HIV-1 infection actually occurs in an unnatural host (having crossed from chimpanzees and other monkey species) and consequently other factors must contribute to the development of AIDS. Two features distinguish the pathogenic (HIV) from nonpathogenic (SIV) infection and may give hints toward understanding the pathogenesis of HIV infection: (1) chronic immune activation and (2) greater levels of infection in varied CD4+ T cell subsets, particularly central memory helper T cells. The former may lead to immune exhaustion and damaged lymphoid architecture and the latter to a loss of critical cells necessary for maintenance of a diverse response to common environmental pathogens. Determining the role of immune activation and infection of various T cell subsets in the causation of AIDS will continue to be an important focus of HIV research.

## 14.3 HIV Evolution After Infection

### *14.3.1 Factors that Effect HIV Genetic Diversity and Evolution*

HIV-1 accumulates very high genetic diversity during the course of infection in a single individual. When first observed, it represented the fastest rate of evolution in any natural eukaryotic system. Today, the record for genetic diversity belongs to the hepatitis C virus. Although the error rate for HIV-1 replication is about average for an RNA virus, its pace of replication, the duration of infection, and the size of the replicating population allow it to accumulate very high levels of diversity and to evolve rapidly. By evolution, we mean the process by which genetic change accumulates during generations of replication. Four basic mechanisms cause these genetic changes:

1. *Mutation*: For HIV-1, mutations are primarily due to the error-prone reverse transcriptase enzyme. However, there are actually two other sources for error: transcription by host RNA polymerase II and ABOBEC3G (or F)-mediated G-to-A hypermutation. The relative contribution of the later is not known but the overall error rate for HIV-1 is around $3 \times 10^{-5}$ mutations per base per replication cycle (Mansky and Temin 1995). Consequently about one in every three genomes contains a mutation after a single round of replication. These mutations come in several forms including point errors, insertions and deletions, duplications, and inversions (Svarovskaia et al. 2003).
2. *Selection*: Selection results from the ability of one viral variant to replicate more effectively than another in a specific environment (also referred to as fitness) leading to higher numbers of infectious progeny with each replication cycle. Differences in fitness occur from variations in the efficiency to perform any step

of the replication cycle. Fitness is often expressed as a selection coefficient, the difference in the relative number of progeny produced by one variant compared to another per generation. For example, a variant that yields 1 % more infectious progeny than another has a selection coefficient of 0.01. In a large population, even variants with a small selective advantage have a large effect on the population after multiple cycles of replication.

3. *Drift*: Drift is the change in the frequency of a variant that is independent of selection or mutation. These variations are only significant in very small stochastic populations where slight changes in the frequency of a variant have a larger impact on the overall diversity of the population. Such small variations have no effect on larger populations since they have no selective advantage or disadvantage.
4. *Linkage*: Linkage refers to the presence of multiple mutations on the same viral variant. Because selection acts on entire genomes, not individual mutations, linkage is a very important consideration, especially in small populations where mutations may be present on only a single or a few variants. In large populations, mutations may be present on many genomes, making linkage less relevant. Muller's ratchet states that, in extreme cases, linkage of mutations in a very small population may lead to declining fitness due to gradual accumulation of deleterious mutations. The effect of linkage on evolution is reduced by recombination which allows mutations to be redistributed among genomes. Recombination between genetically distinct genomes occurs only in cells that are infected with two or more proviruses. The frequency of multiply infected cells in vivo is unknown and contrasting reports exist. It was originally reported that such multiply infected cells are very frequent in HIV-infected individuals (Jung et al. 2002), but more recent results suggest that they are much more rare, constituting less than 10 % of infected CD4+ T cells in blood (Josefsson et al. 2011). Nevertheless, there is ample evidence for the occurrence and importance of recombination in HIV-1 infection, particularly in individuals coinfected with two or more genetically distinct viruses (Keele et al. 2008; Kearney et al. 2009).

### *14.3.2 Accumulation of HIV Diversity*

After transmission, HIV populations expand and accumulate diversity over cycles of replication primarily due to its error-prone reverse transcriptase enzyme. The initial accumulation of diversity is at a rate very close to the RT mutation rate, with little evidence for selection (Kearney et al. 2009). However, over the course of infection, purifying selection dominates and diversity begins to level off at an average pairwise distance around 1–2 % for *pro* and *pol* (Kearney et al. 2009). Diversity reaches higher levels in *env* where there is greater positive selection from the antibody response resulting in immune escape by point mutations (especially at glycosylation sites) and by large insertions and duplications in the hypervariable loops V1–V2 and V4 (Shankarappa et al. 1998; Shankarappa et al. 1999). In *gag*, *pro*, and *pol*, positive selection from the cellular immune response can be observed at sites corresponding to T cell epitopes. In fact, in untreated infection almost all of the nonsynonymous diversity in these genes is due to mutations that confer escape from the CTL response. CTL escape mutations sometimes have a fitness cost to the virus, and, therefore, partial control of the immune response is due not only to immune control but to the loss of fitness from escape mutations.

### *14.3.3 Emergence of Drug Resistance and Virologic Failure*

A consequence of HIV-1 diversity is the selection of drug-resistant mutants in the presence of non-suppressive antiretroviral therapy (ART), for example, in patients on less than optimal regimens, with low adherence, or in women being treated with a single-dose of the nonnucleoside RT inhibitor, nevirapine (sdNVP), during labor to prevent the transmission of HIV from the mother to child. The rapid selection of drug-resistance mutations during monotherapy and from sdNVP suggested that these mutations were present in the replicating population before initiating therapy (Coffin 1995). Indeed, resistance mutations have been reported to be present in treatment-naïve patients. Current research is aimed at determining what levels and types of preexisting drug-resistance mutations are present in vivo and what levels significantly affect future therapy. One study investigating the effect of low-frequency NVP-resistance mutations (that persisted after receiveing sdNVP on subsequent cART contaiing NVP) found that resistance mutations present at frequencies above 0.8 % of the virus population were significantly associated with virologic failure (Boltz et al. 2011). An important, but preliminary, finding suggests that the same level of preexisting mutations found to negatively affect cART in women who previously received sdNVP does not affect treatment in ART-naïve individuals (Boltz et al. 2012b) (Fig. 14.2). This finding suggests that the association between virologic failure and preexisting resistance may be due to linkage of multiple mutations rather than to the level of a single resistance mutation. If true, then it is likely that it was the high frequency of NVP-resistant variants selected shortly after dosing with sdNVP that allowed these mutations to become linked to other preexisting drug-resistance mutations in the population. By contrast, NVP-resistance mutations were most likely never present in high frequencies in women who had not been exposed to sdNVP and, therefore, are not likely to be linked to other drug-resistance mutations, allowing successful treatment with cART (Fig. 14.3). This hypothesis will be tested when assays are available to investigate the linkage of mutations at very low frequencies. Recent studies are also evaluating the effect of total viral diversity and population size on treatment outcome. Preliminary studies using a macaque model for cART, showed that high plasma virus diversities, preexisting drug-resistance mutations, and large virus population sizes at baseline were all associated with virologic failure during ART. Future studies will investigate these factors in large cohorts of HIV-1 infected patients.

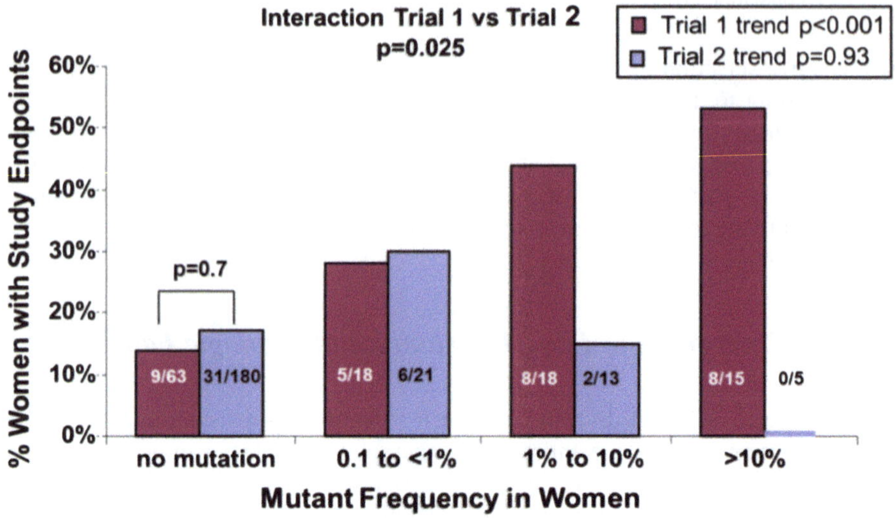


**Fig. 14.2** Proportion of women in Trial 1 (prior treatment with sdNVP) and proportion of women in Trial 2 (no prior treatment with sdNVP) reaching a study endpoint of compared to women with no mutation in each trial. These results show the association of preexisting NVP-resistance mutations in women with prior exposure to sdNVP and the lack of association in women who were treatment naïve (Adapted from Boltz et al. 2012b)

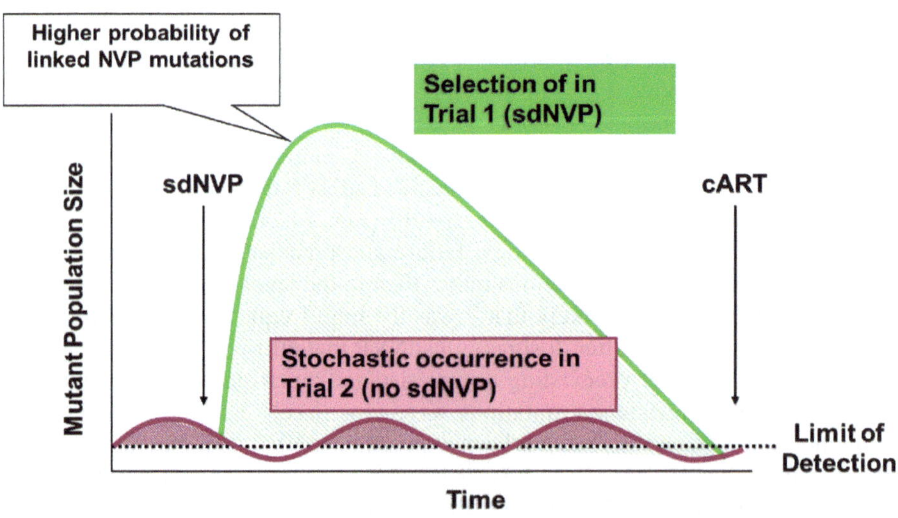


**Fig. 14.3** Model proposed to explain differing impact of preexisting NVP-resistant mutations in women with prior exposure to sdNVP (*Trial 1*) and women without prior exposure (*Trial 2*). The greater area under the curve in *green* represents the greater expansion of mutant populations under drug selective pressure in women in Trial 1 (prior sdNVP) providing more opportunity for linkage of other resistance mutations during ongoing cycles of replication. The smaller area under the curve shown in *red* represents the smaller proportions of mutations occurring stochastically in women in Trial 2 (treatment naïve), providing fewer opportunities for linkage to other mutations on the same genome (Adapted from Boltz et al. 2012b)

## 14.4 Persistent Viremia During cART

Because cART affects the ability of HIV-1 to infect new cells, the steady-state level of viremia decreases following initiation of treatment to levels that are undetectable by standard viral load assays (<50 copies RNA/ml of plasma). However, treatment does not eradicate infection and viremia is still detectable in the majority of patients when measured with ultrasensitive tests that can detect <1 virus copy/ml of plasma (Maldarelli et al. 2007). The kinetics of the decline in viremia on cART reflect the length of time that cells infected prior to initiating treatment live and produce virus after starting treatment (Coffin 1995). HIV RNA levels decay in four phases after starting cART; the first phase includes rapid clearance of virus with a half-life of 1–2 days, followed by a more gradual decline with a half-life of 2–3 weeks, a third phase with a half-life of 6–44 months and a fourth phase with a slope not significantly different from zero (Palmer et al. 2008) (Fig. 14.4). The plateau in the fourth phase of decay suggests that long-term ART fully inhibits HIV-1 replication and that the source of persistent viremia may be a small population of infected cells capable of indefinite self-renewal. The level of persistent viremia during ART is correlated with the level of plasma HIV-1 RNA before initiating cART, further indicating that residual viremia results from cells infected before therapy (Maldarelli et al. 2007).

In contrast to HIV-1 plasma RNA, the kinetics of decay of proviral DNA after initiating treatment typically levels off at about 1–10 % of pretherapy levels and at about 10- to 100-fold higher than RNA, implying that a significant fraction of HIV-1 infected cells persist during treatment but do not produce virus. Furthermore, only

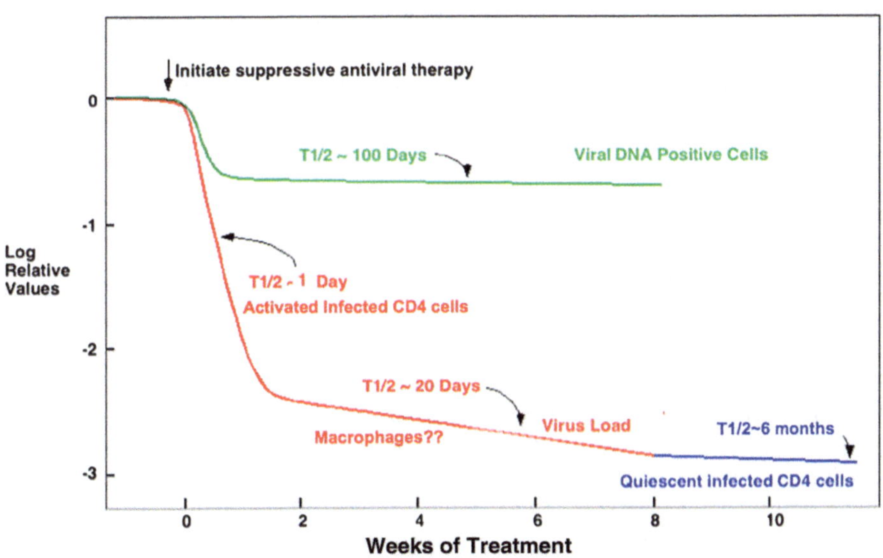


**Fig. 14.4** Phases of decay of plasma virus and infected cells after initiation of cART (Adapted from Fauci and Desrosiers 1997b)

about 1 % of infected cells are able to be induced to produce virus ex vivo (Chun et al. 1997a) perhaps due to mutations in the provirus, to unfavorable sites of integration, or to some undefined cellular factor which prevents expression. Consequently, because of their much longer lifetime than productively infected cells, cells containing latent proviruses accumulate without treatment leaving only a small fraction of the population of infected cells being capable expressing infectious virus.

Although it is well accepted that the large majority of patients on suppressive therapy have low levels of viremia as well as latently infected cells (some being capable of reactivation) the source of the persistent viremia during cART is controversial, with some investigators arguing that the persistent viremia is due to viral expression from long-lived cells infected prior to therapy and others arguing that it is due to low-level ongoing full cycle replication of virus in anatomic sites or cells that are poorly accessible to antiviral therapy (referred to as "sanctuaries"). In support of claims for ongoing replication is the presence of 2-LTR circles (Sharkey and Stevenson 2001) in PBMCs in a fraction of patients (Llibre et al. 2012), a product claimed to be the result of recent replication. However, this claim makes the assumption that 2-LTR circles are inherently unstable and cannot persist in long-lived infected cells, an assumption that has yet to be demonstrated. In contrast, there are several lines of evidence in favor of the idea that the low-level viremia observed during long-term suppressive therapy is derived from cells infected prior to the start of treatment. One important indication is that the level of viremia is not affected by the inclusion of additional drugs to existing regimens. Several studies have been performed in which a suppressive three-drug regimen was intensified by the addition of a fourth inhibitor of a class previously unseen by the patient, (Gandhi et al. 2010; Dinoso et al. 2009; McMahon et al. 2010) showing no change in the level of persistent viremia during the intensification period. It is highly improbable that sanctuaries of replication would be equally inaccessible to all drugs so these results are inconsistent with low-level ongoing replication. Future efforts will inevitably reveal the sources and causes of persistent HIV during suppressive cART.

### *14.4.1 Lack of Evolution During cART*

Supporting a lack of ongoing cycles of replication during suppressive cART is the lack of evidence for new HIV-1 variants emerging during treatment. By sequencing viral populations prior to and during long-term therapy, evolution of viral genomes appears to be halted at the initiation of cART (Ruff et al. 2002; Tobin et al. 2005). In contrast, in patients where therapy is partially suppressive and in patients known as "elite controllers," who naturally control HIV-1 infection at levels similar to those seen in patients on suppressive antiviral therapy (Mens et al. 2010), ongoing evolution can still be detected (Fig. 14.5). A recent study investigated virus populations in samples collected prior to initiating cART, within the first 6 months of cART, and after long-term suppression and found no change in the diversity, divergence, or phylogenetic structure in populations obtained before and after 6 months of treatment but reported clusters of identical sequences during long-term suppression (rather than the diverse quasispecies observed in treatment-naïve individuals or

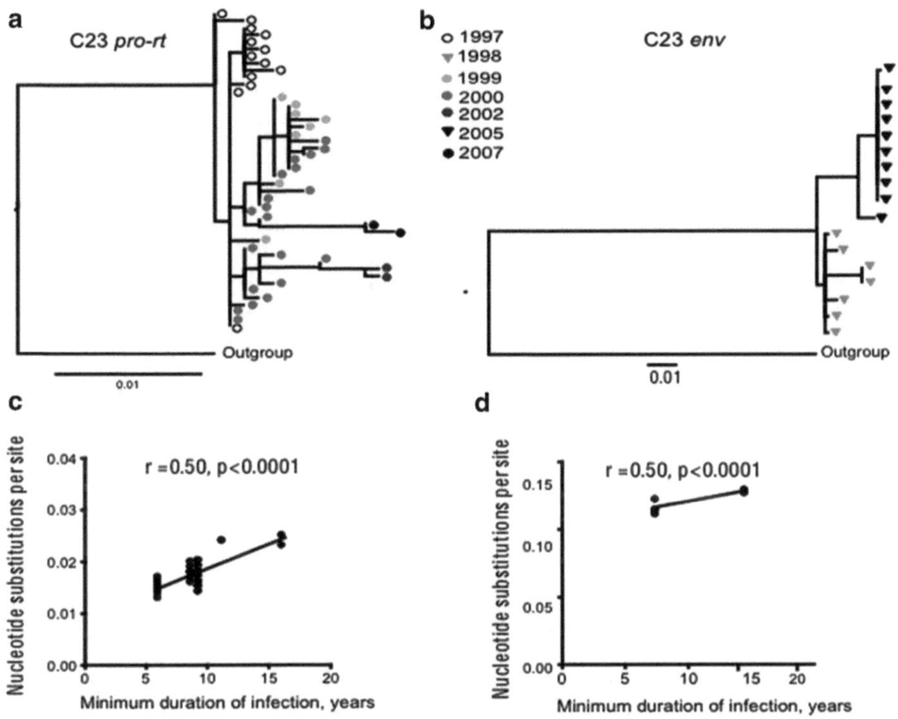


**Fig. 14.5** Evolution in HIV-1 elite controller. Rooted maximum-likelihood phylogenetic trees of pro-pol (**a**) and env (**b**) from one HIV-1 controller C23. The trees were rooted using HXB2 as an outgroup. Root-to-tip distances plotted against sampling time (**c** and **d**) (Adapted from Mens et al. 2010)

during short-term treatment (Kearney et al. 2010)). These results indicate that although the reservoir is seeded prior to treatment with the same virus population as the replicating pool of virus in the plasma, variants are either lost over the course of treatment or are masked by a replicating, infected cell populations (Fig. 14.6). Tests for evolution demonstrated that the populations of identical sequences present during long-term cART fit within the clusters of preexisting phylogenies, giving no evidence for the emergence of new HIV-1 variants during cART (Kearney et al. 2010). The lack of evidence for virus evolution in these patients during prolonged therapy provides further evidence that HIV-1 persists during cART in long-lived cells that were infected before therapy was initiated and is not due to low-level, ongoing virus replication, as occurs in HIV-1 elite controllers with similar levels of viremia. The presence of identical sequences during ART also suggests that virus particles are being produced by an HIV-1-infected clonal cell population, such as hematopoietic stem cells or other proliferating cell types. Additionally, the change in phylogenetic structure after long-term therapy to one or a few populations of identical sequences suggests that variants may be lost over the course of treatment, indicating that long-lived cells may be slowly dying and eventually reducing the size of the reservoir. Future studies will include attempts to identify the reservoirs that express virus during suppressive therapy in order to develop strategies to target and kill those cells.

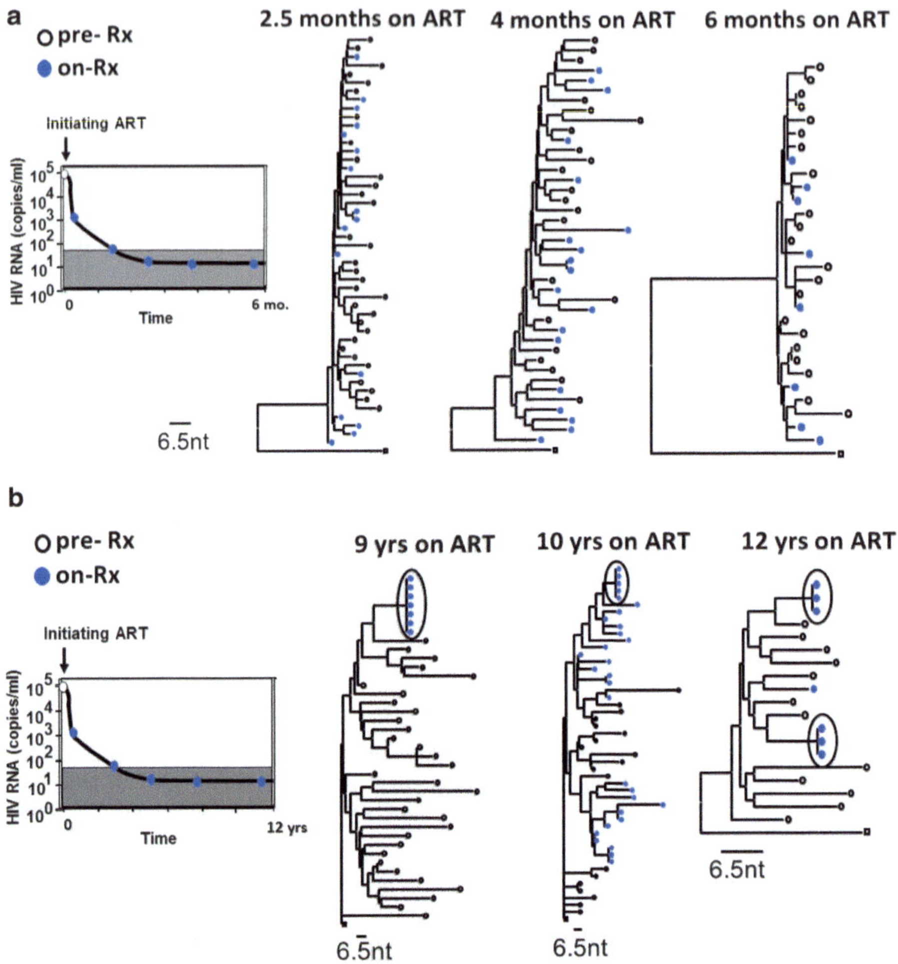


**Fig. 14.6** (**a**) Sequencing from single HIV genomes in plasma samples collected before and after initiating cART. Single-genome *pro-pol* sequences from samples collected before (*open circles*) and within 6 months of initiating cART (*blue circles*) in three patients. There was no change in diversity or population structure after initiating ART. (**b**) Sequencing from single HIV genomes in plasma samples collected before and during long-term cART. Single-genome *pro-pol* sequences from samples collected before (open *circles*) and after long-term treatment with ART (*blue circles*) in three patients. There is a shift in population structure including the appearance of populations of identical sequences after long-term cART. (Adapted from Kearney et al. 2010)

## *14.4.2 Identifying the Reservoir*

Identifying the reservoir (the cellular sources) for persistent viremia during cART is essential for designing effective strategies to eradicate HIV-1 infection. Reactivation of latently infected resting CD4+ T cells have been suggested to be a key HIV-1 reservoir (Chun et al. 1997b; Finzi et al. 1997; Wong et al. 1997). Genetic analyses have revealed that the persistent virus in plasma results from only a small fraction of the infected cell population but the exact sources have not yet been identified. Recent reports showed that HIV-1 can persist in both CD31+ and CD31− populations of naïve CD4+ T cells (Wightman et al. 2010) and in GALT (Lerner et al. 2011), and another study revealed that bone marrow CD34+ cells isolated from four of nine (44 %) patients on suppressive cART had detectable HIV-1 proviral DNA (Carter et al. 2010), indicating infection of hematopoietic stem cells. However, none have been demonstrated to result in the persistent viremia detected in the plasma. Most recently, a study by Yukl et al. showed higher levels of HIV-1 DNA and HIV-1 RNA expression in gastrointestinal mucosal CD4+ cells, especially in the terminal ileum, compared with CD4+ T cells in blood (Yukl et al. 2010). Although informative, the proportion of HIV-1 DNA-containing cells that express HIV-1 RNA in vivo was not determined in these studies. Nor was release of virions from these cells analyzed. As such, HIV-1 reservoirs are still incompletely characterized not only regarding the question of ongoing replication but also as to their anatomic locations, sizes, cell types, levels of HIV-1 expression, and relationship of their virus populations. This fundamental knowledge gap must be filled to appropriately target all reservoirs and ultimately cure HIV-1 infection.

## 14.5 Future Possibilities for HIV Eradication

Despite the success of cART, plasma viremia persists in most patients at very low levels (<1 copy HIV RNA/ml) (Palmer et al. 2008). Continuous cART is essential to maintain HIV suppression; therapy interruption after a few years of suppression often results in prompt rebound of viremia. Long-term cART has unacceptable adverse effects, including dyslipidemia and accelerated cardiovascular disease. New approaches to control or eradicate HIV are needed. Although the majority of patients experience viral rebound after cART interruption, there are reports of spontaneous control of HIV after stopping therapy (Salgado et al. 2011; Van Gulck et al. 2012). Factors contributing to spontaneous control are unknown. It is possible that spontaneous control is associated with reservoirs size or immune response to HIV-1 in patients stopping suppressive therapy. To date, no studies have investigated whether the levels of persistent viremia on cART, the size of the HIV DNA reservoir during cART, or the duration of cART are related to spontaneous control after treatment interruption. Due to such reports of spontaneous control, it is possible that a *functional cure* (control of viremia without treatment) may be achieved in some patients without additional treatments, especially those treated during acute infection. However, in the majority of infected patients, a functional cure will require cART in addition to therapies that target and kill all infected cells that are capable of generating or being induced to generate productive virus. Current research is aimed at discovering such therapies. A *sterilizing cure* (eradication of all HIV-infected cells) has been achieved in a single patient who underwent several treatments for acute leukemia including ablative chemotherapy, radiation therapy, anti-macrophage monoclonal antibody infusions, and stem cell transplants from a donor with a homozygous CCR5 delta-32 genotype (Allers et al. 2011). While the clinical significance of curing HIV infection in this individual is undeniable, bone marrow transplants are not feasible for the treatment HIV which is readily controlled with cART. However, studying the mechanisms for viral eradication in this patient and other HIV-infected patients undergoing cancer therapies may reveal information that will be relevant in developing strategies for curing HIV infection. Which of these interventions was important in leading to eradication of the HIV is currently unknown, but it seems likely that multiple therapeutic interventions will be needed to eliminate the reservoir of HIV infected cells.

## 14.6 Summary and Perspective

HIV is unique among viral infections due to its high rate of replication, high frequency of infected cells, and its high level of genetic diversity in vivo. The features of HIV replication, including the high error rate for reverse transcription, allow the virus to accumulate mutations that confer escape from clearance by the immune response, resulting in life-long infection. Other viral features, most of which are not yet completely understood, allow HIV to kill the majority of infected cells but to persist in some in a latent state. The significance of latently infected cells toward persistence during suppressive cART and toward viral rebound after stopping cART is not yet clear, although it is supposed that a fraction of latently infected cells may become induced to produce viremia during suppression and rebound. Current HIV research is aimed at understanding the mechanisms for viral pathogenesis, the impact of HIV diversity on treatment (especially the clinical significance of preexisting drug-resistance mutations), and mechanisms for viral persistence during cART. A recent report of HIV eradication in a single individual who received a stem cell transplant from a donor with a deleted CCR5 genotype has spurred the possibility of discovering therapies to cure HIV infection. New efforts are being made toward identifying the reservoir for HIV during cART and toward targeting productively infected cells to eradicate HIV infection in the future.

## References

Allers K, Hutter G, Hofmann J, Loddenkemper C, Rieger K, Thiel E, Schneider T (2011) Evidence for the cure of HIV infection by CCR5Delta32/Delta32 stem cell transplantation. Blood 117:2791–2799

Balagam R, Singh V, Sagi AR, Dixit NM (2011) Taking multiple infections of cells and recombination into account leads to small within-host effective-population-size estimates of HIV-1. PLoS One 6:e14531

Batorsky R, Kearney MF, Palmer SE, Maldarelli F, Rouzine IM, Coffin JM (2011) Estimate of effective recombination rate and average selection coefficient for HIV in chronic infection. Proc Natl Acad Sci USA 108:5661–5666

Boltz VF, Zheng Y, Lockman S, Hong F, Halvas EK, McIntyre J, Currier JS, Chibowa MC, Kanyama C, Nair A, Owino-Ong'or W, Hughes M, Coffin JM, Mellors JW (2011) Role of low-frequency HIV-1 variants in failure of nevirapine-containing antiviral therapy in women previously exposed to single-dose nevirapine. Proc Natl Acad Sci USA 108:9202–9207

Boltz VF, Ambrose Z, Kearney MF, Shao W, KewalRamani VN, Maldarelli F, Mellors JW, Coffin JM (2012a) Ultrasensitive allele-specific PCR reveals rare preexisting drug-resistant variants and a large replicating virus population in macaques infected with RT-SHIV. J Virol 86:12525–12530

Boltz VF, Bao Y, Lockman S, Halvas EK, McIntyre JA, Schooley R, Hughes MD, Coffin JM, Mellors JW, Team OA (2012b) The risk of virologic failure associated with low frequency NVP-resistant variants depends on prior exposure to single dose Nevirapine: OCTANE/ACTG 5208 Trial 1 and Trial 2. Oral presentation at the 19th Conference on Retroviruses and Opportunistic Infections

Brown AJ (1997) Analysis of HIV-1 env gene sequences reveals evidence for a low effective number in the viral population. Proc Natl Acad Sci USA 94:1862–1865

Cameron PU, Saleh S, Sallmann G, Solomon A, Wightman F, Evans VA, Boucher G, Haddad EK, Sekaly RP, Harman AN, Anderson JL, Jones KL, Mak J, Cunningham AL, Jaworowski A, Lewin SR (2010) Establishment of HIV-1 latency in resting CD4+ T cells depends on chemokine-induced changes in the actin cytoskeleton. Proc Natl Acad Sci USA 107:16934–16939

Carter CC, Onafuwa-Nuga A, McNamara LA, Riddell J, Bixby D, Savona MR, Collins KL (2010) HIV-1 infects multipotent progenitor cells causing cell death and establishing latent cellular reservoirs. Nat Med 16:446–451

Chakrabarti LA (2004) The paradox of simian immunodeficiency virus infection in sooty mangabeys: active viral replication without disease progression. Front Biosci J Virt Libr 9:521–539

Chun TW, Carruth L, Finzi D, Shen X, DiGiuseppe JA, Taylor H, Hermankova M, Chadwick K, Margolick J, Quinn TC, Kuo YH, Brookmeyer R, Zeiger MA, Barditch-Crovo P, Siliciano RF (1997a) Quantification of latent tissue reservoirs and total body viral load in HIV-1 infection. Nature 387:183–188

Chun TW, Stuyver L, Mizell SB, Ehler LA, Mican JA, Baseler M, Lloyd AL, Nowak MA, Fauci AS (1997b) Presence of an inducible HIV-1 latent reservoir during highly active antiretroviral therapy. Proc Natl Acad Sci USA 94:13193–13197

Coffin JM (1995) HIV population dynamics in vivo: implications for genetic variation, pathogenesis, and therapy. Science 267:483–489

Demma LJ, Jr Logsdon JM, Vanderford TH, Feinberg MB, Staprans SI (2005) SIVsm quasispecies adaptation to a new simian host. PLoS Pathog 1:e3

Dinoso JB, Kim SY, Wiegand AM, Palmer SE, Gange SJ, Cranmer L, O'Shea A, Callender M, Spivak A, Brennan T, Kearney MF, Proschan MA, Mican JM, Rehm CA, Coffin JM, Mellors JW, Siliciano RF, Maldarelli F (2009) Treatment intensification does not reduce residual HIV-1 viremia in patients on highly active antiretroviral therapy. Proc Natl Acad Sci USA 106:9403–9408

Fauci AS, Desrosiers RC (1997a) Pathogenesis of HIV and SIV. In: Coffin JM, Hughes SH, Varmus HE (eds) Retroviruses. Cold Spring Harbor, New York

Fauci AS, Desrosiers RC (1997b) Pathogenesis of HIV and SIV. In: Coffin JM, Hughes SH, Varmus HE (eds) Retroviruses. Cold Spring Harbor Laboratory Press, Cold Spring Harbor, p. 587–635

Finzi D, Hermankova M, Pierson T, Carruth LM, Buck C, Chaisson RE, Quinn TC, Chadwick K, Margolick J, Brookmeyer R, Gallant J, Markowitz M, Ho DD, Richman DD, Siliciano RF (1997) Identification of a reservoir for HIV-1 in patients on highly active antiretroviral therapy. Science 278:1295–1300

Frost SD, Nijhuis M, Schuurman R, Boucher CA, Brown AJ (2000) Evolution of lamivudine resistance in human immunodeficiency virus type 1-infected individuals: the relative roles of drift and selection. J Virol 74:6262–6268

Gandhi RT, Zheng L, Bosch RJ, Chan ES, Margolis DM, Read S, Kallungal B, Palmer S, Medvik K, Lederman MM, Alatrakchi N, Jacobson JM, Wiegand A, Kearney M, Coffin JM, Mellors

JW, Eron JJ (2010) The effect of raltegravir intensification on low-level residual viremia in HIV-infected patients on antiretroviral therapy: a randomized controlled trial. PLoS Med 7(8)

Hirsch VM (2004) What can natural infection of African monkeys with simian immunodeficiency virus tell us about the pathogenesis of AIDS? AIDS Rev 6:40–53

Josefsson L, King MS, Makitalo B, Brannstrom J, Shao W, Maldarelli F, Kearney MF, Hu WS, Chen J, Gaines H, Mellors JW, Albert J, Coffin JM, Palmer SE (2011) Majority of CD4+ T cells from peripheral blood of HIV-1-infected individuals contain only one HIV DNA molecule. Proc Natl Acad Sci USA 108:11199–11204

Jung A, Maier R, Vartanian JP, Bocharov G, Jung V, Fischer U, Meese E, Wain-Hobson S, Meyerhans A (2002) Recombination: multiply infected spleen cells in HIV patients. Nature 418:144

Kearney M, Maldarelli F, Shao W, Margolick JB, Daar ES, Mellors JW, Rao V, Coffin JM, Palmer S (2009) Human immunodeficiency virus type 1 population genetics and adaptation in newly infected individuals. J Virol 83:2715–2727

Kearney M, Spindler J, Shao W, Stephens R, O'Shea A, Rehm C, Poethke C, Mellors JW, Coffin JM, Maldarelli F (2010) Rebound HIV-1 viremia after cessation of long-term suppressive antiretroviral therapy (ART) has low genetic diversity and little divergence from pre-therapy virus populations. Oral presentation at the international HIV & hepatitis virus drug resistance workshop & curative strategies

Keele BF, Van Heuverswyn F, Li Y, Bailes E, Takehisa J, Santiago ML, Bibollet-Ruche F, Chen Y, Wain LV, Liegeois F, Loul S, Ngole EM, Bienvenue Y, Delaporte E, Brookfield JF, Sharp PM, Shaw GM, Peeters M, Hahn BH (2006) Chimpanzee reservoirs of pandemic and nonpandemic HIV-1. Science 313:523–526

Keele BF, Giorgi EE, Salazar-Gonzalez JF, Decker JM, Pham KT, Salazar MG, Sun C, Grayson T, Wang S, Li H, Wei X, Jiang C, Kirchherr JL, Gao F, Anderson JA, Ping LH, Swanstrom R, Tomaras GD, Blattner WA, Goepfert PA, Kilby JM, Saag MS, Delwart EL, Busch MP, Cohen MS, Montefiori DC, Haynes BF, Gaschen B, Athreya GS, Lee HY, Wood N, Seoighe C, Perelson AS, Bhattacharya T, Korber BT, Hahn BH, Shaw GM (2008) Identification and characterization of transmitted and early founder virus envelopes in primary HIV-1 infection. Proc Natl Acad Sci USA 105:7552–7557

Lerner P, Guadalupe M, Donovan R, Hung J, Flamm J, Prindiville T, Sankaran-Walters S, Syvanen M, Wong JK, George MD, Dandekar S (2011) The gut mucosal viral reservoir in HIV-infected patients is not the major source of rebound plasma viremia following interruption of highly active antiretroviral therapy. J Virol 85:4772–4782

Llibre JM, Buzon MJ, Massanella M, Esteve A, Dahl V, Puertas MC, Domingo P, Gatell JM, Larrouse M, Gutierrez M, Palmer S, Stevenson M, Blanco J, Martinez-Picado J, Clotet B (2012) Treatment intensification with raltegravir in subjects with sustained HIV-1 viraemia suppression: a randomized 48-week study. Antivir Ther 17:355–364

Maldarelli F, Palmer S, King MS, Wiegand A, Polis MA, Mican J, Kovacs JA, Davey RT, Rock-Kress D, Dewar R, Liu S, Metcalf JA, Rehm C, Brun SC, Hanna GJ, Kempf DJ, Coffin JM, Mellors JW (2007) ART suppresses plasma HIV-1 RNA to a stable set point predicted by pretherapy viremia. PLoS Pathog 3:e46

Mansky LM, Temin HM (1995) Lower in vivo mutation rate of human immunodeficiency virus type 1 than that predicted from the fidelity of purified reverse transcriptase. J Virol 69:5087–5094

McMahon D, Jones J, Wiegand A, Gange SJ, Kearney M, Palmer S, McNulty S, Metcalf JA, Acosta E, Rehm C, Coffin JM, Mellors JW, Maldarelli F (2010) Short-course raltegravir intensification does not reduce persistent low-level viremia in patients with HIV-1 suppression during receipt of combination antiretroviral therapy. Clin Infect Dis 50:912–919

Mellors JW, Rinaldo CR Jr, Gupta P, White RM, Todd JA, Kingsley LA (1996) Prognosis in HIV-1 infection predicted by the quantity of virus in plasma. Science 272:1167–1170

Mens H, Kearney M, Wiegand A, Shao W, Schonning K, Gerstoft J, Obel N, Maldarelli F, Mellors JW, Benfield T, Coffin JM (2010) HIV-1 continues to replicate and evolve in patients with natural control of HIV infection. J Virol 84:12971–12981

Neher RA, Leitner T (2010) Recombination rate and selection strength in HIV intra-patient evolution. PLoS Comput Biol 6:e1000660

Palmer S, Maldarelli F, Wiegand A, Bernstein B, Hanna GJ, Brun SC, Kempf DJ, Mellors JW, Coffin JM, King MS (2008) Low-level viremia persists for at least 7 years in patients on suppressive antiretroviral therapy. Proc Natl Acad Sci USA 105:3879–3884

Rouzine IM, Coffin JM (1999) Linkage disequilibrium test implies a large effective population number for HIV in vivo. Proc Natl Acad Sci USA 96:10758–10763

Rouzine IM, Coffin JM (2005) Evolution of HIV under selection and weak recombination. Genetics 170:7–18

Ruff CT, Ray SC, Kwon P, Zinn R, Pendleton A, Hutton N, Ashworth R, Gange S, Quinn TC, Siliciano RF, Persaud D (2002) Persistence of wild-type virus and lack of temporal structure in the latent reservoir for human immunodeficiency virus type 1 in pediatric patients with extensive antiretroviral exposure. J Virol 76:9481–9492

Salgado M, Rabi SA, O'Connell KA, Buckheit RW 3rd, Bailey JR, Chaudhry AA, Breaud AR, Marzinke MA, Clarke W, Margolick JB, Siliciano RF, Blankson JN (2011) Prolonged control of replication-competent dual- tropic human immunodeficiency virus-1 following cessation of highly active antiretroviral therapy. Retrovirology 8:97

Schnell G, Price RW, Swanstrom R, Spudich S (2010) Compartmentalization and clonal amplification of HIV-1 variants in the cerebrospinal fluid during primary infection. J Virol 84:2395–2407

Shankarappa R, Gupta P, Learn GH Jr, Rodrigo AG, Rinaldo CR Jr, Gorry MC, Mullins JI, Nara PL, Ehrlich GD (1998) Evolution of human immunodeficiency virus type 1 envelope sequences in infected individuals with differing disease progression profiles. Virology 241:251–259

Shankarappa R, Margolick JB, Gange SJ, Rodrigo AG, Upchurch D, Farzadegan H, Gupta P, Rinaldo CR, Learn GH, He X, Huang XL, Mullins JI (1999) Consistent viral evolutionary changes associated with the progression of human immunodeficiency virus type 1 infection. J Virol 73:10489–10502

Sharkey ME, Stevenson M (2001) Two long terminal repeat circles and persistent HIV-1 replication. Curr Opin Infect Dis 14:5–11

Simmonds P, Balfe P, Peutherer JF, Ludlam CA, Bishop JO, Brown AJ (1990) Human immunodeficiency virus-infected individuals contain provirus in small numbers of peripheral mononuclear cells and at low copy numbers. J Virol 64:864–872

Svarovskaia ES, Cheslock SR, Zhang WH, Hu WS, Pathak VK (2003) Retroviral mutation rates and reverse transcriptase fidelity. Front Biosci J Virt Libr 8:d117–d134

Tobin NH, Learn GH, Holte SE, Wang Y, Melvin AJ, McKernan JL, Pawluk DM, Mohan KM, Lewis PF, Mullins JI, Frenkel LM (2005) Evidence that low-level viremias during effective highly active antiretroviral therapy result from two processes: expression of archival virus and replication of virus. J Virol 79:9625–9634

Van Gulck E, Bracke L, Heyndrickx L, Coppens S, Atkinson D, Merlin C, Pasternak A, Florence E, Vanham G (2012) Immune and viral correlates of "secondary viral control" after treatment interruption in chronically HIV-1 infected patients. PLoS One 7:e37792

Van Heuverswyn F, Li Y, Neel C, Bailes E, Keele BF, Liu W, Loul S, Butel C, Liegeois F, Bienvenue Y, Ngolle EM, Sharp PM, Shaw GM, Delaporte E, Hahn BH, Peeters M (2006) Human immunodeficiency viruses: SIV infection in wild gorillas. Nature 444:164

Wightman F, Solomon A, Khoury G, Green JA, Gray L, Gorry PR, Ho YS, Saksena NK, Hoy J, Crowe SM, Cameron PU, Lewin SR (2010) Both CD31(+) and CD31 naive CD4(+) T cells are persistent HIV type 1-infected reservoirs in individuals receiving antiretroviral therapy. J Infect Dis 202:1738–1748

Wong JK, Hezareh M, Gunthard HF, Havlir DV, Ignacio CC, Spina CA, Richman DD (1997) Recovery of replication-competent HIV despite prolonged suppression of plasma viremia. Science 278:1291–1295

Yukl SA, Gianella S, Sinclair E, Epling L, Li Q, Duan L, Choi AL, Girling V, Ho T, Li P, Fujimoto K, Lampiris H, Hare CB, Pandori M, Haase AT, Gunthard HF, Fischer M, Shergill AK, McQuaid K, Havlir DV, Wong JK (2010) Differences in HIV burden and immune activation within the gut of HIV-positive patients receiving suppressive antiretroviral therapy. J Infect Dis 202:1553–1561

# Part V
# Prevention and Future Approaches

# Chapter 15
# RT Inhibitors as Microbicides

**Jan Balzarini**

## 15.1 Introduction

More than 25 million people have died from the acquired immunodeficiency syndrome (AIDS) since its recognition in 1981 as an emerging novel epidemic disease. The rate of infections with human immunodeficiency virus type 1 (HIV-1), the causative agent of AIDS, is still increasing worldwide, with rapid expansion in Africa, Asia, and Eastern Europe and also a less spectacular but still progressive increase in Western European countries and in the United States. It is clear that the AIDS epidemic is far from over. The development of a prophylactic efficient vaccine to prevent new HIV infections or a therapeutic vaccine to more efficiently control the infection of individuals already harboring the virus is still not in place. It has been estimated that an efficient vaccine will still be awaited for a long time period, if possible at all, to pursue an efficient prevention of the virus infection through immunological intervention.

Instead, the introduction of highly active antiretroviral therapy (HAART) has afforded a marked impact on the replication of the virus and the plasma viral loads for prolonged periods of time, significantly improving the prognosis and life expectancy of HIV-infected individuals. However, therapy failure is often due to drug resistance which can result in direct transmission of such drug-resistant virus strains to uninfected drug-naïve individuals.

Prevention of HIV transmission to uninfected individuals has become an important aim to contain the AIDS epidemic. Most infections with wild-type and drug-resistant virus strains occur through homo- or heterosexual intercourse. As a result, the introduction of microbicides as self-administering prophylactic agents can be of paramount importance to prevent new infections.

J. Balzarini, Ph.D. (✉)
Rega Institute for Medical Research, Laboratory of Virology and Chemotherapy,
KU Leuven, Minderbroedersstraat 10, Leuven 3000, Belgium
e-mail: jan.balzarini@rega.kuleuven.be

S.F.J. LeGrice and M. Götte (eds.), *Human Immunodeficiency Virus Reverse Transcriptase: A Bench-to-Bedside Success*, DOI 10.1007/978-1-4614-7291-9_15,

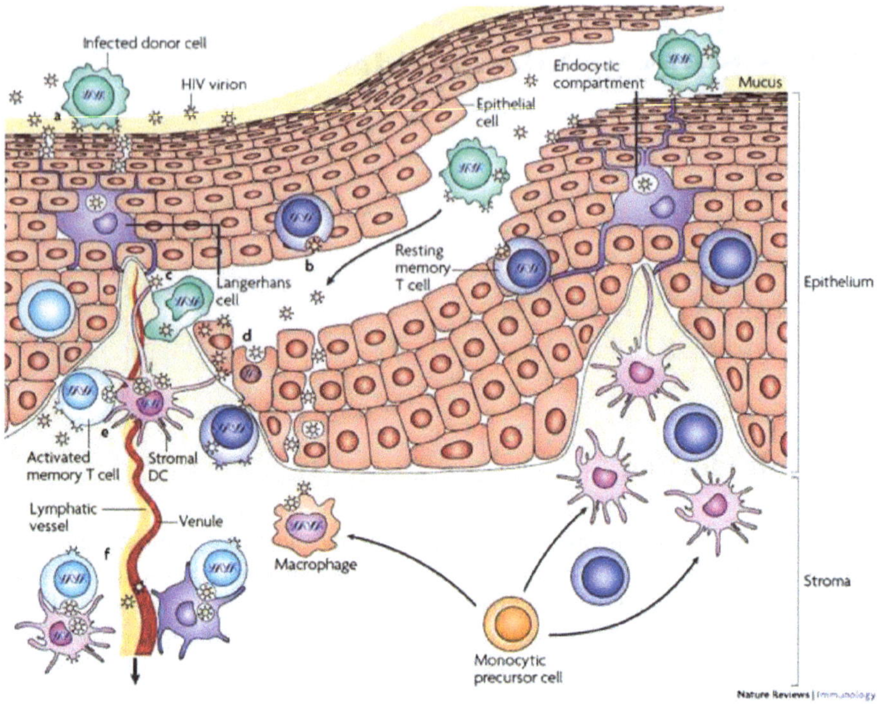


**Fig. 15.1** HIV transmission in women (Published with permission from Hladik and McElrath 2008)

In the genital tract, HIV can enter through the intact stratified vaginal epithelium or the single layer of the columnar cervical or rectal epithelium (Shattock and Moore 2003) (Fig. 15.1). In the genital and rectal submucosal area, $CD4^+$ T lymphocytes but also macrophages and dendritic cells represent the first targets for HIV infection. DC-SIGN, a glycan-binding lectin of the innate immune system and present on the cell membrane of submucosal dendritic cells, can act as an HIV trans-receptor that efficiently transfers HIV to the $CD4^+$ T lymphocytes shortly after exposure of the dendritic cells to the virus. The ideal microbicide should, therefore, be able to prevent the capture of HIV by the dendritic cells (but also macrophages), block the transmission of the virus to the other cells of the immune system, and prevent viral entry in its target cells. However, the ideal microbicide should also be able to block the establishment of HIV in its target cells (as a provirus) and the progressive infection once HIV appeared intracellularly. Therefore, HIV prevention would probably rely on the need to use a combination of several microbicide drugs in one single formulation that also includes at least one or more specific antiretroviral drugs (Balzarini and Van Damme 2007).

After a variety of failures in diverse clinical trials, due to unacceptable side effects (toxicity) and/or infectiveness of the particular drug (Van Damme et al. 2002, 2008; McCormack et al. 2010), the very first double-blind placebo-controlled study was reported in 2010 that showed clear efficacy to prevent HIV-1 transmission. In this trial, designated CAPRISA 004 (Abdool Karim et al. 2010), a 1 % tenofovir topical gel applied twice daily, was demonstrated to diminish HIV-1 transmission by 39 %. Also, the preexposure chemoprophylaxis (PrEP) trial, evaluating Truvada (a combination of tenofovir and emtricitabine) given to men having sex with men as a once-daily oral pill, showed a reduction of the HIV-1 transmission rate by 44 % (Grant et al. 2010). Given the fact that both tenofovir and emtricitabine represent HIV reverse-transcriptase (RT) inhibitors, it could be concluded that RT inhibitors given either as a topical gel formulation or as an oral pill have the potential of microbicidal action against HIV infection if prophylactically administered. In this chapter, HIV RT inhibitors will be further discussed for their potential application as microbicide drugs.

## 15.2 Structural and Functional Different Classes of HIV Reverse-Transcriptase Inhibitors

Since in several other chapters the different classes of HIV-1 RT inhibitors have been discussed in detail, a general but short overview will be presented here on the functionally different type of RT inhibitors that are known today. The HIV reverse transcriptase consists of a heterodimer in which the p66 subunit is non-covalently bound to a p51 subunit. The latter subunit is derived from the p66 subunit after proteolytic cleavage of a p15 moiety from the carboxylic acid end of the p66 subunit. The p15 moiety that is still present in the p66 subunit is endowed with RNAse H activity and hydrolyzes the RNA chain during and after conversion of the genomic RNA to a complementary DNA strand by the reverse-transcriptase activity of the p66 unit. The resulting single-stranded DNA chain will then be converted to double-stranded DNA by the DNA polymerase activity of the HIV RT. Thus, the HIV RT has at least three different functions (RT activity, DNA polymerase activity, and RNAse H activity) to convert single-stranded genomic viral RNA to double-stranded DNA ready to be incorporated as a proviral DNA into the host cell chromosomes of the virus-infected cells (Goff 1990).

At least four different functional classes of RT inhibitors can be distinguished (Table 15.1). The very first specific inhibitor of HIV RT was azidothymidine (AZT, zidovudine) that belonged to the structural class of nucleoside reverse-transcriptase inhibitors (NRTI) (Mitsuya et al. 1985). The NRTIs need to be metabolically converted to their 5′-triphosphate metabolites before being inhibitory against HIV RT (Furman et al. 1986). They not only compete with the endogenous 2′-deoxynucleotide (dNTP) pools for incorporation into the growing viral DNA chain but, upon incorporation, result in DNA chain termination since they do not contain a free 3′-hydroxyl group that supports further DNA chain elongation. The NRTIs do not discriminate between HIV-1 and HIV-2 strains for their inhibitory action.

In 1986, the first antivirally active acyclic nucleotide phosphonate analogues have been reported (De Clercq et al. 1986). These compounds structurally resemble nucleoside analogues in that they contain a purine or pyrimidine base moiety linked to an aliphatic (acyclic) side chain (instead of a pentose sugar). However, these

**Table 15.1** Differences in properties between NRTI, NtRTI, NNRTI, and NcRTI

| Property | NRTI | NtRTI | NNRTI | NcRTI |
|---|---|---|---|---|
| Prototype | Zidovudine [azidothymidine (AZT)] | Tenofovir [(R)-PMPA] | Nevirapine | INDOPY |
| Other members of the functional class | Stavudine (d4T)<br>Didanosine (ddI)<br>Abacavir (ABC)<br>Emtricitabine [(-)FTC]<br>Lamivudine [(-)3TC], etc. | Adefovir (PMEA) | Efavirenz<br>Dapivirine<br>Etravirine<br>Rilpivirine | DAVP |
| Metabolism | 3-step phosphorylation to 5′-triphosphate by cellular enzymes | 2-step phosphorylation to diphosphate by cellular enzymes | No metabolic conversion required | |
| Structures | Nucleoside-like structure | Nucleotide-like structure | Non-nucleoside structure | |
| Antiviral spectrum | Broad antiretroviral spectrum (HIV-1, HIV-2, SIV, …) | | HIV-1 specific | Broad antiretroviral spectrum (HIV-1, HIV-2, SIV, …) |
| RT target | Substrate-binding site | | Non-substrate (allosteric)-binding site | Substrate-binding site |
| Type of inhibition | Competitive to dNTP substrate | | Noncompetitive to dNTP substrate | Competitive to dNTP substrate |
| Mode of action | Incorporation into viral DNA and subsequent DNA chain termination | | No incorporation into viral DNA chain, but slowing down (blocking) the catalytic RT activity | |

compounds have also a phosphonate group linked to the acyclic side chain through a highly stable P–C bond and thus, resemble, in fact, the nucleoside monophosphate structure. They are designated as NtRTI (nucleotide RT inhibitors). Several subclasses of such acyclic nucleoside phosphonates (ANP) have been recognized, including hydroxyphosphonylmethoxypropyl (HPMP), phosphonylmethoxyethyl (PME), phosphonylmethoxypropyl (PMP), and their corresponding 2,4-diaminopyrimidine (DAPym) derivatives (De Clercq 2007). Each subclass of ANP seems to target a distinct and preferred spectrum of viruses to be inhibited. Whereas adefovir (PMEA) is active against most herpesviruses, retroviruses (i.e., HIV), and hepadnaviruses (i.e., HBV), PMPA or tenofovir has a high preference and selectivity for retroviruses and hepadnaviruses. The ANP need to be converted by cellular enzymes to their diphosphate derivatives (the homolog of the active NRTI-triphosphate derivatives) and, like the NRTI, compete with the endogenous dNTP pools for binding to HIV RT and subsequent incorporation into the growing viral DNA chain. The latter incorporation results in DNA chain termination (Balzarini et al. 1991; Balzarini 1994).

In 1989/1990, the first non-nucleoside RT inhibitors, designated NNRTI, were discovered. They were exemplified by the class of HEPT (Miyasaka et al. 1989; Baba et al. 1989) and TIBO derivatives (Pauwels et al. 1990). Later on, a wide variety of structurally diverse NNRTIs were discovered, all of which interact with HIV-1 RT at an allosteric site that is located close to, but is different from, the NRTI (substrate)-binding site on RT (Balzarini 2004). These compounds do not require metabolic conversion but are active as such. They need not to compete with the endogenous dNTP pools for binding to the RT and are not incorporated into the growing DNA chain. Instead, they inhibit the RT reaction in a noncompetitive manner by "freezing" the enzyme in a conformational state in which the substrate active site is somewhat located out of focus towards the primer grip, resulting in the inability of the enzyme to further elongate the DNA chain. The NNRTIs have no structural resemblance with nucleoside analogues but often adapt to a butterfly- or horseshoe-like shape when present in the allosteric NNRTI site of the HIV-1 RT (Das et al. 2004, 2005). A striking property of the NNRTIs is their highly specific inhibitory potential against HIV-1, but not HIV-2 strains. The functional differences between NRTI, NtRTI, and NNRTI are given in more detail in Balzarini (2004).

In 2006, a specific class of indolopyridone non-nucleoside RT inhibitors (designated nucleotide-competing RTI or NcRTI) have been discovered (designated as INDOPY) that block the polymerization reaction in a mode distinct from the NRTIs and NNRTIs (Jochmans et al. 2006). They show competitive inhibition with respect to the nucleotide substrate but do not resemble the nucleoside or dNTP structure. They keep the RT/inhibitor complex in a post-translocational state, preventing binding and incorporation of the next complementary nucleotide into the growing viral DNA chain. Recently, another compound (4-dimethylamino-6-vinylpyrimidine derivative) (DAVP) was reported as a non-nucleoside RT inhibitor that binds close to the polymerase active site (Freisz et al. 2010).

Finally, several classes of inhibitors targeting the RNAse H activity of the HIV RT have been described. These compounds have a non-nucleoside structure and are functionally directed against the RNAse H (p15) site of the p66 unit of RT (Tramontano and Di Santo 2010). They are often less potent inhibitors of RT than the most active NRTI and NNRTI and, together with the NcRTI, are not (yet) subject of clinical trials to investigate their in vivo efficacy. Instead, NRTI, NtRTI, and NNRTIs have been subject of clinical microbicide trials, demonstrating efficacy for the NtRTI tenofovir and the NRTI/NtRTI combination Truvada (tenofovir + emtricitabine) in several studies.

## 15.3 NRTI and NtRTI as Microbicidal Agents

Several microbicide studies have been designed using either tenofovir (Viread) or the combination of tenofovir disoproxil fumarate and emtricitabine (Truvada) as active ingredients (Fig. 15.2).

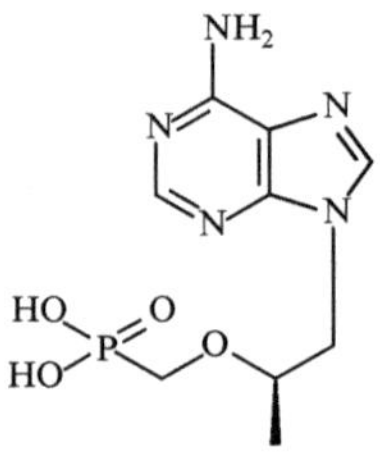

**Tenofovir**

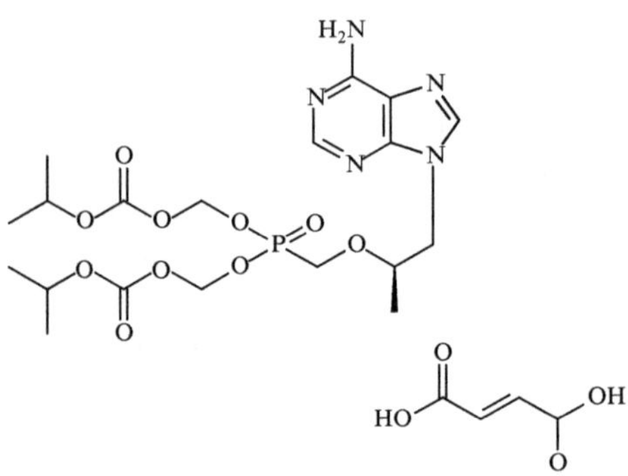

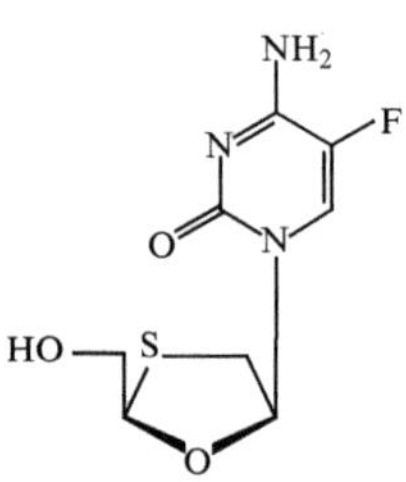

Tenofovir disoproxyl fumarate

Emtricitabine

**Truvada**

**Fig. 15.2** Structural formula of tenofovir and Truvada (tenofovir disoproxyl fumarate + emtricitabine)

The CAPRISA 004 study investigated topical tenofovir gel (1 %) effectiveness in 445 women versus placebo gel applied by 444 women twice daily (Abdool Karim et al. 2010). Monthly follow-up occurred for 30 months. The HIV incidence in the tenofovir gel arm proved 5.6 per 100 women-year versus 9.1 per 100 women-year in the placebo gel arm. The overall efficacy proved 39 %. In addition, an unanticipated reduced risk of herpes simplex virus type 2 (HSV-2) infection by 51 % was observed. Figure 15.3a represents the Kaplan–Meier estimates of cumulative probability of HIV infection in the tenofovir and placebo gel arms. Although an overall efficacy of 39 % was recorded, the efficacy of HIV infection prevention was only 28 % if compliance was less than 50 %, but 68 % if compliance of gel administration was more than 90 %.

The iPrEx Study investigated preexposure chemoprophylaxis by Truvada for HIV prevention in men having sex with men (Grant et al. 2010). Truvada consists of a pill containing the combination of tenofovir disoproxil fumarate and emtricitabine ((-)FTC). The drugs were given once daily as a 300 mg tablet to 2,499 HIV-seronegative men in a placebo-controlled manner. After 30 months of study, an overall reduction of 44 % in HIV incidence was recorded. Figure 15.3b represents the Kaplan–Meier estimates of cumulative probability of HIV infection in the Truvada and placebo arms. No reduced risk of herpes simplex virus type 2 infection was observed (Tan et al. 2011).

The conclusions of these two studies were that an encouraging, yet incomplete (~39–44 %), protection could be afforded against the HIV incidence of infection by prophylactic drug intervention. However, the studies also made clear that the drug adherence (compliance) problem is still significant and likely underestimates the effectiveness of these treatment modalities (Cates 2010). The observed protection data were at least very encouraging and should form the basis of further validation of HIV RT-inhibitory drugs for microbicidal use.

A third study, designated as the FEM-PrEP study that investigated the effect of a daily oral dose of Truvada given to women in Africa, resulted in a surprise during the interim review when 1,951 participants were enrolled (out of the planned 3,900 participants): 28 new infections in the placebo arm and an equal amount of 28 new infections in the treatment arm were recorded (FHI Statement 2011). These findings were the basis of the decision to prematurely stop this trial. The reason of this different outcome compared with the CAPRISA 004 and iPrEP trials was initially rather unclear, and issues such as adherence problems, negative drug–drug interactions, and inadequate drug levels at the local sites were under investigation to explain this lack of activity (van der Straten et al. 2012; Celum and Baeten 2012). Similar disappointing data as obtained in the FEM-PrEP trial were also observed in the oral tenofovir arm of the VOICE study which was also early discontinued (McEnery 2011). To date, of seven trials, from which six were conducted exclusively in sub-Saharan Africa, all found tenofovir and Truvada safe, and four had demonstrated effectiveness. Trial HPTN 052 even demonstrated 96 % HIV transmission reduction in a population of HIV-negative partners in HIV-serodiscordant couples. Yet, the oral Truvada in the FEM-PrEP study and the oral and vaginal tenofovir trial in the VOICE study were stopped for futility. It is now generally

**a**

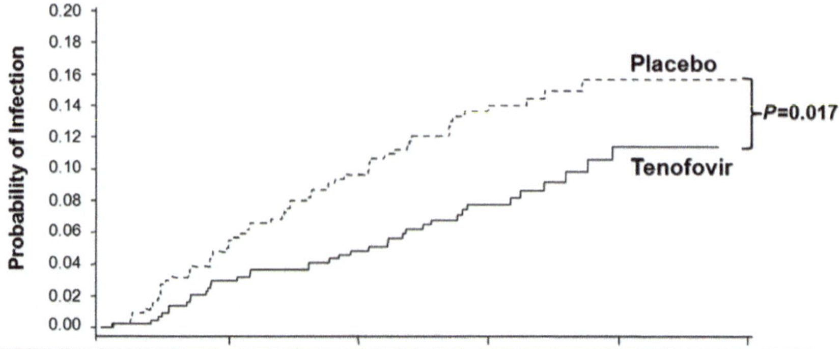


| Months of follow-up | 6 | 12 | 18 | 24 | 30 |
|---|---|---|---|---|---|
| Cumulative HIV endpoints | 37 | 65 | 88 | 97 | 98 |
| Cumulative women-years | 432 | 833 | 1143 | 1305 | 1341 |
| HIV incidence rates (Tenofovir vs Placebo) | 6.0 vs 11.2 | 5.2 vs 10.5 | 5.3 vs 10.2 | 5.6 vs 10.2 | 5.6 vs 9.1 |
| **Effectiveness (*P*-value)** | **47% (0.064)** | **50% (0.007)** | **47% (0.004)** | **40% (0.013)** | **39% (0.017)** |

**b**

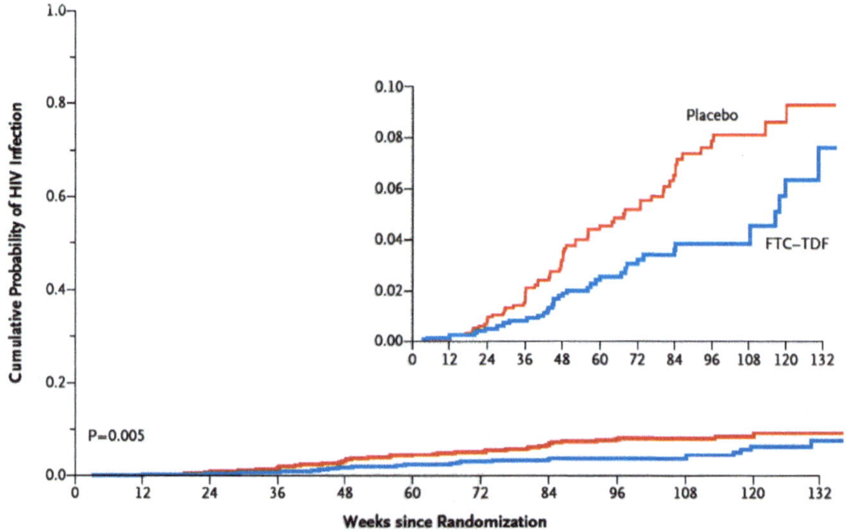


| No. at Risk | | | | | | | | | | | | |
|---|---|---|---|---|---|---|---|---|---|---|---|---|
| Placebo | 1248 | 1194 | 1108 | 1005 | 852 | 647 | 546 | 444 | 370 | 258 | 137 | 60 |
| FTC–TDF | 1251 | 1188 | 1097 | 988 | 848 | 693 | 558 | 447 | 367 | 267 | 147 | 65 |

**Fig. 15.3** (**a**) Kaplan–Meier estimates of cumulative probability of HIV infection in the tenofovir and placebo gel arms. The table provides the cumulative number of HIV end points, corresponding HIV incidence rates, and effectiveness of tenofovir gel for each additional 6 months of follow-up (Published with permission from Abdool Karim et al. 2010). (**b**) Kaplan–Meier estimates of time to HIV infection (modified intention-to-treat population). The cumulative probability of HIV acquisition is shown for the two study groups. The efficacy of preexposure prophylaxis with emtricitabine and tenofovir disoproxil fumarate (FTC–TDF) was 44 %, as compared with placebo ($P$=0.005). The inset graph shows a more detailed version of the overall graph up to a probability of 0.10 (Published with permission from Grant et al. 2010)

believed that adherence is one of the key factors among other additional but often still unclear reasons (van der Straten et al. 2012; Person and Hicks 2012; Van Damme et al. 2012). In fact, it was very recently demonstrated that, notwithstanding tenofovir gel use, those women who acquired HIV had markedly higher systemic innate immune activation prior to infection than women who remained uninfected (Naranbhai et al. 2012). Therefore, it has been suggested that an innate immune activation suppressant may be considered to be added to tenofovir gel, although it should be realized that this may have a negative, rather than positive, impact on other pathogens as well!

To at least partially understand some of the efficacy/non-efficacy observations made in the microbicide trials, it was of interest to consider the pharmacokinetic and pharmacodynamic properties of the different drug formulations and administrations. Whereas cervicovaginal and vaginal tenofovir levels of $10^5$–$10^6$ ng/ml and $10^3$–$10^4$ ng/g were recorded upon 1 % tenofovir gel administration, only 70 ng/ml and 10 ng/g tenofovir levels were observed upon oral administration of tenofovir. Also, upon oral tenofovir treatment, vaginal drug levels were 10 ng/g versus 2,000 ng/g rectal drug recovery (Abdool Karim 2011). Interestingly, tenofovir diphosphate levels as low as 206 fmol/mg rectal tissue were observed versus 2 fmol/mg vaginal tissue upon oral Truvada dosing, whereas 1 pmol/mg tenofovir diphosphate was found in vaginal tissue and 10 pmol/$10^6$ cervical cells upon 1 % tenofovir gel administration. These data, taken together, indicate that oral tenofovir affords markedly higher drug levels in rectal tissue than vaginal tissue which may contribute to the higher effectiveness of the orally administered drug for HIV prevention in the trial of men having sex with men than in the FEM-PrEP trial where women were given oral Truvada treatment. Also, a 1 % tenofovir gel application affords much higher local drug concentrations in the cervicovaginal compartment than an oral drug administration. These findings may provide a rational explanation why the CAPRISA 004 clinical trial revealed both anti-HIV activity and partial protection against HSV-2.

The latter observation was surprising and unanticipated, but most interesting. Tenofovir has been known and developed as a specific antiretroviral drug, not as an antiherpesvirus drug. However, careful analysis of earlier data (Balzarini et al. 1993) and most recent data (Andrei et al. 2011) on tenofovir revealed a rather marginal antiherpetic activity. Tenofovir indeed was endowed with an antiherpetic activity in the higher micromolar range, that is, at drug levels that are ~50- to 100-fold higher than those required to afford anti-HIV activity. The antiherpetic activity of tenofovir at the higher micromolar drug concentrations was confirmed not only against laboratory HSV-2 (and HSV-1) strains but also against a variety of clinical HSV isolates in HEL and keratinocyte cell cultures, in organotypic raft cultures, in monocyte/macrophages (activity at low micromolar drug concentrations), and in ex vivo human tonsil and cervicovaginal tissues (Andrei et al. 2011). Thus, an at-first-glance marginal antiviral activity can turn out to be clinically relevant if sufficiently high drug levels such as those obtained upon topical drug administration can be reached at the local site of infection. These findings are highly relevant, given the fact that HSV-2 is a co-pathogen of HIV and known to increase the transmission rate of HIV (Corey 2007; Tanton et al. 2011). In this respect, it should be mentioned that acyclovir (ACV), the current drug of choice for the treatment of HSV-1 and HSV-2 infections, is also endowed with anti-HIV activity at higher concentrations when activated by herpetic kinases (Lisco et al. 2008). It has been shown that acyclovir prodrugs that deliver the activated ACV-monophosphate derivative intracellularly and independently from the presence of herpesvirus-encoded kinases have intrinsic anti-HIV activity (Derudas et al. 2009; Vanpouille et al. 2010). The observation from several clinical trials that ACV treatment has an effect on plasma HIV RNA levels (Ouedraogo et al. 2006; Mugwanya et al. 2011; Baeten et al. 2008) is in agreement with these data and urges for a more careful consideration of ACV (most likely in its monophosphate prodrug form) as a potential microbicide NRTI. The findings that the ACV triphosphate is inhibitory to HIV RT (Tchesnokov et al. 2009) and, vice versa, tenofovir diphosphate is inhibitory to HSV DNA polymerase (Andreï et al. 2011) further support the development of both tenofovir and/or acyclovir derivatives and combinations thereof as potential microbicide principles with a dual antiviral action.

## 15.4 NNRTI as Potential Microbicide Drugs

Several members of a variety of NNRTI classes have been considered as potential microbicide drug candidates (Fig. 15.4). The chosen compounds are invariably active against HIV in the lower nanomolar range and are considered as tight-binding

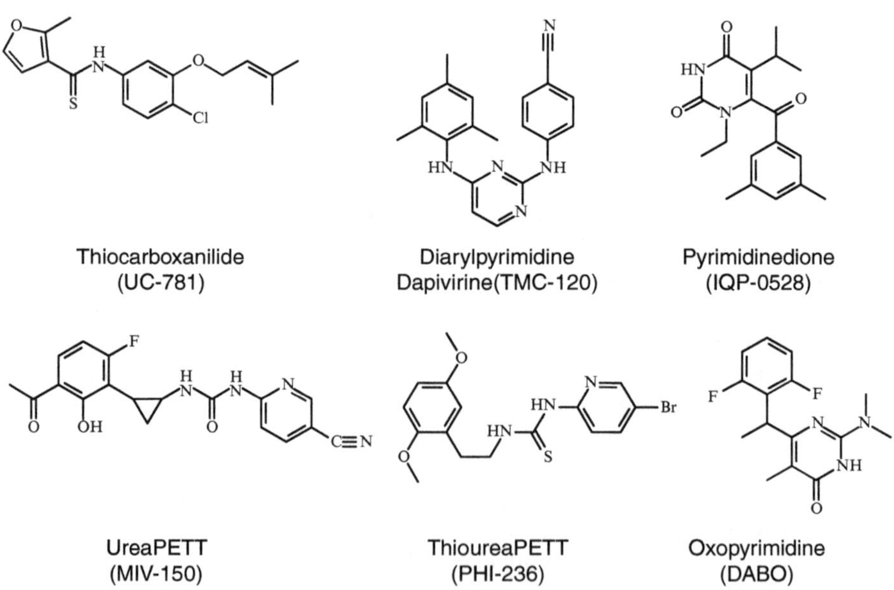


**Fig. 15.4** Structural formulae of NNRTIs

RT inhibitors. Due to their lipophilic nature (a common property of most NNRTIs), the compounds easily penetrate cell layers and may afford a relatively long intracellular half-life. Moreover, several NNRTIs have proven to be endowed with virucidal properties, most likely due to their persistence in the viral particle after budding of the virus from drug-exposed cells. Such properties are favorable for microbicidal application.

### *15.4.1 The Thiocarboxanilide Derivative UC-781*

UC-781 represents the prototype compound among the second-generation thiocarboxanilides and binds rapidly and tightly to HIV-1 RT (Barnard et al. 1997). Infectivity of HIV-1 could indeed be eliminated in a concentration-dependent manner following short exposure of HIV-1 virions to UC-781 (Borkow et al. 1997). Also, endogenous RT activity in UC-781-treated virus particles was markedly reduced by this drug. Treatment of chronically HIV-1-infected cells with UC-781 reduced virus infectivity, and preincubation of uninfected cells with UC-781 rendered these cells refractory to subsequent HIV infection in the absence of the extracellular drug (Borkow et al. 1997). All these properties may be due to the tight-binding and lipophilic properties of UC-781 and are beneficial properties for potential microbicidal application. Moreover, UC-781 proved to block HIV replication in co-cultures of dendritic cells and $CD4^+$ T lymphocytes upon short preincubation of the virus with the drug (Njai et al. 2005; Van Herrewege et al. 2004). The drug is stable at high temperature (i.e., 50 °C) and at low pH. It did not lose antiviral activity upon short exposure to pH 3.5 (mimicking the vaginal pH) (Balzarini et al. 1998). UC-781 is able to suppress a wide variety of mutant HIV-1 strains containing several NNRTI-characteristic resistance mutations in their RT (Balzarini et al. 1998). This has been explained and visualized in crystallographic studies of the HIV-1 RT/UC-781 complex (Ren et al. 1998; Esnouf et al. 1997).

UC-781 was found to afford prolonged protection against HIV in ex vivo cervical and colorectal explant models (Cummins et al. 2007; Fletcher et al. 2006). When administered intravaginally to rabbits as a 5 % replens gel formulation, neither systemic side effects nor local inflammation or damage of the vaginal mucosa or epithelium was observed. Also, there were no adverse effects on the growth of microorganisms such as the commensal vaginal Lactobacillus strains in the presence of the drug (Balzarini et al. 1998). When micronized UC-781 vaginal formulations were applied to rabbits, high vaginal drug concentrations could be reached, exceeding its $EC_{50}$ value by at least 3 orders of magnitude. Interestingly, plasma levels of vaginally applied UC-781 were lower than 2 ng/ml (Clark et al. 2011), making HIV resistance development rather unlikely. Such low, if any, detectable levels of UC-781 in the plasma were also observed in macaques that were vaginally or rectally exposed to water-based formulations of UC-781 (Patton et al. 2007). UC-781 has been subject of phase I/II trials in seronegative women (Schwartz et al. 2008) and found to result in a slight but transient irritation, but mostly undetectable levels in plasma.

Recently, a phase I double-blind, placebo-controlled, rectal microbicide trial using UC-781 gel was performed and reported to be a safe drug application without significant adverse effects and with high acceptability, no significant plasma drug levels, and no mucosal changes. Ex vivo biopsy infections demonstrated marked suppression of HIV infectibility (Anton et al. 2011). So far, no efficacy trials in individuals at risk of HIV-1 infection have been reported.

### *15.4.2 The Diarylpyrimidine Derivative Dapivirine*

Dapivirine (TMC 120) represents the prototype compound among the diarylpyrimidines. It was found to suppress a variety of HIV-1 strains in the lower nanomolar concentration range in several cellular model systems, including co-cultures of monocyte-derived dendritic cells and T cells, human ex vivo cervical tissue and colorectal explants, and HIV-infected cell cultures in a dual chamber model of transepithelial transmission (Harman et al. 2004; Fletcher et al. 2006). In the latter assay model, dapivirine was exposed to cell-associated virus on top of a confluent epithelial cell layer, whereas the target cells (monocyte-derived dendritic cells and T cells) were present in the lower chamber (Van Herrewege et al. 2004). The drug was generally antivirally active at concentrations that were several orders of magnitude lower than its toxicity value. Dapivirine, as also reported for UC-781, is able to penetrate the envelope of HIV-1 particles, likely (partially) due to its high lipophilicity, resulting in inactivation of the virus before target cell infection (Costa et al. 2005; Aguiar et al. 2007; Njai et al. 2005).

Dapivirine showed a preventive potential against vaginal transmission of HIV-1 in hu-SCID mice upon formulation in a gel (Di Fabio et al. 2003), and a long-term controlled release of dapivirine from silicone elastomer vaginal rings could lead to the use of a vaginal ring for the delivery of microbicides such as dapivirine (Malcolm et al. 2005; Woolfson et al. 2006). Despite its rather good oral bioavailability (Lewi et al. 2004), intravaginal dapivirine administration does not result in the appearance of high plasma concentrations, although there has been evidence that the drug is absorbed by the outer mucosal layers.

After the demonstration that dapivirine showed efficacy and safety in hu-SCID mice and low systemic absorption when applied as an intravaginal gel formulation (Di Fabio et al. 2003), phase I/II trials have been initiated and a number of them have been completed already (IPM 2011). It could be generally concluded that upon delivery through a gel or an intravaginal ring, the intravaginally achieved drug concentrations usually exceeded by three to four orders of magnitude those that have been recorded in plasma of such exposed individuals and that intravaginal drug concentrations could be obtained that exceed by three to five orders of magnitude those that afford antiviral activity in HIV-1-infected cell cultures (Nel et al. 2010).

### 15.4.3 Phenylethylthiazolyl(Thio)Urea Derivatives (PETT)

Several members of the compound class of ureaPETTs (i.e., MIV-150/MIV-160/MIV-170) were shown to have low or even sub-nanomolar anti-HIV activity in cell culture models, tight-binding HIV-1 RT properties, a favorable inhibition profile against HIV-1 strains that contain NNRTI-characteristic resistance mutations, and a suppression potential against a variety of clinical isolates (Zhang et al. 2004). Several ureaPETTs have been shown efficacy in the macaque/SHIV model and are currently under further preclinical development as a single agent or combined with other microbicide drugs (Fernandez-Romero et al. 2007; Singer et al. 2011; Selhorst et al. 2011).

The thioureaPETTs are close structural analogues of the trovirdine prototype PETT NNRTI, and several analogues such as PHI-236, PHI-346, and PHI-443 have been investigated for their microbicidal properties (D'Cruz and Uckun 2006). As found for most of the other above-described candidate-NNRTI microbicide compounds, the thioureaPETTs are active against HIV-1 in the low- to sub-nanomolar concentration range, suppress a variety of clinical (wild-type and drug-resistant) HIV-1 isolates (including non-B subtypes), are highly lipophilic, and are endowed with low oral bioavailability. They were shown effective in the HIV-1-infected hu-SCID mice model. PHI-346, but not PHI-236 and PHI-443, was found to display spermicidal activity (D'Cruz and Uckun 2005). The compounds proved rather safe in several ex vivo and animal models and devoid of significant systemic absorption (D'Cruz and Uckun 2005; D'Cruz et al. 2004). It is unclear whether further (pre) clinical microbicide trials are planned with this class of NNRTIs.

### 15.4.4 Pyrimidinedione Derivatives

A wide variety of pyrimidinedione derivatives have been synthesized and evaluated for antiviral activity in cell culture (Buckheit et al. 2007, 2008). Interestingly, several compounds showed antiviral potency against both HIV-1 and HIV-2. This is a rather unusual property of NNRTIs that usually display high selectivity for HIV-1 and lack activity against HIV-2. For the pyrimidinediones, the anti-HIV-2 activity is usually inferior to the anti-HIV-1 activity and could be explained by targeting HIV entry in addition to HIV-1 RT. Thus, the pyrimidinediones are endowed with a dual mechanism of antiviral action. The prototype of this class of compounds is IQP-0528. This compound is highly active and selective against HIV-1 (and HIV-2) strains, including clinical isolates and NNRTI-resistant viruses (Buckheit 2011; Watson et al. 2008). Gel formulations of IQP-0528 turned out to have nanomolar activity in HIV-infected cervical explants without significant side effects (Mahalingam et al. 2011). Also, the compound has been formulated into intravaginal rings as a potential topical microbicide (Ham et al. 2009). When evaluated in macaques, high intravaginal levels of the compound could be reached, without detectable plasma levels (Srinivasan et al. 2010).

### 15.4.5 *Oxopyrimidine Derivatives*

Both oxopyrimidine DABO (MC1220) and thioether-pyrimidine S-DABO derivatives have been investigated as potential microbicidal NNRTIs (Pani et al. 2001; Caron et al. 2010; Stolte-Leeb et al. 2011). Efficacy studies in macaques showed potential protection against SHIV challenge with relatively low appearance of plasma drug concentrations. Although DABO did not show toxicity in rabbit vaginal mucosa, slight vaginal irritation was observed in macaques, and measurable plasma levels were recorded (4.5–30 nM) (La Colla 2007; Caron et al. 2010). Partial protection in SHIV-infected macaques was observed when applied as a gel formulation (Stolte-Leeb et al. 2011).

## 15.5 Conclusion

Although the first clinical successes have been recorded for HIV RT inhibitors as potential microbicide drugs, there is still a long way to go. Activity has not been consistently seen in all microbicide trials, although the nature of the cohorts included in the studies and the formulation modalities and way of administration were often different. Nevertheless, the fact that N(t)RTIs were found so far to be the sole drugs that may show efficacy in HIV prevention points to the HIV RT as a suitable target for microbicide drug development. Several types of NNRTIs are under consideration, and it would not be excluded that, as in HAART, a N(t)RTI should be combined with a NNRTI (and/or additional drugs directed against a different target) to eventually afford the most efficient microbicide activity. Further research will reveal the exact role RT inhibitors may play in microbicide drug development, either as single agents and/or in combination with at least one or two more drugs.

**Acknowledgments** The author thanks Mrs. C. Callebaut for her dedicated editorial help. Funding of the microbicide research of the author has been provided by the European Commission (CHAARM).

## References

Abdool Karim SS, Abdool Karim Q (2011) Antiretroviral prophylaxis: a defining moment in HIV control. Lancet 378:e23–e25

Abdool Karim Q, Abdool Karim SS, Frohlich JA, Grobler AC, Baxter C, Mansoor LE, Kharsany AB, Sibeko S, Mlisana KP, Omar Z, Gengiah TN, Maarschalk S, Arulappan N, Mlotshwa M, Morris L, Taylor D, CAPRISA 004 Trial Group (2010) Effectiveness and safety of tenofovir gel, an antiretroviral microbicide, for the prevention of HIV infection in women. Science 32:1168–1174

Aguiar RS, Costa LJ, Pereira HS, Brindeiro RM, Tanuri A (2007) Development of a new methodology for screening of human immunodeficiency virus type 1 microbicides based on real-time PCR quantification. Antimicrob Agents Chemother 51:638–644

Andrei G, Lisco A, Vanpouille C, Introini A, Balestra E, van den Oord J, Cihlar T, Perno CF, Snoeck R, Margolis L, Balzarini J (2011) Topical tenofovir, a microbicide effective against HIV, inhibits herpes simplex virus-2 replication. Cell Host Microbe 10:379–389

Anton PA, Saunders T, Elliott J, Khanukhova E, Dennis R, Adler A, Cortina G, Tanner K, Boscardin J, Cumberland WG, Zhou Y, Ventuneac A, Carballo-Diéguez A, Rabe L, McCormick T, Gabelnick H, Mauck C, McGowan I (2011) First phase 1 double-blind, placebo-controlled, randomized rectal microbicide trial using UC781 gel with a novel index of ex vivo efficacy. PLoS One 6:e23243

Baba M, Tanaka H, De Clercq E, Pauwels R, Balzarini J, Schols D, Nakashima H, Perno CF, Walker RT, Miyasaka T (1989) Highly specific inhibition of human immunodeficiency virus type 1 by a novel 6-substituted acyclouridine derivative. Biochem Biophys Res Commun 165:1375–1381

Baeten JM, Strick LB, Lucchetti A, Whittington WL, Sanchez J, Coombs RW, Magaret A, Wald A, Corey L, Celum C (2008) Herpes simplex virus (HSV)-suppressive therapy decreases plasma and genital HIV-1 levels in HSV-2/HIV-1 coinfected women: a randomized, placebo-controlled, cross-over trial. J Infect Dis 198:1804–1808

Balzarini J (1994) Metabolism and mechanism of antiretroviral action of purine and pyrimidine derivatives. Pharm World Sci 16:113–126

Balzarini J (2004) Current status of the non-nucleoside reverse transcriptase inhibitors of human immunodeficiency virus type 1. Curr Top Med Chem 4:921–944

Balzarini J, Van Damme L (2007) Microbicide drug candidates to prevent HIV infection. Lancet 269:787–797

Balzarini J, Hao Z, Herdewijn P, Johns DG, De Clercq E (1991) Intracellular metabolism and mechanism of anti-retrovirus action of 9-(2-phosphonylmethoxyethyl) adenine, a potent anti-human immunodeficiency virus compound. Proc Natl Acad Sci USA 88:1499–1503

Balzarini J, Holy A, Jindrich J, Naesens L, Snoeck R, Schols D, De Clercq E (1993) Differential antiherpesvirus and antiretrovirus effects of the (S) and (R) enantiomers of acyclic nucleoside phosphonates: potent and selective in vitro and in vivo antiretrovirus activities of (R)-9-(2-phosphonomethoxypropyl)-2,6-diaminopurine. Antimicrob Agents Chemother 37:332–338

Balzarini J, Naesens L, Verbeken E, Laga M, Van Damme L, Parniak M, Van Mellaert L, Anné J, De Clercq E (1998) Preclinical studies on thiocarboxanilide UC-781 as a virucidal agent. AIDS 12:1129–1138

Barnard J, Borkow G, Parniak MA (1997) The thiocarboxanilide nonnucleoside UC781 is a tight-binding inhibitor of HIV-1 reverse transcriptase. Biochemistry 36:7786–7792

Borkow G, Barnard J, Nguyen TM, Belmonte A, Wainberg MA, Parniak MA (1997) Chemical barriers to human immunodeficiency virus type 1 (HIV-1) infection: retrovirucidal activity of UC781, a thiocarboxanilide nonnucleoside inhibitor of HIV-1 reverse transcriptase. J Virol 71:3023–3030

Buckheit RW Jr (2011) Development of pyrimidine dione NNRTIs with a high genetic barrier to resistance. http://www.ihlpress.com/pd20files/hivdart08_presentations/10_buckheit.pdf

Buckheit RW Jr, Hartman TL, Watson KM, Kwon HS, Lee SH, Lee JW, Kang DW, Chung SG, Cho EH (2007) The structure-activity relationships of 2, 4(1H, 3H)-pyrimidinedione derivatives as potent HIV type 1 and type 2 inhibitors. Antivir Chem Chemother 18:259–275

Buckheit RW Jr, Hartman TL, Watson KM, Chung SG, Cho EH (2008) Comparative evaluation of the inhibitory activities of a series of pyrimidinedione congeners that inhibit human immunodeficiency virus types 1 and 2. Antimicrob Agents Chemother 52:225–236

Caron M, Besson G, Etenna SL, Mintsa-Ndong A, Mourtas S, Radaelli A, Morghen Cde G, Loddo R, La Colla P, Antimisiaris SG, Kazanji M (2010) Protective properties of non-nucleoside reverse transcriptase inhibitor (MC1220) incorporated into liposome against intravaginal challenge of Rhesus macaques with RT-SHIV. Virology 405:225–233

Cates W Jr (2010) After CAPRISA 004: time to re-evaluate the HIV lexicon. Lancet 376: 495–496

Celum C, Baeten JM (2012) Tenofovir-based pre-exposure prophylaxis for HIV prevention: evolving evidence. Curr Opin Infect Dis 25:51–57

Clark MR, McCormick TJ, Doncel GF, Friend DR (2011) Preclinical evaluation of UC-781 microbicide vaginal drug delivery. Drug Deliv Trans Res 1:176–182

Corey L (2007) Synergistic copathogens–HIV-1 and HSV-2. N Engl J Med 356:854–856

Costa L, Aguiar R, Pereira H, Brindeiro R, Tanuri A (2005) Evaluation of nevirapine permeability and retention in HIV-1 cell free and budded virions. In: Proceedings of the 3rd international AIDS society (IAS) conference on HIV pathogenesis and treatment, Rio de Janeiro, 24–27 July 2005

Cummins JE Jr, Guarner J, Flowers L, Guenthner PC, Bartlett J, Morken T, Grohskopf LA, Paxton L, Dezzutti CS (2007) Clinical testing of candidate topical microbicides for anti-human immunodeficiency virus type 1 activity and tissue toxicity in a human cervical explant culture. Antimicrob Agents Chemother 51:1770–1779

D'Cruz OJ, Uckun FM (2005) Discovery of 2,5-dimethoxy-substituted 5-bromopyridyl thiourea (PHI-236) as a potent broad-spectrum anti-human immunodeficiency virus microbicide. Mol Hum Reprod 11:767–777

D'Cruz OJ, Uckun FM (2006) Dawn of non-nucleoside inhibitor-based anti-HIV microbicides. J Antimicrob Chemother 57:411–423

D'Cruz OJ, Samuel P, Uckun FM (2004) PHI-443. A novel noncontraceptive broad-spectrum anti-human immunodeficiency virus microbicide. Biol Reprod 71:2037–2047

Das K, Clark AD Jr, Lewi PJ, Heeres J, De Jonge MR, Koymans LM, Vinkers HM, Daeyaert F, Ludovici DW, Kukla MJ, De Corte B, Kavash RW, Ho CY, Ye H, Lichtenstein MA, Andries K, Pauwels R, De Béthune MP, Boyer PL, Clark P, Hughes SH, Janssen PA, Arnold E (2004) Roles of conformational and positional adaptability in structure-based design of TMC125-R165335 (etravirine) and related non-nucleoside reverse transcriptase inhibitors that are highly potent and effective against wild-type and drug-resistant HIV-1 variants. J Med Chem 47:2550–2560

Das K, Lewi PJ, Hughes SH, Arnold E (2005) Crystallography and the design of anti-AIDS drugs: conformational flexibility and positional adaptability are important in the design of non-nucleoside HIV-1 reverse transcriptase inhibitors. Prog Biophys Mol Biol 88:209–231

De Clercq E (2007) Acyclic nucleoside phosphonates: past, present and future. Bridging chemistry to HIV, HBV, HCV, HPV, adeno-, herpes-, and poxvirus infections: the phosphonate bridge. Biochem Pharmacol 73:911–922

De Clercq E, Holý A, Rosenberg I, Sakuma T, Balzarini J, Maudgal PC (1986) A novel selective broad-spectrum anti-DNA virus agent. Nature 323:464–467

Derudas M, Carta D, Brancale A, Vanpouille C, Lisco A, Margolis L, Balzarini J, McGuigan C (2009) The application of phosphoramidate protide technology to acyclovir confers anti-HIV inhibition. J Med Chem 52:5520–5530

Di Fabio S, Van Roey J, Giannini G, van den Mooter G, Spada M, Binelli A, Pirillo MF, Germinario E, Belardelli F, de Bethune MP, Vella S (2003) Inhibition of vaginal transmission of HIV-1 in hu-SCID mice by the non-nucleoside reverse transcriptase inhibitor TMC120 in a gel formulation. AIDS 17:1597–1604

Esnouf RM, Stuart DI, De Clercq E, Schwartz E, Balzarini J (1997) Models which explain the inhibition of reverse transcriptase by HIV-1-specific (thio)carboxanilide derivatives. Biochem Biophys Res Commun 234:458–464

Fernández-Romero JA, Thorn M, Turville SG, Titchen K, Sudol K, Li J, Miller T, Robbiani M, Maguire RA, Buckheit RW Jr, Hartman TL, Phillips DM (2007) Carrageenan/MIV-150 (PC-815), a combination microbicide. Sex Transm Dis 34:9–14

FHI Statement on the FEM-PrEP HIV Prevention Study (2011) Early end for FEM-PrEP HIV prevention trial. AIDS Patient Care STDS 25:383, www.fhi.org

Fletcher PS, Elliott J, Grivel JC, Margolis L, Anton P, McGowan I, Shattock RJ (2006) Ex vivo culture of human colorectal tissue for the evaluation of candidate microbicides. AIDS 20:1237–1245

Freisz S, Bec G, Radi M, Wolff P, Crespan E, Angeli L, Dumas P, Maga G, Botta M, Ennifar E (2010) Crystal structure of HIV-1 reverse transcriptase bound to a non-nucleoside inhibitor with a novel mechanism of action. Angew Chem Int Ed Engl 49:1805–1808

Furman PA, Fyfe JA, St Clair MH, Weinhold K, Rideout JL, Freeman GA, Lehrman SN, Bolognesi DP, Broder S, Mitsuya H, Barry DW (1986) Phosphorylation of 3′-azido-3′-deoxythymidine and selective interaction of the 5′-triphosphate with human immunodeficiency virus reverse transcriptase. Proc Natl Acad Sci USA 83:8333–8337

Goff SP (1990) Retroviral reverse transcriptase: synthesis, structure, and function. J Acquir Immune Defic Syndr 3:817–831

Grant RM, Lama JR, Anderson PL, McMahan V, Liu AY, Vargas L, Goicochea P, Casapía M, Guanira-Carranza JV, Ramirez-Cardich ME, Montoya-Herrera O, Fernández T, Veloso VG, Buchbinder SP, Chariyalertsak S, Schechter M, Bekker LG, Mayer KH, Kallás EG, Amico KR, Mulligan K, Bushman LR, Hance RJ, Ganoza C, Defechereux P, Postle B, Wang F, McConnell JJ, Zheng JH, Lee J, Rooney JF, Jaffe HS, Martinez AI, Burns DN, Glidden DV, iPrEx Study Team (2010) Preexposure chemoprophylaxis for HIV prevention in men who have sex with men. N Engl J Med 363:2587–2599

Ham A, Watson KM, Yang L et al (2009) Formulation of IQP-0528 into intravaginal rings as a topical microbicide for the prevention of HIV. In: Symposium on advancing prevention technologies for sexual and reproductive health, Berkeley, 24–25 Mar 2009

Ham AS, Rohan LC, Boczar A, Yang L, Buckheit KW, Buckheit RW Jr (2012) Vaginal film drug delivery of the pyrimidinedione IQP0528 for the prevention of HIV infection. Pharm Res 29:1897–1907

Harman S, Watts P, Shattock R, Griffin G, Van Roej J (2004) TMC120 blocks HIV-1 infection in cellular and human cervical tissue models. Microbicides, London (abstr)

Hladik F, McElrath MJ (2008) Setting the stage: host invasion by HIV. Nat Rev Immunol 8:447–457

International Partnership for Microbicides (2011) IPM clinical trials. http://www.ipmglobal.org/our-work/research/clinical-trial

Jochmans D, Deval J, Kesteleyn B, Van Marck H, Bettens E, De Baere I, Dehertogh P, Ivens T, Van Ginderen M, Van Schoubroeck B, Ehteshami M, Wigerinck P, Götte M, Hertogs K, Jochmans D, Deval J, Kesteleyn B, Van Marck H, Bettens E, De Baere I, Dehertogh P, Ivens T, Van Ginderen M, Van Schoubroeck B, Ehteshami M, Wigerinck P, Götte M, Hertogs K (2006) Indolopyridones inhibit human immunodeficiency virus reverse transcriptase with a novel mechanism of action. J Virol 80:12283–12292

La Colla P (2007) Report on the European 6th framework programme. Selection and development of microbicides for mucosal use to prevent sexual HIV transmission/acquisition (SHIVA), June 2007. http://www.edctp.org/fileadmin/documents/Microbicide_Stakeholder_Paolo-La_Colla.pdf

Lewi PJ, Arnold E, Andries K et al (2004) Correlations between factors determining the pharmacokinetics and antiviral activity of HIV-1 nonnucleoside reverse transcriptase inhibitors of the diaryltriazine and diarylpyrimidine classes of compounds. Drugs R&D 5:247–257

Lisco A, Vanpouille C, Tchesnokov EP, Grivel JC, Biancotto A, Brichacek B, Elliott J, Fromentin E, Shattock R, Anton P (2008) Acyclovir is activated into a HIV-1 reverse transcriptase inhibitor in herpesvirus-infected human tissues. Cell Host Microbe 4:260–270

Mahalingam A, Simmons AP, Ugaonkar SR, Watson KM, Dezzutti CS, Rohan LC, Buckheit RW Jr, Kiser PF (2011) Vaginal microbicide gel for delivery of IQP-0528, a pyrimidinedione analog with a dual mechanism of action against HIV-1. Antimicrob Agents Chemother 55:1650–1660

Malcolm RK, Woolfson AD, Toner CF, Morrow RJ, McCullagh SD (2005) Long-term, controlled release of the HIV microbicide TMC120 from silicone elastomer vaginal rings. J Antimicrob Chemother 56:954–956

McCormack S, Ramjee G, Kamali A, Rees H, Crook AM, Gafos M, Jentsch U, Pool R, Chisembele M, Kapiga S, Mutemwa R, Vallely A, Palanee T, Sookrajh Y, Lacey CJ, Darbyshire J, Grosskurth H, Profy A, Nunn A, Hayes R, Weber J (2010) PRO2000 vaginal gel for prevention of HIV-1 infection (Microbicides Development Programme 301): a phase 3, randomised, double-blind, parallel-group trial. Lancet 376:1329–1337

McEnery R (2011) Oral tenofovir arm of VOICE trial discontinued early. IAVI Rep 15:21

Mitsuya H, Weinhold KJ, Furman PA, St Clair MH, Lehrman SN, Gallo RC, Bolognesi D, Barry DW, Broder S (1985) 3′-Azido-3′-deoxythymidine (BW A509U): an antiviral agent that inhibits the infectivity and cytopathic effect of human T-lymphotropic virus type III/ lymphadenopathy-associated virus in vitro. Proc Natl Acad Sci USA 82:7096–7100

Miyasaka T, Tanaka H, Baba M, Hayakawa H, Walker RT, Balzarini J, De Clercq E (1989) A novel lead for specific anti-HIV-1 agents: 1-[(2-hydroxyethoxy)methyl]-6-(phenylthio)thymine. J Med Chem 32:2507–2509

Mugwanya K, Baeten JM, Mugo NR, Irungu E, Ngure K, Celum C (2011) High-dose valacyclovir HSV-2 suppression results in greater reduction in plasma HIV-1 levels compared with standard dose acyclovir among HIV-1/HSV-2 coinfected persons: a randomized, crossover trial. J Infect Dis 204:1912–1917

Naranbhai V, Abdool Karim SS, Altfeld M, Samsunder N, Durgiah R, Sibeko S, Abdool Karim Q, Carr WH, The CAPRISA004 TRAPS team (2012) Innate immune activation enhances HIV acquisition in women, diminishing the effectiveness of tenofovir microbicide gel. J Infect Dis 206:993–1001

Nel AM, Smythe SC, Habibi S, Kaptur PE, Romano JW (2010) Pharmacokinetics of 2 dapivirine vaginal microbicide gels and their safety vs Hydroxyethyl cellulose-based universal placebo gel. J Acquir Immune Defic Syndr 55:161–169

Njai HF, Lewi PJ, Janssen CG, Garcia S, Fransen K, Kestens L, Vanham G, Janssen PA (2005) Pre-incubation of cell-free HIV-1 group M isolates with non-nucleoside reverse transcriptase inhibitors blocks subsequent viral replication in co-cultures of dendritic cells and T cells. Antivir Ther 10:255–262

Ouedraogo A, Nagot N, Vergne L, Konate I, Weiss HA, Defer MC, Foulongne V, Sanon A, Andonaba JB, Segondy M, Mayaud P, Van de Perre P (2006) Impact of suppressive herpes therapy on genital HIV-1 RNA among women taking antiretroviral therapy: a randomized controlled trial. AIDS 20:2305–2313

Pani A, Musiu C, Loi AG, Mai A, Loddo R, La Colla P, Marongiu ME (2001) DABOs as candidates to prevent mucosal HIV transmission. Antivir Chem Chemother 12(Suppl 1):51–59

Patton DL, Sweeney YT, Balkus JE, Rohan LC, Moncla BJ, Parniak MA, Hillier SL (2007) Preclinical safety assessments of UC781 anti-human immunodeficiency virus topical microbicide formulations. Antimicrob Agents Chemother 51:1608–1615

Pauwels R, Andries K, Desmyter J, Schols D, Kukla MJ, Breslin HJ, Raeymaeckers A, Van Gelder J, Woestenborghs R, Heykants J, Schellekens K, Janssen MAC, De Clercq E, Janssen PAJ (1990) Potent and selective inhibition of HIV-1 replication in vitro by a novel series of TIBO derivatives. Nature 343:470–474

Person AK, Hicks CB (2012) Pre-exposure prophylaxis – one more tool for HIV prevention. Curr HIV Res 10:117–122

Ren J, Esnouf RM, Hopkins AL, Warren J, Balzarini J, Stuart DI, Stammers DK (1998) Crystal structures of HIV-1 reverse transcriptase in complex with carboxanilide derivatives. Biochemistry 37:14394–14403

Schwartz JL, Kovalevsky G, Lai JJ, Ballagh SA, McCormick T, Douville K, Mauck CK, Callahan MM (2008) A randomized six-day safety study of an antiretroviral microbicide candidate UC781, a non-nucleoside reverse transcriptase inhibitor. Sex Transm Dis 35:414–419

Selhorst P, Vazquez AC, Terrazas-Aranda K, Michiels J, Vereecken K, Heyndrickx L, Weber J, Quiñones-Mateu ME, Ariën KK, Vanham G (2011) Human immunodeficiency virus type 1 resistance or cross-resistance to nonnucleoside reverse transcriptase inhibitors currently under development as microbicides. Antimicrob Agents Chemother 55:1403–1413

Shattock RJ, Moore JP (2003) Inhibiting sexual transmission of HIV-1 infection. Nat Rev Microbiol 1:24–34

Singer R, Derby N, Rodriguez A, Kizima L, Kenney J, Aravantinou M, Chudolij A, Gettie A, Blanchard J, Lifson JD, Piatak M Jr, Fernández-Romero JA, Zydowsky TM, Robbiani M (2011) The nonnucleoside reverse transcriptase inhibitor MIV-150 in carrageenan gel prevents rectal transmission of simian/human immunodeficiency virus infection in macaques. J Virol 85:5504–5512

Srinivasan P, Johnson T, Martin A et al (2010) Vaginal release kinetics of pyrimidinedione IQP-0258 from a polyurethane intravaginal ring. In: Proceedings of the 28th annual symposium on nonhuman primate models for AIDS prevention – vaccines and microcide, New Orleans, 19–22 Oct 2010

Stolte-Leeb N, Loddo R, Antimisiaris S, Schultheiss T, Sauermann U, Franz M, Mourtas S, Parsy C, Storer R, La Colla P, Stahl-Hennig C (2011) Topical nonnucleoside reverse transcriptase inhibitor MC 1220 partially prevents vaginal RT-SHIV infection of macaques. AIDS Res Hum Retroviruses 27:933–943

Tan DH, Kaul R, Raboud JM, Walmsley SL (2011) No impact of oral tenofovir disoproxil fumarate on herpes simplex virus shedding in HIV-infected adults. AIDS 25:207–210

Tanton C, Abu-Raddad LJ, Weiss HA (2011) Time to refocus on HSV interventions for HIV prevention? J Infect Dis 204:1822–1826

Tchesnokov EP, Obikhod A, Massud I, Lisco A, Vanpouille C, Brichacek B, Balzarini J, McGuigan C, Derudas M, Margolis L, Schinazi RF, Götte M (2009) Mechanisms associated with HIV-1 resistance to acyclovir by the V75I mutation in reverse transcriptase. J Biol Chem 284:21496–21504

Tramontano E, Di Santo R (2010) HIV-1 RT-associated RNase H function inhibitors: recent advances in drug development. Curr Med Chem 17:2837–2853

Van Damme L, Ramjee G, Alary M, Vuylsteke B, Chandeying V, Rees H, Sirivongrangson P, Mukenge-Tshibaka L, Ettiègne-Traoré V, Uaheowitchai C, Karim SS, Mâsse B, Perriëns J, Laga M, COL-1492 Study Group (2002) Effectiveness of COL-1492, a nonoxynol-9 vaginal gel, on HIV-1 transmission in female sex workers: a randomised controlled trial. Lancet 360:971–977

Van Damme L, Govinden R, Mirembe FM, Guédou F, Solomon S, Becker ML, Pradeep BS, Krishnan AK, Alary M, Pande B, Ramjee G, Deese J, Crucitti T, Taylor D, CS Study Group (2008) Lack of effectiveness of cellulose sulfate gel for the prevention of vaginal HIV transmission. N Engl J Med 359:463–472

Van Damme L, Corneli A, Ahmed K, Agot K, Lombaard J, Kapiga S, Malahleha M, Owino F, Manongi R, Onyango J, Temu L, Monedi MC, Mak'Oketch P, Makanda M, Reblin I, Makatu SE, Saylor L, Kiernan H, Kirkendale S, Wong C, Grant R, Kashuba A, Nanda K, Mandala J, Fransen K, Deese J, Crucitti T, Mastro TD, Taylor D, FEM-PrEP Study Group (2012) Preexposure prophylaxis for HIV infection among African women. N Engl J Med 367:411–422

van der Straten A, Van Damme L, Haberer JE, Bangsberg DR (2012) How well does PREP work? Unraveling the divergent results of PrEP trials for HIV prevention. AIDS 26:F13–F19

Van Herrewege Y, Vanham G, Michiels J, Fransen K, Kestens L, Andries K, Janssen P, Lewi P (2004) A series of diaryltriazines and diarylpyrimidines are highly potent nonnucleoside reverse transcriptase inhibitors with possible applications as microbicides. Antimicrob Agents Chemother 48:3684–3689

Vanpouille C, Lisco A, Derudas M, Saba E, Grivel JC, Brichacek B, Scrimieri F, Schinazi R, Schols D, McGuigan C, Balzarini J, Margolis L (2010) A new class of dual-targeted antivirals: monophosphorylated acyclovir prodrug derivatives suppress both human immunodeficiency virus type 1 and herpes simplex virus type 2. J Infect Dis 201:635–643

Watson KM, Buckheit CE, Buckheit RW Jr (2008) Comparative evaluation of virus transmission inhibition by dual-acting pyrimidinedione microbicides using the microbicide transmission and sterilization assay. Antimicrob Agents Chemother 52:2787–2796

Woolfson AD, Malcolm RK, Toner CT et al (2006) Potential use of vaginal rings for prevention of heterosexual transmission of HIV. Am J Drug Deliv 4:7–20

Zhang H, Öberg B, Bottiger D et al (2004) MIV-150 in a vaginal microbicide with superior anti-HIV activities. In: Proceedings of the 15th international conference AIDS 2004, Bangkok, 11–16 July 2004

# Index




S.F.J. LeGrice and M. Götte (eds.), *Human Immunodeficiency Virus Reverse Transcriptase: A Bench-to-Bedside Success*, DOI 10.1007/978-1-4614-7291-9,

## I

## Q



## R

MIX
Papier aus verantwortungsvollen Quellen
Paper from responsible sources
FSC® C105338

If you have any concerns about our products,
you can contact us on
**ProductSafety@springernature.com**

In case Publisher is established outside the EU,
the EU authorized representative is:
**Springer Nature Customer Service Center GmbH**
**Europaplatz 3, 69115 Heidelberg, Germany**

Printed by Libri Plureos GmbH
in Hamburg, Germany